CLUSTERS AND GROUPS OF GALAXIES

ASTROPHYSICS AND SPACE SCIENCE LIBRARY

A SERIES OF BOOKS ON THE RECENT DEVELOPMENTS
OF SPACE SCIENCE AND OF GENERAL GEOPHYSICS AND ASTROPHYSICS
PUBLISHED IN CONNECTION WITH THE JOURNAL
SPACE SCIENCE REVIEWS

VOLUME 111
PROCEEDINGS

CLUSTERS AND
GROUPS OF GALAXIES

INTERNATIONAL MEETING HELD IN TRIESTE,
ITALY, SEPTEMBER 13 – 16, 1983

Edited by

F. MARDIROSSIAN

G. GIURICIN

and

M. MEZZETTI

Osservatorio Astronomico di Trieste, Italy

D. REIDEL PUBLISHING COMPANY

A MEMBER OF THE KLUWER ACADEMIC PUBLISHERS GROUP

DORDRECHT / BOSTON / LANCASTER

Library of Congress Cataloging in Publication Data
Main entry under title:

Clusters and groups of galaxies.

(Astrophysics and space science library ; v. 111)
Sponsored by the European Physical Society, Astronomical Division.
Bibliography: p.
Includes index.
1. Galaxies–Clusters–Congresses. I. Mardirossian, F. (Fabio)
II. Giuricin, G. (Giuliano) III. Mezzetti, M. (Marino) IV. European
Physical Society. Astronomy and Astrophysics Division. V. Series.
QB858.7.C58   1984   523.1'12   84-8389
ISBN 90-277-1772-9

Published by D. Reidel Publishing Company,
P.O. Box 17, 3300 AA Dordrecht, Holland.

Sold and distributed in the U.S.A. and Canada
by Kluwer Academic Publishers,
190 Old Derby Street, Hingham, MA 02043, U.S.A.

In all other countries, sold and distributed
by Kluwer Academic Publishers Group,
P.O. Box 322, 3300 AH Dordrecht, Holland.

This Meeting was sponsored by
the European Physical Society – Astronomical Division

Printed in The Netherlands

TABLE OF CONTENTS

PREFACE

    The large-scale structure of the Universe and systems
like Superclusters, Clusters, and Groups of galaxies are topics
of great interest. They fully justify the meeting on "Clusters
and Groups of Galaxies".
    The topics covered included the spatial distribution
and the clustering of galaxies; the properties of Superclusters,
Clusters and Groups of galaxies; radio and X-ray observations;
the problem of unseen matter; theories concerning hierarchical
clustering, pancakes, cluster and galaxy formation and evolution.
    The meeting was held at the International Center for
Theoretical Physics in Trieste (Italy) from September 13 to September
16, 1983. It was attended by about 150 participants from 22 nations
who presented 67 invited lectures (il) and contributed papers
(cp), and 45 poster papers (pp).
    The Scientific Organizing Committee consisted of F. Bertola,
P. Biermann, A. Cavaliere, N. Dallaporta, D. Gerbal, M. Hack,
J.V. Peach, D. Sciama (Chairman), G. Setti, M. Tarenghi. We are
particularly indebted to D. Sciama, A. Cavaliere and F. Bertola
for their work of coordination. We were acting as the three members
of the Local Organizing Committee. Moreover, we are pleased to
thank the Chairmen of the Sessions (M. Hack, N. Dallaporta, G.
Burbidge, B. Mills, M. Rees, P. Biermann, L.Z. Fang, L. Gouguenheim)
for their valuable help.
    The meeting was sponsored by the European Physical Society
Astronomical Division and supported by: Astronomical Observatory,
Trieste; Azienda autonoma di Soggiorno e turismo di Trieste e
della sua riviera; Banca Cattolica del Veneto, Trieste; Banco
di Sicilia, Trieste; Comune di Trieste; Consiglio Nazionale delle
Ricerche - GNA; Informatica S.p.A., Trieste; International Centre
for Theoretical Physics, Trieste; International School for Advanced

Studies, Trieste; Regione Autonoma Friuli–Venezia Giulia; Università degli Studi di Trieste.

We are very grateful for all the support we reiceved.

In Particular we are pleased to thank M. Hack, A. Salam, and P. Budinich, respectively the Directors of Trieste Astronomical Observatory, of the International Center for Theoretical Physics of Trieste, and of the International School for Advanced Studies of Trieste, and the staffs of their Institutions for their help which greatly contributed to the success of the meeting.

Finally we wish to thank the Authors who followed our editorial suggestions and compressed their texts to six pages for contributed papers and two pages for poster papers.

All questions and answers have been typed from the written version received from Questioners and Authors. In this way we have been able to collect about 60% of the discussion.

Unfortunately the texts of the invited lectures by R. Mushotzky and M. Tarenghi have not arrived.

            Fabio Mardirossian, Giuliano Giuricin, Marino Mezzetti

                              Editors

LIST OF PARTICIPANTS

AHMAD, F., Department of Physics, University of Kashmir, Srinagar, 190006, India.

ANDERNACH, H., Max-Planck-Institut für Radioastronomie, Auf dem Hügel 69, D-5300 Bonn 1, F.R.G.

ANDERSON, E.R., Department of Astronomy, University of Toronto, Toronto, Ontario, M5S 1A7, Canada.

APPLETON, P.N., Department of Astronomy, University of Manchester, Oxford Road, Manchester, U.K.

BAJTLIK, S., Institute of Theoretical Physics, Warsaw University, Hoza 69, Warszawa, Poland.

BARBON, R., Osservatorio Astrofisico, 36012 Asiago, Italy.

BEARD, S.M., Department of Astronomy, Royal Observatory, Blackford Hill, Edinburgh, U.K.

BERTOLA, F., Osservatorio Astronomico, Vicolo dell'Osservatorio 5, 35100 Padova, Italy.

BHAVSAR, S.P., Astronomy Centre, University of Sussex, Falmer, Brighton BN1 9QH, England.

BIERMANN, P., Max-Planck-Institute für Radioastronomie, Auf dem Hügel 69, D-5300 Bonn 1, F.R.G.

BIJLEVELD, W., Sterrewacht, P.O. Box 9513, NL-2300 RA Leiden, Holland.

BOHNENSTENGEL, H.-D., Hochschule der Bundeswehr Hamburg, FB/MB-Mathematik, D-2000 Hamburg 70, F.R.G.

BONOMETTO, S., Istituto di Fisica "G. Galilei", Via Marzolo 8, 35100 Padova, Italy.

BOTTINELLI, L., Observatoire de Meudon, Radioastronomie, 92195 Meudon Principal CEDEX, France.

BRANDUARDI-RAYMONT, G., Mullard Space Science Lab., Holmbury St. Mary, Dorking, Surrey, U.K.

BRESSAN, A., International School for Advanced Studies, Strada Costiera 11, 34100 Trieste, Italy.

BURBIDGE, G., Kitt Peak National Observatory, Box 26732, Tucson, Arizona 85726-6732, U.S.A.

BURNS, J.O., Department of Physics and Astronomy, University of New Mexico, Albuquerque, NM 87131, U.S.A.

CAPACCIOLI, M., Osservatorio Astronomico, Vicolo dell'Osservatorio 5, 35100 Padova, Italy.

CARTER, D., Mount Stromlo Observatory, Private Bag, Woden Post
        Office, ACT 2606, Canberra, Australia.
CAVALIERE, A., Osservatorio Astronomico, Vicolo dell'Osservatorio
        5, 35100 Padova, Italy.
CECCARELLI, C., Università di Roma, Istituto di Fisica, Piazzale
        A. Moro 2, 00185 Roma, Italy.
CIARDULLO, R., Space Telescope Science Institute, John Hopkins
        University, Department of Physics, Homewood Campus, Baltimo-
        re MD 21218, U.S.A.
COLEMAN, C., Mount Stromlo Observatory, Private Bag, Woden Post
        Office, ACT 2606, Canberra, Australia.
CORBYN, P., Physics Department, Queen Mary College, Mile End Road,
        London E1, England.
COUCH, W.J., Physics Department, Durham University, Durham U.K.
CURRIE, M.J., Royal Greenwich Observatory, Herstmonceux Castle,
        Hailsham, East Sussex BN27 1RP, England.
DALLAPORTA, N., International School for Advanced Studies, Strada
        Costiera 11, 34100 Trieste, Italy.
DAVIES, R.D., University of Manchester, Nuffield Radio Astronomy
        Laboratories, Jodrell Bank, Macclesfield, Cheshire SK11
        9DL, England.
DE FELICE, F., Istituto di Fisica "G. Galilei", Via Marzolo 8,
        35100, Padova, Italy.
DE VAUCOULEURS, G., Department of Astronomy, The University of
        Texas, Austin, Texas 78712, U.S.A.
DE ZOTTI, G., Osservatorio Astronomico, Vicolo dell'Osservatorio
        5, 35100 Padova, Italy.
DICKEL, J.R., University of Illinois at Urbana-Champaign, Room
        341, Astronomy Building, 1011 West Springfield Ave.,
        Urbana, IL 61801, U.S.A.
DRESSLER, A., Mount Wilson and Las Campanas Observatories, 813
        Santa Barbara Street, Pasadena, California 91101, U.S.A.
ELLIS, R., Physics Department, Durham University, South Road, Durham,
        England.
FAIRALL, A.P., Department of Astronomy, University of Cape Town,
        Rondebosch, 7700 South Africa.
FANG, L.Z., Astrophysics Research Division, University of Science
        and Technology of China, Hefei, Anhui, Peoples Republic
        of China.
FANTI, C., Istituto di Fisica "A. Righi", Via Irnerio 46, 40126
        Bologna, Italy.
FANTI, R., Istituto di Fisica "A. Righi", Via Irnerio 46, 40126
        Bologna, Italy.
FERETTI, L., Istituto di Radioastronomia, Via Irnerio 46, 40126

Bologna, Italy.

FIALA, M., Institut für Astronomie der Universität Wien, Türkenschanz-
        strasse 17, A-1180 Wien, Austria.

FINZI, A., Department of Mathematics, Technion – Israel Institute
        of Technology, Haifa, Israel.

FLICHE, H.H., Centre de Physique – C.N.R.S., Luminy Case 907, F
        13288 Marseille, Cedex 9, France.

FLIN, P., Obserwatorium Astronomiczne Uniwersytetu, ul. Orla 171,
        30-244 Kracow, Poland.

FOCARDI, P., Dipartimento di Astronomia, C.P. 516, 40100 Bologna,
        Italy.

FONTANELLI, P., Osservatorio Astrofisico di Arcetri, Largo E. Fermi
        5, Firenze, Italy.

FORMAN, W., Harvard-Smithsonian Center for Astrophysics, 60 Garden
        Street, Cambridge, MA 02138, U.S.A.

FORTIERI, C., International School for Advanced Studies, Strada
        Costiera 11, 34100 Trieste, Italy.

FOSTER, P.A., Department of Astronomy, The University, Manchester
        M13 9PL, England.

GAVAZZI, G., Istituto di Fisica Cosmica – C.N.R., Via Bassini 15,
        20133 Milano, Italy.

GELLER, M., Harvard-Smithsonian Center for Astrophysics, 60 Garden
        Street, Cambridge, MA 02138, U.S.A.

GERBAL, D., LAM Observatoire de Meudon, 92195 Cedex, France.

GIOVANELLI, R., National Astronomy and Ionosphere Center, Arecibo
        Observatory, P.O. Box 995, Arecibo PR 00613, U.S.A.

GIOVANNINI, G., Istituto di Radioastronomia, Via Irnerio 46, 40126
        Bologna, Italy.

GIURICIN, G., Osservatorio Astronomico, Via Tiepolo 11, 34131 Trieste,
        Italy.

GØRSKI, K., Copernicus Astronomical Centre, Bartycka 18, 00716
        Warsaw, Poland.

GOUGUENHEIM, L., Observatoire de Meudon, Radioastronomie, 92195
        Meudon Principal Cedex, France.

GREGORINI, L., Istituto di Radioastronomia, Via Irnerio 46, 40126
        Bologna, Italy.

HACK, M., Osservatorio Astronomico, Via Tiepolo 11, 34131 Trieste,
        Italy.

HANISCH, R.J., Radiosterrenwacht Dwingeloo, Postbus 2, 7990 AA
        Dwingeloo, The Netherlands.

HARMS, R.J., Cass C-011, UCSD, La Jolla, CA 92093, U.S.A.

HICKSON, P., Department of Geophysics and Astronomy, University
        of British Columbia, 2219 Main Mall, Vancouver, B.C.
        V6T 1W5, Canada.

HUCHRA, J.P., Center for Astrophysics, 60 Garden Street, Cambridge, MA 02138, U.S.A.

HUCHTMEIER, W.K., Max-Planck-Institut für Radioastronomie, Auf del Hügel 69, D-5300 Bonn 1, F.R.G.

HUNT, L., Osservatorio Astrofisico di Arcetri, Largo E. Fermi 5, 50125 Firenze, Italy.

ILLINGWORTH, G., Kitt Peak National Observatory, P.O. Box 26732, Tucson, AZ 85726, U.S.A.

JAFFE, W., Space Telescope Science Institute, Homewood Campus, Baltimore, Maryland 21218, U.S.A.

JONES, C., Harvard-Smithsonian Center for Astrophysics, 60 Garden Street, Cambridge, MA 02138, U.S.A.

JUSZKIEWICZ, R., Copernicus Astronomical Centre, Bartycka 18, 00176 Warsaw, Poland.

KRISS, G.A., Department of Astronomy, University of Michigan, Ann Arbor, MI 48109, U.S.A.

KRON, R.G., Yerkes Observatory, The University of Chicago, William Bay, Wisconsin 53191, U.S.A.

KOTANYI, C., European Southern Observatory, Karl-Schwarzschild-Strasse 2, D-8046 Garching bei München, F.R.G.

LANZA, A., International School for Advanced Studies, Strada Costiera 11, 34100 Trieste, Italy.

LONGO, G., Osservatorio Astronomico di Capodimonte, Via Moiariello 16, 80130 Napoli, Italy.

LORENZ, H., Zentralinstitut für Astrophysik, DDR-15 Potsdam, R.-Luxemburg-Strasse 17a, German Democratic Republic.

LU, J., International School for Advanced Studies, Strada Costiera 11, 34100 Trieste, Italy.

LUCCHIN, F., Istituto di Fisica "G. Galilei", Via Marzolo 8, 35100 Padova, Italy.

LUCEY, J., Anglo-Australian Observatory, P.O. Box 296, Epping, NSW 2121, Australia.

MACGILLIVRAY, H.T., Royal Observatory, Blackford Hill, Edinburgh, EH9 3HJ, Scotland, U.K.

MARDIROSSIAN, F., Osservatorio Astronomico, Via Tiepolo 11, 34131 Trieste, Italy.

MATARRESE, S., International School for Advanced Studies, Strada Costiera 11, 34100 Trieste, Italy.

MAZURE, A., LAM, Observatoire de Meudon, 92195 Cedex, France.

MCGLYNN, T.A., Institute of Astronomy, Madingley Road, Cambridge CB3 OHA, England.

MCKECHNIE, S.P., WKS, Huygens Laboratorium, Wassenaarseweg 78, Postbus 9504, 2300 RA Leiden, The Netherlands.

MEZZETTI, M., Osservatorio Astronomico, Via Tiepolo 11, 34131 Trieste,

Italy.

MILLS, B.Y., School of Physics, University of Sydney, N.S.W. 2006, Australia.

MORENO, G., Istituto di Fisica "G. Marconi", Piazzale A. Moro 2, 00100 Roma, Italy.

MOSS, C., Specola Vaticana, I-00120 Città del Vaticano.

MUSHOTZKY, R.F., NASA, Goddard Space Flight Center, Greenbelt Maryland 20771, U.S.A.

NANNI, D., Laboratori Nazionali di Fisica Nucleare, Casella Postale 13, 0044 Frascati (Roma), Italy.

NØRGAARD-NIELSEN, H.U., Copenhagen University Observatory, Østervoldgade 3, DK-1350 Copenhagen K, Denmark.

OCCHIONERO, F., Istituto di Astronomia, Università di Roma, Piazzale A. Moro 2, 00185 Roma, Italy.

OKE, J.B., 105-24 California Institute of Technology, Pasadena, CA 91125, U.S.A.

OORT, J.H., Sterrewacht Leiden, Huygens Laboratorium, Wassenasarseweg 78, Postbus 9513, 2300 RA Leiden, The Netherlands.

OSMER, P.S., Cerro Tololo Inter-American Observatory, Casilla 603, La Serena, Chile.

PADRIELLI, L., Istituto di Radioastronomia, Via Irnerio 46, 40126 Bologna, Italy.

PALMER, J.B., Royal Observatory, Blackford Hill, Edinburgh EH9 3HJ, Scotland, U.K.

PALUMBO, G.G.C., Istituto TE.S.R.E./C.N.R., Via de' Castagnoli 1, 40126 Bologna, Italy.

PARKER, Q.A., University Observatory, Buchanan Gardens, St. Andrews, Fife, Scotland, U.K.

PEACH, J.V., Department of Astrophysics, South Parks Road, Oxford, England.

PEEBLES, P.J.E., Joseph Henry Laboratories, Physical Department, Princeton University, Princeton, NJ 08540, U.S.A.; Dominion Astrophysical Observatory, 5071 West Saanich Road, Victoria, B.C. V8X 4M6, Canada.

PEROLA, G.C., Istituto di Astronomia, Università di Roma, Piazzale A. Moro 2, 00185 Roma, Italy.

PERSIC, M., Osservatorio Astronomico, Via Tiepolo 11, 34131 Trieste, Italy.

PHILLIPPS, S., Department of Applied Mathematics and Astronomy, University College, Cardiff, Wales, U.K.

PIETRANERA, L., Istituto di Fisica, Università di Roma, Piazzale A. Moro 2, 00185 Roma, Italy.

PIRRONELLO, V., Istituto di Fisica, Facoltà di Ingegneria, Università di Catania, Corso delle Province 47, Catania, Italy.

PORTER, A., 105–24 California Institute of Technology, Pasadena, CA 91125, U.S.A.

PRESTAGE, R.M., University of Edinburgh, Department of Astronomy, Royal Observatory, Edinburgh EH9 3HI, Scotland, U.K.

PRIMACK, J.R., 34–63, University of California, Division of Natural Sciences, Santa Cruz, California 95064, U.S.A.

RAKOS, K.D., Institut für Astronomie, Universitäts-Sternwarte, Türkenschanzstrasse 17, A–1180 Wien, Austria.

REES, M., Institute of Astronomy, Madingley Road, Cambridge CB3 OHA, England.

REPHAELI, Y., Department of Physics and Astronomy, Tel Aviv University, Tel Aviv 69978, Israel.

RICHTER, O.G., European Southern Observatory, Karl-Schawarzschild-Strasse 2, D–8046 Garching bei München, F.R.G.

RIVOLO, A.R., Space Telescope Science Institute, J.H.U./Homewood Campus, Baltimore, MD, 21218, U.S.A.

SADLER, E.M., European Southern Observatory, Karl-Schwarzschild-Strasse 2, D–8046 Garching bei München, F.R.G.

SALUCCI, P., International School for Advanced Studies, Strada Costiera 11, 34100 Trieste, Italy.

SALVADOR SOLÉ, E., DPTO. Fisica de la Tierra y del Cosmos, Universidad de Barcelona, Av. Diagonal 645, Barcelona 28, España.

SANCISI, R., Kapteyn Laboratorium, P.O. Box 800, 9700 AV Gronigen, The Netherlands.

SANTANGELO, P., Istituto di Astronomia, Università di Roma, Piazzale A. Moro 2, 00187 Roma, Italy.

SARKAR, S., University of Oxford, Department of Astrophysics, South Parks Road, Oxford OX1 3RQ, England.

SCIAMA, D.W., University of Oxford, Department of Astrophysics, south Parks Road, Oxford OX1 3RQ, England; International School for Advanced Studies, strada Costiera 11, 34100 Trieste, Italy.

SEDMAK, G., Osservatorio Astronomico, Via Tiepolo 11, 34131 Trieste, Italy.

SHAFER, R.A., Institute of Astronomy, Madingley Road, Cambridge CB3 OHA, England.

SHAPIRO, P., Department of Astronomy, University of Texas, Austin, TX 78712, U.S.A.

SJOLANDER, N., Uppsala University, Astronomical Observatory, Box 515, S–751 20 Uppsala, Sweden.

SMITH, H., Jr., Department of Astronomy, 211 Space Sciences Research Building, University of Florida, Gainesville, Florida 32611, U.S.A.

SOURIAU, J.M., Centre de Physique Theorique – C.N.R.S., Luminy

Case 907, F 13288 Marseille, Cedex 9, France.

STAVELEY-SMITH, L., Nuffield Radio Astronomy Laboratories, Jodrell Bank, Macclesfield, Cheshire, U.K.

STEVENSON, P., Physics Department, University of Durham, South Road, Durham, OH1 3LE, England.

SUDANAGUNTA, M., International School for Advanced Studies, Strada Costiera 11, 34100 Trieste, Italy.

TAMMANN, G.A., Astronomisches Institut, Venusstrasse 7, CH-4102 Binningen, Switzerland.

TANZELLA NITTI, G., Osservatorio Astronomico di Torino, 10025 Pino Torinese, Italy.

TARENGHI, M., European Southern Observatory, Karl-Schwarzschild-Strasse 2, D-8046, Garching bei München.

THOMPSON, L., Institute for Astronomy, 2680 Woodlawn Drive, Honolulu, Hawaii 96822, U.S.A.

TOLMAN, B., International School for Advanced Studies, Strada Costiera 11, 34100 Trieste, Italy.

TREVESE, D., Osservatorio Astronomico di Roma-Monte Mario, Viale del Parco Mellini 84, 00136 Roma, Italy.

TULLY, B., Institute for Astronomy, 2680 Woodlawn Drive, Honolulu, Hawaii 96822, U.S.A.

TURNER, E.L., Princeton University Observatory, Peyton Hall, Princeton, NJ 08544, U.S.A.

ULMER, M.L., Department of Physics and Astornomy, Northwestern University, 2131 Sheridan Road, Evanston, Illinois 60201, U.S.A.

VALDARNINI, R., International School for Advanced Studies, Strada Costiera 11, 34100 Trieste, Italy.

VAN DEN BERGH, S., Dominion Astrophysical Observatory, 5071 West Saanich Road, Victoria B.C., V8X 4M6, Canada.

VAN WOERDEN, H., Kapteyn Laboratory, Pstbus 800, 9700 AV Groningen, The Netherlands.

VETTOLANI, G., Istituto di Radioastronomia, Via Irnerio 46, 40126 Bologna, Italy.

VITTORIO, N., Istituto di Astronomia, Università di Roma, Piazzale A. Moro 2, 00185 Roma, Italy.

VOGLIS, N., Department of Astronomy, University of Athens, Panepistimiopolis, Athens (621), Greece.

WARMELS, R.H., Kapteyn Astronomical Institute, Postbox 800, 9700AV Groningen, The Netherlands.

WHITTLE, M., Steward Observatory, University of Arizona, Tucson, Arizona 85721, U.S.A.

WILLIAMS, B., National Radio Astronomy Observatory, Edgemont Road, Charlottesville, Virginia 22903, U.S.A.

WILLSON, R.F., Department of Physics, Tufts University, Medford, MA 02155, U.S.A.

YEE, H.K.C., Dominion Astrophysical Observatory, 5071 West Saanich Road, Victoria, BC V8X 4M6, Canada.

ZAMORANI, G., Istituto di Radioastronomia, Via Irnerio 46, 40126 Bologna, Italy.

ZEILINGER, W.W., Institut für Astronomie, Türkenschanzstrasse 17, A-1180 Wien, Austria.

ZITELLI, V., Istituto di Astronomia, Via Zamboni 33, 40100 Bologna, Italy.

SUPERCLUSTERS*

J.H. Oort
Sterrewacht, Leiden, The Netherlands

Galaxies tend to cluster over a wide range of scales, extending from doubles to clusters with thousands of members, and finally to super-clusters, the dimensions ranging from tens of kiloparsecs to hundreds of Megaparsecs. The rich clusters may be the largest bound structures. The superclusters look largely unrelaxed: they have rarely a well-defined centre; their relatively small internal random motions indicate crossing times along the larger diameters of the order of the age of the universe, or longer, so that they cannot have appreciably re-arranged themselves since their birth. They might therefore teach us something about the universe in the era of their formation. For a discussion of their pro-perties it is useful first to look at a few specimens.

The Local (or Virgo) supercluster, to which our Galaxy belongs, can evidently give the best "inside" information. This will be discussed by Professor de Vaucouleurs.

THE PERSEUS SUPERCLUSTER

Two other superclusters which have been studied in some detail are the Perseus and the Coma/Abell 1367 superclusters. The former has a mean velocity of 5000 km s^{-1} and covers a region of about 45° x 45° in the South-galactic hemisphere. It contains a rather concentrated chain of galaxies, between +30° and +42° declination, extending over roughly 4^h in right-ascension; the chain is about 80 Mpc long, has a width of 5-10 Mpc, a velocity dispersion of about 400 km s^{-1}. It contains three dense rich clusters and several dense groups. The supercluster's further structure can be seen from Figures 1-3. The degree to which the chain stands out as an isolated feature in velocity may be best judged from Figure 3, which shows position along the chain against radial velocity:

* For convenience I shall use in this lecture a definite value of the Hubble constant, viz. 50 km s^{-1} Mpc^{-1}. One should keep in mind the great uncertainty of this constant. With H_0 = 100 km s^{-1} Mpc^{-1} dimensions and masses would be halved.

1

F. Mardirossian et al. (eds.), Clusters and Groups of Galaxies, 1–16
© 1984 by D. Reidel Publishing Company.

it indicates a distinct concentration of the velocities between $\sim$ 4500 and 5500 km s^{-1}. The decrease in the number of velocities higher than $\sim$ 7000 km s^{-1} is, however, to a large extent due to the brightness limit of the observations.

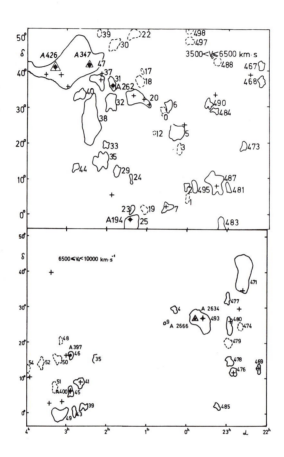

Figure 1.   Perseus supercluster. The contours show outlines of clusters from the catalogue of Zwicky *et al* (1961–68). These clusters are less well-defined than the compacter clusters in the catalogue by Abell (1958) which are discussed in other parts of this report; they are more suitable for studying superclustering because they are much more numerous.
The upper diagram contains the clusters with velocities between 3500 and 6500 km s^{-1}, the lower those with velocities between 6500 and 10.000 km s^{-1}, which are clearly distributed differently. The velocities were mostly estimated from galaxies with measured radial velocities; in some cases (indicated by dotted contours) from magnitudes and cluster diameters. The numbering is from Nilson (1973). Abell clusters are indicated by solid circles and by their numbers in Abell's catalogue; A 426 is the Perseus cluster. Δ: (X-ray sources), +: (radio sources).
Reprinted by courtesy of Einasto *et al* (1980).

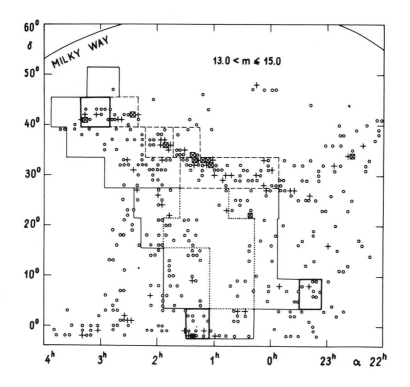

Figure 2. Perseus supercluster. Distribution of galaxies between 13ᵐ0
and 15ᵐ0 in the same region as that shown in Figure 1. Symbols indicate
numbers per square degree (circles 2-3, crosses 4-7, crosses in squares
8-15). From Einasto *et al* (1980).

THE COMA SUPERCLUSTER

        The Coma supercluster is, next to the Virgo supercluster, the most
prominent feature in the distribution of galaxies brighter than 14ᵐ5 in
the North-galactic hemisphere. It was first found as a bridge of galaxies
and groups with velocities between 6000 and 7000 km s^{-1}, extending from
the Coma cluster (Abell 1656) to Abell 1367. The total extent can best be
studied from the data in the large radial velocity survey of galaxies down
to 14.5 Zwicky magnitude made at the Harvard Center for Astrophysics
(Davis *et al* 1982; Huchra *et al* 1983). Figure 5 shows the distribution
of galaxies with velocities between 6000 and 10000 km s^{-1} in the observed
part of the North-galactic polar cap. I have outlined a number of regions
which might contain a supercluster. The solid contour marked COMA indicates the best defined part of the Coma supercluster; possible extensions
are marked by dashes. The total length might be 160 Mpc, the width about
20 Mpc. The part that lies between declinations +20° and +30° is shown
again in a right-ascension velocity plot in Figure 7.

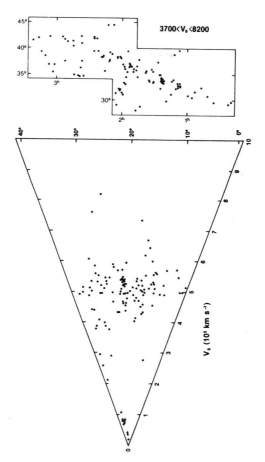

Figure 3.   Perseus supercluster. *(Top)* Distribution of galaxies brighter than m_p= 14.0 in the Zwicky catalogue and having velocities between 3700 and 8200 km s^{-1}. The boundaries of the region surveyed are indicated. The tips of the Perseus supercluster chain lie around ℓ = 150°, b = 12° *(left)* and ℓ = 115°, b = 34° *(right)*; *(Bottom)* "Wedge" diagram for all galaxies in the survey with velocities less than 10 000 km s^{-1}. The positions are measured along the chain.   From Gregory *et al* (1981).

STATISTICS

        The contours in Figures 4-8 are a superficial attempt towards a rough statistic of superclusters in the North-galactic polar cap in the distance range 60-200 Mpc. Evidently there is arbitrariness in outlining such struc-tures. Several may be only chance fluctuations, as stressed by Peebles in a subsequent communication. They are meant mainly to give a general in-dication of sizes, shapes and space frequency of superclusters. In some cases the reality of the structure is supported by the fact that rich clusters in a supercluster feature are elongated parallel to that feature.

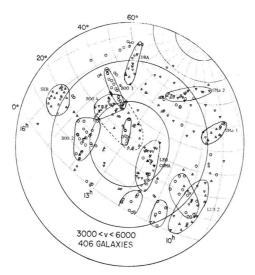

Figure 4. An equal area plot in galactic co-ordinates of galaxies
brighter than m_B = 14.5 above b = +40°. The galactic pole is at the
center, the circles are at b = 30°, 50°, and 70°. Right ascensions and
declinations are indicated by dotted curves. Galaxies whose absolute
magnitudes are fainter than −20.0 (on the H_O = 50 distance scale) have
been omitted. The various symbols denote the following velocity bins:
▽ (3000-4000), ▵ (4000-5000), ○ (5000-6000 km s⁻¹). Reproduced by courtesy of Davis *et al* (1982) and the *Astrophysical Journal*; the contours were
drawn by the present author.

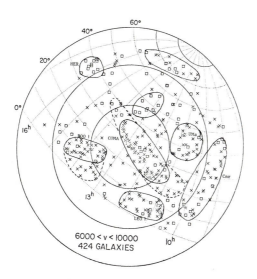

Figure 5. Same as for Figure 4, but for velocities between 6000 and
10 000 km s⁻¹. Symbols: x (6000-8000), ▫ (8000-10 000 km s⁻¹).

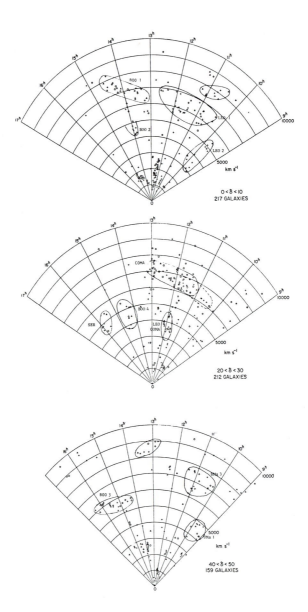

Figures 6, 7 and 8. Distribution of galaxies with m_B < 14.5 in right ascension and velocity for three declination zones. Again, galaxies fainter than M_B = -20.0 were omitted.
Reproduced by courtesy of Davis *et al* (1982) and the *Astrophysical Journal*; contours were drawn in by the present author.

The Coma cluster, for instance, is strongly elongated in a position angle
co-inciding with the direction toward Abell 1367. A similar situation is
observed in the Perseus supercluster where the major axes of the clusters
Abell 426, 347 and 262 are aligned within 10° - 20° of the direction of
the main ridge of the supercluster.

An independent attempt towards statistics of superclusters has recently been made by N. Bahcall & R. Soneira (1983). They studied the distribution of 104 Abell clusters of distance class D < 4, corresponding to
a distance limit of about 500 Mpc. For all of these radial velocities are
available. The distribution in the North galactic hemisphere is shown in
Figure 9. Contours indicate various space density enhancements. Numbers
refer to the catalogue of superclusters which the authors drew up on the
basis of these data. The objects are generally much larger than the structures outlined in Figures 4-8. The largest, No. 12, contains 15 rich clusters, and has a diameter of 360 Mpc. The dashed outermost contours cor-

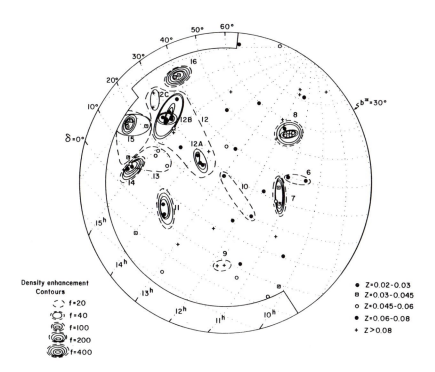

Figure 9. Giant superclusters in the North galactic hemisphere above
30° latitude. The outermost contour is $b = 30^{o}$; the inner contour is the
completeness limit of the sample. Small contours show the space density
enhancement factor over the mean space density at the distance of the
supercluster. Numbers refer to the authors' list of superclusters. The
elongated contour number 10 is the Coma supercluster. Reproduced by
courtesy of N. Bahcall & Soneira (1983).

respond to space density enhancements of 20; in the inner parts the enhancement is often several hundreds. The 16 superclusters listed in the
catalogue contain 54% of all the clusters considered. These fill $\sim$ 3%
of space. The superclusters indicated in the Center for Astrophysics
survey are about 100 times more frequent.

PROPERTIES OF SUPERCLUSTERS

Size

        This ranges from roughly 30 to 300 Mpc.

Mass

        Judging from the numbers of rich clusters contained in them the
masses of the more important superclusters may be estimated to range from
$\sim 10^{15}$ to $\sim 10^{17}$ M$_\odot$.

Shape

        Several appear to have a strongly elongated or flat structure.
Examples are the structural details in the Local supercluster, the main,
"filamentary", branch of the Perseus supercluster, the probably elongated
form of the Coma supercluster, and such objects as for instance BOO 3
marked in Figures 4 and 7. But, as Peebles points out in his talk , the
evidence for such strongly aspherical shapes needs further confirmation.

Internal motions

        From the dispersion in radial velocity these are inferred to be
$\sim$ 300 km s^{-1}, corresponding to $\sim$ 3 Mpc in a Hubble time. As a consequence,
very little mixing can have occurred along the major diameters since the
formation of the superclusters.

Orientation of clusters

        In an important investigation Binggeli (1982) has shown that the
apparent major axes of Abell clusters have a tendency to point in the
direction towards the nearest neighbouring Abell cluster if this lies
within a distance of about 35 Mpc (cf. Figure 10), while there is a distinct correlation between these directions up to distances of 100 Mpc. If
these results would be confirmed by more clusters they would give independent evidence that neighbouring clusters are related with each other
up to large distances, presumably by common membership in a supercluster.

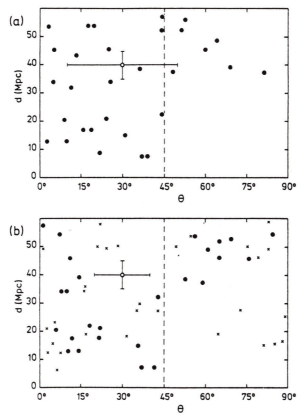

Figure 10. Relation between cluster position angles and directions to the nearest neighbour cluster. (a) Difference in position angle θ of the cluster major axis and the direction to the closest neighbouring cluster as a function of the distance d to this neighbour; (b) same as for (a), except that the cluster position angle is replaced by the position angle of the major axis of the first-ranked galaxy. Crosses refer to clusters for which only the position angle of the first-ranked galaxy is known (from Binggeli 1982).

Orientation of double galaxies

     In an investigation of radial velocities of binary galaxies Tifft (1980) noticed that in several cases the orientations of binaries lying in the same structural feature have a tendency to be parallel, and in a direction which may coincide with that of the feature. Figure 11 shows a few examples.

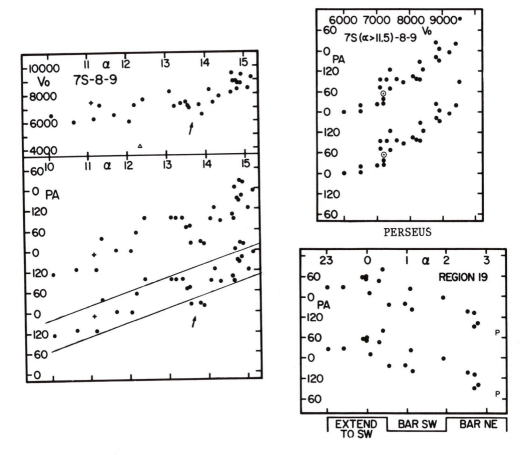

Figure 11 . Orientation of double galaxies in a region comprising the
structures marked LEO 1 and BOO 1 in Figures 5 and 6. The upper left-
hand frame shows the variation of velocity along the region. The lower
righthand frame gives the data for the Perseus supercluster. The data
have been plotted twice to bring out the non-random distribution in
position angles (Tifft 1980). Reproduced by courtesy of the *Astrophysical
Journal*.

Segregation of galaxy types

      Interesting evidence for a segregation of galaxy types in the Perseus
supercluster has been found by Giovanelli *et al* (Figure 12). The upper
left frame has some contamination from fore- and background galaxies, in
the other frames only galaxies with velocities between 3000 and 7500 km s[-1]
have been plotted, and contamination with outside galaxies will be small.
There is a clear progression in the widths of the structural features
from early- to late-type galaxies. A similar phenomenon had earlier been
found by Dressler (1980) in galaxy clusters. Dressler found an outspoken

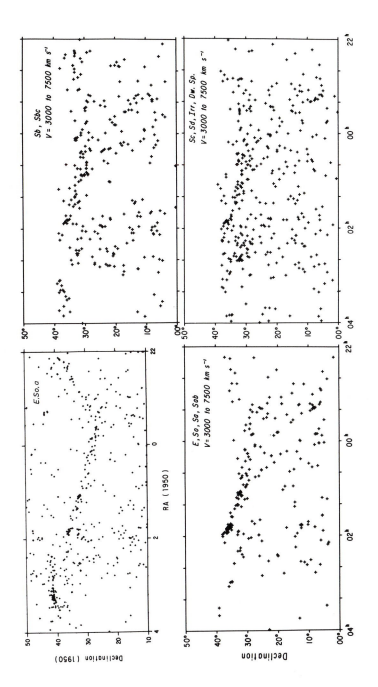

Figure 12. Distribution of galaxies from the catalogue by Zwicky *et al* (1961–1968) in the region of the Perseus supercluster. (Top left) E–SO,Sa for all velocities; the other panels are limited to galaxies with velocities between 3000 and 7500 km s^{-1} (Giovanelli *et al* 1983). I am greatly indebted to the authors for putting their most recent material at my disposal in advance of publication.

correlation between number density of galaxies and their types: where
the density is largest E galaxies prevail, SO galaxies occur most frequently in regions of intermediate density, while the low-density regions
are mainly populated by spirals. These phenomena must in some way be connected with the formation process of galaxies, to which I shall briefly
refer in the last section.

Parallelism of galaxies

Except possibly in the Local Supercluster (cf. MacGillivray *et al*
1982) there is no convincing evidence for parallelism of galaxies in
clusters or superclusters. One remarkable case of parallelism of somewhat
different nature should, however, be mentioned in this connection, viz.
that of the radio galaxy 3C130. This extended source (length 1.9 Mpc) was
found to have two "companions", at about 3 Mpc North and 5 Mpc South,
both equally large as 3C130. The structure of the Northern source is
exactly parallel to that of the central radio galaxy; the Southern source
has a slightly more complicated shape, but its inner region appears again
to be parallel to that of 3C130 (Jägers 1983). Apparently, the axes of
the three "central engines" are parallel.

ORIGIN

It is still unknown how and when galaxies, the "building blocks" of
the Universe, were formed. The only pertinent knowledge we have is that
the bulk of intrinsically bright *quasars* must have been born between
approximately $z = 2.5$ and $z = 3.5$. Quasars are probably galaxies, but
we do not know in how far they are representative of galaxies in general.
There is some evidence that quasars lie in clusters; they *may* have been
born in them. The two very different theories (cf. Zeldovich *et al* 1982,
and references therein) which have been proposed for the origin of galaxies and their clustering illustrate our ignorance. In the "isothermal"
theory it is supposed that galaxies formed first, and that they subsequently clustered in groups, clusters and perhaps superclusters (cf.
Peebles, in Chapter   ). In the alternative, "adiabatic", theory the smal-
lest-size fluctuations which could exist were of supercluster mass. Their
collapse resulted in flat ("pancake") or filamentary structures; galaxies
would have formed as a result of this collapse. In mass as well as shape
the structures formed would resemble superclusters.

In the isothermal scenario the Universe would contain density fluctuations on all scales; it is thought that fluctuations comprising a
mass equal to the Jeans mass at the time of decoupling of matter and
radiation ($\sim 10^6$ $M_\odot$) would be the first to condense, and that by a process of hierarchical clustering these would collect first into galaxies,
and subsequently into larger structures.

Can Superclusters aid in the choice between the two scenario's?

Phenomena which are hard to explain in the pure isothermal picture are, in the first place, the large dimensions of superclusters in general, and, in particular, the extreme sizes found from the distribution of rich clusters; in the second place, their sometimes strongly elongated shapes and the tendency of clusters to be aligned with these shapes. In particular, recent multiparticle simulations made by Dekel *et al* (1983) have indicated that the preferential orientation of rich clusters cannot be produced in the hierarchical scenario, whereas it occurs naturally in the adiabatic scenario.

It is not impossible, however, that the actual universe has a hybrid structure, with isothermal fluctuations superimposed on large-scale adiabatic perturbations, a view favoured by Dekel *et al*, among others. In this model galaxies would form first and collapse subsequently into pancakes or filaments. This collapse would then be largely dissipationless, but can nevertheless lead to flat and thin structures.

Better insight into whether or not the formation of the large structures took place in the gaseous phase can perhaps be obtained from further study of the segregation of different galaxy types, as apparently observed in the Perseus supercluster (cf. Figure 12). A plausible explanation is that the galaxies formed in the dense central parts of the filamentary principal branch would be likely to be born with the smallest angular momentum, and therefore become ellipticals. Equally massive systems formed in the thinner outer regions would have had to collect gas from a larger volume, and were therefore likely to have been endowed with a large angular momentum, and would have become spirals. It should be stressed, however, that there are other plausible ways for explaining the segregation. In particular, Roos & Norman (1979) have suggested that it might be due to the merging of galaxies, whereby spirals would be transformed into ellipticals. Galaxies would then have existed prior to the formation of clusters and superclusters.

The extreme smoothness of the microwave background radiation, with upper limit $\delta T/T < 10^{-4}$ for fluctuations on the scale of superclusters and large clusters, puts interesting constraints on their origin. If density fluctuations corresponding to the above limit started to separate out at the time of decoupling they could not have collapsed in time for the formation of quasars and galaxies. Other phenomena, such as the large virial masses of clusters, the large rotation velocities in the outer region of spirals, the abundance of deuterium, have suggested that most of the mass of the universe consists of invisible, non-interacting particles (for instance, heavy neutrinos). These same particles might also explain the discrepancy between the smallness of the background fluctuations and the epoch of supercluster formation.

For more detailed information on superclusters, see Oort (1983).

14

J. H. OORT

REFERENCES

Abell, G.O.: 1958, Ap.J.Suppl. 3, p. 211.
Bahcall, N.A., Soneira, R.M.: 1983, Ap.J. Submitted for publication.
Binggeli, B.: 1982, Astron. Astrophys. 107, p. 338.
Davis, M., Huchra, J., Latham, D.W., Tonry, J. 1982: Ap.J. 253, p. 423.
Dekel, A., West, M.J., Aarseth, S.J.: 1983, Ap.J. in press.
Dressler, A.: 1980, Ap.J. 236, p. 351.
Einasto, J.,Jôeveer, M., Saar E.: 1980, MNRAS 193, p. 353.
Giovanelli, R., Haynes, M.P. and Chincarini, G.J.: 1983. In preparation.
Gregory, S.A., Thompson, L.A. Tifft, W.G.: 1981, Ap.J. 243, p. 411.
Huchra, J.P., Davis, M., Latham, D.W, Tonry, L.J.: 1983, Ap.J.Suppl.53,
Jägers, W.J.: 1983, Astron. Astrophys, 125, p. 172.   p.89.
MacGillivray, H.T., Dodd, R.J., McNally, B.V., Corwin Jr., H.G.: 1982,
                                                MNRAS 198, p. 605.
Nilson, P.: 1973, Uppsala General Catalogue of Galaxies, Uppsala Obs.
                                                Ann. 6.
Oort, J.H.: 1983, Ann. Rev. Astron. Astrophys. 21, p. 373.
Roos, N., Norman, C.A.: 1979, Astron. Astrophys. 76, p. 75.
Roos, N.: 1981, Astron. Astrophys. 95, p. 349.
Roos, N., Aarseth, S.J.: 1982, Astron. Astrophys. 114, p. 41.
Tifft, W.G.: 1980, Ap.J. 239, p. 445.
Zeldovich, Ya.B., Einasto, J., Shandarin S.F.: 1982, Nature 300, p. 407.
Zwicky, F., Wild, P., Herzog, E., Karpowicz, M., Kowal, C.T. 1961-1968.
        Catalogue of Galaxies and Clusters of Galaxies, Vols. 1-6.
        Pasadena: Calif.Inst.Technol.

DISCUSSION

SHAPIRO: It is true that those fluctuations in the microwave background on the scale of superclusters ($\sim 10'$-$20'$) which would result from the existence of an adiabatic density fluctuation on this scale at recombination can be eliminated by the reionization of the intergalactic gas after recombination. However, this effect will not eliminate the microwave background fluctuations on the very largest angular scales, as follows. If we assume that the primordial spectrum of density fluctuations has a power spectrum $|\delta_K| \propto K^n$, where K is wavenumber and n is some integer, then whatever the primordial amplitude must be on the supercluster scale in order to make superclustersized fluctuations achieve nonlinearity by the present, there is implied some other amplitude on the scales greater than the horizon at recombination. On this scale (which corresponds to $\gtrsim 2°$ in angle) the principal source of microwave background temperature fluctuation is the so-called Sachs-Wolfe effect which results in $(\delta T/T) \sim (\delta\rho/\rho)_H$, the value of $\delta\rho/\rho$ at recombination on the scale of horizon. For a Zel'dovich (n=1) spectrum, for example, if $(\delta\rho/\rho)_H > 10^{-4}$ is required on the supercluster scale when that scale first enters the horizon, then $(\delta\rho/\rho)_H$ has the same value at recombination for the scale which then fills the horizon. Hence, $(\delta T/T) > 10^{-4}$ results for scales which enter the horizon after recombination. This means that measurements of the large angular scales (i.e. $\gtrsim 2°$), scales which are too large to have their $\delta T/T$ eliminated by reionization, actually constrain the value of $\delta\rho/\rho$ on the supercluster scale as well. As a result, even with reionization, the value of $\delta\rho/\rho$ required to make superclusters form by the present in a baryon-dominated universe, is large enough to imply a violation of the microwave background observations unless the primordial density fluctuation spectrum has a steep fall off with wavelength (i.e. n>1), which is somewhat contrived.

OORT: This is a very interesting remark which indicates that reionization is not likely to have been important and the "inos" may indeed be required to explain the absence of background fluctuations.

BURBIDGE: We see that galaxies have preferred axes; binaries have preferred axes; clusters have preferred axes; superclusters have preferred axes as do strings of cluster. Also in this category there are radio sources, jets, etc. which are thought to be short lived phenomena due to non-thermal processes. On the other hand galaxies are thought to be old relaxed systems. But the same geometry is present, as some galaxies lie along radio axes - for example galaxies around Centaurus A (NGC 5128) and the M87 jet. Should we perhaps think about galaxy formation through an explosive process going on continuously as is the case for radio sources, and not attempt to force the picture of all galaxies being formed in the early universe? The conflict between the smooth microwave background and the structures seen in galaxies is very hard to decide within conventional cosmology.

OORT: I agree that new galaxies may well form today and that they may form in an unconventional manner. However, I do not see how this can alleviate conflict between the existence of superclusters and the smoothness of the background radiation.

THREE-DIMENSIONAL STRUCTURE IN THE INDUS SUPERCLUSTER

S.M. Beard,[1] J.A. Cooke,[1] D. Emerson.[1], B.D. Kelly.[2]

1. Dept. of Astronomy, Edinburgh University, Edinburgh, Scotland
2. Royal observatory, Blackford Hill, Edinburgh, Scotland

ABSTRACT

Redshifts of large numbers of galaxies can be obtained to an accuracy of ± 0.01 in Z by measurement of their spectra on UK Schmidt Telescope objective-prism plates using the COSMOS plate-scanning machine. 2294 redshifts have been measured in a 4.7 X 5.2 deg^2 area of the sky which includes part of the Indus supercluster. The redshifts enable the three-dimensional distribution of the galaxies within the supercluster to be investigated, and a subset of this data is presented here. A 6.8 h^{-1} bridge of galaxies connecting the rich clusters 2151-5805 and 2143-5732 is revealed (h = Ho/100 km sec^{-1} Mpc^{-1}). A concentration of galaxies, apparently coincident with 2151-5805 on the sky, is found to lie in the background at a higher redshift, illustrating how the knowledge of redshifts can resolve structure in three dimensions. Much faster and more objective automatic methods of determining the redshifts are currently being developed.

INTRODUCTION

Cooke (1980) and Cooke et al. (1981) have shown it is possible to determine the redshifts of galaxies, to B $\sim$ 19, using spectra obtained from UK Schmidt Telescope (UKST) plates taken using the low-dispersion objective-prism. The objective-prism produces spectra, $\sim$ 1mm in length, between the sensitivity cut-off of the IIIa-J emulsion (5380Å) and the ultraviolet cut-off of the atmosphere and telescope optics (3200Å) at a reciprocal dispersion of 2480Å mm^{-1} at Hγ (4340Å) (Nandy et al. 1977). Redshifts are obtained using the separation of the IIIa-J emulsion cut-off and the 4000Å continuum break found in the spectra of elliptical and early-type spiral galaxies (see Palmer 1984, Parker et al. 1984). Details of the techniques used to extract spectra from COSMOS measurements of UKST plates are described in Cooke et al. (1983a). Because of the low dispersion of the objective-prism spectra, individual redshifts can only be obtained to an accuracy of ± 0.01 in Z.

17

F. Mardirossian et al. (eds.), Clusters and Groups of Galaxies, 17–22.
© *1984 by D. Reidel Publishing Company.*

Visual counts of galaxies to B ∼ 19 by Corwin (1981) revealed an 8°
by 10° annular concentration of galaxies extending over nine SERC/ESO
survey fields in Indus. An apparent filamentary concentration of
galaxies extends southwards from the annulus like the tail of a tadpole.
Beard (1983) has obtained 2294 redshifts in survey field 145 centred at
21 32 -60 00 (1950 coords.). The two richest clusters in the Indus
supercluster annulus (2151-5805) and 2143-5732) lie in the north-east
corner of this field, and the cluster 2131-6215 lies at its southern
edge. Part of the supercluster "tail" extends down the eastern edge of
the field between these clusters.

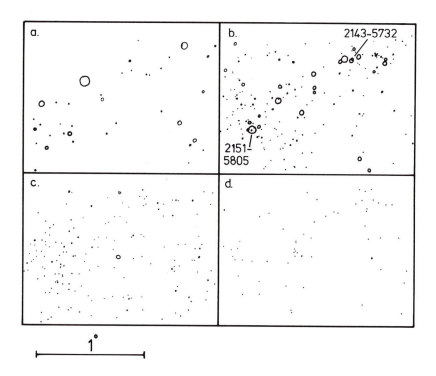

Figure 1. The distribution of galaxies on the sky, in the region
containing clusters 2151-5805 and 2143-5732, sampled in the redshift
intervals: (a) 0.00-0.05; (b) 0.05 - 0.10; (c) 0.05-0.10; (d) 0.15-
0.20. Each galaxy is represented by a circle of radius proportional to
its apparent brightness.

RESULTS

Detailed results from all the redshifts obtained in field 145 are
presented in Beard (1983) and Beard et al. (1983). No evidence was
found for a consistent excess of galaxies with redshifts in the same
range (0.07-0.08) as that of the Indus supercluster on the eastern side

of the field, showing that any supercluster "tail" must exist only at a
low level of significance.

Figs 1a,b,c and d show plots of the two rich clusters in the
north-east corner in four redshift intervals. A "bridge" of galaxies
can be seen extending between the two clusters, which is most prominent
in the same redshift interval occupied by the clusters Fig. 1b). The
surface density of galaxies on the sky in this bridge area is four
times that of the surrounding background. If redshift histograms are
plotted for galaxies in the two clusters, and in the intervening
region (Fig. 2), a peak in the redshifts can be seen near $Z \sim 0.08$ in
all three distributions.

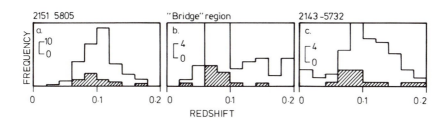

Figure 2. Histograms of the redshift distribution for galaxies in
regions of sky containing the clusters in Fig. 1 and in the intercluster region. Shaded histograms are for redshifts which have been
assigned a higher confidence by Beard (1983).

The 70 arcminute separation of these clusters on the sky, and their
slit spectra redshifts of 0.0760 and 0.0743 given by Corwin & Emerson
(1982) mean the bridge has a length of 6.8 h^{-1} Mpc (where h = Ho/100 km
sec^{-1} Mpc^{-1}). A semi-automatic star/galaxy separation technique, using
the method of MacGillivray & Dodd (1980) has been applied to a COSMOS
measurement of the direct plate of field 145. The distribution of the
galaxies detected confirms the existence of this connecting bridge
(Beard 1983). Filamentary bridges of galaxies between clusters have
been a common feature of recent redshift surveys (e.g. Gregory & Thompson
1978; Chincarini, Rood & Thompson 1981). The bridge found connecting
2151-5805 and 2143-5732 in this paper is much smaller than the
20-50 h^{-1} Mpc lengths of filaments detected typically. It may be part
of Corwin's much larger annulus, but further redshifts in surrounding
fields are needed to verify this.

On Fig. 1c, a concentration of galaxies can be seen at a higher
redshift than 2151-5805, but nearly coincident with the position of
this cluster on the sky. Fig. 3 is a cone diagram of this area, also
showing this background grouping. The images in this grouping are
associated with a group of spiral galaxies seen on the direct plate,

which is very likely a background cluster of galaxies. The cluster
2151-5805 is the richest one in the Indus supercluster, containing over
500 galaxies per square degree in Corwin's (1981) survey. But, in the
objective-prism survey of this cluster, 65 of the 159 galaxies in the
area are found to belong to the more distant cluster. Thus, the richness
of the cluster may have been overestimated by a factor of about 1.5.

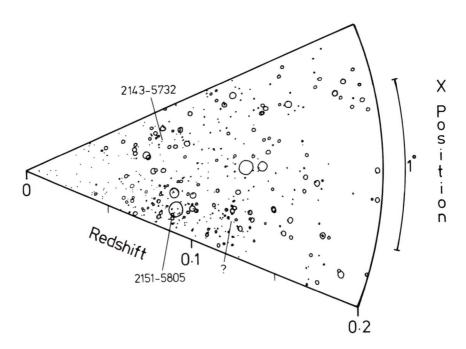

Figure 3. A "cone" diagram of X position on the plate (or R.A. on the
sky) against redshift for the cluster region of Fig. 1 showing the
appearance of a background grouping behind cluster 2151-5805. Each
galaxy is represented by a circle of radius proportional to its luminosity.
Note that the sky angle has been exaggerated by a factor of ~25.

CONCLUSIONS

The use of the redshift information to resolve structure in three
dimensions is demonstrated by the detection of a cluster of galaxies
behind 2151-5805 and the filamentary bridge between this latter cluster
and 2143-5732. These objective-prism redshifts are not of sufficient
accuracy to resolve individual cluster velocity dispersions, but they
are able to detect large-scale structures such as voids, filaments and
superclusters (see Parker et al. 1984). Ways of improving the accuracy
are being investigated.

The results presented here are only a subset of a very tedious manual redshift survey of an entire UKST field by Beard (1983), in which the 4000Å feature was located on each spectrum manually with a cross-hair cursor on a graphical VDU. This manual survey is being used by Cooke et al. (1983bc) to perfect an automated method of redshift determination, which will rely on a pattern-match using the spectrum of an Oke & Sandage (1968) standard elliptical as a template. Automatic redshifts will be very much faster to obtain and will be free of the subjectiveness inherent in any manual measurement. They will enable successive surveys to be made in overlapping UKST fields, and a large-scale survey to be built up in the future.

(SMB was supported by an SFRC studentship).

REFFRENCES

Beard,S.M. : 1983, Ph.D. thesis, University of Edinburgh, in preparation.

Beard, S.M., Cooke, J.A., Emerson, D. & Kelly, B.D. : 1983, Mon.Not.R.astr.Soc, submitted.

Clowes, R.G.: 1984, in "Astronomy with Schmidt-type telescopes", ed. M. Capaccioli, D. Reidel, p. 107.

Cooke, J.A. : 1980, Ph.D. thesis, University of Edinburgh.

Cooke, J.A., Emerson, D., Kelly, B.D., MacGillivray, H.T. & Dodd,R.J.: 1981, Mon.Not.R.astr.Soc, 196, pp.397.

Cooke, J.A., Beard, S.M., Kelly, B.D., Emerson, D. & MacGillivray, H.T. : 1983a, Mon.Not.R.astr.Soc, submitted.

Cooke, J.A., Emerson, D., Beard, S.M. & Kelly, B.D. : 1983b, in "Proceedings of the workshop on astronomical measuring machines, 1982", pp.209, eds. R.S. Stobie, B. Mcinnes, Edinburgh.

Cooke, J.A., Kelly,B.D., Beard, S.M. & Emerson, D. : 1983c, in "Astronomy with Schmidt-type telescopes", ed. M. Cappaccioli, Asiago, in press.

Chincarini, G., Rood, H.J. & Thompson.L.A. : 1981, Astrophys. J. lett., 249, pp.L47.

Corwin, H.G., jr. : 1980, Ph.D. thesis, University of Edinburgh

Corwin, H.G. jr. & Emerson, D.: 1982, Mon.Not.R.astr.Soc, 191, pp.1.

Gregory, S.A., & Thompson, L.A. : 1978, Astrophys. J., 222, pp. 784.

MacGillivray, H.T. & Dodd, R.J. : 1980, Mon.Not.R.astr.Soc, 193, pp.1.

Nandy, K., Reddish, V.C., Tritton, K.P., Cook, J.A. & Emerson, D. : 1977, Mon.Not.R.Astr.Soc, 178, pp.63P.

Oke, J.B., & Sandage, A. : 1968, Astrophys, J., 154, pp.21.

Palmer, J.B.: 1984, in "Clusters and Groups of Galaxies", this volume, p. 71.

Parker, Q.A., MacGillivray, H.T. and Dodd, R.J.: 1984, "Clusters and Groups of Galaxies" this volume, p. 595.

ADDENDUM

Please note that an account of the properties of the Indus supercluster may also be found in:
Corwin, H.G.: 1983, I.A.U. Symposium 104 "Early Evolution of the
        Universe and its Present Structure", eds. G.O. Abell &
        G. Chincarini, Reidel, pp. 293.

DISCUSSION

CARTER: What a proportion of your redshifts do you get wrong because
of misidentification of the continuum break?

BEARD: There are two other features it is possible to mistake for
the 4000 Å continuum break; one at 3600 Å and one at 4470 Å. Mistaking
either of these for the 4000 Å feature will result in redshifts
discrepant by ± 0.1. 283 of the redshifts in field 145 have been
measured a second time to test for repeatability, and in 5% of
these new measurements a different feature was chosen.

PARKER: Because you identified galaxies by their spectra only,
what amount of stellar contamination do you expect in your sample?

BEARD: The spectrum of a late-type star will be misidentified as
that of a galaxy when its 4470 Å feature is mistaken for a 4000 Å
feature at a redshift of about 0.12. In another paper (Beard et
al. 1983) we have plotted a redshift histogram and compared it
with the redshift distribution expected from a universe of randomly-
distributed galaxies. We find an 11% excess of redshifts in the
range 0.09–0.13, which we believe to be due to stellar contamination.

CORBYN: Have you considered, as a possible explanation for the
observed filamentary structures, the effects of gravitational clustering that could take place around supermassive strings in the early
universe?

BEARD: No! The connecting bridge between the two clusters is much
smaller than the large-scale filamentary structure reported elsewhere
in the literature, and the system should be defined more correctly
a binary cluster. We do not yet know if the large-scale annulus
of the Indus supercluster exists in 3-D space.

THE SPATIAL DISTRIBUTION OF GALAXIES IN THE SOUTHERN SKY

A.P. Fairall and H. Winkler
Department of Astronomy, University of Cape Town, Rondebosch,
7700 Cape, South Africa.

Our present understanding of the large scale distribution and superclustering of galaxies rests mainly on the relative wealth of redshifts
obtained in the northern hemisphere. It will be some years before the
quantity and quality of redshifts in the southern hemisphere can comple-
ment the northern data and correct the imbalance. In the meantime, we
wish to offer a preliminary examination using available data - albeit
somewhat heterogeneous.

The region of concern is the sky south of Declination -30°. A
search has been made for all available redshifts and a simple catalogue
prepared (Fairall, Lowe and Dobbie - see also references therein). It
includes references published or available to mid-1983 (but it is not
a definitive work and no guarantees of completeness can be given).

The catalogue shows that some 2600 galaxies (south of -30°)
have been observed spectroscopically, from which reliable redshifts are
available for about 2200 galaxies (the balance having produced near
featureless spectrograms - emission lines can be excluded). The first
author's interest comes partly because he has contributed about a third
of these redshifts and the catalogue and spatial distribution plots serve
to see how his work stands in relationship to other data. Whilst the objects he has observed are morphologically "compact and bright-nucleus
galaxies" and one might question whether they obey the same spatial distribution characteristics as do other galaxies, the advantage is that it
is an overall survey of almost the entire area in question (galaxies are
selected from scanning the ESO and UK Schmidt sky surveys).

The second author has prepared the plots that accompany this
article. The data has been subdivided in Declination zones and plots of
V_o (=$c*z$ + 300 sin ℓ cos b) vs. Right Ascension are shown for each of
the zones. The concentric grid lines are spaced at intervals of 2000
km s^{-1}. The plots convey spatial distribution, except allowance must
be made for intrinsic motions within clusters (thereby elongating clusters in a radial direction). Aside from the distinction between Fairall
galaxies and others, Seyfert galaxies are shown by "sun" symbols.

F. Mardirossian et al. (eds.), Clusters and Groups of Galaxies, 23–28.
© *1984 by D. Reidel Publishing Company.*

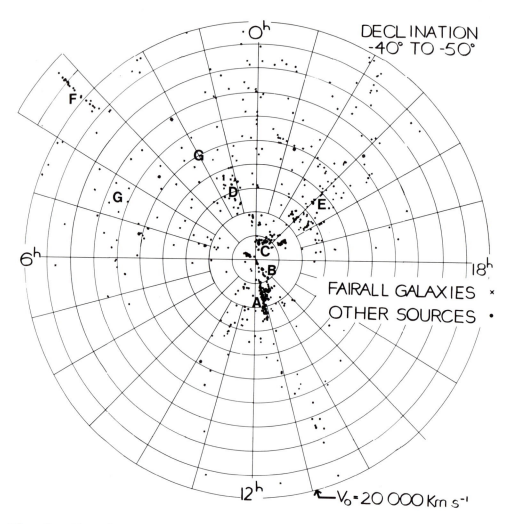

Fig. 1. The -40° to -50° Declination Zone is a suitable starting point.
The most conspicuous concentration (labelled A) is the Centaurus cluster.
Its main body shows a spread of over 3000 km s^{-1} in radial velocity but
it may have a dynamically composite structure - as discussed by Lucey
elsewhere in this volume.   There is also the impression of an intercon-
necting filament (B) to our local supercluster.   The concentration at C
may be a part of our supercluster (Oort 1983).   Elsewhere there is weaker clustering D and E, a part of the Horologium supercluster F, but the
general impression is a wide scattering of galaxies.   It would also appear, at first sight, that the Fairall galaxies tend to fill in the open
spaces (for example at G).   However, considerable caution should be exer-
cised in this interpretation because it is obvious that other observers
have not undertaken general surveys but have concentrated on particular
clusters leaving vast expanses of virgin territory inbetween.   Almost
all the Seyfert galaxies in this zone are Type 2.

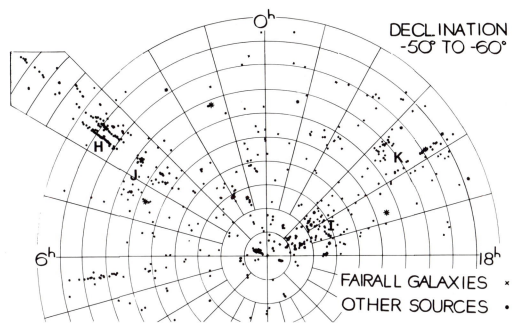

Fig. 2.   This Zone reveals two major structures, the Horologium Cloud
(H) recently discussed by Lucey *et al.* (1983) and a more open concentration in Indus (I) which links to that labelled E in the previous plot.
The three main concentrations of the Horologium supercluster (H in this
plot and F in the previous plot) show a spread of over 3000 km s^{-1} - as
did the Centaurus cluster.   Although the author selected a number of apparent members of the Horologium cloud for observation, almost all turn
out to be foreground objects (J) - a feature also found by Chincarini
(private communication - data not yet published nor included here).   The
concentration of galaxies around K also suggests a possible supercluster.
Most of the Seyfert galaxies here are Type I (longer spikes).

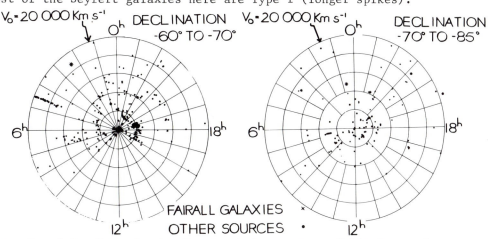

Figs. 3 and 4.   Data in the far south is sparse and little can be sur-
mised about spatial structure.

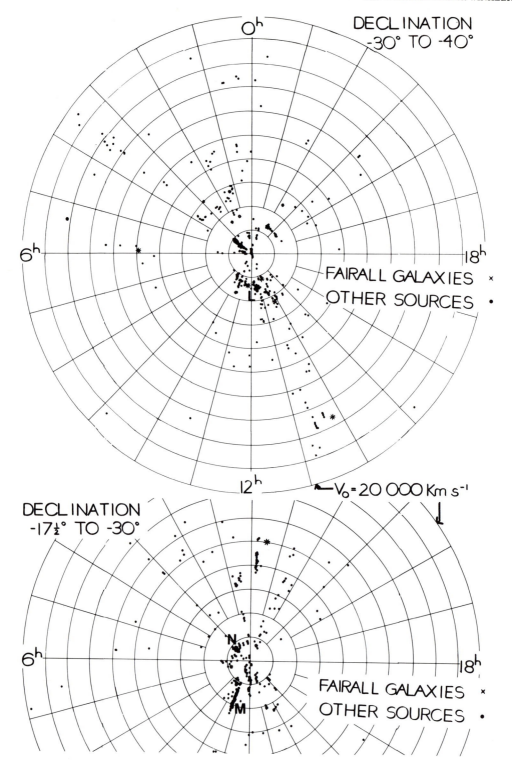

Figs. 5 and 6 (opposite page): Moving northwards (Fairall survey incomplete; catalogue stops at -30°, but some data available to -17½°),
the most dominant structures are the Centaurus-Hydra supercluster (L
and Hydra cluster M) and portions of the local supercluster (including
Fornax cluster N) - the two superclusters seem almost interconnected.
Nearby (low luminosity) Seyfert galaxies favour clusters but distant
ones (higher luminosity) seem to prefer open spaces.

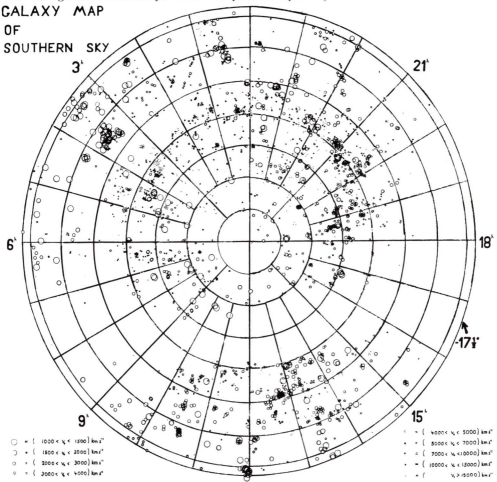

GALAXY MAP OF SOUTHERN SKY

Figure 7 (above): The second author's plot of all the galaxies projected on the sky.

REFERENCES
Fairall, A.P., Lowe, L. and Dobbie, P.J.K., 1983, "A Catalogue of Gala-
xies South of Declination -30° that have been Observed Spectroscopically",
Publ. of the Department of Astronomy, Univ. Cape Town, No. 5.
Lucey, J.R., Dickens, R.J., Mitchell, R.J. and Dawe, J.A., 1983. *Mon.
Not. R. astr. Soc.*, 203, 545.
Oort, J.H., 1983, "Superclusters" to appear in Vol. 21, *Ann. Rev. Astr.
and Astrophys.*

DISCUSSION

LORENZ: Can you quote differences in the properties of the Seyfert galaxies as a function of their position in space relative to clusters of galaxies?

FAIRALL: As mentioned, there seems to be a tendency for the more distant higher luminosity Seyfert galaxies to prefer open spaces; this is certainly the case for three extreme Seyferts in the sample. Obviously, the higher luminosity ones are mainly Seyfert 1. However, all this is tentative since the open spaces may simply reflect the incompleteness of the survey.

WHITTLE: Relating to your statement concerning the distribution of Seyfert galaxies, I have investigated the galactic environment of a sample of 120 Seyfert galaxies and find a weak tendency that the more luminous Seyfert galaxies are found preferentially in regions of lower galaxy density.

RECENT STUDIES OF THE LOCAL SUPERCLUSTER

G. de Vaucouleurs
The University of Texas
Austin, Texas   78712

ABSTRACT

    After a brief historical review of the discovery of the Local
Supercluster, this paper reports on recent research on the space
distribution of nearby galaxies, on the space motions of the Sun,
Galaxy and Local Group relative to these nearby galaxies, and on a
first empirical mapping of the velocity field near the plane of the
Local Supercluster.

1. INTRODUCTION

    The first attempt to search for a local supercluster or "galaxy of
the second order" was made by Charlier (1908, 1922) who noted the
excess density of "nebulae" in the north galactic hemisphere. Both
Reynolds (1921, 1923) and Lundmark (1927) commented on the remarkable
concentration of the largest "extragalactic nebulae" toward a great
circle of the sphere but drew no conclusions from this fact. Shapley
(1930) introduced the word "supergalaxy" in the context of his proposal
that our own Galaxy was a flattened cluster of galaxies, and for a few
years applied the word to various clusters of galaxies.[1] Zwicky (1938)
was probably the first to suggest that the Virgo cluster and its
extensions to the north and south (the UMa cloud and the "southern
extension" of Virgo) formed a vast system, perhaps reaching the Local
Group, although in later years he always opposed the concept of
superclustering, both locally and at large.

    The first definite quantitative indications of a local density
excess on a scale larger than ordinary groups and clusters were
obtained by Holmberg (1937) and Reiz (1941) from their analyses of
galaxy counts at Harvard and Heidelberg. But their results, con-
tradicting the established dogma of statistical uniformity, were
generally ignored at the time. It took nearly forty more years before
the reality of the Local Supercluster became generally accepted.

F. Mardirossian et al. (eds.), Clusters and Groups of Galaxies, 29–42.
© 1984 by D. Reidel Publishing Company.

In 1951 V. Rubin
used the great circle
noted by Reynolds and
Lundmark as a
"universal equator"
in her search for
differential rotation
effects in galaxy
redshifts.   In 1953 I
proposed that the
concentration of the
brighter galaxies
along a great circle
of the sphere was the
trace of a flat
supersystem or "Local
Supergalaxy" centered
in or near the Virgo
cluster and including
the Local Group in an
outlying position.
The growing evidence
in support of this
hypothesis, and of
superclustering in
general, was reviewed
in 1956 and again in
1960 (de Vaucouleurs
1956, 1960), but was
still generally
ignored or dismissed
as sheer speculation.

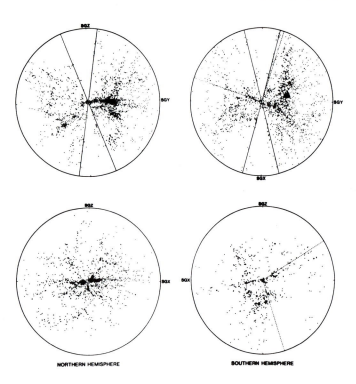

Figure 1
Projections of galaxy distribution in redshift space unto the XY, YZ and ZX supergalactic planes (Tully 1982).

Nevertheless, the evidence continued to grow. An analysis of radial
velocities in the HMS catalogue indicated the presence of non-linearity
and anisotropy in the local velocity field, consistent with a model of
the Local Supercluster in differential expansion and rotation
(de Vaucouleurs 1958). Then, the generality of the superclustering
phenomenon was demonstrated by Abell (1958) through a statistical
analysis of his catalogue of rich clusters. Interestingly the
indicated characteristic scale of superclustering was consistent with
the diameter of the Local Supercluster, about 30-40 Mpc if H = 100
km/s/Mpc.

Some specific objections to the supercluster hypothesis or to the
reality of the velocity anistropy (van Albada 1962, Teerikorpi 1975,
Sandage and Tammann 1975, Bahcall and Joss 1976) were examined and
found to be invalid or implausible (de Vaucouleurs 1964, 1976a,b, 1978).
The supergalactic anisotropy of the velocity field was repeatedly
confirmed as more redshift data became available (de Vaucouleurs 1966,
de Vaucouleurs and Peters 1968, de Vaucouleurs and Bollinger 1979).
By the time of the Tallinn meeting (IAU Symposium 79) in 1977 the

evidence for superclustering in general, and the Local Supercluster in particular, had become overwhelming, and study of this new field has grown rapidly in the past few years.[2]

The recent research efforts have been along three main directions:

(1) Improved mapping of the space distribution of galaxies, whose distances were derived initially from redshifts (de Vaucouleurs 1981, Tully 1982a), but recently from more reliable distance indicators (de Vaucouleurs and Peters 1983b);

(2) Derivation of the space motions of the sun, the Galaxy and the Local Group relative to frames of reference defined by nearby galaxies either empirically with a minimum of assumptions (de Vaucouleur and Peters 1981, 1983a, de Vaucouleurs et $al.$ 1981) or through some theoretical models of the Local Super-cluster (Davis et $al.$ 1980, Hoffman, Olson and Salpeter 1980, Schecter 1980, Aaronson et $al.$ 1982).

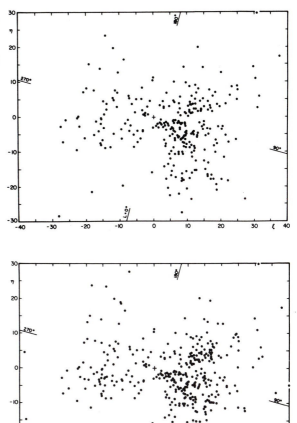

Figure 2

Projection unto XY supergalactic plane of the space distribution of galaxies with known distances within 5 (above) and 10 Mpc from supergalatic plane (de Vaucouleurs and Peters 1983)

(3) Empirical mapping of the velocity field within 30-40 Mpc for galaxies having good distance estimates independent of redshift and independent of any model for the supercluster (de Vaucouleurs and Peters 1983b).

2. THE SPACE DISTRIBUTION OF GALAXIES

In the first approach (Tully 1982a) the 'corrected' redshift V_0 (as defined by IAU Commission 28 in 1973), $V_0 = V + 300 \cos A$, was used

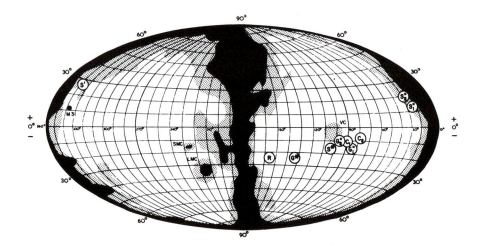

Figure 3

Sky map in supergalactic coordinates showing directions of centroids C_1, C_2 of two samples of 200 and 300 galaxies within $\sim$30 Mpc from Galaxy and of apices of the sun (S) and Local Group (G) relative to different extragalactic frames of reference (': Local Group; ": nearby galaxies; *: background radiation). Note that none points toward Virgo cluster (VC).

as a distance indicator under the assumption that the ideal (linear, isotropic) Hubble law applies. This has the advantage of maximizing the number of galaxies available for the study (only the redshifts need be known), but the disadvantage that the space distribution is distorted in an unknown fashion by departures from the Hubble law. The results can be shown either in projection on the XY, YZ, and ZX planes (in supergalactic coordinates) (Figure 1) or in pseudo-relief graphics. The latter is especially helpful in identifying galaxy groups and clouds, and most of the nearby groups previously known (de Vaucouleurs 1975) can be recognized (Tully 1982a,b).

In the second approach (de Vaucouleurs and Peters 1983a) actual distances derived from a variety of indicators, including the luminosity index (de Vaucouleurs 1979), the HI line width (Bottinelli _et al._ 1980, 1983), the stellar velocity dispersion (de Vaucouleurs and Olson 1982), and inner ring diameters (Buta and de Vaucouleurs 1983a,b), are used to map the space distribution in slabs parallel to the supergalactic plane (Figure 2). This has the advantage of avoiding the assumption of a linear Hubble law, but the disadvantage that the sample is limited to a small fraction (half or less) of the galaxies available, because of the necessity of having good magnitudes and distance indicators. Another, more serious, limitation is that at the present time reliable information of this kind is substantially limited to galaxies in the Shapley-Ames catalogue and suffers from serious incompleteness beyond ∿ 30 Mpc. This limitation will be pregressively reduced by ongoing programs at many observatories.

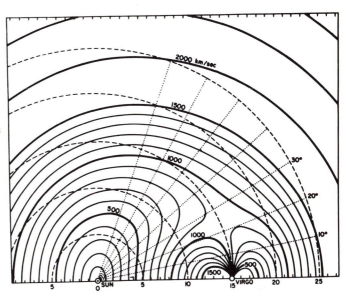

An important result of such studies is that neither a spherical model, nor a flat circular disk model (both centered at the Virgo cluster) are good approximations of the actual space distribution of galaxies. The centroid C of the distribution of 600 galaxies within ∿ 30 Mpc from the Galaxy is in the direction of supergalactic coordinates L = 78°, B = -16°, or nearly 30° to the north-

Figure 4

Theoretical velocity field of a non-rotating "Virgocentric" model of the Local Supercluster (Tonry 1981).

east of the Virgo cluster (Figure 3). Further, there is a large vacuole in space beyond the Virgo cluster in a region where circular symmetry would predict a large density excess (Figure 1). Finally, the galaxy clouds outside the flat disk component which defines the supergalactic plane are not distributed in a spheroidal corona, but tend to be elongated, cigar-shaped concentrations pointing toward the Virgo cluster (or perhaps the center of mass) (Figure 2).

3.   THE SPACE MOTIONS OF THE SUN, THE GALAXY AND THE LOCAL GROUP

In recent attempts to build dynamical models of the LSC combining the distribution and velocity information, a non-rotating spherically symmetric model of the LSC centered at the Virgo cluster and obeying a simple power law of density distribution, such as $\rho \propto r^{-2}$, has been assumed by many authors (Peebles 1976, Hoffman, Olson and Salpeter 1980, Tonry 1980, Schecter 1980, Davis *et al.* 1980). This necessarily results in a fictitious "Virgocentric" component of the total velocity. This model predicts a velocity field, symmetric about the direction of Virgo (Figure 4), which is not supported by observation. As was shown already in our initial study (de Vaucouleurs 1958), a rotational component is required to account for the asymmetry of the velocity field. This has been repeatedly confirmed (de Vaucouleurs 1964, Stewart and Sciama 1967, de Vaucouleurs and Peters 1968, de Vaucouleurs

and Bollinger 1979), and
again in a multi-parameter
fit of a spherical model
(Aaronson *et al.* 1982).
In other words, a simple
dipole component term
allowing for a constant
velocity vector of the
Sun (or Galaxy, or Local
Group) is not an adequate
representation of the
velocity field.

Two recent attempts
to define a "mean"
velocity vector for the
motion of the Sun relative
to the frame of rererence
defined by two all-sky
samples of galaxies in
the distance interval
$2 < \Delta < 30$ Mpc, without
any explicit assumption
of a model for the SLC
(but with the implicit
assumption that the
velocity residuals form
a random vector field
after subtraction of

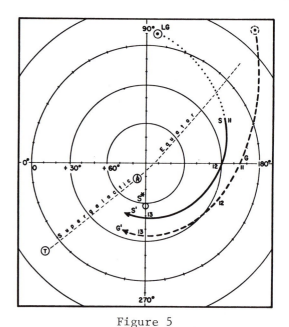

Figure 5
Drift of solar and galactic apices with
respect to galaxies of different apparent
magnitudes (de Vaucouleurs and Peters
1968). S* is solar apex relative to
background radiation.

the reflex solar motion) gave fairly consistent results: the total
sample of 400 galaxies, of which 200 had distances from the optical
luminosity index (de Vaucouleurs and Peters 1981) and 300 from the
radio HI line width (de Vaucouleurs *et al.* 1981) (100 are in common),
gave a solar velocity of 323 km/s toward $L = 26°$, $B = -13°$, and a
Local Group apex close to the direction of the centroid of the space
distribution of galaxies in the sample (Figure 3).

However, the most recent study (de Vaucouleurs and Peters 1983a)
of an enlarged sample of some 600 galaxies having distances by several
indicators confirms that the concept of a simple dipole term
independent of distance is not an adequate representation of reality.
In a previous study (de Vaucouleurs and Peters 1968) of the motion of
the sun and the Galaxy relative to ∿ 1000 galaxies in differnt
intervals of apparent magnitudes ($10 < m < 14$) - the only distance
indicator available at the time - we had found that the direction of
the galactic apex drifted rapidly as the mean apparent magnitude of
the reference galaxies increased from $m \simeq 11$ to $m \simeq 13$, but then seemed
to become stationary toward $L \simeq 140°$, $B \simeq -15°$ when referred to
galaxies fainter than $m \simeq 13$ (Figure 5). At the time this direction
in space had no special significance and most of the displacement could
be explained by the rotating-expanding disk model of the LSC with the

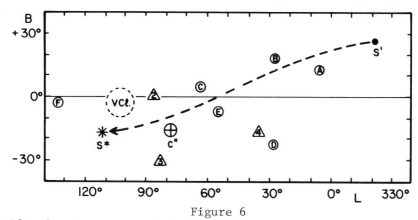

Figure 6
Drift of solar apex relative to different extragalactic frames
of reference (de Vaucouleurs and Peters 1983).

addition of a Z component (de Vaucouleurs 1972, 1978). It is now clear
that this 'asymptotic' direction of the apex with respect to the
fainter and more distant galaxies (<cz> > 3000 km/s) was the first
determination of our 'absolute' motion in space identified today as
the motion with respect to the background radiation.[3]

Our latest solutions for the solar motion with respect to six
samples, each including 100 galaxies, in distance intervals from 6 to
36 Mpc, supplemented by two samples from 21 cm data (Hart and Davis
1982), fully confirm the drift of the apex (Figure 6). All show
clearly the progressive transition from small velocities characteristic
of the local frame to the larger values characteristic of the background frame. For example, the toal velocity V_G of the Local Group
increases steadily from ∿ 0 with respect to the nearby co-moving frame
defined by the nearest galaxies to ∿ 650 km/s with respect to the
background radiation (Figure 7). Simultaneously, the apex moves from
$L \simeq 75°$, $B \simeq -17°$ with respect to galaxies of mean redshift $<cz> \simeq 1000$
km/s, to $L \simeq 110°$, $B \simeq -25°$ with respect to galaxies of $<cz> \simeq 2000$-
2500 km/s and, finally, to $L* = 132°$, $B* = -24°$ with respect to the
background radiation. It is remarkable that most of the variation
takes place in the first 40-50 Mpc; the trend of the curves suggests
that the frame of reference defined by galaxies having redshifts as
small as ∿ 6,000 km/s is essentially at rest with respect to the
background radiation. Conversely, this result suggests that, except
for an improbable coincidence, any intrinsic dipolar anisotropy of the
background radiation is probably $< 10^{-4}$.

4.   THE VELOCITY FIELD NEAR THE SUPERGALACTIC PLANE

With the progress in the development of relatively reliable

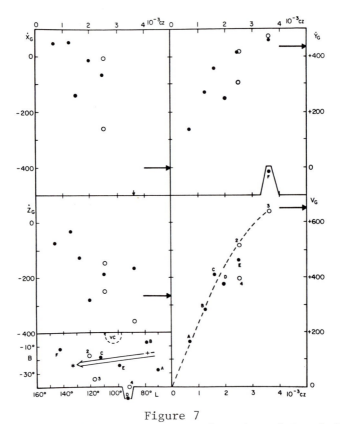

Figure 7
Supergalactic velocity components of motion of Local Group
with respect to galaxies in different distance intervals.

distance indicators (see de Vaucouleurs 1982a,b and references therein)
and the publication of lists of distance moduli independent of redshifts for fairly large samples (several hundreds) of galaxies, it has
become possible to map the departures from the ideal uniform-isotropic
expansion postulated by the Hubble law. Preliminary attempts have
been recently reported by Karoji (1983) and by Brosche, Hövel and
Lentes (1983) who used our sample of 458 galaxies having distances
from the luminosity index (de Vaucouleurs 1979) to map the variations
of the apparent Hubble ratio in space or over the sphere (Figure 8).

Recently we have completed a more detailed empirical study of the
velocity field in or near the supergalactic plane (de Vaucouleurs and
Peters 1983b). This study is based on a larger sample of $\sim$ 500
galaxies selected for the high-quality of their distance moduli and
is independent of any assumption on the shape and density structure
of the LSC, except for the evident fact that it is strongly flattened
toward its equatorial plane (de Vaucouleurs 1976a, 1981; Tully 1982).
This fact allows us to restrict our attention to galaxies within 30°

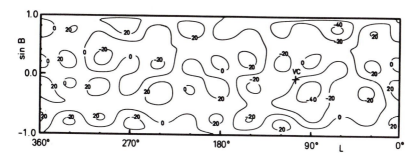

Figure 8

Departures of apparent Hubble ratio V_0/Δ from all-sky average
($<H*> = 107$) for sample of 400 galaxies having distances from
luminosity index, from harmonic analysis in supergalactic
coordinates (Brosche, Hövel and Lentes 1983).

from the supergalactic equator and within a few megaparsecs from its
plane. These restrictions entail only a modest reduction in the
number of galaxies having known distances and redshifts. The galaxies
included in the sample are those of all types having the best determined distance moduli derived from one to four distance indicators
(luminosity index, HI line width, velocity dispersion, inner rings)
in the distance interval $2 < \Delta < 32$ Mpc, in which incompleteness is
not serious and causes no appreciable bias. The linearity of the
distance scales defined by these indicators has been repeatedly
verified by a multiplicity of checks (de Vaucouleurs 1982a,b). A
rectangular slab of space 80 x 60 Mpc in the supergalactic plane, and
either 10 or 20 Mpc thick was considered, with the ξ axis in the
direction $L = 105°$ (essentially the direction of the Virgo cluster),
and the η axis directed toward $L = 195°$. The space distribution of
the galaxies is shown in Figure 2 in projection onto the supergalactic
plane. The Galaxy is at the origin in the center of the maps which
cover only 60 x 60 Mpc.

     The radial velocities were first corrected to the velocity
centroid of the Local Group after de Vaucouleurs, Peters and Corwin
(1977) and projected onto the supergalactic plane. The correction is
always less than 13% and $\sim$ 5% on the average. The projected velocities
were analyzed in several ways by the 'numerical mapping' technique of
Jones et $al.$ (1967) using 6-th order orthogonal polynomials in ξ and η
(28 terms) either unconstrained or constrained to match the ideal
Hubble flow at the outer boundary of the 80 x 60 Mpc field. The
mapping was applied first to the individual galaxies, then to 2 x 2
Mpc averages with equal weights, to produce maps of

(a) the velocity field (Figure 9)

(b) the Hubble ratio (Figure 10)

(c)   the velocity residuals from
      a uniform Hubble flow
      (Figure 11),

for galaxies within 5 and 10 Mpc
from the SG plane.

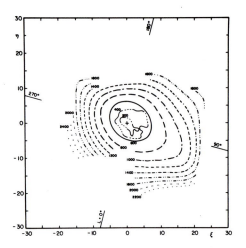

      The velocity maps show
mainly the dominant dipolar
anisotropy reflecting the mean
motion of the Local Group with
respect to the nearby galaxies.
The component of this motion
parallel to the SG plane is
directed toward L ≃ 90° (see
§ 3 above).

      The maps of the Hubble
ratio show the variations of the
mean expansion rate with
direction; lower than average
values are observed in the
general direction of the north
galactic pole, and higher
values in the opposite hemisphere, but with some
irregularities indicating the
presence of terms of higher
order than the simple dipolar
anisotropy. This north–south
asymmetry has been known for a
long time (de Vaucouleurs 1958)
and is, in part, responsible
for the low values of the
Hubble "constant" derived from
samples dominated by galaxies
in the north galactic hemisphere.

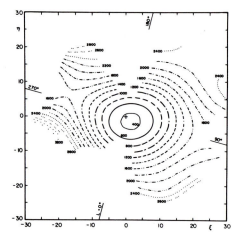

Figure 9
Maps of velocity field for galaxies
within 5 (above) and 10 Mpc from
supergalactic plane.

      Maps of the apparent
Hubble ratio have the dis-
advantage that they tend to conceal local departures from the Hubble
flow, since they represent only the mean expansion rate between the
origin and (ξ, η). Local departures are better shown by maps of the
velocity residuals from an ideal linear, isotropic expansion. Such
maps (Figure 11) show large negative residuals in the general direction
L ≃ 90° at distances of several tens of megaparsecs and positive
residuals toward L ≃ 180° and L ≃ 330°, with a suggestion of another
region of negative residuals toward L ≃ 270°; the latter is poorly
defined because of insufficient data at distances greater than 20 Mpc
in the direction of the south galactic pole. The whole pattern is

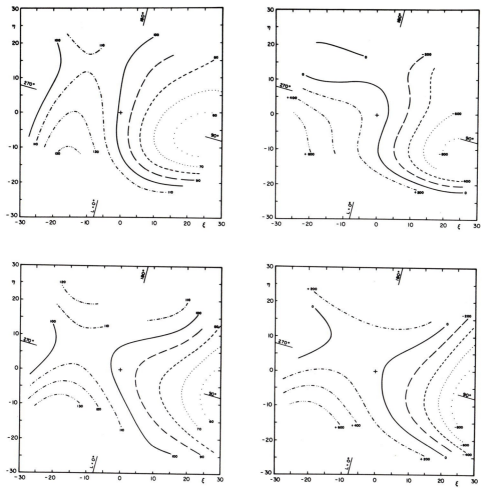

Figure 10
Maps of apparent Hubble ratio
for galaxies within 5 (above)
and 10 Mpc from supergalactic
plane.

Figure 11
Maps of velocity residuals from
ideal Hubble flow for galaxies
within 5 (above) and 10 Mpc from
supergalatic plane.

strongly suggestive of a quadrupole term superimposed on a north-south
dipole term. Current "Virgocentric" spherical models of the LSC cannot
account for this pattern.

We have examined a number of possible sources of systematic errors
in the distance estimates (Malmquist bias, non-linearly, galactic
extinction); none is plausible or even of the sign needed to explain
the observed anisotropies. We can only conclude that (i) the major
part of the departures from the uniform-isotropic expansion illustrated

in Figures 9 to 11 is a real phenomenon, probably related to the structure and dynamics of the Local Supercluster, and (ii) none of the models that have been advanced, from the original rotating-expanding disk, to the currently popular spherical "Virgocentric" rotating or non-rotating models give a satisfactory account of the observed velocity field.

FOOTNOTES

[1] Curiously, when, twenty years later, I adopted the word (which by then had been abandoned) to refer to the local supercluster, Shapley did not conceal his disbelief, but asked that, if the concept turned out to be confirmed, we should remember that the word was invented at Harvard!

[2] For more details of the early work, see previous review papers (Abell 1974, de Vaucouleurs 1971, 1978, 1981).

[3] This remarkable coincidence, first pointed out by Concklin (1969) and by Henry (1971), was poorly received and doubted by many at the time, but has been thoroughly confirmed and refined by all subsequent work. The mean solar apex relative to the background radiation is at $L^* = 122°$, $B^* = -17°$ with $V^* = 336$ km/s.

REFERENCES

Aaronson, M., Huchra, J., Mould, J., Schechter, P.L., and Tully, R.B. 1982, *Astrophys. J. 258*, 64.

Abell, G.O. 1958, *Astrophys. J. Suppl. 3*, 211.

Abell, G.O. 1974, *I.A.U. Symp. No. 63*.

Bahcall, J., and Joss, P.C. 1976, *Astrophys. J. 203*, 23.

Bottinelli, L., Gouguenheim, L., Paturel, G., and de Vaucouleurs, G. 1980, *Astrophys. J. Lett. 242*, L 153.

Bottinelli, L., Gouguenheim, L., Patural, G., and de Vaucouleurs, G. 1983, *Astron. Astrophys. 118*, 4.

Brosche, P., Hövel, W., and Lentes, F.-Th. 1983, *Mitt. Astron. Gesell 58*, in press.

Buta, R., and de Vaucouleurs, G. 1983a, *Astrophys. J. Suppl. 51*, 549.

Buta, R., and de Vaucouleurs, G. 1983b, *Astrophys. J. 266*, 1.

Charlier, C.V.L. 1908, *Ark. Math. Astron. Phys. 4*, No. 24.

Charlier, C.V.L. 1922, *Ark. Math. Astron. Phys. 16*, No. 22 = Medd Lund Observ., No. 98.

Concklin, E.K. 1969, *Nature 222*, 971.

Davis, M., Tonry, J., Huchra, J., and Latham, D.W. 1980, *Astrophys. J. Lett. 238*, L 113.

de Vaucouleurs, G. 1953, *Astron. J. 58*, 30.

de Vaucouleurs, G. 1956, *Vistas in Astron. 2*, 1584.

de Vaucouleurs, G. 1958, *Astron. J. 63*, 253.

de Vaucouleurs, G. 1964, *Astron. J. 69*, 737.

de Vaucouleurs, G. 1966, *Atti del Convegno sulla Cosmologia, Galileo Conf. (Firenze)*, p. 37.

de Vaucouleurs, G. 1971, *Publ. Astron. Soc. Pacific 83*, 113.

de Vaucouleurs, G. 1972, *IAU Symp. No. 44, External Galaxies and Quasi-Stellar Objects*, D.S. Evans, ed. (Reidel: Dordrecht), p. 353.

de Vaucouleurs, G. 1975, *in Galaxies and the Universe, vol. IX of Stars and Stellar Systems*, A. Sandage, M. Sandage, and J. Christian, eds. (Chicago: Univ. of Chicago Press), p. 557.

de Vaucouleurs, G. 1976a, *Astrophys. J. 203*, 33.

de Vaucouleurs, G. 1976b, *Astrophys. J. 205*, 13.

de Vaucouleurs, G. 1978, *IAU Symp. No. 79, The Large Scale Structure of the Universe*, M.S. Longair and J. Einasto, eds. (Reidel: Dordrecht), p. 205.

de Vaucouleurs, G. 1979, *Astrophys. J. 227*, 729.

de Vaucouleurs, G. 1981, *Bull. Astron. Soc. India 9*, 1.

de Vaucouleurs, G. 1982a, *Observatory 102*, 178.

de Vaucouleurs, G. 1982b, *The Cosmic Distance Scale and the Hubble Constant*, Mount Stromlo and Siding Spring Observatories, Australian National University, Canberra.

de Vaucouleurs, G., and Bollinger, G. 1979, *Astrophys. J. 233*, 433.

de Vaucouleurs, G., and Olson, D.W. 1982, *Astrophys. J. 256*, 346.

de Vaucouleurs, G., and Peters, W.L. 1968, *Nature 220*, 868.

de Vaucouleurs, G., and Peters, W.L. 1981, *Astrophys. J. 248*, 395.

de Vaucouleurs, G., and Peters, W.L. 1983a, *Astrophys. J.*, submitted.

de Vaucouleurs, G., and Peters, W.L. 1983b, *Astrophys. J.*, submitted.

de Vaucouleurs, G., Peters, W.L., Bottinelli, L., Gouguenheim, L., and Paturel, G. 1981, *Astrophys. J. 248*, 408.

de Vaucouleurs, G., Peters, W.L., and Corwin, H.G. 1977, *Astrophys. J. 211*, 319.

Hart, L., and Davis, R.D. 1982, *Nature 297*, 191.

Henry, P.S. 1971, *Nature 231*, 518.

Hoffman, G.L., Olson, D.W., and Salpeter, E.E. 1980, *Astrophys. J. 242*, 861.

Holmberg, E. 1937, *Ann. Lund Observ. 6*, 52.

Jones, W.B., Obitts, D.L., Gallet, R.M., and de Vaucouleurs, G. 1967, *Publ. Dept. Astron., Univ. of Texas, Ser. II, Vol. II, No. 8*.

Karoji, N. 1983, preprint.

Lundmark, K. 1927, *Studies of Anagalactic Nebulae, Medd Lund Observ., No. 30*.

Peebles, P.J.E. 1976, *Astrophys. J. 205*, 318.

Reiz, A. 1941, *Ann. Lund. Observ. 9*, 65.

Reynolds, J.H. 1921, *Monthly Notices Roy. Astron. Soc. 81*, 129, 598.

Reynolds, J.H. 1923, *Monthly Notices Roy. Astron. Soc. 83*, 147.

Rubin, V. 1951, *Astron. J. 56*, 47.

Sandage, A., and Tammann, G. 1975, *Astrophys. J. 196*, 313.

Schecter, P.L. 1980, *Astron. J. 85*, 801.

Shapley, H. 1930, *Circ. Harvard Coll. Observ., No. 350*.

Stewart, J.M., and Sciama, D.W. 1967, *Nature 216*, 742.

Teerikorpi, P. 1975, *Astron. Astrophys. 45*, 117.

Tonry, J.L. 1980, *Ph.D. Thesis, Harvard Univ*.

Tully, R.B. 1982a, *Astrophys. J. 257*, 389.

Tully, R.B. 1982b, *Sky and Telescope 63, No. 6*, 550.
van Albada, G.B. 1962, *IAU Symp. No. 15, Problems of Extragalactic Research*, C.G. McVittie, ed. (New York: MacMillan), p. 411.
Zwicky, F. 1938, *Publ. Astron. Soc. Pacific 50*, 218.

DISCUSSION

TULLY: (1) If the vector of solar motion found from studies of the velocity field of nearby galaxies is subtracted from the vector of solar motion implied by the 3K background dipole anisotropy, the residual vector of a couple of hundred kilometers per second is directed toward the Hydra-Centaurus Supercluster, the nearest adjacent large-scale structure. Ed. Shaya discusses this matter in a paper in press. (2) A student at Hawaii has constructed a beads-on-string model of a region around our position, extending out to a distance corresponding to 3000 km/s. All galaxies with greater than a certain intrinsic luminosity were represented, where the cut-off was chosen to assure quasi-completion across the volume in question. As a true believer in the reality of the Local Super-cluster, what was dismaying to me was the relative insignificance of this feature in the model. Future generations may come to ap-preciate that the structure around Virgo is only a modest appendage of the Hydra-Centaurus Supercluster.

PEEBLES: Your observation of a quadrupole anisotropy of the Hubble ratio at small distances is roughly consistent with what would be expected if the peculiar velocity field were due to a mass concen-tration somewhere near the Virgo Cluster. I agree that the spherical model for the local supercluster is at best crude, but your result suggests to me the model may be useful after all!

DE VAUCOULEURS: This may well be the case. We have not yet made detailed comparisons of the observations with several possible models, nor with the velocity fields predicted by the neutrino-dominated models of Centrella and Mellot. We hope to do so in the near future.

A CATALOG OF CANDIDATE SUPERCLUSTERS AND VOIDS

J.O. Burns and D.J. Batuski
Department of Physics and Astronomy
University of New Mexico
Albuquerque, New Mexico 87131, U.S.A.

Most of our impressions of the large-scale structure in the Universe have come from recent magnitude limited, redshift surveys of relatively nearby volumes of space (e.g. Davis et al. 1981; Gregory and Thompson 1978). These surveys have been important in identifying the existence and measuring the sizes of superclusters and voids. However, one must sample much larger volumes to truly map out the three dimensional structure of matter and to differentiate between competing models of cluster formation. To do this with the traditional approach of measuring redshifts for galaxies down to say 17^m or 18^m over the entire sky would require absurdly large amounts of telescope time. However, one might be able to effectively sample structures in a larger volume if a tracer of supercluster structure could be identified and its distance measured. Such a tracer may be the Abell clusters. A number of nearby superclusters (e.g. Coma/A1367, Hercules) are in fact composed of rich cluster knots bridged by low density chains of galaxies.

We have recently re-examined the usefulness of Abell clusters as supercluster markers with the aid of new data and with a new perspective on the sizes of superclusters. A full 72% (196/272) of distance class 4 and closer Abell clusters now have measured redshifts. This third dimension of information is crucial in determining the physical position of a cluster with respect to its neighbors. The new perspective is the realization from nearby superclusters that rich clusters can be separated by 30-40 Mpc and still be physically linked by a chain of galaxies. The total sizes of superclusters can be hundreds of Megaparsecs (Oort 1983).

The need for a larger, deeper sample combined with an appreciation for the enormous extents of superclusters have prompted us to construct a catalog of candidate superclusters composed of Abell clusters. Our idea was to produce a finding list or guide for future observations rather than a statistical sample for correlation analysis. We emphasize that these superclusters should be viewed only as potential groupings since further observations must be performed to search for

F. Mardirossian et al. (eds.), Clusters and Groups of Galaxies, 43–48.

physical bridges between clusters. Using this philosophical approach, we also recognized that all members of the Abell catalog, including richness class 0 clusters, should be used in selecting superclusters. These poor clusters are apparently important links between richer knots of Abell clusters (e.g., A2162 between Hercules and A2197/A2199).

A supercluster in the catalog is composed of Abell clusters with at least one neighbor within 40 $(h=H_0/75)^{-1}$ Mpc. Although still somewhat arbitrary due to the lack of a physically defined scale size, the choice of 40 Mpc separation was based on three considerations. First, the amplitude of the spatial correlation function for Abell clusters (Bahcall and Soneira 1983) becomes quite small (<0.5) for separations $\gtrsim 40$ h^{-1} Mpc, suggesting a much lower probability of physical association at larger separations. Second, $\delta\rho/\rho \lesssim 2$ for two clusters separated by 40 h^{-1} Mpc where the "background density" was assumed to be equal to that given for a uniform distribution of clusters with the same number as Abell's statistical sample. Third, there are several proven superclusters observed to be composed of Abell clusters with separations of ~40 h^{-1} Mpc.

In addition to the core superclusters identified by the above technique, the neighborhoods out to ~80 h^{-1} Mpc around the superclusters were investigated for two purposes. First, there is still a fairly large number of clusters with no measured redshifts within $z<0.1$. An estimated distance for these clusters was made from a new z vs m_{10} (magnitude of 10th brightest galaxy) relationship established from clusters with measured z. Since these z_{est} still have potentially large errors, the unmeasured redshift clusters within ~80 h^{-1} Mpc of the supercluster cores were allowed to vary along the line of sight by $\pm 2\sigma$ in the z vs. m_{10} relation. The fraction of this total variation in which the z_{est} cluster was within 40 h^{-1} Mpc of the nearest member of the core supercluster was then used as an estimate of the probability of membership in the supercluster. Second, the richness of the supercluster environment was established by summing the gravitational force (i.e., $\Sigma N_j/R_{ij}^2$, where N_j = number of galaxies in cluster j which neighbors cluster i, and R_{ij} = distance between clusters i and j) around all Abell clusters. This gravitational weight is then listed in the catalog for all the clusters in superclusters. This weight clearly shows the variation between clusters in and out of superclusters as identified by our algorithm.

As a further verification of the potential of our algorithm for selecting superclusters, we performed Monte Carlo simulations of the clustering using the same number and density distribution, but randomized positions of Abell clusters. In Figure 1, the multiplicity function for the Abell-core superclusters is compared with that for the Monte Carlo simulation. A χ^2-test reveals that the two distributions could only be drawn from the same parent population with a probability of <0.1%.

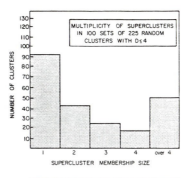

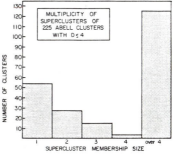

Figure 1. Comparison of number of Abell clusters of all richness classes found in superclusters of given sizes with average memberships for "random cluster catalogs" with the same density-vs-distance profile. Bins are number of clusters per supercluster, with a membership of one representing "isolated" clusters.

Two examples of relatively nearby, prominent superclusters from our catalog are illustrated in stereoscopic projections in Figures 2 and 3. A listing of the member clusters is given in Tables 1 and 2. Figure 2 shows the "revised" Hercules supercluster in which the A2197/2199 and A2147/A2151/A2152 systems are just small portions (in the lower center of the diagram) of a much larger supercluster. The overall diameter of this candidate supercluster is $\sim$200 h^{-1} Mpc. Note the absence of Abell clusters to the left of the main body of the supercluster; this is a part of the larger Böotes void (Kirschner et al. 1981). Note also the location of the Coma/A1367 system (lower left) which is close to Hercules but not formally part of the supercluster according to our selection criteria. Figure 3 illustrates a lesser known supercluster in the Pisces-Cetus region. Like Hercules, it too is extensive, possibly filamentary, and located near several voids of Abell clusters. We have recently completed redshift observations of poor Zwicky clusters which appear in projection to be between Abell clusters in this supercluster. The Zwicky clusters apparently link together several of the Abell clusters. In addition, several of the Zwicky cluster 3-D positions may suggest a bridge of this supercluster towards another less rich supercluster. Our current impressions from these and other superclusters in the catalog suggest that the candidate superclusters are probably not separate entities but merely part of a much larger, interwoven clustering pattern.

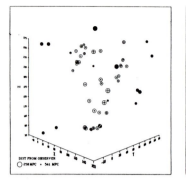 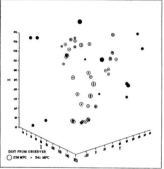

Figure 2. Stereoscopic plots of our supercluster number 237 as viewed by hypothetical observer 500 Mpc from center of cube. Symbols: ● - non-member of supercluster, ○ - rich member (R≥1), ⊕ - poor member (R=0), central dots indicate clusters with measured redshifts. Note large part of Böotes void in left side of figure. Hercules cluster is among tight grouping at bottom. Approximate center of this supercluster is α = 14^h 40^m, δ = 22°, z = 0.055. Largest separation between two member clusters is 195 Mpc.

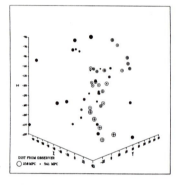 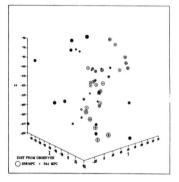

Figure 3. Stereoscopic plots of supercluster number 14 (Pisces-Cetus), with same legend as Figure 1. The approximate center is α = 1^h 15^m, δ = 0°, z = .045. Maximum separation between members is 225 Mpc.

Table 1.   The "Revised" Hercules Supercluster

A1691,	A1749,	A1775,	A1781,	A1785,	A1795,	A1800,	A1813,
A1825,	A1827,	A1831,	A1899,	A1913,	A1927,	A1983,	A1991,
A2019,	A2022,	A2033,	A2040,	A2052,	A2055,	A2056,	A2061,
A2063,	A2065,	A2067,	A2079,	A2089,	A2092,	A2107,	A2124,
A2147,	A2148,	A2151,	A2152,	A2162,	A2197,	A2199	

Table 2.   The Pisces-Cetus Supercluster

A14,	A34,	A71,	A74,	A75,	A76,	A85,	A86,
A102,	A114,	A116,	A119,	A133,	A147,	A150,	A151,
A154,	A158,	A160,	A168,	A171,	A179,	A189,	A193,
A195,	A225,	A240,	A245,	A246,	A279,	A397,	A2572,
A2589,	A2593,	A2634,	A2657,	A2666			

In addition to the candidate superclusters, we also formally searched a volume within z<0.1 and $|b|>30^\circ$ for relatively large regions which are devoid of Abell clusters. In particular, a region was termed a void if no clusters lay within a radius of 50 h^{-1} Mpc. A total of 29 candidate voids, in regions that are not highly obscured (i.e., low HI column density as measured by Heiles, 1975), were identified including, for example, the Böotes void.

Complete listings of the candidate superclusters and voids will appear elsewhere (Batuski and Burns 1984).

Finally the superclusters within z < 0.1 were used to investigate the volume of space occupied by the large-scale clusterings. The volume of a supercluster was estimated by assuming that a cylinder of radius 3 h^{-1} Mpc stretches between all Abell clusters that are within 40 h^{-1} Mpc of each other. By summing up the total supercluster volume, we find that <1% of the universe is filled with Abell cluster-core superclusters. If the matter distribution is represented by these superclusters, it would suggest that the universe is mostly empty space.

J.O.B. wishes to thank NASA, the American Astronomical Society, and the Conference organizers for their support of his travel to Trieste.

REFERENCES
Bahcall, N.A. and Soneira, R.M., 1983, Ap.J., 270, 20.
Batuski, D.J. and Burns, J.O., 1984, in preparation.
Davis, M., Huchra, J., Latham, D.W., Tonry, J., 1981, Ap.J., 253, 423.
Gregory, S.A., and Thompson, L.A., 1978, Ap.J., 222, 784.
Heiles, C., 1975, Astron, Astrophys. Suppl., 1975, 20, 37.
Kirschner, R.P., Oemler, A., Schechter, P.L., Shectman, S.A., 1981, Ap.J., Lett., 248, L57.
Oort, J.H., 1983, Ann. Rev. Astron, Astrophys, 21, 373.

DISCUSSION

DRESSLER: What is the meaning of the figure of 1% for the density of occupied volume?

TULLY: It should be noted that the 1% filling for superclusters is similar to the filling factor of clouds of galaxies in our Local Supercluster: spheroidal shells defined by the 1σ excursions of galaxies from the center of mass of clouds enclose 4% of the volume available.

DRESSLER: You have suggested that space is 99% empty, but your analysis only shows that space is 99% empty of rich clusters. Since rich clusters contain only a few per cent of all galaxies, and rich clusters seem to "cluster" much more strongly than galaxies

(Bahcall and Soneira) it may be incorrect to conclude that space is 99% empty of galaxies.

BURNS: The spatial correlation function for galaxies and rich clusters are not inconsistent with a geometry in which most galaxies lie in low density bridges between clusters within supercluster chains. In fact, a recent 3-dimensional numerical model of the formation of large-scale structure within the pancake scenario by Klypin and Shandarin (1983, MNRAS, 204, 891) suggests just such a geometry and successfully reproduces the observed spatial correlation function. Therefore, we feel that given the present data at hand, our calculation is not an unreasonable estimate of the volume occupied by all galaxies (although the error could still be a factor of 10).

MACGILLIVRAY: How many galaxies are used to establish the cluster redshifts and is it not dangerous to apply too much weight to data based on small numbers of redshifts in any single cluster?

BATUSKI: Of the 365 clusters with z <0.1 (measured or estimated) 125 have no measured redshifts, 123 have one galaxy with measured z, and 117 have two or more. We are very much concerned about the likelihood of error from using cluster redshifts derived from one or two measurements. Our analyses have been done with weightings based on estimates of these uncertainties. All of our supercluster candidates should be considered only as candidates until many redshift measurements of galaxies within and between clusters have been made. Nevertheless, even a single measurement tells us that there is something at a particular location, which can then be considered as part of a supercluster structure.

PEEBLES: Are you (and others!) concerned about completeness of the Abell catalog as a function of position across the sky? Might the density of clusters be in part an effect of variable completeness?

BURNS: Yes, we are quite concerned about the incompleteness of the Abell catalog. This is why we emphasize that the superclusters in our catalog should be viewed only as candidate second order clusterings until further redshift observation confirm the existence of physical bridges between Abell clusters. For the few quasi-statisti cal statements which we make using the catalog, we have utilized clusters that are well above the galactic, plane, z< 0.1, and are not near spurs of galactic HI.

STATISTICAL ANALYSIS OF THE SHAPES OF RICH SUPERCLUSTERS

Robin Ciardullo, University of California, Los Angeles
Space Telescope Science Institute

Richard Harms,   University of California, San Diego

Holland Ford,   Space Telescope Science Institute

ABSTRACT

We outline a technique for estimating the large scale shape of rich superclusters using complete samples of objects extending out to moderately large redshifts. We describe two tests, a principal axes test and a velocity gradient test, which are especially useful for this purpose. We apply these tests to samples of clusters from the Abell catalog and search for evidence of disk-like or filamentary structures.

1. INTRODUCTION

Supercluster shapes are a clue to both the past and future of the universe. Superclusters are unrelaxed, hence the distribution of galaxies today reflects the structure of primordial density perturbations. In addition, the motions of the clusters are affected by the overall supercluster potential. Estimates of large scale matter density are possible from detailed modeling of the slowing of the Hubble expansion. In principle then, supercluster shapes allow us to probe both the physics of the early universe and the cosmological density parameter Ω.

Although extremely important, data on supercluster shapes are difficult to obtain. Surveys of the few nearby superclusters present some evidence for the existence of pancakes or filaments. To understand large scale structure however, we need to sample a larger volume of the universe and observe many more superclusters. Presently, this cannot be done efficiently by measuring redshifts for magnitude limited samples of galaxies. Redshifts of objects in a complete large volume catalog, such as the Abell cluster catalog, are useful for such an investigation, however. By providing data on a large number of superclusters, this type of catalog allows us to estimate the properties of these systems statistically. In this paper we outline a technique for analyzing large volume catalogs and suggest two possible tests for supercluster shape. We apply these tests to the Abell catalog in an attempt to find evidence for large scale disk or filamentary structure.

49

F. Mardirossian et al. (eds.), Clusters and Groups of Galaxies, 49–54.
© *1984 by D. Reidel Publishing Company.*

2. METHOD OF ANALYSIS

The first step in analyzing a large volume catalog is to derive three dimensional positions for each object. This requires the assumption that all redshifts arise from unperturbed Hubble flow. Although this is probably a very good approximation, it is possible that in the richest clusters some internal slowing of the Hubble expansion may have occurred (Ciardullo, et al. 1983). If this is the case, the distribution of derived supercluster shapes will be skewed towards flattened, face-on systems. Although this would make estimating the true supercluster shapes more difficult, it would place interesting constraints on the matter density in these systems.

Once three dimensional positions are established, a list of probable superclusters is produced. This raises the trickiest question in the analysis--what is a supercluster? Objects can be grouped together in any number of ways, and the derived supercluster properties may depend on how one defines these associations. The most popular method for defining superclusters with a small data sample is the single connection, or "friends of friends" approach. In this scheme, an object need only be within a certain distance of any system member for itself to be considered part of that system. Although this method may favor selecting flat or filamentary systems, it is perhaps the best rule to use for an initial search for these structures.

Implicit in this scheme is the concept of "guilt by association". Because we are dealing with small samples, we can only assume that physical proximity denotes physical relationship. For catalogs of galaxy clusters, this may be an acceptable assumption. Studies of the correlation function for galaxies and clusters (Bahcall and Soneira 1983) suggest rich clusters are good tracers of large scale structure. In general however, superclusters defined using this assumption are only probable associations and violations of the rule will limit the effectiveness of any small sample analysis.

The critical parameter in the friends of friends method is the maximum association length, or equivalently, the minimum supercluster density enhancement. Choosing too small a value for this parameter limits associations to rich, high density supercluster cores, while too large an association length links together separate systems and forms enormous, elongated superclusters. Between these two regimes, supercluster shapes are relatively independent of the association length. By varying the parameter and observing the changes in supercluster orientations and multiplicity, reasonable values for the association length can be found.

After generating a list of superclusters for each association length, a shape estimator is applied to all systems containing four or more objects. The distribution of estimator values for the observed superclusters is then compared with that expected from a population of superclusters with a given shape and density profile. It is through this comparison that we obtain the significance of the results.

3. SHAPE ESTIMATORS

Obviously, the goodness of the estimator defines the power of the test. Good estimators must be able to discriminate between a population of spheroids, pancakes, and filaments, and be insensitive to the density profile within these structures. Here we will present two such tests.

3.1 The Principal Axes Test

The moments of N particles about any three axes can be defined through the mass distribution matrix Q, with elements $Q_{i,j} = \sum x_i x_j$. The principal axes of the system are found by diagonalizing the matrix and finding the eigenvalues. The largest eigenvalue defines the scale of the system, while the ratios of the eigenvalues estimate the shape. The direction of the associated eigenvectors provide the system's orientation. If superclusters dynamics are governed by a pure Hubble flow, these eigenvectors will be isotropically distributed.

Figure 1 displays the probability distribution of derived axis ratios from five points drawn at random from a uniform density sphere, 3:3:1 disk, and 3:1:1 filament. The differences in the probability distributions are obvious. If a moderate number of superclusters with multiplicity between 5 and 10 can be identified, statistical estimation of their shapes is possible. The results are not sensitive to the internal supercluster density profile.

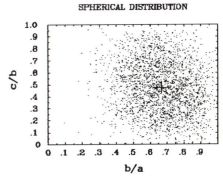

Figure 1. Distribution of principal axis ratios as derived from 5 points drawn from a uniform density volume.

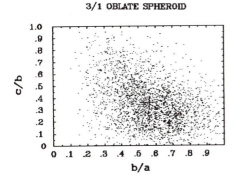

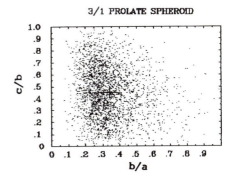

3.2   The Velocity Gradient Test

A different method of examining supercluster shapes uses velocity gradients.   Points drawn at random from quasi-spherical distributions should show no systematic velocity trends.   On the other hand, disk-like structures inclined to the line of sight should have velocity gradients across their apparent minor axes.   Inclined one dimensional structures should show systematic velocity differences along their principal axes.

For superclusters with only a few test particles, the difficulty in defining a minor axis becomes insurmountable, hence gradient tests for pancakes cannot be performed.   A test for strings is straightforward, however.   An apparent major axis is defined using the projected positions of test particles.   Redshift is correlated against position along this axis and the correlation coefficient is used to measure the probability of there being no relation.   The number of systems found to have significant gradients is then compared to that expected from random samples of prolate systems with various axis ratios.

Table 1 lists the probabilities of finding significant (P > 90%) major axis velocity gradients in prolate superclusters of multiplicity 5 and 7.   Once again, a small number of superclusters with 5 - 10 test particles can make a strong statement concerning supercluster shapes.

TABLE 1
Probability of Finding Significant Major Axis Gradient

Multiplicity	Model Supercluster Axis Ratios			
	2:1:1	3:1:1	4:1:1	5:1:1
5	0.23	0.44	0.60	0.70
7	0.35	0.61	0.74	0.78

4.   ABELL CATALOG ANALYSIS

Both tests were performed on the "statistical" sample of Abell clusters with distance class D ≤ 4, all of which have measured redshifts (Hoessel, Gunn, and Thuan 1980).   This is not a true statistical sample; in addition to errors introduced by the semi-qualitative nature of the catalog, this limited sample becomes incomplete at z > 0.08 (Thuan 1980).   Nevertheless, since omissions in the catalog should effect the completeness of the supercluster list much more than the shape estimations, the sample is still very useful for our purpose.

Table 2 lists for various values of the association length, the richest superclusters in the sample, their multiplicities, estimated axis ratios, shape probabilities (normalized to the spherical case), and major axis velocity gradient probabilities.   There is no strong evidence for disks or filaments in the data.

Table 2
Properties of Probable Superclusters

Assoc. Length	Density Contrast	Central Cluster	Mult	a Mpc	b/a	c/b	Prob disk	Prob string	Prob grad
$15h^{-1}$	$57h^{-1}$	2065	4	14	0.64	0.58	0.36	0.19	0.44
$20h^{-1}$	$24h^{-1}$	2065	7	39	0.52	0.24	12.	3.3	0.80
$25h^{-1}$	$12h^{-1}$	1377	4	30	0.27	0.50	2.8	6.5	0.41
		2151	5	27	0.61	0.23	1.6	0.22	0.76
		2065	7	39	0.52	0.24	12.	3.3	0.80
$30h^{-1}$	$7h^{-1}$	1377	4	30	0.27	0.50	2.8	6.5	0.41
		1795	4	37	0.62	0.21	0.94	0.21	0.07
		2151	7	43	0.64	0.66	0.34	0.25	0.70
		2065	7	39	0.52	0.24	12.	3.3	0.80

The maximum and minimum association lengths were set by practical limits. No superclusters with multiplicity $\geq$ 4 were formed with association lengths L $\leq$ $10h^{-1}$ Mpc, while systems formed with association lengths L $\geq$ $35h^{-1}$ Mpc sometimes measured several hundred Mpc across and contained over 15 Abell clusters. The distinctly different nature of these huge, low density contrast systems was obvious from both shape estimators. Highly significant velocity gradients appeared in these systems and their principal axes' lengths and ratios changed dramatically from the higher contrast estimates.

In addition to using the limited statistical sample, we searched for major axis velocity gradients in superclusters formed from the entire Abell catalog with richness class R $\geq$ 1 and known redshifts. Although this is an extremely inhomogeneous and incomplete sample, it is a much larger data base--345 rich clusters now have measured redshifts (Ciardullo, et al. 1984). 11 superclusters with multiplicity $\geq$ 4 were identified using the $30h^{-1}$ association length. None of these, nor any tighter associations had significant velocity gradients. Once again, at lower density contrasts, gradients quickly appeared.

The significance of these results depends on the accuracy of the cluster redshifts. The major uncertainty in the shape estimations arises from the velocity dispersion of each cluster. In rich clusters dispersions of 500 to 1000 km/sec are typical. Since many clusters have only one galaxy with a measured redshift, the large errors in the mean cluster velocities effect our results. If velocity errors were negligible, the lack of observed major axis velocity gradients would be highly significant, and the existence of one dimensional chains of Abell clusters would be ruled out. With a 500 km/sec error, however, our results cannot be considered conclusive.

We wish to thank Dr. A. Rex Rivolo for valuable discussions. The work was supported in part by NASA contract NAS 5-24463.

REFERENCES

Bahcall, N.A., and Soneira, R.M. 1983, Ap.J. 270, 20.
Ciardullo, R., Ford, H., Bartko, F., and Harms, R. 1983, Ap.J. 273, 24.
Ciardullo, R., Ford, H., Harms, R., and Bartko, F. 1984, in preparation.
Hoessel, J.C., Gunn, J.E., and Thuan, T.X. 1980, Ap.J. 241, 486.
Thuan, T.X. 1980, "Proceedings of Les Houches Summer School, Session XXXII, Physical Cosmology," ed. R. Balian, J. Audouze, and D.N. Schramm, (North-Holland Pub. Co.).

DISCUSSION

BURNS: I tend to agree with your conclusions concerning filamentary structure in the Abell clusters distribution. From examinations of many candidate superclusters in our catalog, it would be difficult to define an overall shape to these clusterings.

.A REDSHIFT SURVEY IN THE LINX GEMINI REGION

P. Focardi * , B. Marano, G. Vettolani **
* Dipartimento di Astronomia, Universita' di Bologna,
via Zamboni 33, 40126 Bologna, Italy
** Istituto di Radioastronomia CNR,
via Irnerio 46, 40126 Bologna, Italy

ABSTRACT

The study of the redshift distribution of a complete sample of galaxies brighter than 14.5 mph has been performeded over an area encompassing about 1800 square degrees in Linx and Gemini. The main result is the discovery of a new filament of galaxies in Gemini, at a radial velocity of 4800 km/s, mainly composed of spirals. The possible connection of a cloud of galaxies surrounding the cluster A569, the new filament in Gemini and the Linx-Ursa Major supercluster with the Perseus supercluster is briefly discussed.

THE SURVEY

Evidence for the existence of long filamentary structures (superclus ters) surrounded by regions devoided of bright galaxies (voids) has been growing in the last few years mainly due to redshift surveys of galaxies in large regions of the sky (see Oort 1983 or Chincarini 1982 for a rewiew).

We report here the results of a survey we have performed in a region of about 1800 square degrees from 6 to 10 hours (1950) in right ascension and between +20 and +60 (1950) in declination (Fig 1).

In this region a complete sample of 305 galaxies brighter than the 14.5 magnitude has been extracted from the Uppsala Catalogue (Nilson 1973).

Radial velocities for 223 galaxies were found in the "Catalogue of Radial Velocities of Galaxies" (CRVG) (Palumbo et al 1983) updating it by inspecting the recent literature. We have finally obtained the radial velocities of 59 galaxies, thus reaching the 93 per cent of completeness. The 23 galaxies with radial velocity are uniformly spread over the region and thus their distribution is not biasing the supercluster structures described below.

F. Mardirossian et al. (eds.), Clusters and Groups of Galaxies, 55–61.
© *1984 by D. Reidel Publishing Company.*

Photographic spectra were obtained with the image tube spectrograph at
the 152 cm telescope of the Bologna University in Loiano. They range
from 350 to 700 nm, with a typical resolution of 1.2 nm. Instrumentation and reduction procedure are described in Marano and Vettolani (1982).
The rms velocity error has a typical value of 120 km/s.

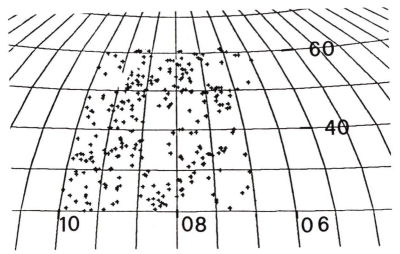

Fig. 1 Equiarea Hammer Aitoff projection of the
305 galaxies of the sample

RESULTS

In the above described area we find evidence for the existence of three
superclusters and of a large cloud of galaxies surrounding A569 (07 06
+49).
Our survey confirms the existence in the region between 40 and 60
degrees in declination of the Linx-Ursa Major supercluster previously
described by Giovanelli and Haynes (1982) as a filament characterized
by a density contrast of about 4 with respect to the surrounding region
and lying between 3000 and 5300 km/s.   However our data indicate that
the Linx Ursa Major higher velocity edge is 4000 km/s only. The
galaxies detected by Giovanelli and Haynes around 5000 km/s are more
likely connected to the cloud of galaxies around A569 which is in the
background of the western portion of the Linx-Ursa Mayor.
Infact the wedge diagram of Fig. 2 shows the western portion of the
Linx-Ursa Major and the existence of the cloud surrounding A569 with a
mean radial velocity of 6000 km/s and a diameter of the order of 10
degrees, corresponding to 12 Mpc (H=100).
Abell 569 (07 06 +49) consists of two main concentrations of galaxies
one with a radial velocity of 5710 km/s the other with 6020 km/s (Fanti
et al 1982).

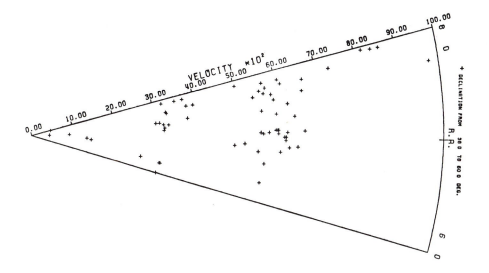

Fig. 2   Wedge diagram obtained projecting in declination
the galaxies in the sample with R.A. between 6 and 8
hours and Dec between 38 and 60 degrees.

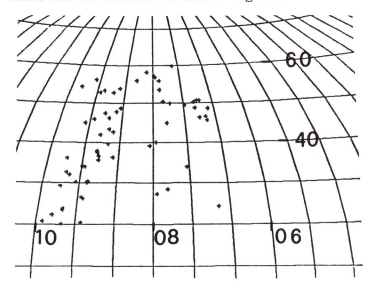

Fig. 3   Equiarea Hammer-Aitoff projection of the galaxies
whith radial velocity between 6000 and 9000 km/s.

Fig 3 shows the existence of a supercluster with a mean velocity of 7200
km/s extending north south across the area studied here. At the center
we find the Abell cluster A779 which has a mean radial velocity of 6811
km/s (Hintzen and others 1978). Maps of the distribution of galaxies
from the CRVG show that this feature extends southern and eastern and
might possibly be connected with the Coma-A1367 supercluster.

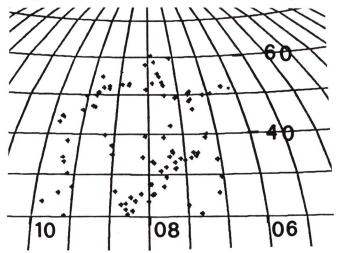

Fig. 4 Equiarea projection of the galaxies
with radial velocity between 3800 and 6000 km/s.

In the southern region a filament of galaxies, running from 06 40 +40 to
08 40 +20 in position angle 135 degrees, stands out (Fig.4) as a
projected density enhancement of a factor about 2.5-3 above the mean.
If we look in detail to the velocity distribution we see that   48
galaxies out of 63 in the filament lie at velocities between 3800 and
6000 km/s (Fig 5).
This filament of galaxies has two striking characteristics which are of
high significance for a better understanding of the structure of the
large scale distribution of galaxies and their formation:
 a) at variance of the best known superclusters such as Coma-A1367 ,
Hercules or the Perseus - Pisces no rich cluster of galaxies is found
along the filament. Low density filaments without rich clusters should
be much more common than this single example. Up to now the study of
superclusters (hence the word itself) has been biased towards regions
containing rich clusters of galaxies. Filaments as the one here
described are difficult to recognize on two dimensional maps of the
galaxy distribution due to their low density contrast, and their
existence can be shown only through redshift surveys in large areas of
the sky.
 b) this filament is characterized by an almost total absence of early

type galaxies: out of 48 galaxies only 2 are ellipticals and 4 SO's. In
this respect Giovanelli and others (1983, see also Giovanelli 1982) have
shown that in the Perseus supercluster there is a regular progression of
the   distribution of morphological types from E/SO to late spirals in
defining the sharpness of the branches of the supercluster. In other
words the ellipticals delineate a narrow filament which is embebbed in a
sea of spirals definig a wider filament.
Since 4 rich clusters of galaxies are found along the Perseus
supercluster whilst none is found along the Gemini filament and that
mean intercluster galaxy density is higher in Perseus than in Gemini, we
are hinted towards the scheme of galaxy formation where galaxies formed
only after the supercluster had collapsed, ellipticals formed in the
densest regions and spirals in regions of lower density. Note that
whilest the morphological segregation could be explained inside rich
clusters via galaxy encounters, this cannot explain the segregation in
superclusters where crossing times are greater than the Hubble time.

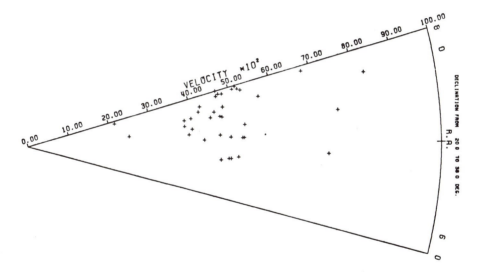

Fig. 5 Wedge diagram obtained projecting in declination
the galaxies in the region of the Gemini filament

A few years ago Chincarini and Rood (1979), studying the redshift
distribution of the Sc I galaxies of Rubin et al (1976) suggested that
the   Perseus   supercluster   could   extend   in   the   northern   galactic
hemisphere emerging from the high galactic absorption region near A 569.
Giovanelli and Haynes (1982) suggested that the Linx-Ursa Mayor filament
could be the extension of the Perseus supercluster.
Focardi et al (1983) have pointed out that aroud l=160 (the point at
which the main axis of the Perseus supercluster intersects the galactic

plane) the survey of galaxies in the galactic plane by Weinberger (1980) shows a large excess of galaxies.

Now the present survey provides a more complicated picture: the cloud around A569 is almost at the same radial velocity and has almost the same density contrast (taking into account the absorption) of the Perseus. Both the Linx Ursa-Mayor and the Gemini filament seems at definitly too low radial velocity to be connected to the Perseus.

It will be extremely interesting to establish in the near future if the A569 cloud is connected to the supercluster around A779 described above and if the latter is really connected to the Coma-A1367 supercluster. If that is true we are in presence of a connected filamentary structure encompassing the whole sky.

REFERENCES

Bothun,G.D., Geller,M.J., Beers,T.C., Huchra,J.P., 1983, Astroph. J., 268, 47

Chincarini,G., 1982, "The Large Scale Structure of the Universe", Lectures at the 3rd Escola de Cosmologia e Gravitacao, Rio de Janeiro, preprint

Chincarini,G., Rood,H.J., 1971, Astroph. J., 168, 321.

Fanti,C., Fanti,R., Feretti,L., Ficarra,A., Gioia,I.M., Giovannini,G., Gregorini,L., Mantovani,F., Marano,B., Padrielli,L., Parma,P., Tomasi, P., Vettolani,G., 1982, Astron. and Astroph., 105, 200

Focardi,P., Marano,B., Vettolani,G., 1983 in "Clustering in the Universe" European Colloquium Clusters of Galaxies, D. Gerbal, H. Mazure edts., Edition Frontiere, Paris, page 335.

Giovanelli, R., 1982, in "The comparative HI content of normal galaxies" Green Bank, p. 105.

Giovanelli,R., Haynes,M.P.,1982, Astron. J., 87, 1355

Giovanelli,R., Haynes,M.P., Chincarini,G., 1983, in preparation

Hintzen, P., Oegerle,W.G., Scott,J.S., 1978, Astron. J., 83, 478

Marano,B., Vettolani,G., 1982, Astron. Astroph. Suppl. 48, 453

Nilson, P., 1973, "Uppsala General Catalogue of Galaxies", Nova Acta R. Soc. Scient. Uppsalensis, Ser. V: A., vol. 1, Uppsala

Oort,J., 1983 "Superclusters" to appear in Annual Review of Astronomy and Astrophysics

Palumbo,G.G.C.,Tanzella-Nitti,G., Vettolani,G., 1983, "Catalogue of Radial Velocities of Galaxies", Gordon & Breach, New York

Rubin,V.C., Ford,W.K., Thonnard,N., Roberts,M.S., Graham,J.A., 1976, Astron. J., 81, 687.

Tifft,W.G., Jewsbury,C.P., Sargent,W.L.W., 1973, Astroph. J., 185, 115

Weinberger,R., 1980, Astron. and Astroph. Suppl., 40, 123

DISCUSSION

HUCHRA: Don't say that there are no rich Abell clusters in this filament you describe because Abell is incomplete for velocities less than about 6000 km/s. Also the Cancer cluster lies at the very end of your filament.

VETTOLANI: The Cancer cluster lies at the very end of the filament but it cannot be defined a rich cluster in the Abell sense on the basis of our counts of galaxies in the region. Moreover Bothum et al. (1983, Ap. J. 268, 47) have shown that the Cancer is not a proper cluster but rather a collection of discrete groups.

GIOVANELLI: 1) Could you please indicate the center of the Cancer in the map with the filament? 2) I wish to support your statement that filaments composed of spirals may be common: for example another extends from the Pegasus cluster to the main ridge of the Perseus Pisces. 3) Can you see an enhancement in the projected density of all galaxies, say, in the CGCG or in the UGC? 4) How do you define the density contrast?

VETTOLANI: 1) The Cancer cluster lies at the extreme south east of the filament. 3) No. 4) There are some problems in defining the density contrast in a region like this where there is an apprecia- ble gradient in absorption due to its low galactic latitude. We have divided the area into squares 3 degrees wide and we have counted the galaxies in the Zwicky Catalogue up to 14.5 where no absorption is present and to fainter magnitude limits where absorption is present (the magnitude limit has been increased of the absorption). We have then compared these counts with the mean counts of Zwicky galaxies at 14.5 in other regions of the sky.

THE ORIENTATIONS OF SPIRAL GALAXIES IN THE LOCAL SUPERCLUSTER

E.R. Anderson, David Dunlap Obs., Toronto, Ontario, Canada.
C. Norman, European Southern Obs., München, West Germany.
G. Illingworth, Kitt Peak National Obs., Tucson, AZ., USA.

ABSTRACT

Studies of the frequency distribution of supergalactic position angles for spiral galaxies of the Local Supercluster show no preferred orientation. However, for galaxies in the magnitude range $14.0 \leq m_B \leq 15.5$ positioned over the whole sky, there is a preferred supergalactic position angle of 94° with a dispersion of 40°.

DISCUSSION

For a sample of 2725 spiral galaxies which have measured position angles, diameters and spiral senses (i.e. S-shaped or Z-shaped), we have calculated the supergalactic position angles and the two possible poles of their angular momentum vectors. This sample was taken from the Morphological Catalogue of Galaxies combined with the Uppsala General Catalogue of Galaxies and the ESO/Uppsala Survey of the ESO(B) Atlas.

The ratio of the number of "S" galaxies to the number of "Z" galaxies, $S/Z=1.13$. Assuming that the chances of "S" or "Z" are equally probable and mutually independent, then the observed number of "S" galaxies deviates from the mean by 3.5σ. It should be noted however, that the ratio does vary from region to region on the sky. We are currently analysing this variation.

Figure 1 shows the frequency distribution of the supergalactic position angles (SGPA) for the whole sample. The distribution is peaked at 94° with a dispersion of 40° significant at the 99% confidence level. This is a curious result. Since the sample covers the whole sky and a large range of magnitude we expected to see a uniform distribution. Further analysis is still in progress. Figure 2 shows the SGPA distribution for galaxies of the Local Supercluster (defined by $m_B \leq 13.5$). This distribution is uniform and thus indicates no preferred orientation in the local Supercluster. The

F. Mardirossian et al. (eds.), Clusters and Groups of Galaxies, 63–64.
© *1984 by D. Reidel Publishing Company.*

galaxies which give rise to the peak in Figure 1 lie in the
magnitude range $14.0 \leq m_B \leq 15.5$.

Results of further investigation will be presented in a future
paper.

REFERENCES

Lauberts, A.:   1982, 'The ESO/Uppsala Survey of the ESO(B) Atlas',
     European Southern Observatory, Munchen, West Germany.
Nilson, P.:   1973, 'Uppsala General Catalogue of Galaxies',
     Publ. Uppsala Observatory, Series V, Vol. 1.
Vorontsov-Velyaminov, B.:   1962-1968,
     'Morphological Catalogue of Galaxies',
     Publ. of Sternberg Observatory, Moscow.

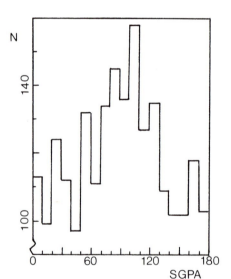

Figure 1.   Frequency
distribution of SGPAs
for the whole sample.

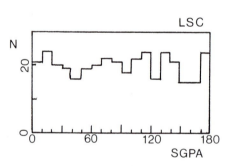

Figure 2.   Frequency
distribution of SGPAs for
the Local Supercluster.

ON THE ORIENTATION OF GALAXIES IN THE LOCAL SUPERCLUSTER

Piotr Flin and Włodzimierz Godłowski

Jagiellonian University Observatory, ul. Orla 171,
30-224 Kraków, Poland

Galaxies of redshifts $V_r < 2600$ km/sec from the Rood Catalogue (1980) with coordinates, position angles (PA) and diameters subsequently taken from UGC (Nilson 1973) served as the observational basis for our study. Four samples were considered: I-all galaxies, II-galaxies with $m < 12^m.5$, III-galaxies with determination of PA and diameters estimated in UGC as certain, IV-the product of sample II and III. Galaxy morphological types were also taken into account.

The coordinates and PA of galaxies given in UGC in the equatorial system were transformed to the supergalactic coordinates (L,B) assuming: $\alpha = 283°.5$, $\delta = +16°$ for the north supergalactic pole and $\alpha = 186°.25$, $\delta = 13°.1$ for Virgo Cluster centre. Allowing for basic meridian of the supergalactic system to pass through Virgo Cluster centre its coordinates are: $L=0°$, $B=3°.19$. The analysis was performed using Hawley and Peebles method (1975) (H-P method), in which only PA are taken into account. The isotropy of the distribution of rotation axes (assumed as perpendicular to PA) was examined by three tests: χ^2-test, Fourier test and correlation coefficient test.

In H-P method samples were divided according to some additional criteria: diameter ratio b/a and supergalactic latitude B. Some of the representative results are presented in the Table, where in columns 1,2 and 3 the number of galaxies, the value of sample statistics from Fourier test (i.e. directly the probability that the distribution is isotropic) and the value of F-coefficient of fitting with its error (for $F > 0$ the rotation axes are parallel to the supergalactic plane) are given respectively. The absence of data in the Table means that the sample is too small to be analysed.

It can be seen that the distribution of PA is isotropic, especially for samples containing galaxies with greater ellipticities. However, when signs indicating preferred orientation are considered, it might be concluded that rotation axes are perpendicular to the supergalactic plane. The very weak anisotropy occurs only for galaxies with b/a<0.7 for samples: I and III. This weak anisotropy is at marginally significant level. Therefore all the following discussion is at the same level of significance. The distribution of rotation axes for galaxies seen face-on and edge-on is different. For bright galaxies the distribution is isotropic. Moreover, the preferred orientation changes drastically for different samples, e.g. for sample I rotation axes are parallel to the supergalactic plane, while for sample II - perpendi-

F. Mardirossian et al. (eds.), Clusters and Groups of Galaxies, 65–66.
© *1984 by D. Reidel Publishing Company.*

TABLE. Testing of isotropy in distribution of rotation axes (H-P method)

Morph. types	Sample	b/a<0.7			b/a<0.45			b/a<0.3			Latitude |B|
		1	2	3	1	2	3	1	2	3	
all	I	820	.08	.05±.00	395	.67	-.25±.01	201	.72	-.29±.02	<90°
all	II	239	.82	.21 .01	99	.67	.25 .00				<90°
all	III	607	.14	-.08 .00	347	.67	-.22 .01	189	.93	-.35 .02	≮90°
all	IV	195	.88	.32 .02	93	.70	.44 .02				<90°
all	I	607	.12	.04 .00	398	.79	-.24 .01	197	.88	-.28 .02	<60°
all	II	238	.82	-.23 .01	99	.67	.25 .01				<60°
all	III	597	.23	.08 .01	341	.76	-.20 .01	185	.96	-.34 .02	<60°
all	IV	194	.92	.34 .02	93	.70	.44 .02				<60°
all	I	672	.20	.13 .00	311	.77	-.06 .01	161	.80	-.15 .01	<30°
all	II	213	.87	.27 .02	86	.90	.51 .01				<30°
all	III	497	.43	-.02 .00	274	.78	-.07 .01	151	.64	-.23 .01	<30°
all	IV	174	.99	.35 .02	81	.85	.58 .02				<30°
Sp	I	629	.15	.01 .00	313	.87	-.08 .01	167	.99	-.12 .01	<90°
non-Sp	I	191	.19	.20 .01	82	.38	-.87 .03				<90°
Sp+Irr	I	723	.11	.09 .00	354	.67	-.13 .01	185	.95	-.15 .01	<90°
non-E	I	799	.06	.01±.00	395	.67	-.25±.01	201	.72	-.29±.02	<90°

cular to the plane. When signs indicating preferred orientation are considered only for samples exhibiting aniosotropy it might be concluded that rotation axes are parallel to the supergalactic plane. The influence of the
sample on the results is the most dominant factor, and the influence of ellipticity criterion is greater than of criteria related to morphological types
and supergalactic latitude.
The same samples of galaxies were analysed using second method (Flin and
Godłowski 1983), based on the Jaaniste-Saar approach (1977), in which both
PA and galaxy inclinations are considered. In this method isotropy of two
angles is studied. The angle γ is between the direction of the rotation
vector and the supergalactic plane, the η angle is between the projection
of this direction on the supergalactic plane and Virgo centre direction.
The analysis, similar to that in the case of H-P method reveals that anisotropy of both angles is statistically significant at a very high level of
confidence, particularly for samples: I and II. The rotation axes are parallel to the supergalactic plane, the η angle points to Virgo Cluster. The
obtained results depend also on the sample and assumed galaxy shape.
The difference of results obtained in these two methods seems to be caused
by highly preferred orientation of the η angle towards Virgo Cluster.
It appears that the sample of galaxies seen edge-on, so essential in H-P
method, is not representative for all galaxies. Previously obtained discordant results also seem to be due to selection effects.

References
Flin, P., Godłowski, W., 1983, in preparation to print
Hawley, D.L., Peebles, P.J.E., 1975, Astron. J. 80, p. 477
Jaaniste, J.Saar, E., 1977, Tartu Preprint A-2
Nilson, P., 1973, Acta Universitatis Upsaliensis Ser. V: A. vol. 1
Rood, H., 1980, A Catalogue of Galaxy Redshifts, Preprint

SUPERCLUSTERS: A NEW INSIGHT INTO THE PROBLEM OF GROUPS WITH DISCREPANT REDSHIFTS

P. Focardi*, B. Marano*, G. Vettolani**
* Dipartimento di Astronomia, Universita di Bologna
  via Zamboni 33, 40126 Bologna, Italy
** Istituto di Radiostronomia, C.N.R.,
   via Irnerio 46, 40126 Bologna, Italy

The Stephan's Quintet is the prototype of compact groups of galaxies with discrepant redshifts. Four members of this group have radial velocity of about 6000 km/s, whilst the fifth, NGC 7320, has $v \sim$ 800 km/s. Several attempts of estimating the density of the "Stephan's Quintet like" groups have been carried on, giving contradictory results about the evidence for a chance projection in the discordant redshift groups. (See e.g. 1,2,3).

We stress that all these studies assumed, as a starting point, the traditional picture of uniform distribution of galaxies in the space, while the recent evidence for the existence of dominant filamentary structures and large voids in the galaxy distribution should be taken into account.

The Stephan's Quintet is projected on a large filament of galaxies belonging to the Perseus-Pisces Supercluster (4),(5),which has a mean radial velocity of 6000 km/s as the four higher velocity galaxies in the quintet.

The analysis of the redshift distribution of a complete sample of galaxies (m $\leqslant$ 14.5) in a wide region surrounding the Stephan's Quintet (6) has shown the presence of a well defined void in the range |1500,4000|km/s and of an extended population of low redshift (1000 km/s), spanning more than 20 degrees, to which NGC7320 appears to be associated (Fig. 1)

This configuration, in which both the high redshift galaxies and the low redshift ones appear to be associated to different large scale structures, compels us to reject any LOCAL interpretation of the redshift discrepancy.

F. Mardirossian et al. (eds.), Clusters and Groups of Galaxies, 67–68.

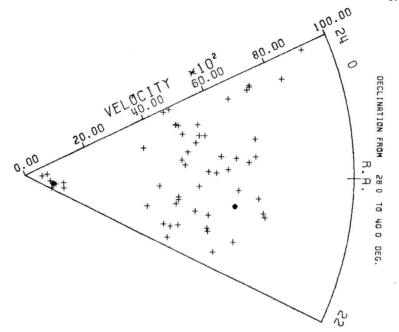

Fig. 1 Redshift distribution of the galaxies in a wide
        region surrounding the Quintet. The two quintet
        galaxies brigther than 14.5 are marked.

REFERENCES

1 Rose,J.A.,1977,Ap.J.,211,311
2 Nottale,L.,Moles,M.,1978,A&A,66,355
3 Sulentic,J.,Ap.J.,1983,270,417
4 Gregory,S.A.,Thompson,L.A.,Tifft,W.G.,1981 Ap.J.,243,411
5 Focardi,P.,Marano,B.,Vettolani,G.,1982 A&A,113,15
6 Focardi,P.,Marano,B.,Vettolani,G,(1983)in preparation
7 Palumbo,G.C.G.,Tanzella-Nitti G.,Vettolani G.,1983,"Catalogue of
   radial velocities of galaxies",Gordon & Breach,New York

THE GEOGRAPHY OF NEARBY SPACE

P. Focardi *, B. Marano *, G. Vettolani **
  * Dipartimento di Astronomia, Universita' di Bologna,
      Via Zamboni 33, 40126 Bologna, Italy
 ** Istituto di Radioastronomia CNR,
      via Irnerio 46, 40126 Bologna, Italy

A series of maps showing the distribution of galaxies in different radial velocity ranges have been prepared.

The sample contains all the galaxies brighter than the 14.5 apparent magnitude in the Uppsala Catalogue (Nilson 1973).

Redshifts have been compiled from the "Catalogue of Radial Velocities of Galaxies" (Palumbo et al 1983) updated with the more recent literature and our own unpublished observations.

Well known superclusters such as the Perseus or the Coma-A1367 as well as new features previously unrecognized clearly show up.

An example of these maps is shown in Fig 1 where all the galaxies with radial velocity between 4500 and 6000 km/s are displayed in an Hammer Aitoff equiarea projection.

REFERENCES

Nilson,P., 1973, Nova Acta R. Soc. Scient. Uppsalensis, Ser V:A, vol 1.

Palumbo,G.G.C., Tanzella-Nitti,G., Vettolani,G., 1983, Gordon and Breach, New York)

F. Mardirossian et al. (eds.), Clusters and Groups of Galaxies, 69–70.
© *1984 by D. Reidel Publishing Company.*

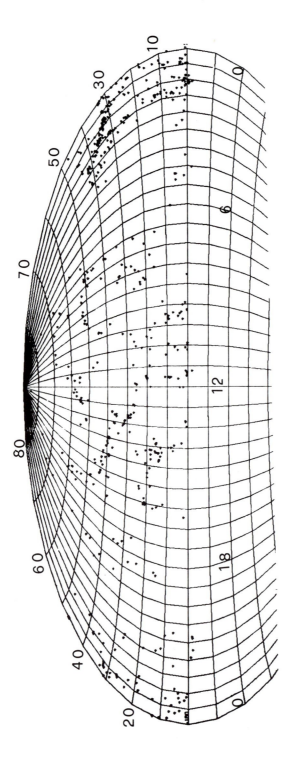

Fig. 1   Hammer Aitoff equiarea projection of the UGC galaxies brighter than 14.5 with heliocentric radial velocity between 4500 and 6000 km/s.

GALAXY REDSHIFTS FROM THE U.K. SCHMIDT OBJECTIVE PRISMS
(MEDIUM DISPERSION)

J.B. Palmer
Edinburgh University Astronomy Department
Royal Observatory, Blackford Hill, Edinburgh EH9 3HJ

Two objective prisms with dispersions of 2480A/mm and 840A/mm (at Hγ) are available on the U.K. Schmidt telescope. Combinations give dispersions of 1240A/mm and 620A/mm at Hγ. This note discusses the accuracy of preliminary galaxy redshift surveys using the two intermediate dispersions.

Previous work using the low dispersion prism has been carried out by, e.g. Cooke et al. 1977, S.M. Beard, Q.A. Parker (this conference). Redshifts are determined either by measuring the separation of the 4000A line blanketing discontinuity and the sensitivity cutoff of the Eastmann-Kodak IIIa-J emulsion or via a coordinate transformation (from the direct plate) and correlation techniques. The latter technique is being applied as an automatic process by J.A. Cooke and B.D. Kelly. Using a IIIa-J sky limited exposure the magnitude range covers > 15.0 to < 19.0 and the accuracy in redshift is +/-3000Km/s. Present work, with the 1240A/mm and 840A/mm (at Hγ) dispersions, at the moment involves an interactive cutoff-feature separation technique for redshift determination. Apart from the 4000A discontinuity we also see the Fe4383 line, Hγ +G-band at 4300A, Ca4227 and the CN-band. Unfortunately Mg-b is too near the cutoff (where the dispersion is very low) to be of any use. The spectra are measured on a microdensitometer with a 16 micron spot and 16 micron stepping increments. The images are then converted from photographic density to relative intensity and checked for any residual variation of cutoff wavelength with image intensity or colour. G-type stars are used to calibrate the cutoff wavelength for each emulsion batch.

PRELIMINARY RESULTS

The limited photometry for faint galaxies (de Vaucouleurs et al. 1976) suggests that the limiting surface brightness of galaxies accessible to this survey (the parameter with the best correlation with image peak intensity) is $m_e' = 17.0$ per □ ". This brighter limit is to

F. Mardirossian et al. (eds.), Clusters and Groups of Galaxies, 71–72.
© 1984 by D. Reidel Publishing Company.

be expected due to the signal to noise characteristics of the
telescope-emulsion setup when comparing the 2480 A/mm plates with
higher dispersions. Figure 1 shows the galaxy spectrum – the result of
the convolution of the dispersion curve, the emulsion sensitivity
curve, the seeing disk, and the image luminosity profile. The sharp
drop at 5400A is the emulsion sensitivity cutoff – nominally 5380A.
Figure 2 schematically illustrates the advantages of the higher
dispersions (2,3) vs. lower dispersion (1) in the range of objects
yielding redshifts. Figure 3 illustrates the magnitude limits (a)
saturation (b) minimum feature detection level (i.e. faintest usable
image) (c) amplitude of random noise. Figure 4 shows the results from
the 1240A/mm (at Hγ) prism compared to slit spectra redshifts from the
literature. The total random error is +/- 1040Km/s and, for a sample
limited to compact ellipticals, approaches +/- 300Km/s. (The dashed
lines show the subset of images for which features have been
mismatched. The correlation technique may remove this ambiguity.) All
results quoted are from measurements taken from plate BJ7612P(2N)
centred on the Virgo cluster.

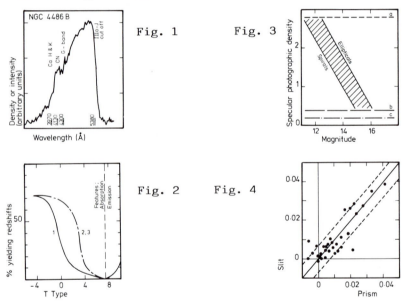

Fig. 1   Fig. 3

Fig. 2   Fig. 4

The author acknowledges the support of an SERC grant, the
unfailing help of the U.K. Schmidt Unit, the use of the RGO PDS and the
support of the ROE.

REFERENCES

Cooke, J.A., Emerson, D., Nandy, K., Reddish, V.C., Smith, M.G., 1977:
    MNRAS 178, pp687

de Vaucouleurs, G., de Vaucouleurs, A., Corwin, H.G., 1976:   Second
    Reference Catalogue of Bright Galaxies.

ON THE LARGE-SCALE DISTRIBUTION OF CLUSTERS OF GALAXIES

K.-H. SCHMIDT
Zentralinstitut für Astrophysik,
Potsdam-Babelsberg, G.D.R.

The redshifts of more than 600 clusters of galaxies have been compiled from the literature in order to study the large-scale distribution of these objects. Because this sample is more complete to larger distances than that of individual galaxies the volume investigated for distribution of extragalactic objects is extended relative to previous papers. For high galactic latitudes the sample is nearly complete for $z \leq 0.06$ and the completeness is sufficient in order to recognize the design up to about z=0.12. In the figure example of the distribution of clusters is shouwn ($\bullet$: ABELL clusters of richness class R $\geq$ 1; $\cdot$: ABELL clusters of richness class R = 0 and non-ABELL clusters).

By examination of this diagram and others many large-scale chains and filamentary structures (besides the well-known superclusters in Perseus, Hercules, and Coma) have been detected. Thus, a large circular filament extending from 10^h to 15^h in right ascension between $- 10°$ and $+ 10°$ in declination may be recognized. This structure is about 200 h^{-1} Mpc in diameter.

Equivalent prominent with the filament and chains of clusters are the large voids detected in previous investigations which are visible in our diagrams, too. The empty regions near the Coma cluster (14.5, $+ 30°$, 1000 km s^{-1}), behind the Virgo cluster ($12^h.5$, $+ 12°$, 5000 km s^{-1}) and at the coordinates (14^h, $+ 60°$, 6000 km s^{-1}) emphasized by M. DAVIS et al. (1982) may be seen as well as the hole in Bootes discovered by R.P. KIRSHNER et al. (1981). Contrary to these coincidences the existence of the large hole proposed by N.A. BAHCALL and R.M. SONEIRA (1982) from a discussion of the distribution of rich nearby ABELL clusters cannot be confirmed certainly from our maps, at least not with

73

F. Mardirossian et al. (eds.), Clusters and Groups of Galaxies, 73–74.
© *1984 by D. Reidel Publishing Company.*

the supposed size. The reason for this fact may partly be caused
by the large dispersion of the redshifts in the group of ABELL
clusters of distance class D = 4.

Further details will appear in Astronomische Nachrichten.

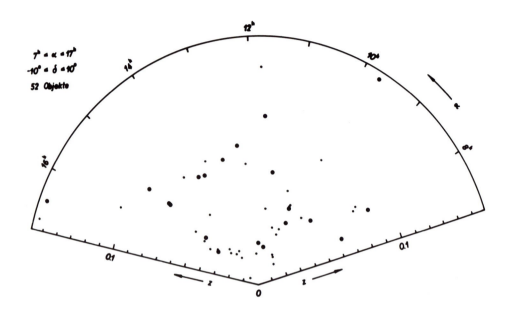

Redshift versus right ascension for clusters in the region
$7^h \leq \alpha \leq 17^h$, $- 10° \leq \delta \leq + 10°$. A large filament extends from $\alpha = 10^h$
to $\alpha = 15.^h5$. Behind this feature which may be part of a giant circular
arrangement of clusters a hole about 200 h^{-1} Mpc in size is visible.

REFERENCES

BAHCALL, N.A., and SONEIRA, R.M.: 1982, Astrophys. J. 262, 419.
DAVIS, M., HUCHRA, J., LATHAM, D.W., and TONRY, J.: 1982, Astrophys.
        J. 253, 423.
KIRSHNER, R.P., OEMLER, A.J., SCHECHTER, P.L., and SHECTMAN, S.A.:
        1981, Astrophys. J. 248, L57.

THE SPATIAL AND PECULIAR VELOCITY DISTRIBUTION OF GALAXIES WITHIN
80 MPC

L. Staveley-Smith and R.D. Davies
University of Manchester
Nuffield Radio Astronomy Laboratories
Jodrell Bank, Macclesfield, Cheshire SK11 9DL
United Kingdom

Strong evidence for the existence of elongated or filamentary structure in the nearby Universe comes from 2-dimensional galaxy distributions supplemented by redshift information (e.g. Davis et al. 1982). The 2-dimensional distribution of 21,000 galaxies taken from the UGC (Nilson 1973) and the ESO/Uppsala (Lauberts 1982) galaxy catalogues is shown in Figure 1 plotted in supergalactic coordinates. This plot shows galaxies with major axis diameters greater than 1 arcmin measured at a brightness level of 25 mag/arcsec^2 in the blue. Galactic obscuration produces the zone of avoidance which passes vertically through Figure 1. The other empty zone is caused by the gap in the catalogues between declinations -17°.5 to -2°.5. The 2-dimensional distribution in Figure 1 gives a strong visual impression of filamentary structure on angular scales up to a radian and more.

Confirmation that these are real features requires redshifts in order to give the third dimension. Joeveer et al. (1978) have established the existence of Perseus-Pisces supercluster extending from SGL = 350°, SGB = -16° to SGL = 317° SGB = 25° while Tully (1982) has found similar elongated structure in the Local Supercluster which is centred on the Virgo Cluster at SGL = 104°, SGB = -3°. These structures can be seen in Figure 1.

The use of redshifts to obtain distances is not appropriate in the vicinity of the Local Supercluster. In this region gravitationally induced velocities can be comparable to Hubble velocities. Optical and IR magnitudes used in conjunction with the Tully-Fisher relation have been used to determine independent distances. We have shown that the λ21 cm neutral hydrogen flux is an equally good luminosity indicator (Hart & Davies 1982).

Observations at λ21 cm of some 460 nearby spiral galaxies have been made with the Jodrell Bank MK IA (76m) radio telescope in a programme to investigate the spatial and peculiar velocity structure in the vicinity of the Local Supercluster. The programme includes galaxies of morphological types Sb, Sbc and Sc with velocities up to 6000 km s^{-1}.

F. Mardirossian et al. (eds.), Clusters and Groups of Galaxies, 75–76.
© *1984 by D. Reidel Publishing Company.*

These data will be used to increase the accuracy of our estimate of the
Local Group motion of 436 km s^{-1} towards SGL = 119°, SGB = -29° (Hart
& Davies 1982) for comparison with the dipole anisotropy in the 3K
cosmic microwave background. They will also be used to derive a model
for the velocity of infall of galaxies towards the Virgo cluster, in a
manner analogous to that obtained by Aaronson et al. (1982) using IR
data. Such a model will provide an indication of the mass distribution
within the Local Supercluster and, by implication, the mean mass density
of the local region of the Universe. We are grateful to Drs. W.H.
Warren Jr and A. Lauberts for providing magnetic tape copies of the UGC
and ESO/Uppsala galaxy surveys.

Aarsonson, M., Huchra, J., Mould, J., Schechter, P. & Tully, R.B., 1982,
        Ap.J., 258, 64.
Davis, M., Huchra, J., Latham, D.W. & Tonry, J., 1982, Ap.J., 253, 423.
Hart, L. & Davies, R.D., 1982, Nature 297, 191.
Joeveer, M., Einasto, J. & Tago, E., 1978, M.N.R.A.S., 185, 357.
Lauberts, A., 1982, The ESO/Uppsala Survey of the ESO(B) Atlas (ESO
        Garching bei Munchen).
Nilson, P., 1973, Uppsala General Catalogue of Galaxies, Uppsala
        Astron.Obs.Ann., Band 6.
Tully   R.B., 1982, Ap.J., 257, 389.

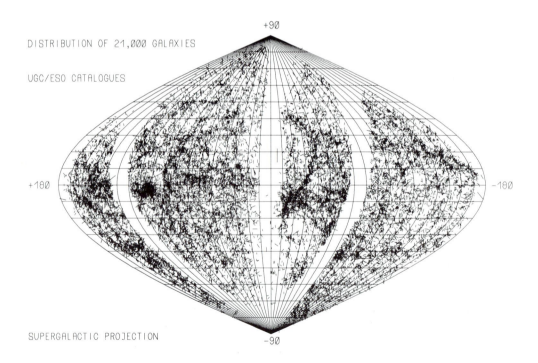

Figure 1.   The distribution of galaxies with D_{25} > 1 arcmin. The length
of each vector is proportional to the galaxy diameter.

LES REGIONS EXTERNES DES GALAXIES NE SONT PAS ORIENTEES AU HASARD

H.H. FLICHE , J.M.SOURIAU
Centre de Physique théorique C.N.R.S - Luminy - Case 907
F- 13288 MARSEILLE CEDEX 9

Des halos (ou couronnes) peuvent être détectées autour d'un certain nombre de galaxies par :
- la présence d'une enveloppe d'hydrogène neutre,
- la présence de galaxies naines satellites (cf. La Galaxie ou M 31) ou de galaxies compagnons (M 51).
Plusieurs études ont montré que, dans le SUPER-AMAS LOCAL, les parties centrales des plans galactiques sont orientées pratiquement au hasard (cf.MAC GILLIVRAY et al., 1982;KAPRANIDIS et al., 1983). La question restait ouverte pour les régions lointaines de ces mêmes galaxies qui, en règle générale, ont une autre direction plane (phénomène de "warping").
Nous avons montré (FLICHE et al., 1982,1983) que la direction d'allongement des enveloppes H I de galaxies est fortement corrélée avec la position dans le ciel. Cela implique un parallélisme statistique très significatif des halos.
L'effet est si net qu'on peut le détecter visuellement (cf.FLICHE et SOURIAU, 1983) :
On choisit un échantillon de galaxies observées dans leurs régions lointaines (enveloppes H I); les images sont allongées, et approximativement elliptiques.
A chaque paire d'images G1 et G2, on associe les points d'intersection P et P' (sur la sphère céleste) des deux grands cercles perpendiculaires aux grands axes de G1 et G2. On effectue cette construction pour TOUTES LES PAIRES de l'échantillon; on reporte tous les points P et P' sur une carte du ciel (figure 1). On constate que les points obtenus sont concentrés en deux taches opposées.
Une figure témoin s'obtient en conservant les POSITIONS des galaxies dans le ciel, mais en faisant tourner aléatoirement le grand axe de chaque image (exemple : figure 2). L'effet de concentration disparaît, ce qui montre que l'effet est bien dû à l'orientation des galaxies, et qu'il ne s'agit pas d'un artefact dû à la répartition des objets dans le ciel.
La concentration observée s'interprète sur la sphère céleste comme un parallélisme statistique des grands axes, autour du pôle ω

$$(6h, 0°) \pm 15°$$

F. Mardirossian et al. (eds.), Clusters and Groups of Galaxies, 77–78.
© *1984 by D. Reidel Publishing Company.*

avec un écart moyen de 20° environ.
DANS L'ESPACE, l'effet signifie que les régions externes des galaxies
sont approximativement planes et parallèles entre elles.
Les galaxies bien connues (M 31, M 33, M 81, M 101, M 83, M 51) sont
bien alignées (FLICHE et al., 1983).
Ce parallélisme est aussi vérifié par le système constitué des nuages H I
à grande vitesse et des galaxies naines "satellites" de la GALAXIE
ainsi que pour le système M 31 (M 31 et ses galaxies satellites).
Le pôle **ω** ne diffère pas significativement d'autres pôles associés au
SUPER-AMAS LOCAL :
- le pôle supergalactique (DE VAUCOULEURS et al., 1976) ;
- le pôle du Nuage Local (DE VAUCOULEURS, 1975)
- le pôle cinématique obtenu par DE VAUCOULEURS (TRIESTE 1983).

Ces faits indiquent une STRUCTURE STRATIFIEE du super-amas local ; stra-
tification qui constitue une contrainte pour les modèles de formation
des galaxies et des amas.

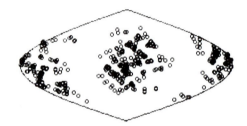 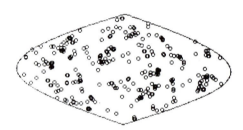

figure 1   figure 2

REFERENCES :
DE VAUCOULEURS, G. : 1975, Astron. Astrophys. 202, 319
DE VAUCOULEURS, G. : 1983, Meeting on clusters and groups of galaxies,
TRIESTE
DE VAUCOULEURS, G., DE VAUCOULEURS, A., CORWIN, H.G. : 1976, "Second
Reference Catalogue of Bright Galaxies", University of Texas Press,
AUSTIN
FLICHE, H.H., SOURIAU, J.M. : 1983, prétirage C.P.T.-83/P.1545, C.N.R.S.,
MARSEILLE
FLICHE, H.H., SOURIAU, J.M., TRIAY, R.: 1982, prétirage C.P.T.-83/P.1402,
C.N.R.S., MARSEILLE
FLICHE, H.H., SOURIAU, J.M., TRIAY, R.: 1983, prétirage C.P.T.-83/P.1502,
C.N.R.S., MARSEILLE
KAPRANIDIS, S., SULLIVAN, W.T.: 1983, Astron. Astrophys. 118, 33
MAC GILLIVRAY, H.T., DODD, R.J., MAC NALLY, B.V., CORWIN, H.G. : 1982,
Monthly Notices Roy. Astron. Soc. 198, 605.

THE CORE OF THE VIRGO CLUSTER

J. P. Huchra, R. J. Davis and D. W. Latham
Harvard-Smithsonian Center for Astrophysics
60 Garden Street
Cambridge, MA 02138

I. INTRODUCTION

We are now extending the CfA Redshift Survey (e.g. see Huchra et al. 1983) to cover a large area of the sky and to reach fainter limits. One of the logical first steps was a more detailed study of the core of the Virgo cluster. Although it is not terribly rich, the Virgo cluster is several times nearer than any other rich cluster in the northern celestial hemisphere. Thus it is well suited for studies of the faint end of the cluster luminosity function. In addition, its internal dynamics are of great interest in x-ray studies, and for the determination of the mean mass density of the cluster and the mean mass-to-light ratio for cluster galaxies. The Virgo cluster also plays an important role in the determination of the Hubble constant, due to its distortion of the local Hubble flow and the use of cluster members as calibrators of the tertiary distance indicators. Because of its unique position as the core of our own Supercluster, the Virgo cluster has been the subject of many previous studies (e.g. Sulentic 1977, Sandage and Tammann 1976, de Vaucouleurs and de Vaucouleurs 1973, Kraan-Korteweg 1982, and Eastmond 1977). Most of these used substantially fewer velocities than the present study.

We have accumulated velocities for 92% of the 473 galaxies brighter than m_{pg} = 15.5 in the Zwicky (1961-68) and Nilson (1973) catalogs and within 6° of the luminosity weighted center of the Virgo cluster, RA = $12^{h}27^{m}.8$ and DEC = $+12^{o}56'$ (Huchra et al. in preparation). Hereafter this will be called the Zwicky Core Sample (ZCS). There are also 78 fainter galaxies with measured redshifts, which brings the total number of galaxies with measured velocities in the 6° circle to 513. Of these, 317 have velocities less than 3000 km/s and are good candidates for cluster membership. Of the remaining 38 galaxies in the ZCS, approximately 30 are low surface brightness dwarfs and are also presumably cluster members. Of the 513 velocities, 343 were measured at Mt. Hopkins (with the 1.5m and MMT). These velocities have typical external accuracies of 30 to 50 km/s. The remaining velocities come from the Second Reference Catalog, the Revised

79

F. Mardirossian et al. (eds.), Clusters and Groups of Galaxies, 79–87.
© *1984 by D. Reidel Publishing Company.*

Shapley-Ames Catalog, Fisher and Tully (1981), Karachentsev and
Karachentsev (1982), Eastmond and Abell (1977), Giovanelli and Haynes
(1983), and Helou and Salpeter (1983).

II. THE LUMINOSITY FUNCTION

     Roughly three quarters (303/435) of the galaxies in the ZCS
have velocities between -700 and 3000 km/s. There are a very few
galaxies between 3000 and 5000 km/s, while most of the background
galaxies are concentrated in a spur of the Coma Supercluster at
velocities between 5000 and 9000 km/s. The area of the sky covered
by the Virgo core (36 pi square degrees) would contain only about 20
galaxies in the velocity range < 3000 km/s if there were no rich
cluster present. We therefore assume that to a first approximation
all of the galaxies in this velocity range are cluster members, and
furthermore that the 30 low surface brightness dwarves without measured
velocities are also members. Adopting 15.7 Mpc for the distance to
the cluster core (Mould, Aaronson and Huchra 1981), we derive the
cluster luminosity function shown in Figure 1.

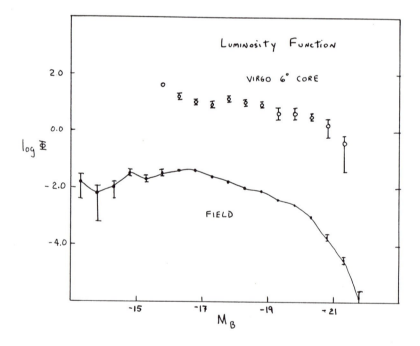

Figure 1. The differential luminosity function of the galaxies in the
Virgo core as compared to the field luminosity function derived from
the whole CfA Redshift Survey.

     The luminosity function is fairly flat over the range $-16 > M_B > -20$,
similar to the field luminosity function. There is no steep rise in

the number of dwarf galaxies fainter than -17, and there is only a weak turnover at the bright end. The Schechter function parameters for this sample are α = -1.18 and M^* = -20.1, essentially the same as the field (-1.30 and -19.4 respectively). Note that the Virgo core is about 1000 times as dense as the field.

III. VELOCITY AND SPATIAL DISTRIBUTIONS

The velocity histogram for the Virgo core is shown in Figure 2. The distribution is decidedly non-Gaussian. It may be better described by two overlying distributions, one broad and one narrow.

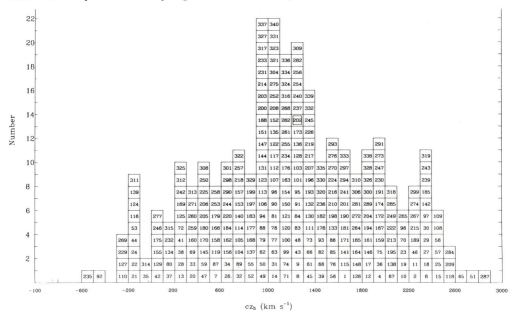

Figure 2.   The velocity histogram for the Virgo core.

An interesting view of the Virgo core is seen in the surface density plots of Figure 3 a-d. Even when all the velocities < 3000 km/s are included (Figure 3 a), it is apparent that the cluster core cannot be represented by a single smooth condensation. There is a large central lump centered nearly on M87, another large and slightly more compact lump centered on M49, and weaker features to the north and west. These features stand out more clearly when viewed in velocity windows (fig. 3 b-d). The central condensation extends through the full velocity range, while the one around M49 dominates the middle velocity range (500 to 1500 km/s). The condensation to the west of center in the integrated distribution (Virgo West) is actually two independent structures, well separated in velocity space.

It is apparent that the Virgo "core" is not a simple, smooth distribution. The galaxy surface density does not fall off uniformly

N

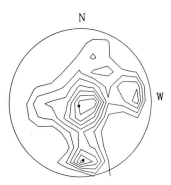

W

Cluster: VIRGO CORE
Bin size: 120.0 arcmin
Scale: 4.72 arcmin/mm

VIRGO −1000 TO 500

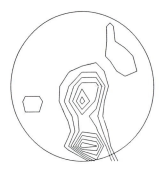

VIRGO 500 TO 1500

VIRGO 1500 TO 3000

Figure 3.   The spatial distribution of galaxies in the Virgo core - (a)
All galaxies with velocities less than 300 km/s.   The dots mark the
positions of M87 and M49.   (b) Galaxies with velocities less than
500 km/s.   (c) Galaxies with velocities between 500 and 1500 km/s.
(d) Galaxies with velocities between 1500 and 3000 km/s.   Contours
are 90%, 80% etc. of the peak galaxy surface density in each map.

From the center; it is incorrect to treat this region as an
object with a single core radius.   The Virgo core is undoubtedly
bound, but it is incorrect to assume that it is virialized.   This has
important implications for the dynamical analysis of the cluster.   Its
line-of-sight velocity dispersion is 729 km/s, so with the bound
assumption, its mass is 4.5×10^{14} $M_\odot$, and its mass-to-light ratio
$M/L_{B(0)}$ is 500 in solar units.   The dispersion of the M87 clump is
760 km/s, higher than that of the core taken as a whole while the
dispersion of the M49 clump is only 490 km/s.   This perhaps explains

why M87 is the center of the x-ray emission even though M49 is the
brightest galaxy in the cluster.

Table 1   Cluster Mean Velocity

m_{lim}	# gal	V_H km/s	σ_v km/s
13.0	86	1153 +/- 75	689
14.0	143	1149 57	681
15.0	217	1211 48	710
ALL	325	1172 41	732
Spirals	136	1158 70	816
Ellipticals	88	1237 63	592

Correction to center of Local Group (Yahil et al.
   1977) = 117 km/s.

IV.   THE CLUSTER VELOCITY AND MORPHOLOGICAL DISTINCTIONS

     We now address two long-standing and important questions:
(1) what is the mean velocity of Virgo (and does is change as fainter
members are included), and (2) is there any difference between the
velocity distributions of the spiral and the elliptical galaxies?
Table 1 contains the mean heliocentric velocities and velocity
dispersions for several subsamples of galaxies. Note that we have
made no attempt to remove foreground or background galaxies - there
is good independent distance information for only 10% of the sample.
The mean cluster velocity is not dependent on the limiting magnitude
to within the statistical errors (we quote the standard deviation of
the mean). The cluster mean velocity relative to the Local Group is
1055 km/s. This value is higher than the result of Mould, Aaronson
and Huchra (1980), although still within the one sigma bounds. The
present sample includes twice as many galaxies.

     The spiral and elliptical galaxies in the Virgo core have the
same mean velocity (to 0.8 sigma). However, the detailed distributions
appear to have significant differences. Figure 4 a and b show the
velocity histograms for the spirals and ellipticals separately.
Figure 5 a and b present their respective surface density distributions.
The spirals are in several clumps and tend to avoid the cluster center.
The densest spiral concentration is the Virgo West clump. The
ellipticals, on the other hand, are clumped symmetrically around M87.
The velocity dispersion of the spirals is 1.4 times higher than that
of the ellipticals.

     Although we have not yet constructed a detailed dynamical model
for the core of the Virgo cluster, a reasonable and simple description
is that the ellipticals form a relaxed, collapsed cluster core while
the spirals are still infalling.

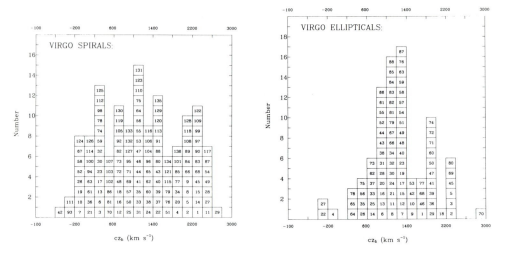

Figure 4. The velocity histograms for (a) spiral and (b) elliptical and SO galaxies in the Virgo core.

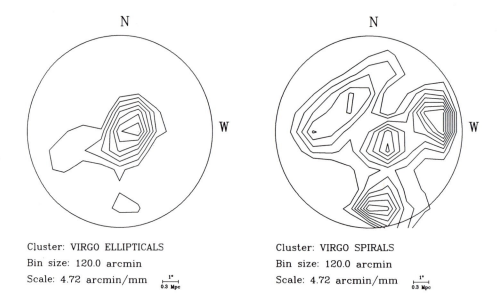

Cluster: VIRGO ELLIPTICALS
Bin size: 120.0 arcmin
Scale: 4.72 arcmin/mm

Cluster: VIRGO SPIRALS
Bin size: 120.0 arcmin
Scale: 4.72 arcmin/mm

Figure 5. The spatial distribution for (a) spiral and (b) elliptical and SO galaxies in the Virgo core.

V. THE HUBBLE CONSTANT

We can derive a quick and dirty value of the Hubble constant using 15.7 Mpc for the distance to the Virgo core (Mould, Aaronson and Huchra 1980), 250 km/s for the "infall" of the Local group towards

Virgo (Aaronson et al. 1982), and 1055 km/s for the actual cluster
velocity relative to the Local Group. The value is (1055 + 250)/
15.7 = km/s/Mpc.

We wish to thank the many Smithsonian staff members who have
contributed to the success of this extension to the CfA Redshift
Survey, especially Ed Horine and Jim Peters for their work at the
telescope; Bill Wyatt for major improvements to the software systems;
John Geary, Charlie Hughes, and Bas van't Sant for keeping the
detectors running; and Susan Tokarz for much of the data reduction.
Part of this work was supported by the Smithsonian Scholarly Studies
program.

REFERENCES

Aaronson, M., Huchra, J., Mould, J., Schechter, P. and Tully, R. B.
    1982, Ap. J. 258, 64.
de Vaucouleurs, G. and de Vaucouleurs, A. 1973, Astron. and Ap. 28, 109.
Eastmond, T. 1977, Ph.D. Thesis, University of California: Los Angeles.
Eastmond, T. and Abell, G. 1978, P.A.S.P. 90, 367.
Fisher, R. and Tully, R. B. 1981, Ap. J. Suppl. 47, 139.
Giovanelli, R. and Haynes, M. 1983, private communication.
Huchra, J., Davis, M., Latham, D. and Tonry, J. 1983, Ap. J. Suppl.
    52, 89.
Helou, G. and Salpeter, E. 1983, private communication.
Karachentsev, I. 1982, Astrofiz. 18, 501.
Karachentsev, I. and Karachentsev, V. 1982, Pis'ma Astron. Zh. 8, 198.
Kraan-Korteweg, R. 1982, Astron. and Ap. Suppl. 47, 505.
Mould, J., Aaronson, M. and Huchra, J. 1980, Ap. J. 238, 458.
Nilson, P. 1973, Uppsala General Catalog of Galaxies.
Sandage, A. and Tammann, G. 1976, Ap. J. 207, L1.
Sulentic, J. 1977, Ap. J. 211, L59.
Yahil, A., Tamann, G. and Sandage, A. 1977, Ap. J. 217, 903.
Zwicky, F. et al. 1966, Catalog of Galaxies and of Clusters of Galaxies.

DISCUSSION

VAN WOERDEN: Your average velocity has gone up from 1019 ± 51 km/s in Mould, Aaronson, Huchra (1980) to 1094 ± 41 km/s now (IAU correction applied), if I remember correctly. Can you explain the difference?

HUCHRA: Part of the change is due to better velocities for a lot of the old galaxies that only had Humason, Mayall and Sandage velocities, part is due to me not making any effort to remove possible background objects (such as IC769). The two values are within their joint uncertainty − in fact, this is a good example of how a sample mean can change if you double the sample size. I don't know, off hand, the exact details of what has caused the change.

REPHAELI: Is it meaningless to ask if you can say anything about the spatial variation of the velocity dispersion in Virgo? Perhaps that of the subgroup of ellipticals around M87?

HUCHRA: Sorry, about that, I have the data to do it in my briefcase but haven't yet. It probably would not make much sense for the cluster as a whole, but should be possible and meaningful for the biggest subgroups.

TURNER: I noticed that the ratio of spiral to elliptical velocity dispersions you quoted is very close to $\sqrt{2}$. Do you think the ellipticals are virialized with $T/|W|=\frac{1}{2}$ but that the spirals are falling into the same potential well and so have $T/|W| \approx 1$?

HUCHRA: Give that man a gold star. That's exactly what I think is the best (or most probable) explanation of what we have seen.

ILLINGWORTH: The M/L in Virgo can best be established by getting a length scale for the M87 subclump. There seems to be little to be gained by establishing the galaxy distribution at larger radii ($< 6° \lesssim r \lesssim 12°$) because of the velocity subclumping. Can one get a good core length-scale for the galaxies around M87?

HUCHRA: If we try to do it with a velocity selected sample − probably a good idea given the existence of a background cluster − it's going to be hard because the dwarf galaxies we now need to get redshifts for are very low surface brightness. You're right though, we need to get good scales for the potentials of the subclumps, the one around M87 has only ~ 90 galaxies in it.

RIVOLO: Your velocity distribution for spirals within 6° of Virgo is distinctly trimodal implying systematic streaming. Is it possible for you to fit a luminosity function to the two "side lobes" of the distribution, and thereby confirm the streaming flow?

HUCHRA: There are about 40 galaxies in each "lobe" — it might be possible, I haven't tried it.

VETTOLANI: Do you think that not having seen an increase in the number of dwarf galaxies in the luminosity function is a fact or could be due to a bias in the sample.

HUCHRA: Zwicky did pretty-well checked against Gibson Reeves' catalog of dwarfs. The bias is not likely to be more than 20 or 30% in the last bin of the luminosity function (much less in the brighter bins) which would hardly be visible on the log-log plot. Afterall, we're only talking about -15.5, this isn't really low surface brightness dwarf territory yet.

THE STRUCTURE AND DYNAMICS OF THE COMA CLUSTER

J.V. PEACH
Department of Astrophysics, Oxford, England.

1. INTRODUCTION

There has recently been a renewal of interest in the dynamics of clusters
of galaxies and a number of studies have used the observed distributions
of mass and radial variation of velocity dispersion to give information
on the distribution function, in the first place of the Coma cluster
(Kent and Gunn 1982, Materne and Fuchs 1982) and more recently of the
Perseus cluster (Kent and Sargent 1983). The major limitation on this
work is in the relatively small number of available velocities, as the
distribution function is more critically dependent on the velocity
structure than on the density distribution in the cluster. Nevertheless
the surface density profile is an essential element in the analysis and
partly as a contribution to providing a definitive profile for Coma we
have recently published a catalogue of magnitudes, colours,
ellipticities and position angles for more than 6000 galaxies in the
central 2.62 degree area of the cluster (Godwin, Metcalfe and Peach
1983).

The surveyed area covers rather more than half the area of the cluster
as determined by Abell (1977) and reaches the effective edge of the
cluster in the SE and NW quadrants although the elongation of the
cluster in the NE and SW quadrants takes it outside our area in these
directions. The photometry is complete to 20B and also includes
galaxies to 21B.

2. DISCRIMINATION BETWEEN CLUSTER MEMBERS AND FIELD GALAXIES

The limit of the photometry is considerably fainter than in previous
surveys (Godwin and Peach 1977) which leads to a large field component
in the sample, and the major problem in deriving the surface density
distribution is the removal of this field. Radial velocity is not a
realistic discriminant of cluster membership for such a large sample as
only about 300 velocities are available in the sample area. Galaxy
colour can, however, be used to give a sample from which much of the

F. Mardirossian et al. (eds.), Clusters and Groups of Galaxies, 89–92.
© *1984 by D. Reidel Publishing Company.*

field component has been removed, as we have described in detail
elsewhere (Peach 1983).

For early-type galaxies more distant than Coma the K-shift moves galaxies
out of the narrow sequence occupied by the cluster ellipticals and
lenticulars in the colour—magnitude diagram and all galaxies with
colours redder than the cluster sequence may be confidently rejected
as field galaxies. Spirals are less easy to apportion between cluster
and field because their K-shift tends to move them closer to (or into)
the cluster early-type sequence. The limitation of the sample to
those galaxies falling in a 0.25 mag. band centred on the cluster
sequence removes the bulk of the distant field. The elimination of
galaxies bluer than the sequence removes a further component of the
field at the expense of also rejecting that small (about 10 per cent)
proportion of cluster members that are spirals.

3.   THE SHAPE OF THE CLUSTER

The cluster is clearly elongated with its major axis along a position
angle of about 65° but there is considerable clumping making
elliptical isopleths only a poor approximation. The line joining the
two central giant ellipticals NGC4874 and NGC4889 has a similar
position angle (about 85°). The ellipticity of the cluster varies
erratically with radius between 0.2 and 0.5, with the lowest
ellipticity values occuring at a radius of about 1500 arc seconds
(0.95 Mpc).

Hawley and Peebles (1975) have claimed the possible presence of a
general radial alignment in the major axis position angles of the
cluster galaxies. There is no evidence for this in any subset of the
new data nor is there evidence for any preferred orientation with
respect to any direction on the sky, in particular the cluster major
axis. The position angles were calculated using the moments of the
galaxy brightness distribution and not by an eye measurement of the
principal axes of the images. This is clearly superior in
eliminating many of the subjective biases that seem to have troubled
previous work.

4.   LUMINOSITY AND TYPE SEGREGATION

Power laws can be fitted to the surface density of galaxies classified
by either luminosity or type. Bearing in mind the ellipticity of the
cluster and the subclustering this procedure is only a crude way of
looking for general trends. The table below shows the power law index
α, where (surface density) = const.(radius)$^{\alpha}$, for sequence galaxies,
that is, for a sample consisting overwhelmingly of early type cluster
members, and for blue galaxies of which only a small fraction are
cluster members. The value of α is given from fits to the data at all
radii, and for the data within a radius of 1300 arc seconds.

	Spectral index, α	
	All radii	Within a radius of 1300 arcsec
Sequence galaxies	-1.3 ± 0.1	-0.7 ± 0.1
Blue galaxies	-0.8 ± 0.1	0.0 ± 0.2

It is clear from these figures that the early type cluster members are
very much more compactly clustered than the spirals, and that the
density distribution in the cluster core is much less sharply peaked
than would be expected from an extrapolation to the core of the
distribution fitting the outer parts. This strong type segregation
confirms previous results and is consistent with Dressler's (1980)
general result on the spiral/early type ratio as a function of galaxy
density.

We have considered the problem of luminosity segregation elsewhere
(Peach 1983). The effect is weak compared with that expected from
equipartition, but similar to that predicted by the n-body calculations
with merging of Roos and Aarseth (1982).

5. DYNAMICS

Approaches to the dynamics of clusters of galaxies have been either
through n-body calculations simulating the cluster collapse (e.g.
Peebles 1970), which although they produce results roughly resembling
actual clusters do not give detailed insight into the dynamics, or
through self-consistent analytical dynamical models. In the latter
plausible assumptions as to possible distribution functions are made and
predictions are calculated for the variation of surface density and
velocity dispersion with radius. Comparison with the observations then
leads to the elimination of unsatisfactory distribution functions (Kent
and Gunn 1982). Alternatively the observed surface density profile has
been used to infer the volume density and the gravitational potential.
Then differing assumptions on the anisotropy of the velocity
dispersions give differing predictions for the observed variation of
velocity dispersion (Fuchs and Materne 1982). The latter approach
depends very sensitively on the observed surface density distribution
as this has to be inverted to obtain the space density distribution,
and during this process observational errors or noise on the observed
profile multiplies rapidly.

Using a surface density distribution based on the photometry of Godwin
and Peach (1977), Kent and Gunn concluded that only models with
distribution functions in which the ratio of tangential to radial
velocity dispersion is not a function of radius could give acceptable
fits to the observations of velocity dispersion, and that the data rule
out models which have isotropic orbits in the cluster core and
predominantly radial motions in the halo. As they demonstrate, the

radial dependence of the velocity dispersion is the crucial factor
in choosing between the models, it being possible to construct models
with quite dissimilar dynamics and virtually identical surface density
profiles. Our new data give essentially the same density profile as
our earlier work so these conclusions remain unchanged.

REFERENCES

Abell, G.O., 1977. Astrophys.J., 213, 327.
Dressler, A., 1980. Astrophys.J., 236, 351.
Fuchs, B. and Materne, J., 1982. Astron.Astrophys., 113, 85.
Godwin, J.G., Metcalfe, N. and Peach, J.V., 1983. Mon.Not.R.Astron.Soc.,
                                                                        202, 113.
Godwin, J.G. and Peach, J.V., 1977. Mon.Not.R.Astron.Soc., 181, 323.
Hawley, D.L. and Peebles, P.J.E., 1975. Astron.J., 80, 477.
Kent, S.M. and Gunn, J.E., 1982. Astron.J., 87, 945.
Kent, S.M. and Sargent, W.L.W., 1983. Astron.J., 88, 697.
Peebles, P.J.E., 1970. Astron.J., 75, 113.
Peach, J.V., 1983. In Proceedings of Paris Conference on Clusters of
                                                                     Galaxies.
Roos, N. and Aarseth, S.J., 1982. Astron.Astrophys., 115, 41.

DISCUSSION

PEEBLES: Is there a danger that our estimates of the surface density
near the edge of the sample are seriously affected by uncertainty
in the background correction?

PEACH: Yes, but we do not think that the uncertainty due to an
error in the background is great enough to seriously affect our
conclusions.

FIBRE OPTIC SPECTRA AND THE DYNAMICS OF SERSIC 40/6

D. Carter[1], P. F. Teague[1] and P. M. Gray[2]
[1]Mount Stromlo and Siding Spring Observatories, Australia.
[2]Anglo-Australian Observatory, Australia.

For the study of the dynamics of rich clusters of galaxies it is important to know rather precise radial velocities of fairly large numbers of cluster galaxies. Kent and Gunn (1982) have analysed the surface density and velocity dispersion profiles of the Coma cluster, where more than 200 radial velocities are known, in terms of a series of models with various assumptions about the form of the distribution function. Obtaining large numbers of redshifts to an accuracy of order 50-100 km/s takes a lot of time on a large telescope, but here we show how redshifts of substantial numbers of cluster galaxies might be obtained simultaneously, making it possible to obtain 100 or more precise redshifts in a cluster per night.

EXPERIMENTAL ARRANGEMENT

Following the example of Hill et al. (1980) at Steward Observatory, a fibre optic multiple object coupler between the 12 arcmin diameter auxiliary focus field of the Anglo-Australian Telescope and the RGO intermediate dispersion spectrograph has been constructed. Details of the fibre optic coupler are given by Gray (1983), and results from a preliminary observing run with the device are discussed by Ellis et al. (1983).

The device consists of 50 individual all silica fibres three metres long and with core diameter 200 μm (1.35 arcsec at the f/8 auxilliary focus at the AAT). The input ends of the fibres are carefully epoxied into brass ferrules, which fit tightly into holes drilled into a 1.6 mm thick brass plate at the positions of the galaxies to be observed. The brass plate locates in the auxiliary focus camera at the AAT. This arrangement provides precise location of the input ends of the fibres, but enables fields to be changed rapidly. The output ends of the fibres are epoxied in a line into another brass plate which is located at the focus of the f/8 collimator of the spectrograph, a position normally occuped by the spectrograph entrance slit. Acquisition and guiding in three coordinates (Right Ascension, declination and the

F. Mardirossian et al. (eds.), Clusters and Groups of Galaxies, 93–98.
© *1984 by D. Reidel Publishing Company.*

position angle of the instrument) are provided by two bundles of seven individual fibres which feed light from the positions of two guide stars in the field to the acquisition TV camera. The spectrograph has a 25 cm f/1.6 camera; we used a 600 lines/mm grating, blazed at 5000 Å; giving a reciprocal dispersion of 66 Å/mm. The detector was the University College London Image Photon Counting System, which is a two dimensional noise free photoelectric detector.

The fibre optic coupler enables us to observe up to 50 objects simultaneously at the expense of a loss of flexibility, and a loss of light in the fibres of about a factor of two. This is caused by Focal Ratio Degradation; the slow f/8 beam input to the fibre emerges more spread out at the output end, and about half the light thus misses the f/8 collimator mirror. Focal ratio degradation is caused by bending and stress of the fibres, measurements of this effect in fibres of this type are discussed by Gray (1983).

A further source of light loss in our experiment is caused by the small (1.35 arcsec diameter) input aperture of the fibres. For large galaxies and in the poor seeing which plagued our observing run for all objects, this results in the acceptance of at least a factor of two to four less light than slit spectroscopy.

OBSERVATIONS

Here we describe observations of the very rich southern cluster Sersic 40/6 (0430-6130). A substantial number of published redshifts (Melnick and Quintana 1981; West and Fransden 1980; Green 1978; Ellis, Inglis, Green and Godwin 1983); photographic photometry of several hundred galaxies (Green 1978); and X-ray data (e.g., Quintana and Melnick 1982) are all available for this cluster. Existing red-shifts suggest a high velocity dispersion for this cluster, compatible with the high X-ray luminosity. However, the main reason for choosing this cluster was that long slit spectroscopy of the central supergiant dumbell galaxy (Carter, Inglis, Ellis, Efstathiou and Godwin 1984 in preparation) shows a marked rise in the stellar velocity dispersion into the cD type halo (Fig. 1). This can be interpreted as the influence of a high mass to light ratio component with a much larger scale length than the luminous component of the galaxy, perhaps comparable with the scale length of the cluster, or as evidence that the material in the cD halo has been accreted from other galaxies in the cluster. One purpose of our observations is to establish the distributions of mass within the cluster, and to see whether this can give rise to the observed properties of the central supergiant galaxy.

Our observations were obtained in about four hours on a night of poor seeing and some thin cloud at the 3.9 metre Anglo-Australian Telescope. Because of the geometrical distortion in the magnetically focussed image tube the spectra of the galaxies, and the associated calibration lamp spectra, were curved in the two dimensional data frame.

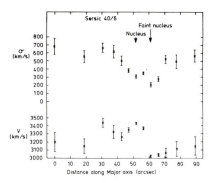

Fig. 1   Velocity dispersion and rotation along the major axis of the cen-
         tral dumb-bell galaxy. Velocity dispersion is the upper panel
         and increases from just over 300 km/s at the main nucleus to
         600 km/s in the halo. Velocity is the lower panel and the zero
         point here is arbitrary. The positions of the main and faint
         nucleii of the dumb-bell are marked.

The spectra were straightened and separated using a 2D distortion
correction routine developed by C.D. Pike at the Royal Greenwich
Observatory for use within the SPICA data reduction package of Starlink.
Care was taken that exactly the same correction was applied to the
calibration frames as to the galaxy frames. The spectra were then
rebinned to a logarithmic wavelength scale by fitting a fifth order
polynomial to the relation between channel number and wavelength. Sky
and continuum were removed, and velocities were obtained by cross
correlating with two template star spectra in the manner described by
Tonry and Davis (1979).

    Of the 44 individual spectra that we obtained 2 could not be used
because the calibration lamp was obscured from the position in the focal
plane of that fibre; of the rest 33 yielded reliable redshifts of which
29 were consistent with cluster membership. Thirteen of these had
previously published redshifts, the scatter of the differences about
zero is consistent with the formal errors on both sets of measurements,
and any systematic errors must be very small.

    The velocity histogram of the cluster (Fig. 2) shows a single
component Gaussian form, with a mean CZ = 18,010 km/s, and a dispersion
of 1590 km/s, which when corrected for cosmological effects (Harrison
1974) and observational errors gives a projected radial velocity
dispersion at the cluster of 1410 km/s, and a three dimensional velocity
dispersion of 2440 km/s. Thus we confirm the high velocity dispersion
of the cluster, and in this case there is no question of the high
dispersion being due to the superposition of two clusters in the line of
sight.

    In Figure 3 we plot the velocity against radial distance from the

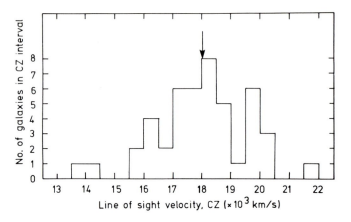

Fig. 2   Velocity histogram of Sersic 40/6

giant dumb-bell galaxy. The velocity dispersion perhaps drops slowly
with radius, but this is consistent with a number of the models of Kent
and Gunn (1982), including the isotropic King models. As with the Coma
cluster we can rule out models in which motions in the core are
isotropic and motions in the halo are predominantly radial, because in
these models the velocity dispersion drops sharply with radius.

     Our conclusions are not as definite as we would like them to be,
and maybe four times the number of velocities are required before we
can reach any definite conclusions about the dynamical state of the
cluster. The high velocity dispersion, together with the apparent
increase of velocity dispersion into the halo of the supergiant dumb-bell

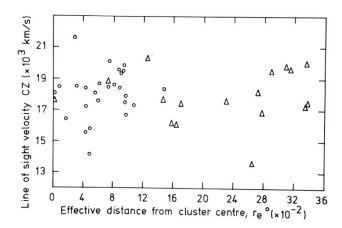

Fig. 3   Velocity against distance from the giant dumb-bell galaxy. The
circles are weighted means are of our data and previous data,
the triangles are previous data only.

galaxy, provide strong evidence for a substantial component of dark matter in the cluster, and the form of the velocity dispersion profile perhaps suggests that it is distributed on the scale of the cluster. Motions of the galaxies could be isotropic, although the models of Kent and Gunn with constant anisotropy also fit the form of the velocity dispersion profile.

Since the observations described here were made two major improvements have been made to the fibre optic system. Firstly the 200 μm core diameter fibres have been replaced by fibres with core diameter 400 μm (2.7 arcsec). Secondly an f/6 collimator mirror can now be used in place of the normal f/8 collimator, reducing light losses due to focal ratio degradation at the expense of a slight loss of resolution. These two changes should result in a gain in speed of a factor of about six. Thus the observations described here could now be made in about an hour, alternatively in a four hour period one could obtain sufficient signal to measure internal velocity dispersions for all of the cluster galaxies, and investigate in detail the luminosity-velocity dispersion relation in the cluster.

REFERENCES

Carter, D., Inglis, I., Ellis, R.S., Efstathiou, G., and Godwin, J.G.: 1984, in preparation.
Ellis, R.S., Gray, P.M., Carter, D., and Godwin, J.G.: 1983, Mon. Not. Roy. astr. Soc., in press.
Ellis, R.S., Inglis, I., Green, M.R., and Godwin, J.G.: 1984, in preparation.
Gray, P.M.: 1983, Proceedings S.P.I.E., Vol 374.
Green, M.R.: 1979, Thesis, Oxford.
Harrison, E.R.: 1974, Astrophys. J. (Letters), 191, L51.
Hill, J.M., Angel, J.R.P., Scott, J.S., Lindley, D., and Hintzen, P.: 1980, Astrophys. J. (Letters), 242, L69.
Kent, S.M., and Gunn, J.E.: 1982, Astron. J., 87, 945.
Melnick, J., and Quintana, H.: 1981, Astron. J., 86, 1567.
Quintana, H., and Melnick, J.: 1982, Astron. J., 87, 972.
Tonry, J., and Davis, M.: 1979, Astron. J., 84, 1511.
West, R.M., and Fransden, S.: 1980, Astron. Astrophys. Suppl., 44, 327.

DISCUSSION

BURNS: Have you used the Cross-Correlation tecnique on poorer signal/ noise galaxy spectra to allow the computation of accurate velocity dispersions in clusters?

CARTER: All of the redshifts I showed were derived by a cross correlation technique. To derive an internal velocity dispersion for a galaxy we use the Fourier quotient technique or some variant of it. To get a reliable dispersion we usually need about three times as many photons as to get a reliable velocity. For perhaps half a dozen galaxies in this sample we do have sufficient counts to do this.

CLUSTERS OF GALAXIES WITH VERY LARGE REDSHIFTS

J. B. Oke
Palomar Observatory, California Institute of Technology

1. INTRODUCTION

There are a number of ways by which the deceleration parameter q_o can be estimated. By assuming objects are bound and by studying gravitational motions one can measure the total mass, both seen and unseen, in galaxies, groups of galaxies, clusters of galaxies or perhaps even in superclusters. Each of these methods at best measures the mass within the particular system and this may not give a true picture of the total mass in the universe. The only way to measure the total mass in a substantial volume of the universe is to use the Hubble relation to determine q_o.

It has been shown that at least for relatively nearby clusters of galaxies the brightness of the first-rank galaxy has an impressively low spread of no more than 0.25-0.35 mag (Sandage 1972; Gunn and Oke 1975). First-rank galaxies therefore provide a potentially suitable standard candle for plotting the Hubble diagram and determining q_o. Unfortunately, to determine q_o by this means it is necessary to extend the observations into the domain where the redshift z is above 0.40 to obtain sufficient sensitivity of the Hubble relation to q_o. At these values of z, the look-back time is large enough so that one can no longer rely on the constancy of the first-rank galaxy. Because of stellar and galactic evolution, galaxies at large look-back times should be brighter than they are now by significant amounts (Tinsley and Gunn 1976). Not only should the galaxies be brighter, but they should also be slightly bluer; therefore it may be possible observationally to actually measure color differences and hence estimate differences in luminosity.

Galaxies in clusters are also subject to dynamical effects over long intervals of time and these must also be allowed for. For instance galaxies may merge with each other to form brighter single galaxies or a single large galaxy may swallow up many small galaxies, becoming brighter and perhaps bluer in the process. These processes may show up as changes in the core radius of the first-rank galaxy, or the

99

F. Mardirossian et al. (eds.), Clusters and Groups of Galaxies, 99–107.
© *1984 by D. Reidel Publishing Company.*

first-rank galaxy may have multiple nuclei. Since these are both observable quantities it may be possible to estimate the rate of change
of luminosity with time.

Although the problem of determining q_0 from the Hubble law is formidable
it is not overwhelming and therefore it should be pursued vigorously.
To solve the problem involves at least four lines of research.

(1) Produce a catalog of faint very distant clusters of galaxies where
the selection criteria are well understood.

(2) Derive redshifts for the clusters in the catalog.

(3) Study the evolution of the brightest cluster galaxies by looking
for changes in spectral features, overall energy distributions and
colors.

(4) Study the dynamical problems of clusters by measuring structural
indices of brightest cluster members, and analyzing multiple nuclei
cases.

2.   AN OBSERVATIONAL PROGRAM

At Caltech 10 years ago Gunn and Oke began an extensive program to carry
out the studies outlined above. Several graduate students, including
J. Hoessel and D. Schneider, worked on the project; it is being continued at present by Gunn, Hoessel and Oke. I will talk about some of
the results particularly with reference to points (1), (2), and (3),
above.

Most well-defined searches for clusters such as those carried out with
large Schmidt telescope plate surveys do not extend to redshifts larger
than about z = 0.35. More clusters have been discovered, but they have
usually been individual clusters or clusters around radio galaxies
(for example, the clusters around 3C 295 [Minkowski 1960]). It was
therefore decided to carry out an extensive optical survey for clusters
where the limiting magnitudes would be clearly defined.

A first survey was carried out in six 1.2-meter Palomar Schmidt fields
at high galactic latitudes in both northern and southern galactic caps.
Deep IIIa-J or IIIa-F plates were obtained and were searched over
roughtly the central 5°x5° areas. In total 144 square degrees were
searched and 232 clusters were found. For this sample the redshifts
are not likely to be greater than 0.4. A second survey was carried out
using a 90-mm magnetically focused image intensifier mounted at the
prime focus of the 5-meter Hale telescope. The incident light was confined to a 6000-7000 Å wavelength range and the green phosphor output
was recorded on IIIa-J emulsion. Exposure times were 5 minutes. A
total of 12 square degrees within the confines of the original Schmidt
survey areas was searched; 82 clusters were discovered. Redshifts of

this sample can be as high as 0.7.

Finally, Gunn and Hoessel carried out a survey using the Kitt Peak
Mayall telescope at prime focus. Hypersensitized IIIa-F or IV-N plates
were used. The IIIa-F plates covered 6 square degrees and yielded 22
clusters. The IV-N plates covered 21 square degrees and provided 90
clusters. These clusters can have redshifts up to 0.9.

Since the search plates give virtually no information except that a
cluster is present, a substantial amount of observing time has been used
during the last two years to obtain high quality deep pictures of all
candidate clusters. These pictures were taken with a Gunn R filter
(6000-7000 Å) and an 800x800 TI CCD which covered a 320x320 arcsecond
field. Exposure times were 150 sec for 1.2-meter Schmidt clusters and
300 sec for the others. These pictures are being used to (1) verify
the reality of the cluster, (2) obtain accurate coordinates, and (3)
plan further observations. The complete catalog will be published very
shortly.

The only way to make a survey which goes fainter than the present one is
to use a large telescope and a CCD detector. One approach which is
being pursued is to use our existing CCD pictures and look for faint
clusters away from the field center where a cluster already exists.
This procedure will allow a survey of about 3 square degrees to be made.
We also plan to make a new survey using a four 800x800 CCD camera which
has recently been built for the 5-meter telescope by Gunn.

3. REDSHIFTS

When the redshift program began the only instrument which was capable
of determining high redshifts was the Multichannel Spectrometer on the
5-meter telescope. It produces absolute spectrophotometric measurements
over a very broad wavelength range and with enough spectral resolution
so that features such as the Ca II H and K line break, the G-band, and
the Mg b-band can usually be seen. Up to the end of 1978 a total of 61
first-rank galaxies were observed. These yielded 56 redshifts. Of the
five failures, two were almost certainly stars, two were flat spectrum
galaxies with no detectable features, and one had very poor data. Examples of fits of the multichannel data to a standard giant elliptical
galaxy energy distribution (Yee and Oke 1978) are shown in Figure 1.

In 1976-77 attempts were made to obtain redshifts using a low resolution
spectrograph with a cooled SIT vidicon detector mounted on the 5-meter
telescope. This instrument proved to be inferior to the Multichannel
Spectrometer, but 16 redshifts were determined.

From late 1979 to the present the spectrographic mode of Gunn's Prime
Focus Universal Extragalactic Instrument (PFUEI) which uses a 500x500,
or more recently on an 800x800 TI CDD, has been used for all redshift
determinations. With it 78 clusters have been observed. Redshifts have

been determined for 57 of these, while 17 are still being processed.
There have been four failures in which the observed objects are either
stars or blue galaxies with no measurable features. Twenty-one of the
78 clusters have been observed with a multislit technique which permits
4 to 8 galaxies to be observed per exposure. Examples of PFUEI spectra
are shown in Figures 2 and 3.

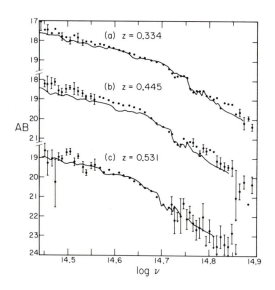

Figure 1.   Multichannel observations of 2154.2+0508 (a),
2158.0+0429 (b), and 2142.6+0348 (c).   The solid curve
is a standard galaxy energy distribution.

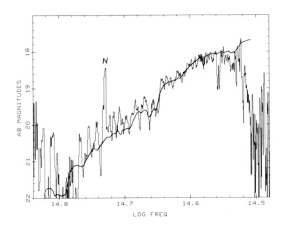

Figure 2.   PFUEI spectrum of 1322.5+3155 at a
redshift z = 0.714.   NS marks a strong night sky
line.   The smooth curve is a standard galaxy
energy distribution.

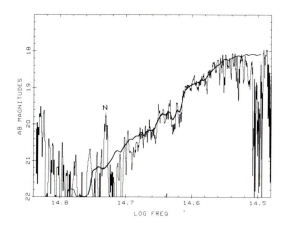

Figure 3.   PFUEI observation of 0021.3+0406 at
a redshift of z = 0.830.

When all the present data are reduced we will have about 130-140 cluster
redshifts in the range z = 0.15 to 0.92; most of them are in the range
0.20 to 0.75.   This should be a large enough sample so that the uncertainty in q_o due to the limited sample size should be no more than 0.1
to 0.2.

4.   EVOLUTIONARY CHANGES

When one looks at the spectra of first-rank cluster galaxies over the
whole range of z covered one is impressed by the fact that the vast
majority are very similar to each other and to nearby ellipticals. A
few percent of the galaxies have such unusual spectra that it is unlikely that they represent simple evolutionary effects. To quantify the
galaxy spectra we will define two colors $c_{vis} = AB'_{4050} - AB'_{6600}$ and
$c_{uv} = AB'_{3360} - AB'_{4050}$. Here AB' is the value of AB at the redshifted
position of the subscripted wavelength and $AB = -2.5 \log f_\nu -48.60$ where
f_ν is the flux in ergs cm^{-2} s^{-1} Hz^{-1}. Thus, c_{vis} and c_{uv} are intrinsic
colors independent of the redshift. c_{vis} is defined over about twice
the wavelength baseline of B-V. c_{uv} is a much more discrminating color
than U-B. We also define an index K which measures the size of the
break at the H and K lines of Ca II. It is the difference in magnitudes
at a rest wavelength of 3935 Å between the extrapolation of the visual
energy distribution derived from AB'_{4050} and c_{vis} and the actual flux in
100 Å band centered at 3935 Å.   K is about 0.5 mag in normal elliptical
galaxies.

The above defined colors are plotted in Figure 4 as a function of redshift z for the galaxies which have been observed with the multichannel
spectrometer; they are derived from the measured energy distributions
which are photometric in quality. The arrows on the left show the location of our standard galaxy. If we omit the four galaxies with

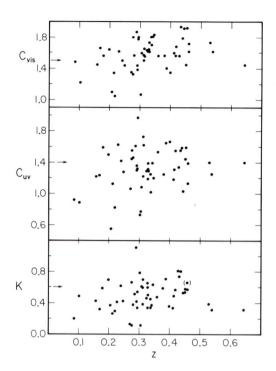

Figure 4. Colors c_{vis} and c_{uv} and index K (see text) plotted against redshift z for the multi-channel observations.

$c_{vis} \leq 1.20$ (discussed below) there is no significant change in either color or K with z. If we derive average values we find $\langle c_{vis} \rangle$ = 1.51 ± 0.16, $\langle c_{uv} \rangle$ = 1.35 ± 0.21, and $\langle K \rangle$ = 0.40 ± 0.19. Since the measuring uncertainties are approximately ±0.10, ±0.15, and ±0.10 respectively, it appears that there is some intrinsic scatter in the colors of the first-rank galaxies but it is quite small. That there are real but small differences are also obvious in Figure 1. Variations in c_{vis} and c_{uv} are not correlated, but there is some correlation between c_{uv} and K in the sense that when c_{uv} is too blue, K is smaller than average. This suggests that a moderately hot uv source of radiation is present at a level which varies from galaxy to galaxy. The four objects with $c_{vis} \leq 1.20$ show a correlation between c_{vis} and c_{uv} in the sense that when c_{uv} is abnormally blue c_{vis} is also abnormally blue.

In the case of the PFUEI spectra it is also possible to derive c_{vis}, c_{uv}, and K, but in this case a slit is used and the photometric quality of the color data is not assured. However, since the wavelength range covered is only from 5000 to 8500 Å and since observations are not made at large zenith distances, atmospheric effects should be small. The value of K should be accurate. The colors and K are shown in Figure 5. They have been derived by simply comparing by eye the spectrum of the

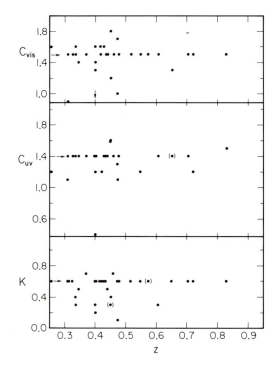

Figure 5.   Same as Figure 4 but for PFUEI data.

unknown with that of the standard galaxy. The results are very similar
to those from the Multichannel observations, i.e., most galaxies are
very similar to the standard and show no trend with z. Again, there
are a few abnormally blue objects but they are not correlated with z.

To summarize, there is no evidence in our data for changes of colors or
the K index with z.

5.   BLUE OBJECTS IN THE SAMPLE

It has been found from IUE observations of nearby elliptical galaxies
(see, for example, Bertola, Capaccioli, and Oke [1982]) that there is a
hot component, corresponding to a blackbody or model atmosphere at a
temperature of $30,000°$ K, which contributes varying amounts of flux in
the uv. If one makes models for the observed flux which correspond to
the sum of a standard galaxy and a varying $30,000°$ K source contribution
one finds excellent fits to the energy distributions of the very blue
objects referred to above. Furthermore, those objects with modest color
changes in c_{uv} can also be fitted with the same type of model. This
possibility that a hot source, perhaps horizontal-branch stars or OB
stars produced in a recent burst of star formation, may be present at a
significant flux level in first-rank cluster galaxies should be pursued.

6. OTHER GALAXIES IN CLUSTERS

So far we have discussed what has been judged to be the one brightest
member of a cluster. In this sample one finds that at most 10 percent
of the objects are peculiar, usually in the sense that the objects are
abnormally blue. Strong emission-line objects such as Seyferts are
extremely rare in this sample although [O II] $\lambda 3727$ is occasionally
present.

When one looks at less luminous objects in a cluster the situation may
be very different. Butcher and Oemler (1978) found large numbers of
blue objects in the 3C 295 ($z = 0.46$) cluster and in 0024+1654 (0.37).
Dressler and Gunn (1983) have obtained spectra of 26 objects in the
3C 295 cluster. Twenty-three objects yield redshifts. Six of these,
including 3C 295 itself, are normal galaxies which look like standard
ellipticals. Six others are blue, three with active galactic nuclei
and three with evidence for recent bursts of star formation. Nine
galaxies are foreground galaxies. There are no normal spirals.

Dressler and Gunn (1982) have also looked at six galaxies in the
0024+1654 cluster. Four are normal ellipticals and two are blue with
strong emission lines. More recent data which they have obtained
reveal that 40 percent of the galaxies observed are blue and have
spectra characteristic of normal spirals but with high surface brightness. Bautz, Loh, and Wilkinson (1982) have done photometry of two
clusters Abell 370 ($z = 0.37$) and 2244-02 ($z = 0.33$). Their colors are
consistent with all the galaxies being normal ellipticals although there
are some problems with the 7500-8500 Å colors.

In our own recent work we have been using PFUEI in a multislit mode.
So far we have reduced spectra of 50 galaxies in 21 clusters. Thirty-
seven galaxies give a value of z and their spectra look like those of
normal giant ellipticals. Six others have poor quality spectra but
they appear to be normal ellipticals if they are cluster members. Six
others are blue objects which could be foreground galaxies, stars, or
peculiar galaxies. One object has strong hydrogen lines and probably
is a galaxy which has undergone a recent burst of star formation.

To summarize, the vast majority of first-rank galaxies in clusters have
spectra similar to those of giant ellipticals. When less luminous
cluster members are observed one is more likely to fine active galactic
nuclei, spirals, or galaxies with recent bursts of star formation.

REFERENCES

Bautz, M., Loh, E., and Wilkinson, D. T.: 1982, Astrophys. J. **225**, 57.
Bertola, F., Capaccioli, M., and Oke, J. B.: 1982, Astrophys. J. **254**, 494.
Butcher, H. and Oemler, A.: 1978, Astrophys. J. **219**, 18.
Dressler, A. and Gunn, J. E.: 1982, Astrophys. J. **263**, 533.
Dressler, A. and Gunn, J. E.: 1983, Astrophys. J. **270**, 7.
Gunn, J. E. and Oke, J. B.: 1975, Astrophys. J. **195**, 255.
Minkowski, R.: 1960, Astrophys. J. Letters, **132**, L908.
Sandage, A.: 1972, Astrophys. J. **178**, 1.
Tinsley, B. M. and Gunn, J. E.: 1976, Astrophys. J. **203**, 52.
Yee, H. K. C. and Oke, J. B.: Astrophys. J **226**, 753.

DISCUSSION

MUSHOTZKY: Any new information on z vs. M diagram?

OKE: Not yet, still in progress.

CARTER: Because you are using the continuum shape, as well as the 4000 A° break to determine the redshift, are you not constraining the colours to be more or less constant simply by the criterion of whether you accept the redshift or not?

OKE: We always require that we see spectral features, otherwise peculiar energy distributions would lead to incorrect values of z and eliminate the peculiar cases. We also keep track of the observations which do not lead to values of z because these are the cases where the energy distributions are sometimes peculiar. These comprise less than 10 per cent of the whole sample.

CCD OBSERVATIONS OF DISTANT CLUSTERS OF GALAXIES

Richard S. Ellis
Durham University
England

1. Spectral Energy Distributions

By imaging clusters of galaxies through a sequence of 7 narrow-band
filters we have been able to reconstruct individual spectral energy
distributions from 0.4 microns to 0.9 microns for large numbers of
galaxies in clusters at redshifts between 0.22 and 0.31.

The technique offers many advantages over direct spectroscopy. Firstly
we have a very large multi-object capability over a wide field of view;
typically 50 - 75 objects within 7 arcmin^2. Secondly we can maintain
a photometric accuracy of better than 10% at R = 21 by using fast CCD
detectors. Finally we have a wide wavelength coverage with the RCA CCD
which is essential for unscrambling the effects of evolution, morphology
and foreground reddening.

Our methods and results are described in detail by Couch et al (1983a)
and will not be repeated here. In our analysis of 3 rich clusters we
identified 3 kinds of galaxy:-

  i) the majority of galaxies have colours expected of an elliptical
    at the cluster redshift but effects of foreground reddening are
    apparent and there is some room for evolution at the shortest
    wavelengths studied. After correction for a colour-luminosity
    relation, the brightest member is typical of this class.

 ii) Some galaxies are bluer than the cluster mode and these remain
    significantly bluer at all wavelengths. The average spectral
    class is consistent with that of a Sbc galaxy at the cluster
    redshift.

iii) A few galaxies are redder than the cluster mode, particularly at
    the longest wavelength studied. The nature of these objects is
    unclear at present.

F. Mardirossian et al. (eds.), Clusters and Groups of Galaxies, 109–111.
© *1984 by D. Reidel Publishing Company.*

2. Selecting Very Distant Clusters

It is well-known that selecting the apparently brightest cluster
member for evolutionary or cosmological tests may introduce redshift-
dependent biasses. Studying the evolution of ensembles of galaxies in
clusters,via methods discussed above, overcomes some of these problems
but the method of selecting the clusters themselves becomes important
at high redshifts.

For example, the near infrared IV-N photographic band appears to offer
a sensitive way of finding distant clusters. Calculations show
(Couch et al 1983b) that this passband should reveal a greater contrast
between distant cluster and field than any other visible band -
provided elliptical galaxies were not significantly bluer in the recent
past. On the other hand, a search in the blue IIIa-J passband would
be the most productive were ellipticals to evolve along the models
produced by Bruzual (1980). Since we can see a wide variety of cluster
types at low redshift it follows that the evolutionary picture
determined by careful studies of distant clusters may still be directly
related to the waveband used to first find them.

Using a large number of prime focus plates taken in various passbands
and applying the photographic amplification technique developed by
Malin (1978), we have scanned a total area of about 40 $\deg^2$ to limits
equivalent to $B_J = 24.5$ and found many distant cluster candidates.
We are now embarking upon an observational programme to verify these
candidates and to determine the redshift distribution as a function of
the detection passband. Our technique is described in more detail by
Couch et al (1983b) and an example of the success of this method viz.
a cluster at z = 0.59, is discussed by Couch elsewhere in these
proceedings.

References:

Bruzual, G. 1980 Ph. D. thesis, University of California at Berkeley.
Couch, W.J., Ellis, R.S., Carter, D. and Godwin, J. 1983a
          Mon. Not. R. astr. Soc., 206,000
Couch, W.J., Ellis, R.S., Kibblewhite, E.J., Malin, D.F. and Godwin, J.
          1983b Mon. Not. R. astr. Soc., in press.
Malin, D.F. 1978 Nature 276,591

DISCUSSION

HUCHRA: At what redshift will spirals in clusters become the dominant members? How will this affect your ability to study evolutionary effects?

ELLIS: The answer is statistical in nature because it depends on the luminosity function and K-correction differences between el- lipticals. for clusters with z > 0.5 there is a very good chance that the apparently brightest blue galaxy is a spiral member. It is precisely by studing ensembles of galaxies that such difficulties can be avoided.

PEEBLES: Are there enough unidentified radio sources that complete blanketing of distant clusters by dust is an appreciable effect?

ELLIS: Infrared photometry of such radio sources is being attempted by Walsh, Longair and others. I understand that their 2 micron fluxes are compatible with their non-detection in the visible without any complications from dust. However what fraction of such sources are actually seen at 2 micron I don't know.

RAKOS: We are using narrow band filters, actually v, r, y Strömgren filters and even with only three filter we have very promising results. What is the typical exposure time for one of your filters and let say 21 magnitude object?

ELLIS: Using the 3.9 metre AAT and filters with widths $\sim$500 $\overset{\circ}{A}$ (FWHM) exposures of 2-3000 sec yield 5-10% photometry to R=21.

GALAXY MORPHOLOGY IN DISTANT CLUSTERS

Laird A. Thompson[1]
Institute for Astronomy, University of Hawaii

Abstract: Direct information on the evolution of galaxies and of rich clusters may be available in the images of clusters at high redshift. While this type of investigation will flourish once Space Telescope is operating, ground-based observations can provide a valuable starting point. A program is underway at Mauna Kea Observatory to obtain subarcsecond images of distant clusters. The program is facilitated by the recent completion of a precise, high-frequency autoguiding system capable of removing image motion from the telescope focal plane. CCD images of clusters out to z = 0.4 have been obtained under good seeing conditions through a filter that mimics standard photographic bandpasses used in studies of nearby clusters (IIIa-J + GG 385).

The Butcher-Oemler effect—the claim that distant clusters contain an overabundance of blue galaxies—remains the center of controversy five years after it was first reported (Butcher and Oemler 1978). Associated with the solution to this problem is the question of whether or not there were more spiral galaxies in rich clusters in the past than are present there today. While an overabundance of spiral galaxies relative to other morphological types may provide a sufficient explanation of the Butcher-Oemler effect, it is by no means the necessary explanation. Quite recently, Dressler and Gunn (1982, 1983; see also Henry et al. 1983) have raised the alternative suggestion that there may be two other types of galaxies contributing to the blue population: (1) Seyferts and (2) galaxies that have experienced impulsive bursts of star formation. Until now, only photometric and spectroscopic data have been available for these studies. Clearly, high-resolution images will provide a new insight into the investigation.

While Space Telescope will be of prime importance with regard to

---

[1]Visiting Astronomer, Canada-France-Hawaii Telescope, operated by the National Research Council of Canada, the Centre National de la Recherche Scientifique of France, and the University of Hawaii.

F. Mardirossian et al. (eds.), Clusters and Groups of Galaxies, 113–116.
© 1984 by D. Reidel Publishing Company.

this problem, a significant head start can be made from Mauna Kea Observatory by utilizing a recently completed Image Stabilizing Instrument System (ISIS) (Thompson and Ryerson 1983). The ISIS is capable of removing image motion at the telescope focal plane to an accuracy of 0.06 arcsec at frequencies as high as 500 Hz. In addition, the instrument can monitor image quality at a similarly high frequency and decide whether to open or close an active shutter (at a frequency of 100 Hz) to preferentially select the moments of best seeing.

The goal of the ISIS development is to produce images down to a resolution limit of FWHM = 0.3 arcsec. Under such conditions it is fair to ask: To what limit in redshift can galaxy morphology be studied? Consider the angular diameter versus redshift relation (c.f., Peebles 1971); assume that $q_o=0.5$ and that a cluster of galaxies lies at $z=0.4$. Then 0.3 arcsec resolution corresponds to a rod of length 1.0 h^{-1} at the cluster, where h = Hubble parameter/100 km s^{-1} Mpc^{-1}. For giant galaxies, there will be 10^2 to 15^2 resolution elements if a galaxy is viewed face-on. This is adequate for the detection of strong spiral structure.

Because morphological studies of nearby galaxies rely on blue sensitive photographic plates (most often IIIa-J emulsion with a GG 385 filter), it is best to use a similar bandpass in the high redshift rest frame when imaging distant clusters. For the new studies with ISIS, a special broadband filter is being used (5190A-7070A) in conjunction with the Institute for Astronomy/Galileo CCD system (Hlivak et al. 1982, 1983). For clusters in the range 0.35 < z < 0.40, this combination provides an excellent match to the photographic studies with IIIa-J plates. The high quantum efficiency of the CCD is another necessary part of the investigation. The surface brightness of any extended object decreases as $(1 + z)^{-4}$, so the outer, low surface brightness regions of a galaxy quickly fade into the light of the night sky. The broad bandpass of the filter and the high quantum efficiency of the CCD both help to ameliorate the effects of the cosmologically induced $(1 + z)^{-4}$ effect.

While the best seeing occurs only on some nights, at the present time it is impossible to predict when these nights will occur. Quite obviously, the bonus of observing on the best nights is very great, not only because the images are good, but also because the effectiveness of the ISIS is optimized under the best seeing conditions (Fried 1966, Thompson and Ryerson 1983). It is incumbent on those of us who observe on Mauna Kea to develop a strategy of observing to insure that the best nights are, indeed, available for high-resolution imaging. So far, the ISIS has been used on two telescope runs for a total of nine nights. In this time the best images obtained have been FWHM = 0.55 arcsec.

Figure 1 shows one of the first results from this study of distant clusters, a 40-minute exposure of the original Butcher-Oemler cluster 0024+16, which lies at a redshift of 0.395 (Dressler and Gunn 1982). The image was produced at the Cassegrain focus of the 3.6-m Canada-France-Hawaii Telescope with our 500x500 Texas Instruments three-phase

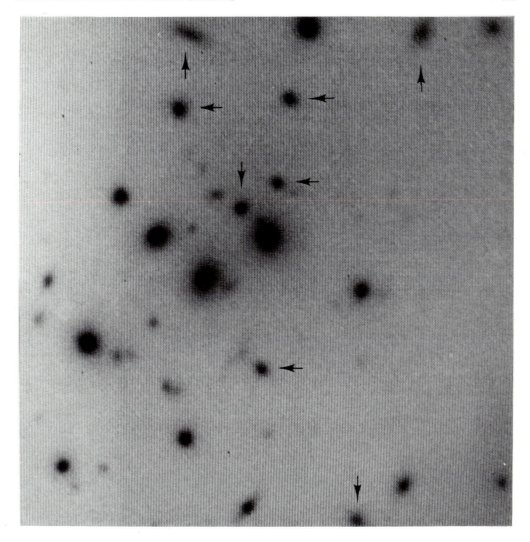

Figure 1. The central core of the cluster of galaxies 0024+16 is shown
in this CCD image, which spans a 37 x 37 arcsec area. The 40-minute
exposure was taken behind a broadband filter that mimics a IIIa-J
photographic plate + GG 385 filter commonly used in studies of nearby
galaxy clusters. Pointers mark those objects that Butcher and Oemler
(1978) designated as blue galaxies. North is to the top and east to the
right.

CCD camera after UV flooding (Hlivak et al. 1983) . The ISIS was
operated with the active shutter open at all times, and the frequency
response for active autoguiding was limited at 40 Hz (repositioning time
25 ms). While there is no star in the 0024+16 field of view from which
to measure the image quality, the CCD frame that immediately preceded
this one was of the same exposure time and produced star images in the
range FWHM = 0.5 to 0.6 arcsec.

While the image quality in Figure 1 is not quite up to the ultimate
goal of 0.3 arcsec FWHM, there are interesting conclusions to be drawn
from it. All galaxies in the field of view that were designated by
Butcher and Oemler to be blue have been marked with pointers. The first
conclusion is that the blue galaxies have a wide range of surface
brightness. This runs contrary to a suggestion by Dressler and Gunn
(1983) that most blue galaxies have an exceptionally high surface
brightness, a conclusion probably biased by the lack of a sufficiently
deep image of the cluster. Cluster 0024+16 lies near the ecliptic plane
where the night sky is quite bright. When the high sky brightness is
combined with the $(1 + z)^{-4}$ cosmological effect, the outer parts of
galaxies become difficult to detect. A second conclusion comes from a
qualitative comparison of Figure 1 with a photograph of the central 8.5
x 8.5 arcmin area of the Coma cluster (a comparable metric field of
view). Within this central area of Coma, the galaxy population is dominated by smooth, early-type galaxies. With the present limited resolution 0024+16 is similar.

The investigation of distant clusters with high-resolution imaging
has just started at Mauna Kea. Within the next few years, it should be
possible to push the quality of the observations to 0.3 arcsec FWHM to
help bridge the gap between conventional ground-based imaging and the
future studies by Space Telescope.

REFERENCES

Butcher, H. and Oemler, A. Jr.: 1978, Astrophys. J. 219, p. 18.
Dressler, A. and Gunn, J.E.: 1982, Astrophys. J. 263, p. 533.
Dressler, A. and Gunn, J.E.: 1983, Astrophys. J. 270, p. 7.
Fried, D.L.: 1966, J. Opt. Soc. Am. 56, p. 1372.
Henry, J.P., Clark, J., Bowyer, S., and Lavery, R.: 1983, Astrophys. J.
    272, p. 434.
Hlivak, R.J., Pilcher, C.B., Howell, R.R., Colucci, A.J., and Henry,
    J.P.: 1983, S.P.I.E. Proc. 331, "Instrumentation in Astronomy IV,"
    p. 96.
Hlivak, R.J., Henry, J.P., and Pilcher, C.B.: 1983, S.P.I.E. Proc. 445,
    "Instrumentation in Astronomy V," in press.
Peebles, P.J.E.: 1971, " Physical Cosmology" (Princeton Univ. Press),
    p. 186.
Thompson, L.A. and Ryerson, H.R.: 1983, S.P.I.E. Proc. 445, "Instrumen-
    tation in Astronomy V," in press.

THE EVOLUTION OF GALAXIES IN CLUSTERS

Alan Dressler
Mount Wilson and Las Campanas Observatories of the
Carnegie Institution of Washington

The following text is an abbreviated version of an article for Volume 22 of <u>Annual Reviews of Astronomy and Astrophysics</u>. More detail and complete references can be found there.

ABSTRACT

Clusters of galaxies are laboratories for the study of galaxy evolution. Predictions of the effects of tidal stripping, merging, and accretion are compared to optical observations, particularly in connection with the formation of cD galaxies in rich clusters. The development of the different morphological types is discussed in terms of models that depend entirely on later evolution and those that rely on initial conditions.

1. INTRODUCTION

In the 1970's a new course of research in clusters of galaxies began to emerge with the resurgence of interest in galaxy evolution. It was recognized that certain clusters contain "super-galaxies" unlike any objects seen in the general field. How had these objects evolved in the unique environment of clusters? The general population of galaxies in clusters is highly skewed toward elliptical and S0 galaxies, a population quite unlike the spiral-dominated field where most galaxies are actively forming new stars. What could this obvious difference tell us about the influences of environment both in the formation and evolution of the different morphological types? It was discovered that many rich clusters contain a pervasive, hot intergalactic gas. Might interactions between this gas and galaxies be strong enough to alter galaxy properties such as their own gas fraction and rates of star formation? Clusters, it was realized, are laboratories for the study of galaxy evolution and may become and useful as star clusters are in the study of stellar evolution.

F. Mardirossian et al. (eds.), Clusters and Groups of Galaxies, 117–132.
© *1984 by D. Reidel Publishing Company.*

The difference between the populations of high- and low-density regions of space is a clear indication that galaxy evolution is very sensitive to environment. The principal question is whether it is the very early environment (i.e., initial conditions) or much later evolution that plays the key role.

2. THE LABORATORY

A small fraction, about 5%, of the galaxies in the present-epoch universe are collected into groups and clusters whose space density is larger than 1 galaxy per cubic megaparsec (about two orders of magnitude greater than the average). Some of these aggregations are both dense and populous. These are the rich clusters cataloged by Abell (1957) during the original Palomar Sky Survey. Typically they contain 100 galaxies with a luminosity spread of two orders of magnitude (and perhaps a larger number of fainter systems) in a volume several megaparsecs across. A roughly equal number of galaxies inhabit relatively poor groups with less than 30 galaxies which also have these relatively high densities (Bachall 1980).

The Virgo cluster, the rich cluster nearest to our Galaxy, is a representative example, with the typical number and density of galaxies. Its structure is also quite typical, for although it has the beginnings of a well-developed core, a general lack of symmetry and significant subclustering indicates that dynamical relaxation on a large scale has just begun. Its population of galaxies is normal, with all morphological types, ellipticals, S0's, spirals (and irregulars) more or less equally represented. Thus in all these characteristics measured by optical techniques, the Virgo cluster is a very average cluster, as one might hope for our nearest neighbor.

It has been perhaps a mixed blessing that another of the nearest clusters, the Coma cluster, is a very rare specimen. Its proximity has allowed detailed studies of a cluster that is unusually populous and rich in early-type galaxies, but has contributed to a widespread notion that Coma is an archetypical rich cluster. In fact, it is no more typical of rich clusters than M87 is typical of galaxies. The Coma cluster, with its population of several hundred bright galaxies, is richer than 95% of the clusters cataloged by Abell. Its galaxy distribution is highly concentrated and very symmetrical, with little or no sign of subclustering, probably indicating that it virialized many cluster crossing times ago (several x 10^9 years, Gunn and Gott 1972). The population of its core is extreme, as it contains nearly equal numbers of elliptical and S0 galaxies, and virtually no spirals or irregulars. The great advantage in having a cluster like Coma so nearby is that one can study an extreme environment where galaxy evolution may have been most influenced by the formation and evolution of the cluster. To the extent that the formation and evolution is a function of the local environment, for example the local density and characteristic velocity of the galaxies, the processes that have influenced galaxy development in the Coma cluster are a smooth extrapolation of what has happened in the less populous but more numerous groups of galaxies.

3. MERGERS AND TIDAL STRIPPING: THE EVOLUTION OF cD GALAXIES

3.1 Criteria for Identifying cD Galaxies

Matthews, Morgan, and Schmidt (1964) called attention to the
existence in some clusters of bright elliptical galaxies surrounded by
extensive, amorphous stellar envelopes (Morgan and Lesh 1965). These
they called D galaxies. The largest (and thus the most luminous) of
these D galaxies, those with an extent 3-4 times larger than the largest
lenticulars (S0's) in the cluster, were dubbed supergiant D: cD galaxies. These are the only primary attributes of cD galaxies--large size
and extensive stellar envelopes ($\sim$100 kpc[1]), and are the proper criteria
to use when classifying clusters (Struble and Rood 1982).

These and later studies have called attention to other secondary
characteristics that are neither sufficient nor necessary for a galaxy
to be considered a cD. For example, cD galaxies usually reside in the
cores of rich, regular clusters, have very extensive ($\sim$1 Mpc) stellar
envelopes of low surface brightness (Oemler 1976), and are frequently
found to be at the kinematical centers of their clusters (Quintana and
Lawrie 1982). They are often very flat and aligned to a flattened
distribution of cluster galaxies. Most of the archetype cD's have low
central and average surface brightness and large core radii of many
kiloparsecs (e.g., A2029, Dressler 1979). A significant fraction (about
25-50%) have multiple nuclei (Hoessel 1980, Schneider, Gunn, and Hoessel
1983).

The evolution of cD galaxies is the best place to start to analyze the influences of cluster environment on galaxy evolution because
cD galaxies do not occur in the low density field ($\rho \lesssim 1$ galaxy/Mpc^3).
The fact that objects with the primary characteristics of cD galaxies
have been found very poor but high-density groups (Morgan, Kayser, and
White 1975, Albert, White, and Morgan 1977) is a clear indication that
local conditions such as density and velocity dispersion, rather than
global parameters like cluster richness or size, are the key to understanding the formation of these unique systems.

Unfortunately, much of the discussion about how cD galaxies might
form is often muddled by a careless application of the original definition by Matthews et al. Thus, two quite dissimilar objects like the cD
in A2029 and NGC 1316, the brightest member of the Fornax cluster
[labeled a cD by Schweizer (1980), but only a D galaxy in Mattews et al.]
are compared to models of cD formation. The cD in A2029 is about an
order of magnitude more luminous than NGC 1316. In this early stage of

---

[1]Linear scales have been determined assuming a value of
$H_0 = 50$ km s^{-1} Mpc^{-1}. This value has been adopted for convenience
throughout this paper.

comparison of observations with models of formation, it is preferable to restrict one's attention to the largest and most luminous examples of the class, where the predictions made by the models should be manifest in the extreme.

3.2 Simulations of Cluster Evolution and cD Models

     The long-standing debate over whether the brightest members of galaxy clusters are simply the tail of the bright end of the luminosity function or the products of special formation and/or evolution is not fully resolved, but it seems certain that the most luminous cD galaxies are indeed the special products of dense environments. They are not found in the low density field, despite the very similar luminosity functions for field and cluster galaxies, and they occur at local density maxima even in already dense environment of clusters (Beers and Geller 1983). Perhaps the clinching piece of evidence that cD galaxies are unique is the high incidence (about one half) of multiple nuclei, compared to the order-of-magnitude lower frequency for the second- and third-ranked ellipticals in the clusters (Schneider, Gunn, and Hoessel 1983).

     Variations of three basic models have been proposed to explain the properties of cD galaxies. (1) Mergers of the brightest cluster members and/or accretion of smaller galaxies is facilitated by dynamical friction between the galaxies and the stellar envelopes of their neighbors (Lecar 1975). This is the so-called "cannibalism model" (Ostriker and Tremain 1975, White 1976). (2) The central galaxy is an ordinary luminous elliptical surrounded by the stellar envelopes stripped from cluster galaxies (Richstone 1976, Merritt 1983b). (3) Vast amounts of cooling gas from the intercluster medium (ICM) builds a cD "in situ" at the cluster center (Mushotzky et al. 1981).

     The first two schemes can be studied by comparing the results of computer simulations to observed parameters for cD galaxies and clusters. Early attempts to model cluster evolution (Richstone 1976, Roos and Norman 1979, Aarseth and Fall 1980) demonstrated that tidal interactions could whittle down cluster galaxies, releasing mass and luminosity into the intercluster medium (ICM), while mergers could build a significant number of more massive galaxies.

     The work of Malumuth and Richstone (1983), Merritt (1983a, b) and Miller (1983) have significantly expanded the predictive ability of computer simulations of clusters. In the Monte Carlo simulation by Malumuth and Richstone, the Bautz-Morgan (1970) type is used to characterize the prominence of the brightest cluster member. Allowing the rich, relaxed clusters in their simulations to evolve for 10^{10} years they find a $\sim$20% occurrence of cD galaxies, and that the development of a cD galaxy is independent of cluster richness, but the brightness of the envelope of tidal debris is a strong function of cluster richness, all in agreement with observations (Leir and van den Bergh 1977, Thuan and Romanishin 1981).

   In spite of these successes, however, the simulations by Malumuth
and Richstone fail to produce a high enough <u>frequency</u> of cD galaxies.
Although 20% is close to the Leir-van den Bergh figure, most clusters in
the Abell catalog have not been relaxed for 10^{10} years. Probably only
∿10% of present-epoch clusters are as evolved as the clusters produced
in the simulations, which suggests that the predicted cD frequency is
actually an order of magnitude too low.

   Merritt's (1983b) simulations are even less efficient in producing cD galaxies from dynamical interactions. This appears largely to be
due to his adopted setup, in which a strong tidal field produced by the
cluster mass distribution pares down the galaxies to a maximum size of
only 30 kpc. With cross sections this small, mergers between galaxies
become very rare, and thus they are neglected in Merritt's simulations.
Merritt therefore concludes that if mergers build cD galaxies, they must
occur <u>at cluster collapse</u> and only for galaxies on bound orbits (Aarseth
and Fall 1980). Little evolution is expected after cluster collapse.

   The strong tidal limits imposed in Merritt's simulations may be
too severe. For example, recent determinations of core radii for clusters find values nearly twice the 250-kpc value used by Merritt, which
reduces the strength of the mean tidal field. Similarly, Merritt's
assumption that each galaxy receives the maximum tidal force for an
essentially infinite time may be too stringent. Miller (1983) also
points out that if the outer stars have large tangential velocity components they should be much more difficult to remove.

   If one or more of these factors results in an easing of the tidal
cutoff, then merging of cluster galaxies cannot be ignored, and significant evolution <u>after virialization</u> can be expected. If Merritt's parameters are correct, however, the effects of merging and accretion are
weakened to the point that they will only be important in the context of
special "initial" conditions, for example, bound orbits and low velocity
dispersions. Merritt, consequently, suggests an alternative interpretation of a cD galaxy, in which it is the chance survivor of a number of
large, massive galaxies which were ripped apart by the mean tidal field
of the cluster. If one of these comes to rest in the center of the
cluster, it will <u>not</u> be torn apart due to its symmetrical placement
within the gravitational field. It will retain its envelope, and
additionally will be identified with the more extensive envelope of
tidal debris belonging to the cluster as a whole (Richstone 1976).

   Miller's (1983) simulations are unique in their use of "collision
rules" taken from N-body simulations of galaxy encounters (see references in Miller 1983). He models a variety of initial conditions including collapsing clusters from isothermal perturbations and pancake-
like models. A number of runs covering a range of tidal stripping
efficiency show that as little as 10% to as much as 70% of the initial
mass in galaxies is released into the cluster potential. Though mergers
are important in some of Miller's simulations, he agrees with Merritt
that most of the action takes place during cluster collapse, when

densities are high and velocity dispersions are still relatively low. Miller emphasizes the importance of measuring the intracluster light that is predicted by all three studies for both cD and non-cD clusters. The intracluster light has been measured in a few clusters at a typical level of 10-40%, but the measurements are difficult and the results obtained so far are very uncertain. These types of measurements are much needed, and should be facilitated by the new generation of linear area detectors.

Reviewing these cluster simulations, it is apparent that various techniques produce very similar results, and that the main point is that all of the simulations discussed here are relatively <u>inefficient</u> in producing cD galaxies, even given a longer time than is available to most clusters. Unless tidal stripping is much less efficient than thought, the high frequency of cD occurrence must be telling us that the primary evolution of such systems occurs in subcondensations (small groups), <u>before virialization</u>, when tidal stripping is less effective and velocity dispersions are significantly lower. The idea that most of the evolution took place in the "small-group" phase may also be more compatible with the morphology/density relation for clusters (discussed below).

A completely different mechanism for forming cD galaxies has been suggested by X-ray observations which indicate that large amounts of gas are flowing from the IGM into certain centrally located cluster galaxies (Fabian and Nulsen 1977, Canizares <u>et al.</u> 1979, Mushotzky <u>et al.</u> 1981, Fabian <u>et al.</u> 1981a, b). In the context of simple models, the cooling flows that are detected imply an accretion of from 10's to 100's of solar masses of material each year. Continuation of this process for anything approaching a Hubble time will obviously result in the accumulation of mass comparable to a cD galaxy. Since the insides of cD galaxies are much like normal ellipticals (Thuan and Romanishin 1981), one would conclude that either (1) the accretion is taking place onto a preexisting galaxy and has not significantly altered its structure, or (2) the accretion is directed to the center of dark matter binding the cluster ($R_{core} \sim 500$ kpc) but dissipation in the gas causes it to form a much more condensed structure, or (3) a condensed structure in the dark matter, such as the "black pit" suggested by Blandford and Smarr (1983), forms the skeleton over which the cD forms from the inflowing gas.

Blandford and Smarr's model attempts to explain the large number of fast-moving companion ellipticals found within ~ 10 kpc of many cD galaxies. They propose that the dark binding material of the cluster actually has a much smaller radius, ~ 10 kpc, than the $R_{core} \sim 500$ kpc) of the galaxy distribution. Observations of the velocity dispersion as a function of radius (Dressler 1979, Tonry 1983) do not show, however, the rise in dispersion that is expected in the black pit model.

Let us return to the cases of accretion flows onto a preexisting galaxy or into a dark matter distribution with a large core ($\sim$500 kpc). It is well-established that the interiors of cD galaxies are similar to normal ellipticals (Thuan and Romanishin 1981), including velocity dispersions, mass-to-light ratios, stellar populations (optical spectra and colors) and densities (Malumuth and Kirshner 1981, Faber, Burstein, and Dressler 1977, Dressler 1979). Valentijn's (1983) report to the contrary of strong color gradients and extremely red nuclei in cD's is in strong disagreement with all previously published data on the color and color gradients of cD galaxies (e.g., see Lugger 1983, Wirth and Shaw 1983) which imply that cD galaxies have the colors of normal ellipticals. These colors are compatible with a single burst of star formation some 10^{10} years old, as is usually assumed for an elliptical galaxy. The signs of continuing star formation might be hidden if only low-mass stars are being formed (Fabian, Nulsen, and Canizares 1982, Sarazin and O'Connell 1983, Valentijn 1983), but this would imply that the mass-to-light ratio of cD galaxies should be significantly higher than normal ellipticals, and this does not seem to be the case (Malumuth and Kirshner 1981, Thuan and Romanishin 1981).

Because of the direct evidence for cooling flows in the form of temperature inversions and optical filaments (Heckman 1981, Fabian et al. 1981b), it appears that gas accretion rates of order 10 solar masses per year are occurring in the centers of some clusters. However, there are no optical observations that support the cases of accretion of hundreds of solar masses per year, which should be capable of building a giant galaxy in a Hubble time. This suggests that the models of such huge accretion flows are incorrect, or that accretion on such a scale has not persisted for such long periods of time. It is worth pointing out, however, that if cD galaxies were built primarily from inflow of this type, then it is probable that all galaxies formed in a similar way, since the properties of cD's and other ellipticals are so similar.

3.3 Optical Observations Relevant to Merging, Accretion, and Tidal
     Stripping

ARE cD GALAXIES BUILT FROM HOMOLOGOUS MERGERS? Ostriker and Hausman (1977) and Hausman and Ostriker (HO, 1978) have proposed that mergers and accretion build giant galaxies with radial profiles that scale simply as R α M. This assumption allows them to predict the surface brightnesses and metric magnitudes of cD galaxies as a function of their total luminosities. This approach has been sharply criticized by Schweizer (1981) who claims that both N-body simulations and observations show that mergers are not homologous, but rather build cores with larger central concentrations than the progenitors.

Schweizer points out that atmospheric seeing could be hiding very small cores in most ellipticals that have been studied (e.g., Hoessel 1980). This is probably not relevant to the question of homologous growth on a large scale. Measurements of the logarithmic intensity

gradient α by Hoessel (1980) and Schneider, Gunn, and Hoessel are very
insensitive to seeing, and they show α increasing with luminosity in a
way consistent with the HO models. Also consistent with the cannibalism model are the observations that giant cD's have much lower surface
brightnesses within their effective radii (Thuan and Romanishin 1981,
Morbey and Morris 1983), and that the archetype cD's, like A 2029,
A2199, and A1413, have very low surface brightness.

Perhaps the best piece of evidence that the merger process is
responsible for building the insides of cD galaxies is the observation
that a large fraction (25-50%) of such objects have multiple nuclei.
The frequency of multiple nuclei for the second- and third-ranked
cluster galaxies is an order of magnitude less, which seems to confirm
the special place of the cD in the bottom of the cluster potential well,
and its unique evolution compared to other bright galaxies in the
cluster.

If none of this evidence is compelling for the case that cD
galaxies grow by "cannibalism," it is sufficient to say that most
optical observations are consistent with the general features of this
model.

4. THE EVOLUTION OF DIFFERENT MORPHOLOGICAL TYPES

The previous section reviewed the major processes that can alter
the mass functions and spatial distributions of galaxies in regions of
high galaxy density. In this section the level of complexity is raised
to include the forms of the luminous matter, described by the morphological type of the galaxy. This requires an additional knowledge of
how the processes described above, combined with initial conditions,
will vary the density, angular momentum and gas content in a way that
produces the observed range and distribution of morphological types.
Again, clusters are ideal laboratories in which to study these effects,
since they contain a different mix of galaxy types than is seen in the
low-density field.

4.1 A Basic Description of Galaxies

The following is a list of the primary attributes of galaxies
and their distributions that a comprehensive theory of formation and
evolution should explain.

1. Different morphological types exist. There are spheroidal
galaxies and galaxies with both disk and spheroid. Most gas-poor
systems have had little star formation within the last several billion
years.

2. All morphological types are found in both low- and high-
density environments, but there is a clear morphology/density relation
in the sense that the fraction of spiral galaxies decreases as the
fraction of S0's and E's increases with local galaxy density (Dressler
1980b).

3. The sizes, colors (metal abundances?), and characteristic internal velocities of elliptical galaxies scale monotonically with luminosity (e.g., see Faber and Jackson 1976, Tonry and Davis 1981). There is some evidence of a second parameter which could be galaxy ellipticity, surface brightness, or mass-to-light ratio. The average surface brightness of ellipticals rises slowly with increasing bright-ness, but levels off and begins to fall again for $M_V < -21$. Most elliptical galaxies can be represented by models of oblate isotropic rotators, however, with rising luminosity $M_B < -21$, an increasing fraction owe their flatness to anisotropic velocity dispersions (Binney 1981, Illingworth 1983, Davies et al. 1983).

4. S0 galaxies almost always have small D/B ($\sim$1) and appear to have a "thick disk" component in addition to the thin disks found in both S0's and spirals.

5. Spirals are classified by their D/B and by the detailed form of their spiral pattern: openness of the arms, arm thickness, contrast of the arms to the underlying disk (Sandage 1961). Their forms do not correlate well with luminosity, but the spiral pattern and gas content seem related to D/B (Strom 1980). At least a two-parameter family is indicated (Brosche 1973, Whitmore 1983).

6. Along with these characteristics, it is important to remember that the dominant structure in some or all types may be a massive, un-seen halo, and interactions among halos may have a critical effect on the luminous matter within them.

4.2   Three Classes of Models

The models that will be reviewed fall roughly into three classes. The first type assume similar initial conditions for all galaxies (such as in hierarchical clustering models) regardless of their destiny to be cluster or field galaxies. The aim is to reproduce all the morpholog-ical variations with fairly late evolution (after clusters became important), for example, producing S0's from stripping and ellipticals from merging an initial population of spiral galaxies.

In the second class, later evolution is retained as the primary modifier of galaxy type, in particular through the truncation of disk development, but initial conditions or very early evolution are added to account for the prominence of galaxies with luminous spheroids in regions of high galaxy density. For example, mergers in the early evolution of clusters (the small-group phase) may build up a population of more massive spheroids.

Finally, there is the possibility that initial conditions were primarily responsible for galaxy morphology. The challenge for this third class of models is to understand how galaxies "knew" at the time of formation about their eventual environments. Recent cosmological models may provide a solution to this problem.

CLASS 1. Models which explain the existence of SO galaxies as "normal spirals" which have lost their gas due to environmental influences are attempts to explain variation in morphological types by recent environment alone. Spitzer and Baade (1951) originally proposed that spiral galaxies in dense clusters might strip gas from each other in direct collisions, but today, ram-pressure stripping (Gunn and Gott 1972) and gas evaporation (Cowie and Songaila 1977) by a hot (T ~ 107-8°K) intracluster gas are considered to be more important. Strong evidence that this is not the primary mechanism for the production of SO galaxies is presented by Dressler (1980b) who points out that most (~80%) SO's are found in low-density environments (the outskirts of clusters and the field) where these processes are ineffective.

This is not to say that gas ablation is unimportant in the evolution of spiral galaxies. Recent observations which show a tendency for spirals in the intermediate-density environments to be gas-poor by factors of 2-3 (relative to their field counterparts at the same Hubble type) may be cases of gas removal.

Gas "deficiencies" of spirals are modest (factors of 2-3) in these environments where SO's are quite common. Since SO's are deficient in gas by factors of 100 or more compared to spirals, this is further evidence that this type of environmental influence does not explain the existence of most SO galaxies. Only in the Coma cluster has a deficiency of order 10 been found, and Bothun (1981) claims that in this one case there are some small-bulge SO's in the very core, possibly the true remains of stripped spirals. It is doubtful that a spiral could survive in such an environment, thus the rarity of small bulge, gasless systems is probably a sign that (1) few large D/B spirals have plunged into these conditions up to the present epoch, or (2) disk formation is slowest for these systems, so their development will be inhibited by the cluster virialization (see below).

Toomre and Toomre's (1972) suggestion that ellipticals form by mergers of disk galaxies is another attempt at a class-1 model. This idea has been more fully developed by Roos (1981), who attempts to build both ellipticals and the bulges of disk galaxies from mergers of what are initially purely disk systems. The attractions and problems of such merger models are briefly discussed in the next section.

CLASS 2. Kent (1981) has provided a prescription for altering disk luminosity as a function of density that reproduces the morphology/density relation. Kent has proposed that a "fading" of galactic disks that is proportional to the collapse time for local density enhancement can reproduce the morphology/density relation. This is a class-2 model since Kent requires an ad hoc "initial condition" that large bulge systems preferentially "become" SO's while small bulge systems remain spirals. This scheme fails only for the reason that Dressler (1980b) rejected the ~2 magnitude fading that is required in the densest regions: the LF (luminosity function) of the "faded" galaxies should be quite different from the LF of the galaxies that have finished building their disks (the low-density clusters or field). But, in fact, it is not. This issue is discussed in more detail below.

Larson, Tinsley, and Caldwell (1980) concentrate their efforts on a different aspect of the same scheme by suggesting a mechanism for the "fading." Their calculations indicate gas exhaustion times for spirals of only a few billion years. Therefore, they suggest that spirals must be refueled by infall from tenuous gas envelopes that would be easily "stripped" by tidal encounters, so a morphology/density relation is expected. Like Kent's model, it cannot easily explain the similarity of spiral and S0 LF's. Larson et al. postulate that bulges in denser regions are more luminous, so S0's will come preferentially from large-bulge spirals.

Comparing luminosity functions to test disk truncation and initial conditions. The evidence that brighter spheroids, in addition to fainter disks, are necessary to explain the morphology/density relation rests on a comparison between LF's of high- and low-density regions. For example, from Dressler's (1980a) data, Hercules (Abell 2151) has a 55% population of spirals and a $<D/B> = 3.70$, and Coma has a 12% spiral population and a $<D/B> = 1.18$. Therefore, a "fading" of the disks by a factor of 3.14 would be required to bring Hercules to the same $<D/B>$ as Coma, which implies a $\Delta M^* = 0.85$ for a comparison of the LF's.

Schechter's (1976) analysis of Oemler's (1974) data gives $M^*(Coma)-M^*(Hercules) = 0.26$, considerably smaller than predicted in the disk fading model. Indeed, for the average of Oemler's 4 "spiral-rich" clusters compared to the 5 "cD" clusters (lowest proportion of spirals), $\Delta M^* = -0.21$, which goes in the wrong sense for the disk fading model.

These data suggest that a brightening of spheroids by factors of 2-3 in dense regions is needed to compensate the lower luminosities of disks and thus leave the LF basically unchanged. It is perhaps encouraging that early attempts to model cluster evolution (Malumuth and Richstone 1983, Miller 1983) do show an approximate balance between the opposing effects of mergers and tidal stripping.

Larger spheroids in denser regions could be the product of initial conditions or early mergers in the "small-group phase" when velocity dispersions were low. It seems very natural that mergers might account for the most luminous ellipticals, which have low surface brightnesses and are often flattened by anisotropic velocity fields. But in order to account for the larger spheroids in disk galaxies in dense regions, mergers would have to be responsible for them as well. This introduces a number of additional problems for merger models, for example, explaining why some mergers produce gas-free systems (ellipticals), while others leave gas in the system (large bulge spirals). There are also the questions of whether the merger would leave the angular momenta of bulges and disks so well-aligned, and whether the

chemical histories of galaxies can be understood if disks formed <u>before</u>
spheroids (Ostriker and Thuan 1975). A merger model that can <u>explain</u>
these observations, together with a limitation on disk building as a
function of density, could probably account for all of the basic data
on morphological types, and the morphology/density relation as well.

CLASS 3. The models described above rely on later evolution to
produce the range in morphological types observed and their dependence
on present environment. It is also possible that these later influences
are <u>minor</u> compared to the role played by initial conditions. The first
attempts to identify initial conditions that would lead to different
galaxy types identified angular momentum (Sandage, Freeman, and Stokes
1970) and density (Gott and Thuan 1976) as the variables responsible
for the differentiation. The models have grown in complexity in recent
years, but a dependence of galaxy concentration, and possibly angular
momentum, on the amplitude of the initial density perturbation is still
expected. If density perturbations of high amplitude lead consistently
to the conversion of most of the available gas into stars before large-
scale dissipation forms a global disk, then the basic structures of
present-epoch galaxies could be established at very early epochs. The
small fraction of gas in field S0's and early-type spirals suggests
that the presence of a large spheroidal component <u>alone</u> can induce increased rates of star formation that lead to gas exhaustion, <u>independent</u>
<u>of external environment</u> (Roberts, Roberts, and Shu 1975). Thus the
Larson <u>et al.</u> type of model of truncating the formation of disks may be
superfluous, and the fraction of gas "left over" in the Coma cluster
(roughly 2 x the luminous matter) may be typical of <u>all</u> environments.

Unfortunately, models that employ only initial conditions provide
no obvious explanation for the observed dependence of morphology on
later environment. To achieve this, the perturbation that becomes the
proto-galaxy must be affected by the larger-scale perturbation that
grows into a group or cluster. In a crude sense, they must form at the
same time. For the isothermal perturbation model (hierarchical clustering, Peebles 1980) this will occur if the initial spectrum contained
more power at large scales (pink noise), or was phased non-randomly.
The adiabatic perturbation models (Doroskevich <u>et al.</u> 1980 and references therein) have a built-in preferred scale of large mass, so it will
be important to the final amplitude of a proto-galaxy fluctuation
whether it is in a proto-cluster or field region. The recent emphasis
on cosmologies with exotic particles that dominate the mass of the
universe has produced several models where these preferred mass scales
for galaxies and clusters are expected (see Primak and Blumenthal 1983
for a review), and within these types of models such coupling of gal--
axies to clusters is likely, if not unavoidable. High-energy physics,
then, could supply the extra ingredient that enables galaxy formation
models to be built which rely primarily on initial conditions.

Alternatively, it has been suggested that differences might arise
in the angular momentum gained through tidal torques (Peebles 1969) in
proto-cluster and proto-field regions (diFazio and Vagnetti 1979, Shaya

and Tully 1983). Large scale variations in the L distribution would provide a correlation with environment for the model of galaxy differentiation first suggested by Sandage, Freeman, and Stokes (1970).

At the present time it might be best to consider these types of models "last resorts." Our ignorance of initial conditions is greater than our ignorance of later evolutionary effects, therefore, such models are rather ad hoc and offer few predictions or tests. Until they are able to do so, it seems advisable to embrace this alternative only if the models that stress later external influences fail to explain the data on morphological types and their environments.

5. SUMMARY

In the last decade a large amount of data on the distributions, brightnesses, colors, and morphology of galaxies in clusters has been collected. Theoretical models and simulations with greater predictive ability have also become available. Nevertheless, we can point to few, if any, definitive cases where the models and observations converge to a unique explanation. The good news is that our newly acquired data and models have sharpened our questions. At the moment, the primary one seems to be: Given that the morphology, stellar content, and structure of galaxies are clearly linked to environment, is it the environment at formation, or that during much later evolution, that plays the key role?

REFERENCES

Aarseth, S. J., and Fall, S. M.: 1980, Astrophys. J. <u>236</u>, 43.
Abell, G. O.: 1957, Astrophys. J. Suppl. <u>3</u>, 211.
Albert, C. E., White, R. A., and Morgan, W. W.: 1977, Astrophys. J. <u>211</u>, 309.
Bahcall, N. A.: 1980, Astrophys. J. Lett. <u>238</u>, L117.
Bautz, L. P., and Morgan, W. W.: 1970, Astrophys. J. Lett. <u>162</u>, L149.
Beers, T. C., and Geller, M. J.: 1983, Preprint.
Binney, J.: 1981, in The Structure and Evolution of Normal Galaxies, eds., S. M. Fall and D. Lynden-Bell (Cambridge: Cambridge Univ. Press), p. 55.
Blandford, R. D., and Smarr, L.: 1983, Preprint.
Bothun, G. D.: 1981, PhD thesis, University of Michigan.
Brosche, P.: 1973, Astron. Astrophys. <u>23</u>, 259.
Canizares, C. R., Clark, G. W., Markert, T. H., Berg, C., Smedira, M., Bardas, D., Schnopper, H., and Kalata, K.: 1979, Astrophys. J. Lett. <u>234</u>, L33.
Cowie, L. L., and Songaila, A.: 1977, Nature 266, 501.
Davies, R. L., Efstathiou, G., Fall, S. M., Illingworth, G., and Schechter, P. L.: 1983, Astrophys. J. <u>266</u>, 41.
diFazio, A., and Vagnetti, F.: 1979, Astrophys. Space Sci. <u>64</u>, 57.
Doroshkevich, A. G., Khlopov, M. Yu., Sunyaev, R. A., Szalay, A. S., and Zeldovich, Ya. B.: 1980, in Tenth Texas Symposium on Relativistic Astrophysics, eds., R. Ramaty and F. C. Jones, (New York: New York Academy of Sciences), p. 32.

Dressler, A.: 1979, Astrophys. J. 231, 659.

Dressler, A.: 1980a, Astrophys. J. Suppl. 42, 565.

Dressler, A.: 1980b, Astrophys. J. 236, 351.

Faber, S. M., Burstein, D., and Dressler, A.: 1977, Astron. J. 82, 941.

Faber, S. M., and Jackson, R. E.: 1976, Astrophys. J. 204, 668.

Fabian, A. C., Hu, E. M., Cowie, L. L., and Grindlay, J.: 1981a,
        Astrophys. J. 248, 47.

Fabian, A. C., Ku, W. H.-M., Mahlin, D. F., Mushotsky, R. F., Nulsen,
        P. E. J., and Stewart, G. C.: 1981b, Mon. Not. R. A. S. 196, 35P.

Fabian, A. C., and Nulsen, P. E. J.: 1977, Mon. Not. R. A. S. 180, 479.

Fabian, A. C., Nulsen, P. E. J., and Canizares, C. R.: 1982, Mon. Not.
        R. A. S. 201, 933.

Gott, J. R., and Thuan, T. X.: 1976, Astrophys. J. 204, 649.

Gunn, J. E., and Gott, J. R.: 1972, Astrophys. J. 176, 1.

Hausman, M., and Ostriker, J. P.: 1978, Astrophys. J. 224, 320.

Heckman, T. M.: 1981, Astrophys. J. Lett. 250, L59.

Hoessel, J. G.: 1980, Astrophys. J. 241, 493.

Illingworth, G.: 1983, in Internal Kinematics and Dynamics of Galaxies,
        ed., E. Athanassoula (Dordrecht: Reidel), p. 257.

Kent, S. M.: 1981, Astrophys. J. 245, 805.

Larson, R. B., Tinsley, B. M., and Caldwell, C. N.: 1980, Astrophys. J.
        237, 692.

Lecar, M.: 1975, in Dynamics of Stellar Systems, ed., A. Hayli
        (Dordrecht: Reidel), p. 161.

Leir, A. A., and van den Bergh, S.: 1977, Astrophys. J. Suppl. 34, 381.

Lugger, P. M.: 1983, Preprint.

Malumuth, E. M., and Kirshner, R. P.: 1981, Astrophys. J. 251, 508.

Malumuth, E. M., and Richstone, D. O.: 1983, Preprint.

Matthews, T. A., Morgan, W. W., and Schmidt, M.: 1964, Astrophys. J.
        140, 35.

Merritt, D.: 1983a, Astrophys. J. 264, 24.

Merritt, D.: 1983b, Preprint.

Miller, G. E.: 1983, Astrophys. J. 268, 495.

Morbey, D., and Morris, S.: 1983, Preprint.

Morgan, W. W., and Lesh, J. R.: 1965, Astrophys. J. 142, 1364.

Morgan, W. W., Kayser, S., and White, R. A.: 1975, Astrophys. J. 199,
        545.

Mushotsky, R. F., Holt, S. S., Smith, B. W., Boldt, E. A., and
        Serlemitsos, P. J.: 1981, Astrophys. J. Lett. 244, L47.

Oemler, A., Jr.: 1974, Astrophys. J. 194, 1.

Oemler, A., Jr.: 1976, Astrophys. J. 209, 693.

Ostriker, J. P., and Hausman, M. A.: 1977, Astrophys. J. Lett. 217, L125.

Ostriker, J. P., and Thuan, T. X.: 1975, Astrophys. J. 202, 353.

Ostriker, J. P., and Tremaine, S. D.: 1975, Astrophys. J. Lett. 202,
        L113.

Peebles, P. J. E.: 1969, Astrophys. J. 155, 393.

Peebles, P. J. E.: 1980, The Large Scale Structure of the Universe
        (Princeton: Princeton University Press).

Primak, J. R., and Blumenthal, G. R.: 1983, Preprint.

Quintana, H., and Lawrie, D. G.: 1982, Astron. J. 87, 1.

Richstone, D. O.: 1976, Astrophys. J. 204, 642.

Roberts, W. W., Roberts, M. S., and Shu, F. H.: 1975, Astrophys. J.
     196, 381.
Roos, N.: 1981, Astron. Astrophys. 95, 349.
Roos, N., and Norman, C. A.: 1979, Astron. Astrophys. 76, 75.
Sandage, A.: 1961, The Hubble Atlas of Galaxies (Washington, D.C.:
     Carnegie Institution of Washington).
Sandage, A., Freeman, K. C., and Stokes, N. R.: 1970, Astrophys. J.
     160, 831.
Sarazin, C. L., and O'Connell, R. W.: 1983, Astrophys. J. 268, 552.
Schechter, P. L.: 1976, Astrophys. J. 203, 297.
Schneider, D. P., Gunn, J. E., and Hoessel, J. G.: 1983, Astrophys. J.
     268, 476.
Schweizer, F.: 1980, Astrophys. J. 237, 303.
Schweizer, F.: 1981, Astrophys. J. 246, 722.
Shaya, E., and Tully, R. B.: 1983, Preprint.
Spitzer, L., and Baade, W.: 1951, Astrophys. J. 113, 413.
Strom, S. E.: 1980, Astrophys. J. 237, 686.
Struble, M. F., and Rood, H. J.: 1982, Astron. J. 87, 7.
Thuan, T. X., and Romanishin, W.: 1981, Astrophys. J. 248, 439.
Toomre, A., and Toomre, J.: 1972, Astrophys. J. 178, 623.
Tonry, J.: 1983, Preprint.
Tonry, J., and Davis, M.: 1981, Astrophys. J. 246, 680.
Valentijn, E.: 1983, Astron. Astrophys. 118, 123.
White, S. D. M.: 1976, Mon. Not. R. A. S. 174, 19.
Whitmore, B. C.: 1983, Preprint.
Wirth, A., and Shaw, R.: 1983, Astron. J. 88, 171.

QUESTIONS

J. Peebles: If disks of spirals were built by accretion over a Hubble time, then to keep disks thin we would have to assume that the direction of the angular momentum vector has not changed much as mass is added. Is that reasonable?

A. Dressler: It is probably only reasonable if the gas is intimately connected with the galaxy, for example, falling in from the galaxy halo. I agree that if the gas were falling in from a very extended distribution that received its angular momentum much later than the galaxy did, the problem of alignment could be a serious one.

L. Thompson: On the topic of cD galaxy formation, it is important to recall that Oemler identified the galaxy, NGC 4839, which shows all the photometric properties of a cD, yet it is not located at the center of a cluster. NGC 4839 may provide an interesting test-point for the various models of galaxy formation.

A. Dressler: NGC 4839 is an interesting counterexample, though because it is such a rare one, it might be considered the exception that proves the rule. I wonder if this galaxy might be the remains of a small, compact group that has coalesced to form this cD-like object?

A. Porter: I've done the same kind of surface photometry on
cD's that Hoessel did. My results agree with his, and at the same time,
any seeing corrections behave the way Schweizer says they should.
Furthermore, these core radii correlate extremely well with Einstein
cluster X-ray luminosities. Incidentally, cluster luminosity increases
significantly more slowly with galaxy core radius than a simple dimensional analysis (based on homologous merger theory) would lead you to
expect. This might be used to argue against homology, but it might
also be a sign that tidal stripping and gas inflow are at least as
prevalent as mergers.

A. Dressler: The seeing disk for Hoessel's data corresponds
typically to 1-2 kpc in radius. I think it entirely possible that
most giant ellipticals have core radii this large.

P. Shapiro: You have emphasized the point that the existence of
S0 galaxies outside of clusters makes an explanation for the origin of
S0's inside clusters in terms of the interaction of spiral galaxies
with the hot, intracluster medium (e.g., ram pressure or evaporative
stripping to remove the interstellar medium) unlikely. It is worth
noting, however, that in the pancake scenario for galaxy formation,
both in a baryon-dominated and in a neutrino-dominated universe, a
great deal of hot gas is generated by the pancake shocks which cannot
cool in the age of the universe. As a result, there exists in this
scenario a hot, intergalactic gas at substantial pressures outside of
clusters. This intercluster medium may be capable of transforming
some of the spirals outside of clusters by the same processes as have
been suggested for removing gas from spirals within clusters.

A. Dressler: If this were the case, we would expect to see a
class of small bulge S0 galaxies which were the late-type spirals (the
most common type in the field) which lost their gas through interaction
with this putative hot intergalactic gas. They're just not there!

THE RADIO LUMINOSITY FUNCTIONS OF cD GALAXIES;
THE FUELLING OF RADIO SOURCES

Willem Bijleveld
Sterrewacht Leiden, The Netherlands

Edwin A. Valentijn
European Southern Observatory, Garching bei München, Germany

INTRODUCTION

cD galaxies are often found to be centered exactly on an extended X-ray source (e.g. Abramopoulos and Ku, 1983; Jones and Forman, 1984 JF) establishing a special class of extended extra-galactic X-ray sources associated with cD galaxies. We report on a correlative study of four observational parameters of cD galaxies, viz. the X-ray luminosity (L_x), radio power ($P_{1.4}$), absolute visual magnitude (M_v) and richness (R) of the cD cluster or group. Since cD galaxies are the most luminous galaxies in all three wavelength regimes, correlations amongst the observational parameters can be studied best for these systems. A physical process, which might give rise to a correlation is the accretion of hot X-ray emitting gas by the cD from the general cluster reservoir by means of a "radiative cooling flow" (Gunn and Gott, 1972). If only a small part of the accreted gas reaches the nucleus it would be sufficient to power a nuclear radio source. In the two family evolutionary cluster scheme of JF (1984) the special class of X-ray cD galaxies is referred to as XD systems. These XD systems are characterized by a small X-ray core radius, which indicates that the X-ray gas is trapped in the potential well of the galaxies themselves, rather than that of the total clusters.

H_o = 50 km s^{-1} Mpc^{-1} is used throughout this paper. Details of the study presented here can be found in Valentijn and Bijleveld (1983).

THE SAMPLE

We compiled a list of 104 cD galaxies. In order to trace selection effects three different categories (O, R and X) have been distinguished, indicating whether the cD was found and classified for the first time in an optical, radio or X-ray survey. The optically selected cDs have again been separated into two homogeneous subsamples: cDs in Abell clusters and cDs in poor groups of galaxies.

133

F. Mardirossian et al. (eds.), Clusters and Groups of Galaxies, 133–138.
© 1984 by D. Reidel Publishing Company.

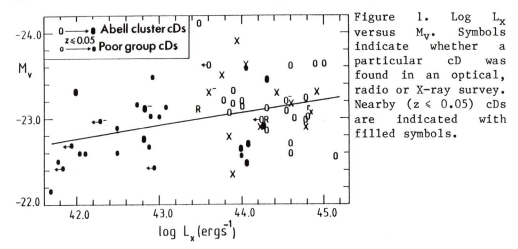

Figure 1. Log L_x versus M_v. Symbols indicate whether a particular cD was found in an optical, radio or X-ray survey. Nearby ($z \leqslant 0.05$) cDs are indicated with filled symbols.

For most of the cD galaxies the radio power at 1.4 GHz ($P_{1.4}$) or an upper limit to the radio power is available, partly based on the survey described by Bijleveld and Valentijn (1983). Optical magnitudes (M_v) from different authors have been transformed to a uniform system. We have chosen the metric absolute visual magnitude, within a galaxy diameter of 38 kpc, and a standard K-correction and correction for visual absorption. The X-ray luminosities are mainly from data from Kriss et al. (1983), Abramopoulos and Ku (1983) and JF (1984). The X-ray luminosities refer to the 0.5-3.0 keV energy band, within a 0.5 Mpc radius around the galaxy.

CORRELATIONS

In studying correlations one might analyse luminosity-luminosity diagrams; e.g. $P_{1.4}$ versus M_v. However, a diagram which involves $P_{1.4}$ is problematic since a radio source varies its output on a 10^6-10^8 yr timescale which is most clearly demonstrated by the irregular brightness distribution in the tails of head-tail radio sources. This implies that a non-detection does not say that the particular object cannot be very powerful at a different epoch. This problem can be circumvented by the use of radio luminosity functions, in which the fraction of non-detections and the detection limits of the radio observations are taken into account.

An L_x-M_v diagram (fig. 1) is, however, informative. From this diagram we draw three conclusions:
1) The mean M_v of cD galaxies in poor groups ($-22^m.73 \pm 0.46$) is $0^m.42$ fainter than the mean M_v of cluster cDs ($-23^m.15 \pm 0.36$).
2) Poor group cDs have a much lower X-ray luminosity than cluster cDs. If we restrict ourselves to the subsample of optically selected cDs with $z \leqslant 0.05$ (filled symbols) this conclusion is strengthened.
3) From a least square fit to all the data points a relationship is found: $M_v = -16.44 - 0.15$ (± 0.05) log L_x, which implies a very strong dependence of L_x on M_v.

Figure 2. The XRLF: the
integral radio luminosity
function of cD galaxies for
three intervals of L_x. The
XRLF of A1367 elliptical
cluster galaxies is indicated as well. These luminosity
functions give the fraction of
galaxies with a radio power
larger than a specific value.
E.g. 50±9% of $L_x > 10^{44}$ erg s^{-1}
cDs have a central radio
source with $P_{1.4} > 10^{24}$ WHz^{-1},
while 12^{+12}_{-5} % of $L_x < 10^{43}$ erg s^{-1}
cDs have a radio source of
that power.

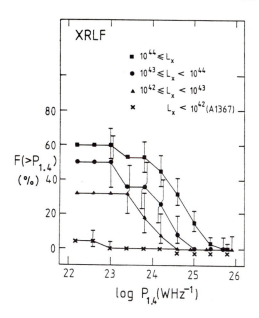

RADIO LUMINOSITY FUNCTIONS

     The best technique to study correlations of a sample of radio
sources is to construct a radio luminosity function (RLF), which is
described in detail in Valentijn and Bijleveld (1983). We constructed
RLFs for various subsamples, both as function of M_V (the optical
bivariate RLF: ORLF), as function of L_x (the X-ray bivariate RLF:
XRLF), and as function of M_V and L_x (the trivariate RLF: TRLF). The
ORLF of cDs in clusters conforms very closely to the ORLF (Auriemma et
al., 1977) for the brightest elliptical galaxies, while the ORLF of
cDs in poor groups follows the Auriemma et al. functions for the
optically next brightest ellipticals. Consequently, from a radio point
of view cD galaxies are similar to giant ellipticals and
straightforwardly extrapolate the properties of normal elliptical
galaxies. In fig. 2 the XRLF of all optically and X-ray selected cD
galaxies is reproduced. The three brightest L_x intervals refer to cDs
in our sample while the lowest interval refers to the elliptical
galaxies in A1367. The important conclusion to be drawn from this
diagram is that if a cD galaxy has a bright X-ray halo, it has a large
probability to contain a central radio source! In addition the TRLFs
(not shown here) reveal that the L_x-$P_{1.4}$ relation holds independent of
M_V in different magnitude intervals and emphasizes this correlation on
the 4σ level.

DISCUSSION

     The correlations described above can be expressed in analytical
expressions, which are reviewed in fig. 3. We also included cluster
richness, since this parameter is known to correlate with L_x and M_V.

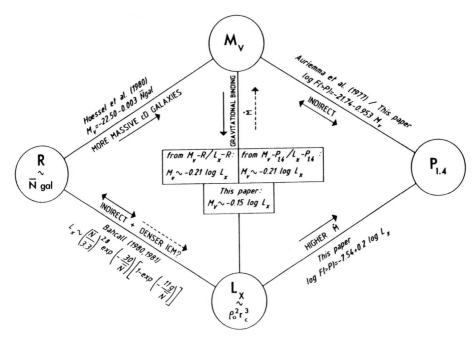

Figure 3. Diagram indicating the relations between average properties of cD galaxies: cluster richness R, cD optical luminosity M_v, X-ray luminosity L_x and 1.4 GHz radio power $P_{1.4}$. Indicated are both the observed regressions and their physical interpretation in terms of basic processes and indirect correlations.

All correlations presented in fig. 3 are observationally determined and internally consistent.

     We presume that rich clusters just have more material to build a cD galaxy; consequently, the R-M_v relationship is a causal one. If the R-L_x relation would be causal than that would imply that richer clusters contain a higher intracluster gas density and therefore produce brighter X-ray sources. However, from several independent radio observations of head-tail radio sources it is known that cluster richness is not a direct indicator of intracluster gas density. Probably the R-L_x relation is not causal and hence we consider whether the M_v-L_x relation is causal and reflects an on-going process. The formalism that predicts the observed relationships is Bondi's (1952) solution of the steady state, gravitationally driven, spherical accretion of an infinite gas cloud onto a central mass concentration. This solution predicts $M_v \sim \left(-2.5/(7+7\alpha)\right) \log L_v$, with α defined by $M_{gal}/L_v \sim L_v^\alpha$. The general accepted value of $\alpha = 0.5$ results in the observed slope in the M_v-L_x relation. Therefore we conclude that the X-ray gas is probably gravitationally bound to the cD galaxies and subject to a gravitationally driven cooling flow eventually fuelling the central radio source.

This conclusion is further strengthened from the analysis of the TRLFs that show that the $P_{1.4}-L_x$ correlation holds independent of M_v and thus must be causal and in our interpretation reflects that cD galaxies with the brightest X-ray haloes have cooling flows at the highest rate that provide the material for the nuclear radio activity. JF (1984) claim a correlation between the central X-ray excess (L_x^e) and the radio power: $L_x^e \sim P^{1.4\pm0.4}$. Their L^e is however strongly correlated with the total L_x: $L_x^e \sim L_x^{1.67\pm0.2}$ (regressions determined by us). Thus one would predict from their data $L \sim P^{0.84\pm0.45}$. Our XRLF's give however $L_x \sim P^{1.7\pm0.1}$ on the basis of a much larger sample. We have already pointed out the problematics in interpreting power-power diagrams, due to the nature of radio sources. We suspect that JF's findings are either hampered by this problem and are affected by observational bias. Alternatively, the 4 to 5 cD galaxies with both a high L_x and a high P that 'make' their correlation could represent a group of objects for which in the X-ray band a non-thermal component has been detected of the same source that emits in the radio band. This would imply a significant point source contribution in the quoted X-ray excess luminosity. The radio-X-ray spectral index of $\alpha = -0.55$ ($S_\nu \sim \nu^\alpha$) of these galaxies supports this interpretation.

CONCLUSIONS

The following scheme emerges from our analysis:
1) Richer clusters form more massive cD galaxies with brighter M_v.
2) More massive cDs accrete more gas from their environment and thus produce brighter X-ray sources.
3) Richer clusters contain very little denser cluster-pervading gas and the $R-L_x$ relation is a result of 1) and 2).
4) High mass galaxies accrete more gas, hence have a high L_x and fuel their central radio sources at the highest rate: $P_{1.4}-L_x$ reflects a principal proccess.
5) Optically bright galaxies have a higher probability to become a radio source, which results indirectly from 2) and 4). Finally we note that especially the continuity of the RLFs from normal ellipticals to cD galaxies strongly suggests that a similar process, but operating at a lower rate, fuels the radio sources in less massive galaxies.

Abramopoulos, F., and Ku, W.H.M.: 1983, Astroph. J. 271, 446
Auriemma, C., Perola, G.C., Ekers, R., Fanti, R., Lari, C.,
    Jaffe, W.J., and Ulrich, M.H.: 1977, Astron. Astroph. 57, 41
Bahcall, N.A.: 1981, Astroph. J. 247, 787
Bondi, H.: 1952, Monthly Notices Roy. Astron. Soc. 112, 195
Bijleveld, W., and Valentijn, E.A.: 1983, Astron. Astroph. 125, 217
Gunn, J.E., Gott III, J.R.: 1972, Astrophys. J. 176, 1
Hoessel, J.G., Gunn, J.E., and Thuan, T.X.: 1980, Astroph. J. 241, 486
Jones, C., and Forman, W.: 1984, Astroph. J. in press; JF
Kriss, G.A., Cioffi, D.F., and Canizares, C.R.: 1983 (in preparation)
Valentijn, E.A., and Bijleveld, W.: 1983, Astron. Astroph. 125, 223

DISCUSSION

BURNS: To follow-up on your result, Owen et al. (poster paper, this meeting) find that cluster with cooling X-ray cores tend to have radio sources but they are low power and confined to the cores of the dominant galaxies. The powerful, wide-angle tailed sources are associated with central dominant galaxies that are not in cooling core clusters. Do you think that this is consistent with your results?

BIJLEVELD: Yes. Our results have been obtained for an optically selected sample of cD galaxies. For this sample we concluded that 90% of the radio sources are unresolved, while the powers of the unresolved radio sources range from low to high. WATs are in general not included in our study since they form part of the radio selected sample. Moreover we note that the detection of a cooling X-ray core depends strongly on the X-ray flux of a particular cD galaxy.

WHITTLE: To what extent have the correlations been artificially enhanced by plotting luminosity against luminosity in a flux limited sample ?

BIJLEVELD: We carefully checked whether the correlations, as found in plots of luminosity against luminosity, are present in a distance limited sample ($Z \lesssim 0.05$) as well. As can be seen in fig. 1 the correlation is confirmed for this sample.

GLOBULAR CLUSTERS AND GALAXY FORMATION IN CLUSTERS OF GALAXIES

Sidney van den Bergh
Dominion Astrophysical Observatory
Herzberg Institute of Astrophysics

ABSTRACT

Taken at face value globular cluster counts indicate that: (1) Most elliptical galaxies in rich clusters were not formed by merging spirals and (2) central galaxies in rich clusters were already very special at the time when they were forming globular clusters. These results are difficult to reconcile with models in which clusters of galaxies are formed entirely by gravitational clustering of pre-existing galaxies.

I. INTRODUCTION

Globular clusters are the oldest known objects in galaxies. They should therefore provide valuable information on the earliest phases of galactic evolution. Studies of their frequency, spatial distribution and composition might throw some light on the distant era wher the first clusters of galaxies were formed.

To investigate this possibility Dr. William Harris of McMaster University and I have started a large program of globular cluster counts in galaxies (with special emphasis on the Virgo cluster) using the 3.6 m Canada-France-Hawaii telescope. To transform the observed cluster counts (which reach $J \sim 23.5$, corresponding to $M_J \sim -7.5$ at the Virgo distance) to total number of globulars per galaxy we assume that the globular cluster luminosity function is universal and similar to that in the Galaxy and M31 (Harris and Racine 1979, Racine and Shara 1979). The limiting magnitude of each plate is estimated from the observed background star and galaxy counts. This introduces some uncertainty because the detection limit for slightly extended galaxies is somewhat brighter than that for stars and globular clusters (which at the distance of Virgo appear star-like). To derive the specific globular cluster frequency, S, (i.e. the number of globulars per $M_V = -15$ of parent galaxy light) the Hubble parameter has to be known. Intercomparisons of the specific globular cluster frequencies in different distant galaxies are, however, insensitive to the assumptions that are made regarding the

F. Mardirossian et al. (eds.), Clusters and Groups of Galaxies, 139–143.

numerical value of the Hubble parameter and the velocity of the Local Group towards Virgo.

II. SPECIFIC CLUSTER FREQUENCY IN E AND SO GALAXIES

To date (cf. van den Bergh 1982) specific globular cluster frequencies are available for two dozen E, SO and Dwarf Spheroidal galaxies. These data show a real range in specific cluster frequency with S ranging from 1.7 ± 0.8 in NGC 3379 to 18.2 ± 4.5 for NGC 4486. Relying on the classifications of Sandage and Tammann (1981) there appears to be no significant systematic difference between the specific cluster frequencies in E and in SO galaxies. Nor is there, over a range in excess of 10^3 in parent galaxy luminosity, any indication of a systematic dependence of S on parent galaxy luminosity. The data (see Table 1) may, however, show marginal evidence for a dependence of the specific cluster frequency on environment in the sense that S is larger in rich clusters than it is in poor ones.

TABLE 1
Dependence of Specific Globular Cluster
Frequency on Environment*

Region	n	< s >	Remarks
Virgo Cluster	11	6.6 ± 0.6	Excludes M87
Other Groups	8	4.7 ± 0.9	
Local Group	3	3.1 ± 1.2	Excludes Fornax, M32

* E and SO galaxies only

III. FREQUENCY OF GLOBULARS IN CENTRAL GALAXIES OF RICH CLUSTERS

It has been known for many years (cf. Harris and Racine 1979, Harris and van den Bergh 1981) that NGC 4486 = M87, which is located near the bottom of the Virgo cluster potential well (e.g. van den Bergh 1977), has an unusually high specific globular cluster frequency. Recent observations by Harris, Smith and Myra (1982) show that NGC 3311, which is a D galaxy that occupies a similar position in the Hydra I cluster (Smyth, 1982, Richter, Huchtmeier and Materne 1983) also has a much higher S value than do other bright early-type galaxies in this cluster. Finally Dawe and Dickins (1976) find that NGC 1399, which is the central galaxy of the Fornax I cluster, is particularly rich in clusters on an AAT plate. I have confirmed this conclusion by inspection of the Fornax cluster on an SRC Sky Survey negative. In summary it appears that the central galaxies in all three rich clusters that have so far been studied have an unusually high specific globular cluster frequency.

IV. DISCUSSION

The observations quoted above show that either (1) central galaxies in rich clusters were special _ab initio_ or (2) they have been able to collect globular clusters subsequent to their formation because of their priveleged position.   The second alternative seems improbable for a number of reasons:

a.   Central galaxies (and cD galaxies in particular) are thought to have grown by cannibalism and/or mergers.   Since normal galaxies have a lower specific cluster frequency than central cD galaxies, cannibalism will lead to a _reduction_ of the S value of the central galaxy with time.

b.   If the excess of globulars in central galaxies is due to stripping from other cluster galaxies then the globulars acquired in this way would be expected to be relatively metal-poor.   This is so because (1) clusters would be stripped pre ferrably from the (presumably metal-poor) outer parts of passing galaxies and (2) stripped galaxies will, on average, be much fainter than the central galaxy so that their globular clusters would (on average) be expected to be metal-poorer than those initially associated with the luminous central galaxy (van den Bergh 1975).   This conflict with observations by Racine, Oke and Searle (1978) and Brodie and Hanes (1981), which appear to indicate that the most luminous M87 globulars are, in fact, quite metal-rich.

c.   It is difficult to see how any stripping scenario can account for the enormous number of globulars presently associated with central galaxies.   According to Racine and Harris (1979) M87 contains ≥10,000 more globular clusters than it would have had with a normal S value.   To account for this excess popula-tion M87 would have had to strip (and retain) the total cluster populations of ~ 100 galaxies similar to our Milky Way System!

These considerations suggest that the central galaxies in rich clusters were formed with an unusually large globular cluster population i.e. _they were already very special at the time when they first collapsed_.   This indicates that the present central galaxies in rich clusters may have been the kernels around which these clusters formed.

REFERENCES

Brodie, J.P., and Hanes, D.A. 1981, in Astrophysical Parameters for
    Globular Clusters.   A.G. Davis Philip and D.J. Hayes Eds.
    (Schenectady L. Davis press) p. 381.
Dawe, J.A., and Dickins, R.J. 1976, Nature 263, 395.
Harris, W.E., and Racine, R. 1979, Ann. Rev. Astr. Ap., 17, 241.
Harris, W.E., Smith, M.G., and Myra, E.S. 1982, preprint.
Harris, W.E., and van den Bergh, S. 1981, Astron. J., 86, 1627.
Racine, R., Oke, J.B. and Searle, L. 1978, Ap.J., 223, 82.
Racine, R., and Shara, M. 1979, Astron. J., 84, 1694.
Richter, O.G., Huchtmeier, W.K., and Materne, J. 1983, Astr. Ap., in
    press.
Sandage, A., and Tammann, G.A. 1981, A Revised Shapley-Ames Catalog of
    Bright Galaxies (Washington: Carnegie Institution).
Smyth, R.J. 1982, preprint.
van den Bergh, S. 1975, Ann. Rev. Astr. Ap., 13, 217.
van den Bergh, S. 1977, Vistas in Astronomy, 21, 71.
van den Bergh, S. 1982, Pub. A.S.P., 94, 495.

DISCUSSION

BURBIDGE: How reasonable is it to impose that the globular cluster luminosity function is the same in all galaxies? Secondly, how sure are you that the globular cluster candidates are really globular clusters?

VAN DEN BERGH: Our results could also be interpreted by assuming that the luminosity function of globular clusters associated with central galaxies of rich clusters differs from that in other ellipticals. This would also lead to the conclusion that these central galaxies must already have been very special at the time that they were formed.

DRESSLER: You use the assumption that the number of globular clusters is conserved to argue against a merger origin for giant ellipticals. Is there observational data that rule out the possibility that many of the globulars are young (like those in the Magellanic Clouds), and that they formed during the merger?

VAN DEN BERGH: The following arguments suggest that globular cluster are not formed during galaxy mergers: i) Observations by Strom et al. show that, at any radius, the globulars associated with Virgo ellipticals are blues than their parent galaxies. These obser- vations show that either (1) all globulars associated with Virgo ellipticals have ages $\lesssim 1 \times 10^9$ years or (2) globulars are metal – poor compared to their parent galaxies. Since the first alternative is very improbable it follows that globular clusters are older than their parent galaxy and hence not a product of collisions and subsequnet mergers. It would clearly be very important to confirm the photographic photometry of Virgo globulars using modern CCD photometry. ii) The radial gradient in mean cluster colors is difficult to understand if they were formed during violent collisions between galaxies which would probably result in rather chaotic cluster orbits. iii) It is hard to understand the rather uniform properties of globular cluster if they were formed in mergers; which must have taken place with rather a wide range of collision velocities.

ILLINGWORTH: What is the ratio of the specific frequencies of globular clusters in SO galaxies and spiral galaxies when spheroid (and not total) luminosities are used?

VAN DEN BERGH: For the integrated light spirals are deficent in globulars by a factor of ~10 as compared to E and SO galaxies. The corresponding factor for bulge light only is ~3, which is probably of only marginal statistical significance in view of the fact that data are presently only available for 4 spirals.

PARKER: Going back to an earlier question with regard to globular cluster identification, did the 21 spectra of candidate globular clusters confirming them as globulars cover the full range of the limited part of the luminosity function that you could get data for?

VAN DEN BERGH: Only in the case of the brightest globular cluster suspects near NGC 5128 (=Cen A) was it possible to obtain spectra with the CTIO 4-m telescope.

INFALL INTO THE VIRGO CLUSTER AND SOME COSMOLOGICAL CONSTRAINTS

R. B. Tully and E. J. Shaya
Institute for Astronomy, University of Hawaii

A family of mass models has been developed to describe the observed infall of galaxies in the Virgo Southern Extension toward the Virgo Cluster. The requirement that the mass models also explain the motion of our Galaxy with respect to the Virgo Cluster provides some constraints of cosmological interest. If the age of the universe is $10 < t_0 < 15$ Gyr, then there is a rough coincidence between the mass required to explain the infall pattern of galaxies near to the cluster, the mass required to explain the motion of our Galaxy, and the mass implied for the central cluster by the virial theorem. In this case, most of the mass of the supercluster must reside in the Virgo Cluster. Since 80% of the light of the supercluster lies at larger radii, the mass-to-light ratio must drop by perhaps an order of magnitude going out from the central cluster. If $t_0 < 10$ Gyr, models can be formulated which require additional mass at large radii, such that there need not be marked variations in M/L with radius. However, it may be difficult to accept that the universe is so young. If $t_0 > 15$ Gyr, there is already too much mass at small radii, as implied both by the virial analysis of the cluster and the infall model for the galaxies close to the cluster, to explain the motion of our Galaxy. If it is insisted that the universe is older than this limit then a tractable conclusion is that the cosmological constant is positive.

With the mass-age models and a census of the distribution of galaxies near the Virgo Cluster, it is possible to estimate the near-future accretion rate of galaxies into the central cluster. The influx of gas-rich systems is sufficiently large that it is plausible that all of the spirals and irregulars presently observed in the Virgo Cluster have arrived on a timescale of a third to half the age of the universe. By contrast, the influx of gas-poor systems is insignificant compared with the number of these kinds of galaxies already in the cluster. There are several indirect indications that the Virgo Cluster was composed mainly of ellipticals and lenticulars until recently, but is in the process of being modified as the consequence of a merger with a large cloud of spiral galaxies.

F. Mardirossian et al. (eds.), Clusters and Groups of Galaxies, 145–146.
© *1984 by D. Reidel Publishing Company.*

DISCUSSION

REPHAELI: Can you briefly compare your results with those of Hoffman and Salpeter?

TULLY: We are in very good agreement.

CORBYN: You said that if the age of the universe is in the range $10 < t_0 < 15$ Gyr you expect a very low Ω. What value did you have in mind.

TULLY: The implicit value of Ω is coupled to the age of the universe. So the constraints are very poor. We have some preference for models which would imply $\Omega \lesssim 0.4$.

MATTER DISTRIBUTIONS OF VARIOUS COMPONENTS IN THE COMA CLUSTER

Gerbal,D.,Mathez,G.,Mazure,A.,Monin,J.L.
Laboratoire d'Astrophysique
Observatoire de Paris Meudon

SUMMARY : Making use of the hydrostatic equilibrium equation observed density and velocity dispersion profiles of galaxies lead to obtain the distribution of total matter responsible for the gravitational potential. Applied to ICM the equivalent equation gives its temperature and density from the X-ray surface brightness profile.
At 3° the dynamical mass of Coma is $1.3\ 10^{15}$ $M_\odot$. Unseen and luminous matters have not the same distribution. ICM temperature is about 13 kev in the center and drops to 5 kev at 20'. ICM mass is about $2,5\ 10^{14} M_\odot$ at 3°.

1-INTRODUCTION

In this communication, we report an attempt to describe the spatial distributions of the various components (luminous matter, X-ray emitting matter or ICM and the so-called unseen matter) of the Coma cluster. We also derive the temperature profile of the ICM.

This has been done by relying on all the available data (galaxy density and velocity dispersion profiles, X-ray surface brightness and ICM temperature) in the frame of the simplest hypothesis.

In particular our main hypothesis are the following :
a) spherical symmetry (see discussion in Peach's contribution)
b) both ICM and galaxy gas are perfect fluids in hydrostatic equilibrium within the gravitational well

c) matter and energy exchanges between components are negligible.

Then we can make use of the hydrostatic equation :

$$(1) \quad \frac{T(r)}{m} \frac{d}{dr} \log_e(NT) = -\frac{dU}{dr}$$

147

and of the Poisson equation :

$$(2) \quad M(r) = r^2 \frac{dU}{dr} / G$$

expressing the total mass $M(r)$ (whatever its nature is) in terms of the acceleration it generates. $N(r)$ and $T(r)$ are the number density and the temperature of a given component and m the corresponding particle mass.

Equations (1) and (2) can be used in several manners, for instance, using a further equation given by the expression of the "bremsstrahlung" emissivity White and Silk (1980) obtain the ICM distribution and temperature in the cluster A576. The same method has been applied to Perseus center (Fabian et al. 1981). Conversely, supposing that ICM is isothermal, Fabricant et al. (1980) derive the corresponding distribution of dynamical matter in the galaxy M87 .

With the new hypothesis that the velocity dispersion of galaxies $Vg(r)$ is nearly isotropic and depends little on the mass of galaxies (i.e. mass segregation is neglected), equation (1) may be applied to the galaxy fluid. In this case $2Vg'/Vg$ is substituted to T'/T. Fuchs and Materne (1983) made use of equation (1) to test the isotropy of the velocity dispersion in the Coma Cluster.

There is a great amount of data of various types on Coma which, in this sense, appears as unique. On this ground we derive the distibutions of all the components : galaxies, ICM, unseen matter.

2 - EQUATIONS.

In a first step we admit that the behavior of the galaxy density Ng and velocity dispersion Vg shows the tracks of the potential generated by the dynamical matter. We fit the density profile derived from data of Godwin, Metcalfe & Peach (1983) by the modified Hubble law :

$$Ng(r) = Ng(0) \left[1 + (r/7')^2 \right]^{-0.6}$$

The 3-velocity dispersion profile derived in Capelato et al. (dashed curve of fig.3, 1982) from data of Gregory and Tift (1976) is fitted by the simple analytical expression :

$$V_g^2(r) = V_g^2(0) \left[\frac{1 + (r/450')^2}{1 + (r/20')^2} \right]^{0.4}$$

From equations (1) and (2) the distribution of total matter is then given by:

$$(3) \quad M(r) = 2.7 \; 10^{13} \; (r/1') \left[\frac{1 + (r/450')^2}{1 + (r/20')^2} \right] \left[\frac{1.1}{1+(\frac{7'}{r})^2} + \frac{0.4}{1+(\frac{20'}{r})^2} - \frac{0.4}{1+(\frac{450'}{r})^2} \right] h_{50}^{-1} M_\odot$$

In a second step, the now known acceleration (right hand-side of (2)) is input into the hydrostatic equation (1) applied to ICM (N=Nx, T=Tx). Moreover we fit the 0,5-4,0 keV surface brightness profile of Abramopoulos et al. (1981) by the Hubble law :

$$Bx(r) = Bx(0) \left[1 + (r/14')^2\right]^{-2.75}$$

By deprojection, and using the expression of the emissivity :

$$\epsilon_x = 1.5\ 10^{21}\ \rho_x^2\ T_x^{\frac{1}{2}}\ [e^{-E1/T} - e^{-E2/T}]\ \text{erg}\ s^{-1}\ cm^{-3}$$

we obtain a system of two equations for Tx and Nx :

$$N_x^2\ T_x^{\frac{1}{2}}\ [e^{-E_1/T} - e^{-E_2/T}] = 4.79\ 10^{-6}\ [1 + (r/14')^2]^{-3.25}\ h_{50}\ cm^{-6} keV^{\frac{1}{2}}$$

(4)

$$\frac{T_x}{\mu m_p}\ \frac{d}{dr}\ log_e(N_x T_x) = - G\ M(r)\ /\ r^2$$

Numerical integration leads to the ICM density and temperature profiles and to its total mass (out to a given radius).

3- RESULTS

The run of total dynamical mass with radius is given in table (1). The values given correspond to extreme values of the parameters adopted to fit the data. Note that the results do not differ very much. At 3^0, the total mass is $M = 1,3\ 10^{15} M_\theta$.

R	M(R)/M$_\odot$		
1'	6.4 10^{11}	to	3.3 10^{12}
30'	6.3 10^{14}	to	7.1 10^{14}
1^0	8.0 10^{14}	to	9.7 10^{14}
2^0	8.7 10^{14}	to	1.3 10^{15}
3^0	9 10^{14}	to	1.5 10^{15}

Table 1 : Distribution of the total dynamical mass.

In table (2) we compare this value to previous ones encountered in the litterature relating more than 10 years of dynamical analysis of Coma. The mass we obtain is small and close enough to those of Smith(1983) and Des Forets et al.(1983). Note that other results neglect the dynamical role of ICM which may be important due to the Limber effect (Smith 1980).

Years	Mass 10^{15} $M_\odot$	References
1970	2.9	Peebles
1972	4-6.2	Rood et al.
1974	1.4-3	H. Smith Jr.
1976	2.3	S. White
1977	2±1.5	Abell
1979	2.6	Struble and Bludman
1982	2.9	Kent and Gunn
1983	1.5	H. Smith
1983	1.4	Des Forêts et al.

Table 2 : The calculated mass of Coma.

An interesting information on the location of the unseen matter is obtained. In fig(1) we show the function M(r). The integrated luminous mass of galaxies M_L (r) is plotted too. This profile has been obtained from the integrated blue luminosity assuming a mass-to-light ratio M/L = 30. It is interesting to compare the behaviors of both profiles. They are not paralleles, so that the differences between both curves indicate the location of unseen matter. M(r) first grows faster than M_L (r). On the contrary, at large enough r, it is M_L (r) which grows faster. The more probable location of unseen matter seems to be between 30'and about 2°.

Concerning the ICM we find that it is non isothermal : its temperature decreases from the central value Tx(0) = 13 kev to about 5 keV at 20'and then to an asymptotic value of 3kev. The gradient is everywhere negative : there is no central cooling process. Temperature and density profiles are shown in figure (2). The integrated ICM mass is also plotted in fig.(1). This mass is 2.5 10^{14} $M_\odot(3^o)$, but notice that X-ray data are reliable only out to 20'from center.

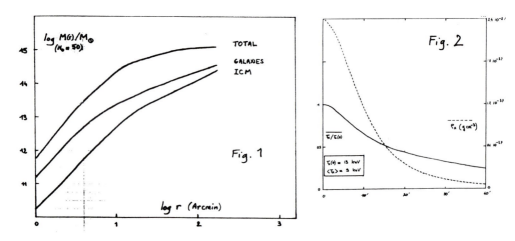

Fig. 1

Fig. 2

We have calculated the energy spectrum emitted between 1 and 30 kev by the ICM with such temperature and density profiles. The resulting energy spectrum cannot be distinguished from that of an isothermal

ICM at 9 kev. This ensures definitely the compatibility of our results with X-ray data.

REFERENCES

Abell,G.O. 1877 Ap.J.213,327

Abramomopoulos,F.,Chanan,G.A. & Ku,W.H.-M. 1981 Ap.J.248,429

Capelato.H.W.,Gerbal,D.,Mathez,G.,Mazure,A.& Salvador-Sole,E.1982
         Ap.J.252,433

Des   Forets,   G.,Dominguez-Tenreiro,R.,   Gerbal,D.,   Mathez,G.,   Mazure,A.   &
Salvador-Sole,E. 1983 submitted to Ap.J.

Fabian,A.C.,Hu,E.M.,Cowie,L.L. & Grindlay,J. 1981 Ap.J.248,47

Fabricant,D.,Lecar,M. & Gorenstein,P.1980 AP.J.241,552

Fuchs,B.& Materne,J. 1982 Astron.Astroph.113,85

Godwin,J.G.,Metcalfe,N. & Peach,J.V. 1983 MNRAS 202,113

Gregory,S.A. & Tift,W.G. 1976 Ap.J.206,934

Kent,S.M.& Gunn,J.E. 1982 A.J.87,945

Peebles,P.J.E. 1970 A.J.75,13

Rood,H.J.,Page,T.L.,Kustner,E.C. & King,I.R. 1972 Ap.J.175,627

Smith,H.Jr.1980 Ap.J.241,63

Smith,H.Jr.1983 Ap.J.270,422

Struble,M.F. & Bludman,S.A. 1979 Ap.Sp.Sci.64,301

White,S.D.M. 1976 MNRAS 177,717

White,S.D.M. & Silk.J.,1980 Ap.J.241,864

DISCUSSION

REPHAELI: It was not clear to me what you get for the relative distribution of dark matter and galaxies.

GERBAL: Comparing the distribution of dynamical matter (C1) to the distribution of integrated number of galaxies (C2), we note that 2 curves are not parallel essentially for $R \gtrsim 40'$. At $R \sim 3°$, C2 growths more quickly than C1: then the region in which unseen matter is located is between 40' and 3'. However, the true amount of this matter is not known without a hypothesis on the mass of luminous matter per galaxy in view of transform number of galaxies into integrated luminous mass.

THE TWO-COMPONENT CLUSTER IN CENTAURUS

J. R. Lucey, Anglo-Australian Observatory
M. J. Currie, Royal Greenwich Observatory
R. J. Dickens, Rutherford Appleton Laboratory
J. A. Dawe, Royal Observatory Edinburgh

In a previous paper (Lucey et al 1980) we reported the existence of a double-peaked velocity distribution for the rich, nearby Centaurus cluster (α = 12h 46.9m, δ = -40° 57'). We have now obtained an enlarged data set for Centaurus consisting of 203 redshifts and GF photometry (g = IIIa-J + GG495, F = IIIa-F + RG630) for 329 galaxies which lie within 3° of the cluster's centre. We have redshifts for all but 2 of the 99 galaxies brighter than $G_{26.5}$ = 16.5 that lie within 1° of the cluster's centre. The velocity histogram for the enlarged data set is shown (figure 1). The velocity distribution is bimodal and has best fit double gaussian parameters of: n_1 = 123, v_1 = 3041 km s^{-1}, σ_1 = 577 km s^{-1} and n_2 = 57, v_2 = 4570 km s^{-1}, σ_2 = 262 km s^{-1}, where the velocity dispersions have been corrected for a measuring error of 100 km s^{-1}. We have labelled the lower velocity and the higher velocity components Cen30 and Cen45 respectively. The sky distributions of the two components overlap completely. The galaxy distribution in the core region of the cluster is shown in figure 2.

A crucial question is the spatial location of the two velocity components. If Centaurus's velocity structure was caused by a mere chance projection of unrelated clusters at different Hubble velocities their corresponding separation would be $\sim$ 15 h^{-1} Mpc (h = H_0/100 km s^{-1} Mpc^{-1}). In this case we would expect a distance modulus difference, $\Delta\mu$, between the components of 0.9 magnitude. However, if the velocity structure was caused by the internal motions within one cluster, $\Delta\mu$ would be close to zero.

A major conclusion of our study is that colour-magnitude data and luminosity function data independently provide evidence that the two velocity components lie at the same distance. This is illustrated in figures 3 and 4. Thus the bimodal velocity distribution is caused by the relative motion of two components within one cluster. N-body studies have indicated that clusters grow by the amalgamation of subcondensations (White 1976). Centaurus is readily interpreted as a cluster (Cen30) which is currently accreting such a subcondensation

153

F. Mardirossian et al. (eds.), Clusters and Groups of Galaxies, 153–158.
© 1984 by D. Reidel Publishing Company.

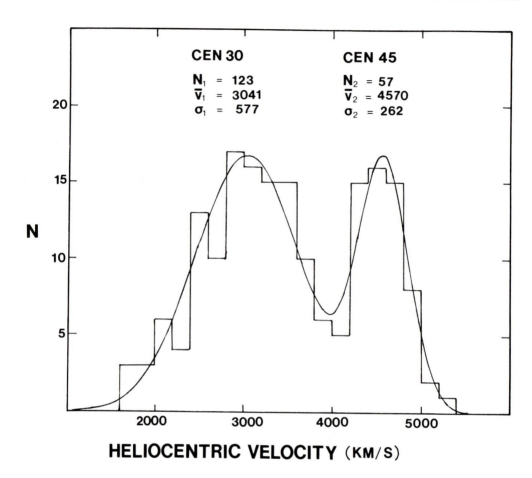

Figure 1. The velocity histogram.
The histogram includes all galaxies within 3° of the cluster centre
with heliocentric velocities less than 6000 km s^{-1}.

(Cen45). As the components have similar centres on the sky, the
relative motion of the components is predominatly along the
line-of-sight. Thus the $\sim$ 1500 km s^{-1} velocity difference between the
components is principally one of infall. Dynamically such an
interpretation is quite plausible because the escape velocity of the
main cluster, Cen30, is $\sim$ 2000 km s^{-1}. If the two-particle model (see
Beers et al 1982) is used to describe the system's motion, a current
separation of 0.7 h^{-1} Mpc is derived.

    The sky distribution of galaxies in Centaurus is visibly clumpy
with several subcondensations being apparent. Most of these subunits
have small velocity dispersions indicating that they are distinct
physical units. With hindsight, we might be surprised if such a clumpy
cluster as Centaurus did not show velocity substructure. Additional

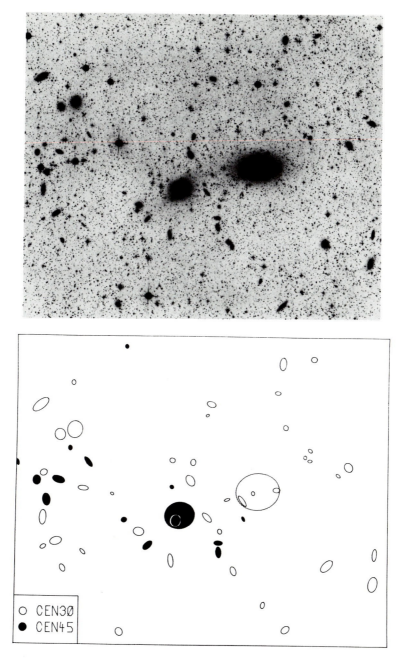

Figure 2. The sky distribution of galaxies in the cluster core.
The photograph is a composite made from four Schmidt J plates by David
Malin (AAO) and shows the central 1°1 x 0°9 region of the Centaurus
cluster. The chart below shows the membership of the two velocity
systems - open ellipses for Cen30 and filled ellipses for Cen45.

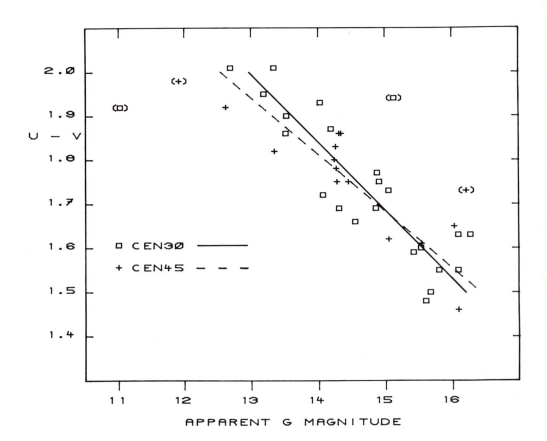

Figure 3. The colour-magnitude data
The colour-magnitude relation for elliptical and S0 galaxies has been
used to measure the relative distances of the two components. The
majority of the galaxies' colours were derived from the IPCS spectra
used for the redshift determinations. These colours have been
calibrated and supplemented by CCD UBV image data. We have used colours
derived from the equivalent of a 12 square arcsec aperture. There is a
well-defined colour-magnitude relation for each of the components, with
a formal $\Delta\mu$ between the components of -0.08 ± 0.17 magnitude. If the
components were at their Hubble distances, colour-aperture effects would
revise this difference, because of the fixed aperture nature of the
colours, to 0.23 ± 0.17. Thus colour-magnitude data clearly indicate
that the two components lie at the same distance.

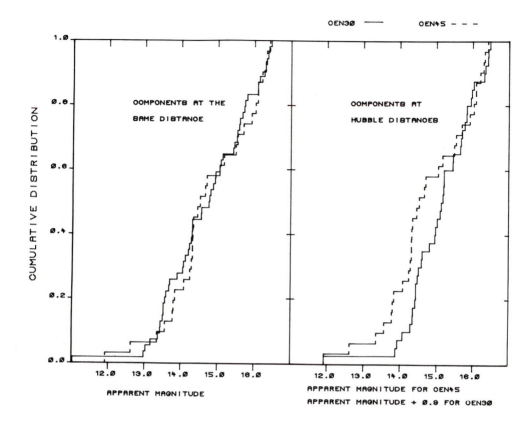

Figure 4.   Luminosity function data
The luminosity functions of the two components have been used to
estimate their relative distances.   We used our most complete deep
galaxy sample (r < 1° and $G_{26.5}$ < 16.5) and examined how well the
cumulative luminosity functions match for each hypothesis.   For the
Hubble-distance case the luminosity function of the lower velocity
component, Cen30, has been shifted to +0.9 magnitude and restricted to
those galaxies brighter than 15.6 magnitude to remove the distance
effect.   The solid curves represent Cen30 and the dashed curves Cen45.
The better fit of the luminosity functions occurs when the components
lie at the same distance.   For the Hubble-distance case the mis-match
of the two luminosity functions is different at the 10% significance
level (Kolmogorov-Smirnov test).

velocity substructure could exist within Centaurus, but both the limited number of galaxies involved and the probable overlapping nature of their velocity distributions make its detection very difficult.

Is Centaurus an unusual galaxy cluster? Substructure in the surface density distribution of galaxies within clusters is well-known. This spatial substructure will have an associated velocity substructure. However the detection of velocity substructure will be limited to clusters where i) a large number of redshifts is available, and ii) the relative velocity of the components is significantly greater than their velocity dispersion. These considerations leads us to regard Centaurus as a well-studied rather than an unusual cluster.

REFERENCES

Beers, T.C., Geller, M.J. and Huchra, J.P., 1982. Astrophys. J., 257, 23.

Lucey, J.R., Dickens, R.J. and Dawe, J.A., 1980. Nature, 285, 305.

White, S.D.M., 1976. Mon. Not. R. astr. Soc., 177, 717.

REDSHIFT ESTIMATION BY COLORS

D.C. Koo [1], R.G. Kron [2], D. Nanni [3], D. Trevese [3], A. Vignato [3]
[1]) Carnegie Institution of Washington
[2]) Yerkes Observatory, Williams Bay
[3]) Astronomical Observatory of Rome and National Laboratories of Frascati

1. INTRODUCTION

Redshift estimation by colors could prove to be valuable for cosmological investigations, since colors for faint galaxies can be produced in large quantities by automatic techniques.

In this contribution we explore this idea in practice by analysis of the colors of galaxies in the field of a rich, compact cluster at known redshift. The main points to be determined are the size of the random errors in the redshift estimation, the size of the systematic errors, and whether the scientific conclusions are sensitive to these errors. Thus a reasonably large amount of spectroscopic calibration must be done before the technique can be fully evaluated and applied (e.g. Koo 1981a). As an example of the technique, we consider here the isolation of the cluster II Zw 1305.4+2941 (z = 0.241) by colors. A similar discussion for a cluster at higher redshift was given by Koo (1981b).

2. OBSERVATIONS

Photometry is done by PDS scans of fine-grained plates obtained with the Kitt Peak 4m prime focus camera. The characteristics of these plates are given in Table 1. The reduction of the data (image detection, background evaluation, intensity calibration, star/galaxy image classification, and magnitude evaluation) have for the most part been described in previous publications (Agnelli et al. 1979; Di Chio et al. 1982). These techniques in the present instance yield a catalog of 600 galaxies brighter than J = 22.5 or F = 21.0 in a square area 18.6 arc min on a side, centered on the cluster. Each galaxy is characterized by its position, the J magnitude derived from the average of two plates, and the colors.

F. Mardirossian et al. (eds.), Clusters and Groups of Galaxies, 159–162.

Mayall PF	3314	1053	1561	1571	3315
Emulsion	IIIaJ	IIIaJ	IIIaJ	127.D2	III N
Filter	UG5	GG385	GG385	GG495	RG695
Exp. (minutes)	135	45	45	60	60
Band	U	J	J	F	N
λeff	3650	4650	4650	6100	8000

Table 1. Plate characteristics

3. ANALYSIS

For any pair of wavebands, the color-redshift relation for a particular type of galaxy spectrum can be computed in a straightforward way. In general a given color can correspond to more than one redshift, and of course there are different types of intrinsic galaxy spectra, so at minimum we require two colors to obtain a redshift estimate. According to Koo (1981a), for redshifts less than about 0.4 the color-color diagram formed from only the three bands UJF is the best redshift discriminator. Therefore in this work we do not explicitly consider the N band data. The evolving, synthetic galaxy energy distribution models of Bruzual (1981) form a convenient grid of spectra for calculation of the iso-redshift loci in the U-J, J-F diagram (Figure 1). Redshifts may then be estimated by simple interpolation between these loci. The random errors in the redshift depend on the random errors in the colors, which will be a function of apparent magnitude. The systematic errors depend on how accurately Bruzual's set of synthetic galaxy spectra represent the spectra of real galaxies, and how accurately the photometric system of the data has been modeled.

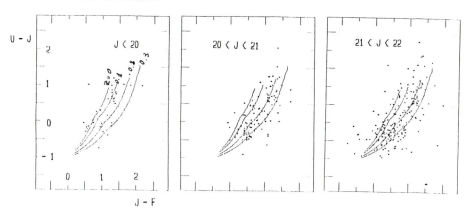

Figure 1) U-J versus J-F for various intervals of J magnitude, as indicated.

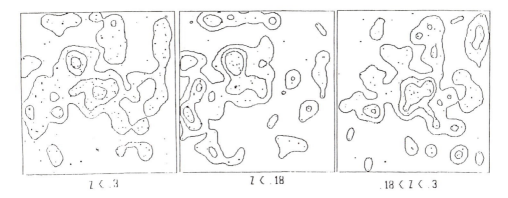

Figure 2) The distribution of galaxies (side of square = 18.6 arc min) when selected according to color criteria suggested by the groupings of points in Figure 1. The cluster II Zw 1305.4+2941 is centered. The lower isophlets correspond to the mean galaxy densities.

The absolute accuracy of neither the models nor the photometric zero points can be guaranteed; therefore spectroscopic redshifts are required to establish an empirical color-redshift calibration. In this way the models can be refined. In summary, the model evolving energy distributions of Bruzual provide an essential tool for the interpretation of the color distribution of faint galaxies when combined with spectroscopic checks. We have a small number of spectroscopic redshifts in this field which indicate clusters at $z = 0.241$ and $z = 0.125$. Therefore a good test of the principle is whether or not these clusters are apparent in the two-color diagram (Figure 1), and, when isolated by colors into groups nominally in the interval $0.06<z<0.18$ and $0.18<z<0.30$, whether or not the clusters are distinguishable spatially (Figure 2).

4. CONCLUSIONS

The distribution of points in Figures 1 and 2 does indeed suggest the existence of clusters at different redshifts. The cluster at $z = 0.241$ is comprised mainly of red galaxies, consistent with its compact, rich (= high density) appearance. The cluster or clusters at $z = 0.125$ are looser, not nearly as rich, and contain many blue galaxies. Since the model redshift loci converge at the blue end (Figure 1), unfortunately the redshift discrimination is relatively poor for these galaxies. The technique appears to have promise for a variety of programs, among which is the better determination of the membership probabilities for galaxies in clusters, thereby allowing better profiles and luminosity functions to be derived (see for example the discussion in this context by Peach in this volume of the use of just one color). Also, the technique appears (Koo 1981a) to have applications for the study of the large-scale distribution of field galaxies.

REFERENCES

Agnelli,G.,Nanni,D.,Pittella,G.,Trevese,D.,and Vignato,A.: 1979,Astron. Astrophys. 77,pp.45-52
Bruzual,A.G.: 1981,PhD dissertation, University of California, Berkeley.
Di Chio,P.,Nanni,D.,Pittella,G.,Trevese,D.,Vignato,A.: 1982,"Multicolor photometry and classification of galaxies: clustering in color diagram", Clustering in The Universe,D. Gerbal, and A. Mazure Eds.
Koo,D.C.: 1981a,PhD dissertation, University of California, Berkeley.
Koo,D.C.: 1981b,Astrophys.J. 251, pp.L75-L79.

THE ELLIPTICITY OF GALAXY CLUSTERS

Piotr Flin

Jagiellonian University Observatory, ul. Orla 171,
30-244 Kraków, Poland

This paper deals with changes of ellipticity with radial distance from the cluster centre. The Dressler sample of galaxy cluster is analysed. It is shown that the ellipticity changes are statistically significant. In many clusters they are not caused by subclustering. For this reason ellipticity changes ought to be regarded as an intrinsic independent cluster feature.

1. INTRODUCTION

During the preparation of one of the previous papers (Flin 1982) it was noticed that ellipticities of galaxy clusters change with radial distance from the cluster centre. Significance estimation of this effect together with some implications was also discussed (Flin 1984). The change of ellipticity has been observed also by several authors, e.g. Schipper and King (1978) and Peach (1982) for Coma Cluster, Carter and Metcalfe (1980) for several clusters and Burgett (1982) in the case of analysed by him clusters from Dressler sample (clusters containing more than 100 objects).

2. OBSERVATIONAL DATA

Dressler sample (1980) of 55 rich clusters was analysed. From this survey data (i.e. galaxy positions in mm on the plates) were taken without any correction. The equal weight to each galaxy was assigned. There is no background correction. In the case of present study such correction is not neccesary. The comparison of Burgett (1982) surface number density maps, where background correction is disregarded, with those presented by Geller and Beers (1982), where the statistical background correction is taken into account, demonstrates that in the considered sample this correction does not change the general appearence of the cluster delivered from isopleths. It is assumed throughout this paper that H_o=50 km/sec·Mpc. The redshifts of Abell clusters were taken from Sarazin et al. (1982) work, for remaining cluster from Dressler paper (1980). The distances, in Mpc, were calculated with the help of formula, in which the angle distance Θ = $0.573 \cdot (1+z^2)/(z(1+z)/2)$ corresponds to the linear distance of 1Mpc.

F. Mardirossian et al. (eds.), Clusters and Groups of Galaxies, 163–168.
© 1984 by D. Reidel Publishing Company.

3. METHOD OF ANALYSIS

3.1. The Determination of Cluster Centre

Cluster centre was determined for each cluster by an interative procedure,
slightly modificated in comparison to the previous one (Flin 1982). The
arithmetic mean of galaxy positions $\{x_o, y_o\}$ served as the first approximation of the centre. At the second step of iteration from this centre the
circle with radius $r_m = \max \{|x_\kappa - x_o|, |y_\kappa - y_o|\}$, where x_κ, y_κ are the coor-
dinates of the K-th galaxy, was descibed. From galaxies lying inside the
circle the new center $\{x_o^1, y_o^1\}$ was determinated. The procedure was repeated after diminishing the radius of circle described from the new centre.
Ellipticities $e = 1 - b/a$ and position angles were calculated for galaxies lying
inside the circles with radius being the multiple of 0.25 Mpc. This was
performed for each iteration step, that is for each centre $\{x^i, y^i\}$, by
application of dispersion ellipse method (Trumpler and Weaver 1953). The
procedure was terminated when:
1. the number of galaxies lying inside the innermost ring with radius
   equal to 0.25 Mpc was maximal,
2. the solution was stable, that is the repetition of the iteration step
   would change neither the cluster centre nor the computed values of
   ellipticities at all distances.
This method is increasing the number of galaxies in the cluster centre.
The all sample mean difference of number of galaxies falling into the
innermost rings between the adopted centre $\{x_o^f, y_o^f\}$ and the centre
$\{x_o^o, y_o^o\}$ is 4 galaxies for the ring with the radius 0.25 Mpc, and 1.5 galaxy
for the ring with the radius 0.5 Mpc.
Difficulties occuring here are similar to those met in all other methods.
Particularly difficult is the determination of the cluster centre for loose
groups with subcondensations. Moreover, in some cases there are more than
one solution fulfilling the criteria. The additional considerations, as e.g.
the smoothness of density profile, comparison with surface number density
map, presently not performed, might be useful in choosing the best solution.

3.2. The Analysis of the Changes of Ellipticities

The method described in the previous section allowed to find the changes
of ellipticities with radial distance from the cluster centre. An example

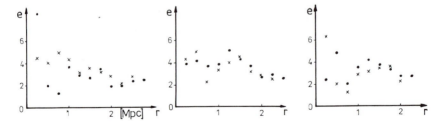

Fig. 1 The ellipticity determined at different radial distances from the
cluster centre for various centres (Cluster DC 0559-40).

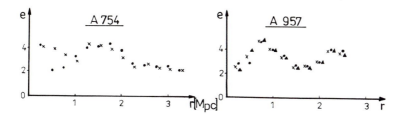

Fig. 2 The change of the mean ellipticity with radial distance for cluster
with several possible centres (for better visualization symbols for
different centers are shifted).

of these changes is presented in Fig. 1. However, such diagrams are not
convenient for further analysis. Therefore, the mean ellipticity at the
given distance from the cluster centre averaged over centres was calculated. Subsequently the relation mean ellipticity versus radial distance
from the cluster centre was investigated. Some examples of this relation
are presented in Fig. 2 and Fig. 3.

An important question arises: are these changes statistically significant? In order to answer the question several attempts were made.
Firstly, the difference between the mean ellipticities at the same distance from the centres: $\{x_0^o, y_0^o\}$ and $\{x_0^f, y_0^f\}$ was calculated (Flin 1984).
The all sample mean is $\Delta e' \approx 0.05$.
Secondly, the differences among the ellipticities at the same distance from
the centres: $\{x_0^i, y_0^i\}$ and $\{x_0^{i+1}, y_0^{i+1}\}$ were calculated. The all sample
mean $\Delta e''$ is smaller than $\Delta e'$.
These two error estimation attempts permitted me to conclude that some
changes of ellipticities must be statistically significant.

The mean ellipticity at given distance and the standard deviation of
the mean was used as the first statistical test. Obtained results demonstrated that on the 3σ level some changes of ellipticities are statistically
significant. The 3σ level error bars are marked in Fig. 3. However this statistics is meaningful only when the distribution of ellipticities in cluster is

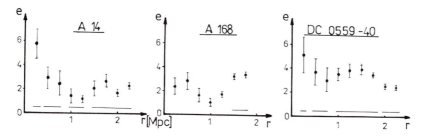

Fig. 3 The change of mean ellipticity with radial distance.

Gaussian. This requirement is not fulfil in several clusters. The histogram
of ellipticities for some clusters reveal the existence of two maxima, which
can be connected with both the influence of the procedure of cluster
centre finding and internal structure of the cluster.

From the described method of investigation it is clear that for various
centres ellipticities at different distances from the centre has been determined. This permit me to apply the analysis of variance for two-way
classification (Ahrens 1967, Oktaba 1976). The correct application of the
analysis of variance requires the absence of interaction between classifications, and that is the case. The analysis of variance applied to each
cluster revealed that:
a. in almost all clusters the change of mean ellipticity with radial distance
is statistically significant at the level of confidence $\alpha = 0.01$.
b. in only few cases the influence of the cluster centre position was statistically significant. This occurs mainly in cases of clumpy structures with
subcondensations lying far away from the main condensation, where the
final centre was adopted. The further analysis shows that removal of a
few first approximations permits to avoid this dependence.

The fact that analysis of variance demonstrated the significance of
ellipticity changes permit one to ask the question which changes are
important. This was checked using the new test of multipled range
(Oktaba 1976). This is combination of Keuls test of studentized range with
Duncan test of multipled comparisons.
After performing this test the answer to the important for us question
whether differences at consecutive distances from the center are statistically significant is possible. The applied level of confidence is $\alpha = 0.05$.
The grups of insignificant ellipticity changes at consecutive radial distances
are marked by segments in the lower part of Fig. 3.

4. DISCUSSION OF THE RESULTS

The statistical analysis shows that ellipticity changes are significant.
Therefore one can ask: what causes them? The simplest, and a priori
probably the most common explanation is subclustering. The 36% of clusters
exhibit the significant, in the sense of Geller and Beers (1982) subclustering, while almost all clusters show significant change of ellipticity. On
the other hand in the majority of clusters this structural feature can be
noticed. Due to the fact, that a priori it cannot be assumed that only significant subclustering causes ellipticity changes, the possible influence of
subclustering was checked for all clusters.
In order to do that the comparison has been made between the distance
from the cluster centre at which the statistically significant change of
ellipticity occurs with distance of subclustering. The distances of subclusters was read out from original Dressler maps, as well as surface
number density contour maps published by Geller and Beers (1982) and
Burgett (1982). All possible combinations of effects were noticed. This
means that there are clusters with smooth distribution (no subclustering)
in which significant ellipticity change occurs (e.g. A 14), clusters with
weak, insignificant subclustering in which subclustering does cause the

ellipticity change (e.g. A 400) or not (e.g. A 957). Also the strong sub-clustering could be responsible for ellipticity change (e.g. A 838) or not (e.g. DC 0317-54).
The analysis of position angles presently is not performed due to difficulties with application of adequate statistics.

5. CONCLUSIONS

The performed analysis shows that almost all clusters in Dressler sample exhibit the significant change of ellipticity with radial distance from the cluster centre. The statistical analysis is performed at the cost of smoothing obtained ellipticities and averaging them over not too small regions of the cluster. Two factors have the influence on the obtained results:
a. the number of galaxies increases with the radial distance, which causes the smaller reliability of changes closer to the centre than those at greater distance (as can be easily seen from standard deviation of the mean),
b. averaging over the centres smooths ellipticities over neighbouring regions, which is diminishing the obtained ellipticity changes; therefore the results given here are rather the lower limit of changes.
Taking these factors into account the ellipticity changes occuring further from the cluster center could be regarded as real phenomenon. On the other hand I share the opinion expressed by Peach (1984), that neither spherical nor axial symmetry are the best description of the cluster shape. It is shown, that subclustering causes some of the ellipticity changes, but it is not responsible for each change. Hence, some changes of ellipticity ought to be regarded as intrinsic independent property of the cluster structure itself.

Acknowledgments
I would like to express my deep gratitude to all, who kindly send me preprints and reprints of their papers.
I thank very much the organizing Committee for financial support permitting me to attend this Meeting.

References
Ahrens, H., 1967, Varianzanalyse, Akademie-Verlag, Berlin (Polish translation: PWN, Warszawa, 1970)
Burgett, W.S., 1982, Master Thesis, University of Oklahoma
Carter, D., Metcalfe, N., 1980, MNRAS 191 pp. 325-337
Dressler, A., 1980. Astroph. J.Suppl. 42, pp. 565-609
Flin, P., 1982 in "Clustering in the Universe" (eds. D.Gerbal, A.Mazure), Eds. Frontieres, Gif-sur-Yvette
Flin, P., 1984 in "Formation and Evolution of Galaxies and Large Structures in the Universe" (eds. J. Audouze and J. Tran Thanh Van) D.Reidel, Dordrecht, p. 137.
Geller, M.J., Beers, T.C., 1982, PASP 94, pp. 421-439
Oktaba, W., 1976, Elements of mathematical statistics and methodics of experimentation, PWN, Warszawa (in Polish)
Peach, J.V., 1982 in "Clustering is the Universe" (eds. D.Gerbal and A. Mazure), eds. Frontieres, Gif-sur-Yvette

Peach, J.V., 1984, this volume, p. 89.
Sarazin, C.L., Rood, H.J., Struble, M.F., 1982, Astron.Astroph. 108,pp. L7-L10
Schipper, L., King, I.R., 1978, Astroph. J. 220, pp. 798-808
Trumpler, R.J., Weaver, H.F., 1953 Statistical Stronomy, University of California Press, Berkeley and Los Angeles

DISCUSSION

MACGILLIVRAY: The method of dispersion ellipse is not a very good method for the determination of cluster ellipticities. Firstly, it systematically underestimates the ellipticity and secondly it is highly sensitive to contamination of field objects. The latter is even more serious for cluster of higher redshifts.

FLIN: I agree that the method underestimates ellipticities. Therefore I am speaking about the lower limit. The comparison of surface number density maps presented by Geller and Beers, where the background correction is taken into account, with these without correction published by Burgett shows that this problem is not important here. Also redshifts of investigated clusters are small.

GERBAL: Does ellipticity changes for different magnitude of member galaxies?

FLIN: This was not investigated. However Burgett shows that for various magnitude bands different ellipticities are obtained.

STRUCTURE OF THE HYDRA-I CLUSTER OF GALAXIES

F.W. Baier and H. Oleak
Zentralinstitut für Astrophysik,
Potsdam-Babelsberg, GDR

In a previous paper Baier (1983) showed that the secondary maxima in the radial number density distribution of some clusters of galaxies are not due to density shells around the cluster centres but to substructures.

Richter et al. (1982) discussed the density distribution and the distribution of the velocity dispersion for the Hydra-I cluster. By introducing a two component model with exponential decrease they obtained a good fit with their data. Further, they concluded that there is no pronounced substructure in the cluster.

Our investigation of the galaxy distribution and velocity distribution of the cluster leads us to an o p p o s i t e   c o n c l u s i o n . We found a p r o n o u n c e d   s u b s t r u c t u r e . From counts of galaxies on red palomar prints within a field of 4^0 x 4^0 around the cluster centre we found two week concentrations of galaxies in the northern cluster area and three pronounced galaxy concentrations in the southern cluster area up to radius values r = 30' and between r = 40' and r = 50', respectively. There is a strong correlation between the positions of these concentrations and the occurence of secondary maxima in the radial number-density distributions for the southern cluster area.

Mean values of the radial velocities were calculated for successive ring zones (with widths of 15') from the cluster centre up to a maximum radius of 90'. We found an excess of radial velocities in zone II (r = 15' to r = 30') and in zone IV (r = 45' to r = 60'). There is a coincidence between the positions of galaxies with higher radial velocities and the positions of the above mentioned galaxy concentrations. We consider this coincidence in connection with a statistical test for the

F. Mardirossian et al. (eds.), Clusters and Groups of Galaxies, 169–170.
© *1984 by D. Reidel Publishing Company.*

differences of the mean radial velocities in different
zones as a serious argument in favour of the multiple
structure of the Hydra I-cluster.

Furtheron, there is a coincidence between the
position of the secondary maxima in the radial number –
density distribution and a raised number of E – and SO –
galaxies according to Melnick and Sargent (1977) and Wirth
and Gallagher (1980), indicating the influence of another
cluster or group of galaxies.

We consider the existence of substructures in this
cluster as a normal phenomenon in connection with the
results of Geller and Beers (1982) and Baier (1983) that
a large percentage of clusters of galaxies shows significant substructures.

Details of this investigation are given by Baier
and Oleak (1983).

REFERENCES

Baier, F.W.: 1983, Astron. Nachr. 304, 5, 211-220.
Baier, F.W., and Oleak, H.: 1983, Astron. Nachr. 304, 6,
        273-279.
Geller, M.J. and Beers, T.C.: 1982, Publ. Astron. Soc.
        Pacific 94, 421.
Melnick, J. and Sargent, W.L.W.: 1977, Astrophys. J. 215,
        401.
Richter, O.G., Materne, J. and Huchtmeier, W.K.: 1982,
        Astron. Astrophys. 111, 193.
Wirth, A. and Gallagher, J.S.: 1980, Astrophys. J. 242,
        469.

REDSHIFTS FOR ZWICKY'S NEAR CLUSTERS

G.C Baiesi-Pillastrini[1], G.G.C. Palumbo[2], G. Vettolani[3]
1, Via Garzoni 2II, 40138 Bologna, Italy
2, Istituto TE.S.R.E./CNR Via De' Castagnoli 1, 40126,
Bologna, Italy
3, Istituto di Radioastronomia/CNR, Via Irnerio 46,
40126 Bologna, Italy

Studies of the structure of the Universe on scales larger than
clusters of galaxies may provide a test for alternative theoretical
models for the evolution of the Universe itserlf. Abell's clusters
(Abell 1958) and, to a lesser extent, Zwicky's clusters (Zwicky
1961-68) have recently been used to evidence large structures (Superclusters) and voids. (See for instance Oort, this conference). These
studies become possible when some estimate of distance is available
and three- dimensional pictures can be drawn. For Zwicky's clusters
the redshifts available have been scarce and distances have been
estimated mostly from the magnitude of the brightest galaxies or
similar indicators. With a complete coverage of the literature we
have compiled the Catalogue of Radial Velocities of Galaxies (Palumbo
et al. 1983) which we constantly update. From it, as a first step,
here we present the results obtained in assigning redshift to the 498
near clusters in the Zwicky catalogue using all the available redshifts of UGC galaxies (Nilson, 1973).

Total number of near clusters with some redshift
information (i.e. at least one galaxy with measured   121   (24%)
redshift)

of the above: unusable because only one galaxy
              had a measured redshift   93
              > 1 galaxy had a measured redshift
              but the values are widely different   28

Therefore the number of "usable" near cluster
(i.e.) with > 1 galaxy with measured redhisft is   170   (34%)

However only 39 of the usable clusters have > 10 galaxies with measured redshift and therefore may be considered reliable estimate.
Furthermore only 9 of these have a unique redshift. The figure summa-
rizes the spatial distribution of usable near clusters; simbols are
explained in the caption. A complete search for redshifts of all
galaxies in Zwicky's near clusters is in progress and will be published in a forthcoming paper.
It is already apparent that: a) the available redshift information
about Zwicky's clusters is very incomplete, b) it is risky to use
Zwicky's clusters to map large scale structures in view of the high

F. Mardirossian et al. (eds.), Clusters and Groups of Galaxies, 171–172.
© 1984 by D. Reidel Publishing Company.

contamination by chance superposition of unrelated groups.

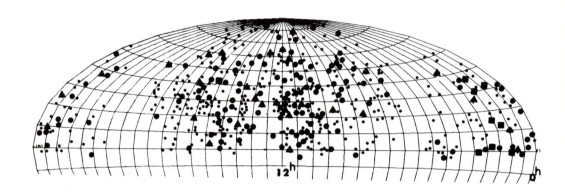

498 Zwicky's near clusters. Large dots: > 1
galaxy with measured redshift; triangles: > 10
galaxies; squares: unique redshift.

References:
Abell, G.O., (1958), Astrophys. J. Suppl., 3, 211.
Nilson, P., (1973), "Uppsala General Catalogue of Galaxies" Nova Acta
Reg. Soc. Sci. Upps., Ser. V,: A, Vol. 1, Uppsala.
Palumbo, G.G.C., Tanzella-Nitti, G., Vettolani, G., (1983) "Catalogue
of Radial Velocities of Galaxies", Gordon & Breach, New York.
Zwicky, F., Herzog, E., Wild, P., Karpowicz, M., Kowal, C.T., (1961-
1968) "Catalogue of Galaxies and Clusters of Galaxies" 6 Volumes,
Caltech., Pasadena.

TIDAL STRIPPING OF GALAXIES IN CLUSTERS REVISITED

Malcolm J. Currie
Rutherford Appleton Laboratory, Chilton, Didcot,
Oxon OX11 0QX, UK.

The Stroms (1978a, b, c, 1979) claimed that they had found evidence for stripping of galaxies in cluster cores because galaxian sizes varied with environment, however this work has not been corroborated.

Observations of radius-luminosity relationships by Currie (1983), Michard (1979) and Dixon (1979) have shown that single linear fits to these, as adopted by the Stroms, are poor approximations to reality. There are two parallel sequences of early-type galaxies in the logarithmic effective radius (RE) versus magnitude space, separated by 0.35 in RE. The lower surface brightness sequence is only present at faint magnitudes, typically $> m_* + 1$, hence it is called the faint sequence. Further, galaxies belonging to it have a steeper colour-magnitude relationship than those of the other ('bright') sequence, also they tend to reside in the outer parts of their cluster. These facts coupled with the luminosity segregation in the Coma cluster (Capelato et al. 1980) makes it inevitable that on average the core galaxies will be brighter and therefore smaller for their luminosity than those beyond the core. Similar luminosity segregation is found in Cl 0342-538 and Cl 0329-527 and therefore a similar effect to that of the Stroms was obtained when a linear regression was made to each of three annular samples in Cl 0342-538. This also explains why the Stroms' regressions were not parallel. The relative number ratios of the bright to faint sequence galaxies grows rapidly with increasing radial distance, and therefore the mean effective radius also increases radially, which accounts for the erroneous conclusion of the Stroms. However, they did not notice the presence of the faint sequence in their colour-magnitude diagrams, or in their RE-magnitude plots, and therefore computed combined regressions (see Figure 15 and 16 of Strom and Strom 1978a for example). This explains the difference in slope between their fits and the Kormendy relation.

This argument also accounts for the fact that the Stroms (1979) observed a much smaller difference between the mean radii of the core and halo of the clusters Abell 401 and Abell 2670. Since these clusters are at least twice as distant as the previous clusters

F. Mardirossian et al. (eds.), Clusters and Groups of Galaxies, 173–174.
© *1984 by D. Reidel Publishing Company.*

described by these authors, the absolute limiting magnitude was
brighter, and therefore included a much smaller fraction of faint
sequence galaxies. The blueness within and outside the core, as noted
by the Stroms, can also be explained by this selection effect.
Comparing the relations for both the bright and faint sequences in
three 1.2 Mpc (H = 100 km/s/Mpc) concentric annuli, a diminution in RE
of the galaxies of ~ 0.05 ± 0.02 was observed for both sequences in the
central circle compared to the outer annulus in Cl 0342-538 and
Cl 0329-527. The isophotal radii show no difference within the errors
for both clusters.

REFERENCES

Capelato, H.V., Gerbal, D., Mathez, G., Mazure, A., Salvador-Solé, E.
     and Sol, H.: 1980, Astrophys. J. 241, p.521.

Currie, M.J.: 1983, D. Phil. Thesis, Sussex.

Dixon, K.L.: 1979, D. Phil. Thesis, Oxford.

Michard, R.: 1979, Astron. Astrophys. 74, p.206.

Strom, K.M. and Strom, S.E.: 1978a, Astron. J. 83, p.73.

Strom, S.E. and Strom, K.M.: 1978b, Astron. J. 83, p.732.

Strom, K.M. and Strom, S.E.: 1978c, Astron. J. 83, p.1293.

Strom, S.E. and Strom, K.M.: 1979, Astron. J. 84, p.1091.

CLUSTER ELLIPTICAL AND LENTICULAR COLOUR-MAGNITUDE RELATIONSHIPS

Malcolm J Currie
Rutherford Appleton Laboratory, Chilton, Didcot,
Oxon OX11 0QX, UK.

Although galaxy colour-magnitude relationships (CMR) have been known to exist for twenty-five years, there is still debate about the underlying physical processes and whether they are universal for given wavebands, environment and morphological type. Here only the recent results concerning the E+S0 sequences will be described in outline. (Further details are given in Currie (1983) and in a forthcoming paper.) Visvanathan and Sandage (1977, 1978) (VS, SV) claimed that they are linear and universal in visual bandpasses. This is important to check before such relationships can be utilised for accurate relative distance determinations.

The relationship is not linear. Several authors have noted the presence of a flattening of the CMR for the brightest galaxies (e.g. Frogel et al. 1978; Lugger 1979), indeed it is evident in SV's Virgo data. This may be due to galactic cannibalism (Hausman and Ostriker 1978). In addition to this there is a second and more significant deviation from linearity at fainter magnitudes. The sequence steepens by a factor of about three faintward of an absolute ('break') magnitude which varies from cluster to cluster, and seems to be apparent only for certain colours. This form is seen in the data of de Vaucouleurs (1961), Dixon (1979), Metcalfe (1983), Butcher et al. (1983) and in other clusters observed by the author. In the following the shallower and steeper sequences will be called the bright and faint sequences respectively.

An analysis of four clusters, for which the author has obtained photometry, shows that the logarithmic isophotal radius (RI) vs. magnitude relationship shows no appreciable difference for the two sequences, but it has a shallower slope faintward of the 'break' magnitude in the CMR. However, the logarithmic effective radius (RE) vs. magnitude appears to show two parallel 'sequences', which overlap in magnitude, but have a difference in RE of 0.35 - the 'larger-RE' sequence if plotted in a CM diagram matches the faint sequence and the smaller sequence corresponds to the bright sequence with a CMR slope as predicted by VS. While this work was in progress a paper by Michard

175

F. Mardirossian et al. (eds.), Clusters and Groups of Galaxies, 175–176.
© *1984 by D. Reidel Publishing Company.*

(1979) was found in the literature which substantiates the above conclusions.

Morphological types in Cl 0342-538 were determined from a deep AAT plate taken in 0.5 arcsec seeing to V ~ 18.5. The faint end of the bright sequence consists almost exclusively of ellipticals while the bright end of the faint sequence consists mostly of lenticulars - a K-S test on the colour distributions of E's vs SO's at magnitudes where the sequences overlap gave a probability of 1.6% that they could be drawn from the same population. The spatial distributions of both sequences highlight supercluster bridges and filaments, but only in the bright sequence are the rich clusters apparent. This might explain why luminosity segregation is observed in some clusters, but there is no evidence for mass segregation. Galaxies in the faint sequence are, on average, flatter by 0.05 in ellipticity.

The RI for the fainter sequence do not form a parallel sequence with magnitude, as do the RE, it is concluded that there is a trend for RI/RE to increase as a power law for the faint sequence, but to be a constant factor for the bright sequence. If the majority of faint sequence members are lenticulars this can be interpreted as a decrease of bulge-to-disc ratio with increasing magnitude.

REFERENCES

Butcher, H., Oemler, A. and Wells, D.C.: 1983, Astrophys. J. Supp. 52, p.183.

de Vaucouleurs, G.: 1961, Astrophys. J. Suppl. Ser. 6, p.213.

Currie, M.J.: 1983, D. Phil. Thesis, Sussex.

Dixon, K.L.: 1979, D. Phil. Thesis, Oxford.

Frogel, J.A., Persson, S.E., Aaronson, M. and Matthews, K.: 1978, Astrophys. J. 220, p.75.

Hausman, M.A. and Ostriker, J.P.: 1978, Astrophys. J. 224, p.320.

Lugger, P.M.: 1979, Astron. J. 84, p.1677.

Metcalfe, N.: 1983, D. Phil. Thesis, Oxford (submitted).

Michard, R.: 1979, Astron. Astrophys. 74, p.206.

Sandage, A. and Visvanathan, N.: 1978, Astrophys. J. 223, p.707 (SV).

Visvanathan, N. and Sandage, A.: 1977, Astrophys. J. 216, p.214 (VS).

ON THE PERCEPTIBILITY OF DISTANT CLUSTERS OF GALAXIES

H.-E. FRÖLICH, K.-H. SCHMIDT, and R. SCHMIDT
Zentralinstitut für Astrophysik,
Potsdam-Babelsberg, G.D.R.

The near launch of the large Space Telescope will make accessible the realm of very faint and distant galaxies. The question arises wether distant clusters of galaxies will be recognizable against a "noisy" background of numerous field galaxies or not. We have therefore estimated the limiting magnitude one must reach to detect a cluster of given redshift (up to $z \approx 1$) and richness in 90 per cent of all cases (at a level of significance of one per cent).

Particularly the following assumptions have been made:
(i) Kron's (1980) J system magnitudes are used together with the colour relations for converting B and V magnitudes into J if necessary.
(ii) The number-magnitude relation for the background holds for a $q_0 = 0.02$ cosmology ($H_0 = 50$). It is derived from Tinsley's (1980) differential count predictions without luminosity evolution. It should be noted that the cosmological model only marginally influences the count predictions. Our calibration differs from that prefered by Tinsley for we rely on the counts performed by Butcher and Oemler (1978).
(iii) Regarding the cluster population the richness criteria defined by Abell (1958) are fulfilled. The surface density declines with increasing distance from the centre according to a modified Hubblelaw with shape parameter $\alpha = 0.5$ (cp. Fuchs and Materne 1982). An outer out-off of 30 times the core radius (≈ 50 kpc) is used. The expected number of visible cluster members is computed taking the richness criteria of Abell and a Schechter-type luminosity function with $M_j^* = -21.4$ and slope 1.24 (Kirshner, Oemler, and Schechter 1979). No luminosity evolution has been taken into account. In order to compute the K corrections the tables of Coleman, Wu, and Weedman (1980) are used. An old stellar population comparable to the bulge

177

F. Mardirossian et al. (eds.), Clusters and Groups of Galaxies, 177–178.

of M 31 is assumed.

(iv) In order to decide if a cluster should be recognizable or not the maximum deviation of the expected distribution function (the total number of galaxies within a given distance to the centre) from an uniform distribution is considered. A detection probability of 90 per cent means that at such a high percentage the maximum deviation exceeds the critical limit resulting for a level of one per cent. To optimize our treatment an area of three times the area covered by the cluster is considered.

An our estimates show distant clusters do not "drown" in a huge background of field galaxies for the number—magnitude relation flattens considerably in an expanding universe (see figure).

Further details will appear in Astronomische Nachrichten.

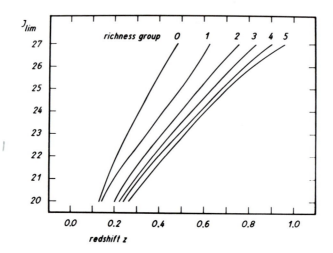

Depending on redshift and richness the limiting J magnitude is shown above which one can detect a cluster in more than 90 per cent of all cases.

REFERENCES

Abell, G.O.: 1958, Astrophys. J. Suppl. 3, 211 – 288.

Butcher, H. and Oemler, A.: 1978, Astrophys. J. 226, 559 – 565.

Coleman, G.D., WU, C.-C., and Weedman, D.W.: 1980, Astrophys. J. Suppl. 43, 393 – 416.

Fuchs, B. and Materne, J.: 1982, Astron. Astrophys. 113, 85 – 93.

Kirshner, R.P., Oemler, A., and Schechter, P.L.: 1979, Astron. J. 84, 951 – 959.

Kron, R.G.: 1980, Astrophys. J. Suppl. 43, 305 – 325.

Tinsley, B.M.: 1980, Astrophys. J. 241, 41 – 53.

LUMINOSITY SEGREGATION AND ANTI-SEGREGATION IN CLUSTERS OF GALAXIES

A. Mazure[*], H. Capelato[**], G. des Forêts[*]
* Observatoire de Meudon, 92190 Meudon, France.
** Instituto Astronomico e Geofisico, Saõ Paulo, Brasil.

Gravitational systems and in particular clusters of galaxies seem to undergo various transient evolutionary phases : violent relaxation, two-body relaxation, secular evolution (Lightman & Shapiro, 1978) with different time scales. However uncertainties still exist concerning the estimate of these time scales as well as on the details of the evolution of these systems. Thus it seems well-advised to look for indications on their dynamical stage in observational data. Furthermore the presence of a large mass spectrum in a cluster of galaxies and the non-ponctual character of its constituents lead also to modify the conventional scenario of the evolution of stellar systems.

From the observational point of view, the dynamical stage of a cluster of galaxies can be displayed by analyzing the velocity dispersion of galaxies versus magnitude or radius. Another technic is the detailed study of the projected distribution of galaxies with respect to their magnitudes.

Here such an analysis is made by the simultaneous study of the projected density, the distribution of interparticles separation and their moments with respect to the magnitude (Capelato et al., 1980). This allows to exhibit the luminosity (mass) segregation and the presence of eventual substructures.

These technics are applied on 5 clusters : Coma, Perseus (A426), A2147, A2151 and 0004.8-3450, (Capelato et al., 1980; Capelato, 1983; Capelato et al., 1983). The results may be summarized as follows :

a) Luminosity segregation is common to the 5 clusters analysed.
b) It was suggested that for the Coma cluster there is a critical magnitude m_c above which no significant luminosity segregation may be found. There is some indications that this result is also valid for Perseus (A426) and 0004.8-3450 clusters. For the A2147 and A2151 clusters no conclusions can be drawn because of the limited range of magnitudes covered by the samples. (Capelato et al., 1980b).

179

F. Mardirossian et al. (eds.), Clusters and Groups of Galaxies, 179–180.
© 1984 by D. Reidel Publishing Company.

However, although undeniable, the observed mass (luminosity) segregation among the heaviest galaxies appears weak compared to the results of N-body simulations (White, 1976). A possible explanation of such a disagreement is that these calculations do not take into account inelastic phenomena.

Roos & Aarseth (1982) have shown, using a model similar to that of White (1976), how the inclusion of such effects leads to recover the segregation observed in Coma. Besides, a more detailed analysis of the brightest groups of galaxies shows an anti-segregation phenomenon for these galaxies, in agreement with the preceding simulations. This is due to the fact that collisions between massive galaxies can lead to their merging. There is thus a progressive depletion of the proportion of the massive objects reducing the segregation, even leading to anti-segregation.

REFERENCES

Capelato, H.V., Gerbal, D., Mathez, G., Mazure, A., Salvador-Solé, E.,
    Sol., H., Ap. J. 241, 521.
Capelato, H.V., Gerbal, D., Mathez, G., Mazure, A., Roland, J., Salvador
    -Solé, E., 1980b, Astron. Astrophys. 87, 32.
Capelato, H.V., Gerbal, D., Mathez, G., Mazure, A., Roland, J., Salvador
    -Solé, E., 1981, Astron. Astrophys. 96, 235.
Capelato, H.V., 1983, Thèse d'Etat Université Paris 7.
Capelato, H.V., des Forêts, G., Mazure, A., 1983, unpublished work.
Roos, N., Aarseth, S.J., 1982, Astron. Astrophys. 114, 41.
White, S.M.D., 1976, M.N.R.A.S. 177, 717.

AN OBJECTIVE PRISM SURVEY OF EMISSION LINE GALAXIES IN NEARBY ABELL CLUSTERS

C. Moss *
Vatican Observatory, Castel Gandolfo, Italy.
J.F. Goeke
Present address: St. Louis Univ. H.S., St. Louis, U.S.A.
M. Whittle *
Steward Observatory, Univ. of Arizona, U.S.A. and
Institute of Astronomy, Cambridge, U.K.

ABSTRACT

A survey is being undertaken for $H\alpha$ emission line galaxies in nearby Abell clusters (cz < 9500 km/s) using the Burrell Schmidt telescope on Kitt Peak equipped with the high dispersion 10 deg. prism. A preliminary investigation of three clusters has shown that of the galaxies listed in the CGCG, approximately 15% are detected in emission, and that this sample comprises approximately half the detected emission line galaxies on the plates. Ultimately, the results of this homogeneous survey will be used for a statistical investigation into the effect of both local and cluster environment upon the production of $H\alpha$ emission in galaxies.

* Visiting astronomers, Kitt Peak National Observatory, operated by the Association for Research in Astronomy, Inc., under contract with the National Science Foundation. Observations made with the Burrell Schmidt of the Warner and Swasey Observatory, Case Western Reserve University.

181

F. Mardirossian et al. (eds.), Clusters and Groups of Galaxies, 181.
© *1984 by D. Reidel Publishing Company.*

STUDIES OF GALAXY CLUSTERS USING APM

Steven Phillipps
Department of Applied Mathematics and Astronomy
University College Cardiff
Wales UK

The Automatic Plate Measuring System (APM) at the Institute of Astronomy, Cambridge, has been used to obtain surface photometry of a large sample of galaxies. Several studies involving galaxies in the Fornax cluster are currently being undertaken.

INTRODUCTION

Automatic measuring machines have been used for some time for studies of large numbers of objects on astronomical plates. Mostly such studies have involved counting objects (i.e. stars or galaxies) to various magnitude or apparent size limits.

To extend this work a collaboration has been set up between University College, Cardiff (M.J. Disney and S. Phillipps) and the APM group at Cambridge (E.J. Kibblewhite and M.G.M. Cawson) to study automated surface photometry of galaxies in clusters.

The Automated Photographic Measuring system (Kibblewhite et al 1975) has been used to obtain raster scans of all galaxies above a certain isophotal size limit (though smaller galaxies included in the same scans were also analysed) on a UKSTU plate in the region of the Fornax cluster (see e.g. Jones and Jones 1980). A pixel size of 8 μm ($\sim$ 0.5") is employed in the rasters.

ANALYSIS

From the raster scans we can obtain best fitting elliptical contours (see Cawson 1983 for details) and hence radial profiles. Satisfactory results are obtained for galaxies $\sim$30" in diameter or even less (note that even a 20" diameter circle contains over 1000 pixels). Standard profile laws can then be fitted to the data to attempt an automatic classification of the galaxies by morphological type. The outermost isophotes used are somewhat larger than the diameter measured by eye by Holmberg et al (1978), indicating the faint limits attainable with the present technique.

F. Mardirossian et al. (eds.), Clusters and Groups of Galaxies, 183–184.
© *1984 by D. Reidel Publishing Company.*

GALAXY AND CLUSTER PROPERTIES

A number of interesting studies can be made using the same data, for
example, using just the classifications the spiral to elliptical ratio
can be estimated, the luminosity function can be obtained for the different morphological types and the distribution of types across the
cluster can be determined.

If we use the actual profile data, or more particularly the parameterized
fits to the profiles, then we can study the distrubutions of characteristic surface brightness and size among the cluster galaxies, and their
possible variation with distance from the cluster centre. One particular
current study is outlined below.

Note that automated measurements are doubly valuable in this respect
since the large number of galaxies measured obviates the need for separate galaxy counts in order to determine the background (field) contamination and the extent and centre of the cluster.

LOW SURFACE BRIGHTNESS OBJECTS

One of the motivations for the overall project was to search for low
surface brightness galaxies that might be missed from normal photographic surveys. The preliminary analysis of the Fornax cluster data
reveals that such galaxies are indeed quite common, and that some galaxies of extremely low surface brightness are present. By comparison
with previous photometry of the outer parts of some cluster galaxies
we estimate the central surface brightness of one such object near the
cluster to be $\sim$ 25.5 Bμ. It is $\geqslant$ 1' in diameter and its profile is well
fitted by an exponential. It would, therefore, appear to be an extremely
low surface brightness irregular. Its integrated magnitude B $\sim$ 19 is
not particularly faint and implies $M_B \sim -12$.

REFERENCES

Cawson, M.G.M., 1983, Workshop on Astronomical Measuring Machines,
Occasional Reports, R.O.E., 10, 173.
Holmberg, E.B., Lauberts, A., Schuster, H-E. and West, R.M., 1978,
Astron.Astrophys.Suppl., 34, 285.
Jones, J.E. and Jones, B.J.T., 1980, Mon.Not.R.Astr.Soc., 191, 685
Kibblewhite, E.J., Bridgeland, M.T., Hooley, T., and Horne, D., 1975,
Image Processing Techniques in Astronomy p245 (Reidel; Dordrecht).

NON THERMAL RADIO SOURCES IN CLUSTERS OF GALAXIES

R. Fanti
Dipartimento di Astronomia, Università di Bologna e
Istituto di Radioastronomia del CNR, Bologna, Italia.

Pluralitas
non est ponenda
sine necessitate.

W. Ockham

1) Introduction

This paper deals mostly with properties of non thermal radiosources in rich Abell clusters, with particular attention to radio morphology and to the Radio Luminosity Functions (RLF). The following analysis is based on the large bulk of data existing in the literature, the effort of several researchers who, with different scientific purposes and different methods of work, have lead to the accumulation of an impressively large but also incoherent amount of information. The main risk in dealing with these data is to incurr in not well understood biases or selection effects. On the other hand the advantage is to have at one's disposal a set of hundreds of radio sources instead of a few tens. I believe that the following analysis will be reasonably free of the aforementioned risks, because biases and incompleteness have been taken into account through a painful and time consuming but hopefully useful work.

Only high resolution data from aperture synthesis observations have been used. They come mostly from the various Westerbork surveys and from the work of Owen and collaborators. The bulk of these data, containing more than two hundred radiosources in Abell clusters over more than three orders of magnitude in radio power, give a representative picture of the source morphology without any significant bias.

As far as the RLFs are concerned, much more care was taken and smaller subsamples with well known selection criteria have been used. Up to distance class 3, $\delta > 18°$, there are 64 Abell clusters, of which 59 have measured red-shift. About 80% of them have been studied with aperture synthesis techniques, with different resolutions (ranging from $\sim 2''$ to $\sim 50''$), different sensitivities (from ~ 2 to ~ 400 mJy at 1.4 GHz) and

F. Mardirossian et al. (eds.), Clusters and Groups of Galaxies, 185–201.
© *1984 by D. Reidel Publishing Company.*

different fractions of cluster coverage. For the remaining 20% upper
limits have been set to the radio flux density of any present radio-
galaxy using the pencil beam survey by Owen et al. (1982). This sample
is the 'flower' of the data used here. It allows the use of 69 radio
galaxies, of which 54 have confirmed membership (owing to radial velocity measures). Also the remaining ∿ 20% of radio galaxies are very
likely to be cluster members owing to various kinds of considerations
(specifically : proper range of magnitude for the suggested optical
identification and often small distances from cluster center as compared
to Abell radius). The different limits on radio luminosity and on fractions of cluster coverage are taken into account by well known renormal-
isation methods. In order to improve the study of the RLF at high radio
luminosities, distance class 4 and 5 clusters, with rather high radio
flux density limits, have also been used. This allows the addition of
some 20 extra radio galaxies, typically at log P > 24.4, (W/Hz).

A comparison is also made with radio properties and RLFs of radio
galaxies not belonging to rich Abell clusters. One would hope to get
from it some information on: a) the surrounding ambient whose interaction with the radio sources influence the radio morphology; b) probability of galaxies with different "social status" to become radio sources.

A value of the Hubble constant of 100 Km/sec Mpc is used.

2) Radio Source Structures in Abell Clusters.

The most striking difference between radio galaxies in and out of
Abell clusters is the radio morphology. The generally dominant double
structure (D), aligned with the parent galaxy, exhibited by most sources
in radio catalogues is here noticeably less or absent. The largest
majority of radio sources with dimensions > 40 Kpc (larger than about
the galaxy size) show morphology distortion from the typical double
structure of the types generally called "Head Tail" (HT) or "Wide angle
tail" (WAT). Table 1 presents a concise summary of the source morphological types found in Abell clusters, based on more than 200 radio galaxies. HT's and WAT's represent about 30% of the whole sample, but they
account for up to 70% of the large size sources. There are trends of
morphological composition with radio power, the HT's and WAT's being
more frequent in the intermediate power range. At low radio luminosities, radio galaxies tend to have small linear sizes (this is found also
in radio galaxies outside Abell cluster) and the angular resolutions
they have been observed with are generally not good enough to map them
properly. At high radio luminosities double aligned structure seems to
occure more often. For comparison (see section 3) radio galaxies outside
Abell clusters show a much larger proportion of aligned double structure
and a definitely lower proportion of HT's and WAT's.

The percentages of (D's + HT's + WAT's) in and out of Abell clusters
are similar to each other, suggesting that the D missing in the Abell
clusters sample are sources of HT or WAT type, whose morphology is dramatically affected by ambient effects.

TABLE 1. Distribution of radio morphological
types.

log P 1.4		L < 40 Kpc	L > 40 Kpc			
			N.C.	D	WAT	HT
logP<23	Abell cl.	84 %	5 %	3%	6 %	--
	Comp.sample	100 %	-	-	---	
23<logP<24	Abell cl.	52 %	10 %	3 %	11 %	14 %
	Comp.sample	57 %	-	33 %	(- 10 % -)	
24<logP<25	Abell cl.	31 %	16 %	23 %	11 %	19 %
	Comp.sample	36 %	2 %	53 %	(- 9 % -)	
logP>25	Abell cl.	27 %		27 %	45 %	
	Comp.sample	18 %	18 %	63 %	---	

N.C. = Not Classified ; D = Double ; WAT = Wide Angle Tail
HT = Head Tail

It is generally assumed that the main effect is a drag action on radio components due to the galaxy's motion through the intergalactic medium (Miley et al. 1972; Jaffe and Perola 1974). Such medium is now currently seen by means of X-ray observations. Other sometimes suggested explanations for the distorted structure are: a) buoyancy effects on the radio emitting regions; b) asymmetric ram pressure effects on moving components; c) bulk motion of the intracluster gas. The first explanation can be tested by measuring the velocity of the parent galaxy (in a statistical sense) with respect to the mass center of the cluster. HT sources are expected to be associated to the faster galaxies; WAT's to less fast galaxies and D sources to galaxies almost at rest. Furthermore, if the radio morphology is due to motion of individual galaxies, it is expected that no systematic orientation effects of the source axis should be found. The contrary may be expected in the case of the other suggestions.

First of all, it appears (see also Owen and Rudnick 1976) that the WAT and D sources are generally associated with the brightest galaxies (Mpg < -20.), while the HT's are equally divided among bright and less bright galaxies. This is consistent with the drag model, since the less bright galaxies (assumed also to be less massive) are expected to have larger velocities.

Analysis of the velocities of small samples of HT's (Baggio et al. 1978; Ulrich 1978) have shown indeed that these objects are moving with respect to the cluster rest frame. At present those previous analysis can be expanded to a sample about twice as large (21 HT's and 8 WAT's). The results are shown in Fig. 1. Both the HT's and WAT's have velocities in the range of few hundred to few thousand Km/sec. WAT's however are slower by about a factor 1.5 - 2.0. For both types the velocities are comparable to those of typical cluster galaxies, as evidentiated in the lower panel of fig. 1. The distribution of the quantity $Vr/\sigma(V)$, (Vr being the radial velocity of the galaxy with respect to the cluster average velocity and $\sigma(V)$ the cluster velocity dispersion) for HT's and WAT's together, is consistent with a normal distribution with a standard deviation of 1. However, WAT's alone have a standard deviation of ~ 0.7, indicating that they are slightly slower than typical cluster galaxies. Whether these velocity differences are large enough to produce the two types of distortions is not obvious, since one has to rely on specific models (e.g. Valentjin 1978).

The data for double sources are very limited. Only in 8 cases the parent galaxy's velocity, with respect to the cluster, can be estimated from literature data. In three cases this velocity is small ($V_r < 200$ Km/sec) and consistent with no motion. In at least three other cases significicnat velocities are found. However it should be realised that these three sources have been studied with rather poor angular resolution and an eventual WAT structure could have escaped detection. It would be very important to reobserve these objects at radio wavelengths with higher resolution to check their radio morphological type. Alternatively it would be important to determine whether these clusters have an intergalactic gas density too low to produce the drag action necessary to modify the radio structure.

Fig. 2 shows the distribution of the HT and WAT radio axis orientations, with respect to the line joining the parent galaxy to the cluster center. While the HT orientations are distributed more or less at random, the WAT's seem to point away from the cluster center in the large majority of cases. The effect was noted first by Rudnick and Owen (1977). Such an effect, if real, would raise serious doubts on the explanation for WAT morphology based on galaxy motion and would rather support mechanisms invoking large scale coherent effects, however not such as to influence the random orientation of HT's. Objection to the galaxy motion as a cause of the WAT structure has been raised by Burns (1981), mostly on the basis of the assumed small or null velocities of the giant galaxies associated to these sources, coupled to the sometimes very large physical sizes. Other alternative explanations for the WAT class do not however appear more convincing (Burns 1983).

Effects a) and b) cannot be a general explanation for the WAT since at least in the case of 3C 465 (Abell 2634) the radio trails appear to bend toward regions of enhanced X-ray emission and therefore of higher densities (as in Burns et al. 1982).

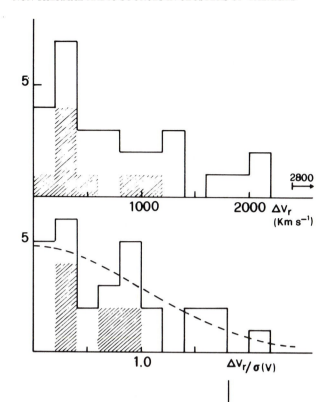

Fig. 1. Distribution of
radial velocities of HT's
and WAT's (hatched squa-
res) with respect to cl-
uster mean. Top: absol-
ute values. Bottom: val-
ues normalised to veloc-
ity dispersions.

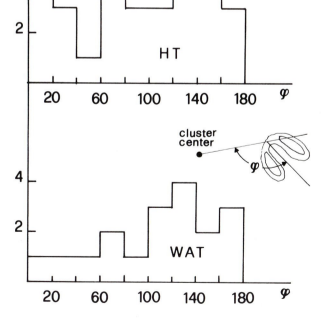

- Fig 2. Distribution of
HT's and WAT's as function
of the angle between radio
axis and cluster axis. Only WAT's more distant than
about 2' from cluster center have been considered.

Buoyancy however has been suggested by many authors to explain details in the sources morphology, as the curvatures, sometimes sharp, seen in the lower brightness regions of these sources (see e.g. Feretti et al., this volume). From a statistical point of view, an analysis of HT and WAT curvatures based as well on X-ray maps of clusters, allowing to trace the hot gas distribution, has been presented by Harris (1982). He shows that the low brightness regions of these sources usually bend toward regions of lower gas density, consistently with the buoyancy model.

The low brightness regions of HT's and WAT's are generally assumed to be in static pressure equilibrium with the external medium. Typical equipartition internal energy densities (derived on the basis of usual equipartition assumptions and of a ratio proton/electrons of 1) are in the range of 10exp-11 - 10exp-14 erg/cm^3, requiring for the density Nig and temperature Tig of the intracluster medium values such that NigxTig $\sim$ 10exp3 - 10exp4. The now more and more frequent information on the intracluster medium of Abell clusters from X-ray observations has led several authors (see, e.g. Harris 1982) to compare equipartition internal pressure with external gas pressure. It is generally found that the last one is able to produce the required confinement.

3) Comparison of Properties of Radio Galaxies in and out of Abell
   Clusters.

As a comparison sample, radio galaxies are used from two B2 sources samples identified with bright galaxies (m < 16) outside Abell clusters (Colla et al. 1975; Fanti et al. 1978). Such sources span a radio power interval similar to the one examined here for Abell clusters radio galaxies. Table 1 shows the radio morphological composition of the comparison sample. It contains 6 well defined and not questionable cases of HT and WAT sources. They are in the range of powers 10exp23 - 10exp24 W/Hz and account for $\sim$ 10% of the sources in this power range, or 15% of the sources larger than 40 Kpc. (This figure has to be considered as a lower limit however, since often the angular resolution of the radio observations is such that some WAT's might have been misclassified as normal Ds). A further attempt was made in order to divide the comparison sample into two subclasses according to membership (or not) to Zwicky, non Abell, clusters. Only about half of the comparison sample radio galaxies have been tentatively classified in either class, due to uncertainties in defining the membership. The radio morphologies of radio galaxies inside and outside Zwicky, not Abell clusters, are not significantly different. Of the 6 HT's and WAT's, two fall in the Zwicky clusters subclass. The remaining four are among the non classified objects.

Comparison of physical sizes of radio sources inside and outside Abell clusters have been made by several authors (e.g. Hooley 1974; Burns and Owen 1977) with questionable results. A crucial point, in doing this, is how to compare sizes of sources of such different morpho-

logical types as D's on the one hand and HT's and WAT's on the other.
Specific models have to be assumed for that (e.g. Ekers et al. 1981) and
the results would depend on those assumptions.

As a final point one may examine the situation of confinement of low
brightness regions (like in relaxed doubles or on the few HT's and
WAT's). Assuming that also in this case the confinement is due to the
static pressure of the outer gas one gets some information on its density and temperature. Bearing in mind the possibility of biases introduced by the present finite resolution and sensitivity of radiotelescop-
es, which tend to confine the detectable brightness levels, and therefore
minimum equipartition energy densities, to certain ranges, it is nevertheless surprising that the ranges of fainter brightness observed in
radio galaxies both in and out of Abell clusters are remarkably the same
(see, e.g. Ekers et al. 1981). It is tempting to conclude that the differences in the ambient density and temperature are, on average, smaller
than a factor 10, and perhaps even smaller, in the two cases.

The largely different proportion of doubles and HT and WAT radio
galaxies in and out of Abell clusters is therefore likely to be due to
differences in the typical velocities of the parent galaxies in the two
types of "society". This would be consistent with the notion that the
velocity dispersion in groups of galaxies is an increasing function of
richness. It is important to obtain X-ray observations of the regions
around radio galaxies not in Abell clusters in order to test such expected uniformity of the outer medium. A limited number of X-ray observations of such objects, presented by Burns et al. (1981) seem to
confirm such uniformity.

4) Radio Luminosity Functions

An important item is whether galaxies belonging to different aggregates have different probabilities of being radio sources. This item is
generally examined by constructing the RLF.
We begin now examining the properties of the RLF in Abell clusters
and discuss at the end the comparison with other aggregates. Determinations of the bivariate RLF in Abell clusters have been made by several
authors, mostly using data from the Westerbork surveys of rich nearby
Abell clusters (see e.g. Fanti et al. 1982 and references therein).
These determinations are based on limited numbers of detections and the
statistical errors are high.
Here a new determination is presented, based on the samples mentioned in the introduction. Since the lack of optical counts of ellipticals
for the "average cluster" studied makes it difficult to determine the
RLF in terms of percentage of radio sources per galaxy, a different approach is followed (as also in Lari and Perola 1978). A RLF is computed
which represents the number of radio galaxies per average cluster, as a
function of radio power.
This cluster RLF is shown in fig.3. Its similarity with the general

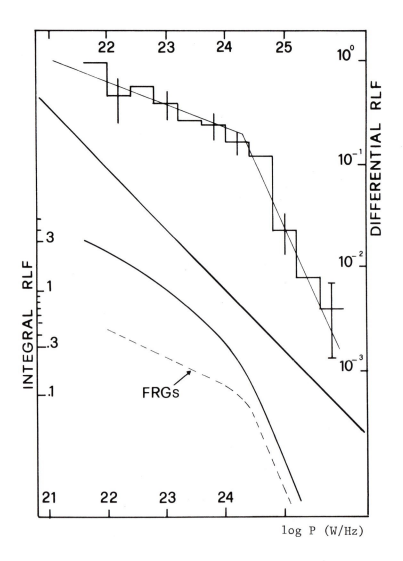

Fig. 3. The Cluster RLF. Top right : differential RLF, per interval
        of log P = 0.4. The RLF from Auriemma et al. is shown as
        light curve (see text). Bottom left : integral RLF.
        It is shown also the contribution of FRG's. Units in ordinate
        are : number of galaxies /cluster. In abscissa log of radio
        power, in Watts/Hz, at 1.4 GHz.

slope of the RLF of radio galaxies, irrespective of the social status,
as in Auriemma et al. (1977) or in Meier et al. (1979) is remarkable.
Also the proportion of bright (Mpg < -20) to less bright galaxies (-18.5
> Mpg > -20), a factor $\sim$ 1.5 - 2, is remarkably similar to that in the
previous papers (see fig. 4). Comparison in absolute value (namely in
terms of sources per galaxy) would require the knowledge of the average
number of elliptical galaxies per cluster in the above ranges of optical
luminosities, and this is not accurately known. In order to have agreement in absolute units as well, one would need to have $\sim$ 3 - 4 ellipticals per cluster brighter than Mpg = -20.0 and 10 - 12 in the fainter
optical luminosity interval, figures which are not unlikely.

The next point of investigation is the RLF of the first ranked galaxies (FRGs). In order to avoid subtle biases due to an "a posteriori"
classification of FRG's, the sample used is that of Hoessel et al.(1980)
containing Abell clusters FRGs up to distance class 4, complete for richness > 1, supplemented by FRGs of richness 0 clusters with distance class
1 and 2, which are easily established on the basis of magnitudes from
the Zwicky catalogue. The sample obtained in this way contains 12.6.
FRGs for all of which flux densities (or upper limits) were found in the
literature. The so obtained RLF is shown in fig. 5. It is seen that
FRGs contribute by 30% - 50% to the RLF at log P $\sim$ 24.0, but
their RLF flattens off more rapidly than that of other cluster galaxies.
Both the shape and absolute value of the FRG's RLF are in excellent agr-
cement with the RLF of the brighter ellipticals in Auriemma et al.

A further point to be investigated is the dependence of the RLF on
cluster richness. The result is shown in fig. 6, where the RLF is shown
for three richness classes (R = 0, 1, 2). The average number of sources
per cluster scales with richness class by $\sim$ a factor 1.5 - 2.0, in excellent agreement with the ratio of galaxies contents between consecutive
richness classes as defined by Abell. A similar result was obtained by
Lari and Perola (1978), for log P> 24.4. Therefore the number
of radio sources per cluster is proportional to the number of galaxies
in the cluster.

Correlations with the BM class have been searched for as well. No
effect is found on the RLF. Actually a trend, contrary to any expectation, consisting in a larger number of sources/clusters for late (III)
BM types is clearly due to different proportions in richness among different BM types. A clear cut effect with BM class appears when analysing
the FRG's RLF (see fig. 7). FRGs in BM 1 and I-II clusters appear to
have a significantly higher probability of radioemitting. Note that their
RLF is identical to that from Auriemma et al. for galaxies brighter than
Mpg = -21 , while the FRGs of later BM type clusters have a RLF close to
that from Auriemman et al. for the range -20 < Mpg < -21. Since the
average absolute optical magnitude of FRGs depends on BM type (Sandage
and Hardy 1973), it is likely that the differences among the RLFs are
due to differences in the absolute optical luminosity content. If so,
it would appear that FRGs do not have significantly different probability
of producing a radio source as compared to other equally optically bright

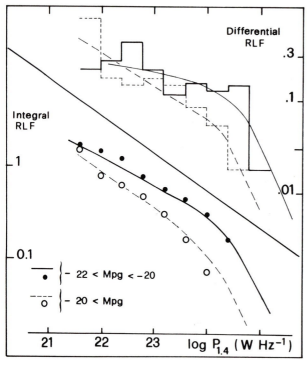

Fig. 4. Bivariate cluster
RLF. Continuous and broken
curves are from Auriemma et
al. (see text). Top right:
differential RLF. Bottom
left : integral RLF. Units
are as in fig. 1.

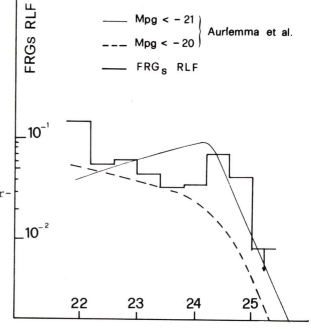

Fig. 5. FRG's differential
RLF and comparison with Auriemma et al. Units are as
in previous figures.

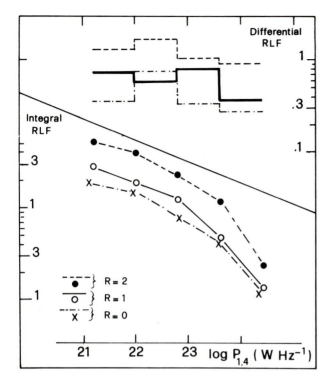

Fig. 6. Clusters RLF as
function of richness class.
Top right : differential.
Bottom left : integral.
Units are as in previous
figures.

Fig. 7. FRG's integral
RLF as function of BM
type and comparison with
Auriemma et al. (see text)

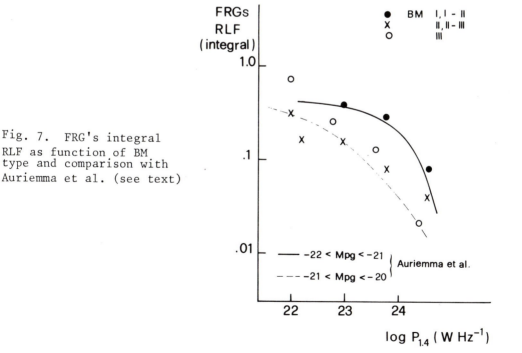

galaxies. (Note that from the Hoessel et al. photometry the differences
in optical luminosity among FRGs of different BM types are minor ones,
smaller than is given in Sandage and Hardy, 1973. This may be due to
the fact that Hoessel's et al. photometry refers only to the inner 19 Kpc
region of galaxies and therefore the luminosity of large cDs, which account for ∿ 70% of FRGs of BM types I and I-II could be largely underestimated). See however Feretti et al., this volume.

Sixty percent of sample D < 3 clusters used to compute the RLF and
∿ 50% of the sample used for studying the RLF of FRG have been studied
and detected at X-ray wavelengths. A search has been made for correlation between radio emission probability and X-ray luminosity. Such a
search is motivated by the already mentioned importance of the intergalactic gas and its possible contribution to fuel the radio sources. Effects on the RLF could be expected.

As far as the cluster R L F is concerned there is not any effect of
this kind. The RLF of clusters with log Lx > 43.4 (ergs/sec) does not
differ by more than 20% from that of the weaker X-ray emitters. A marginally significant effect is found in the RLF of FRGs, where in strong
X-ray clusters FRGs have higher levels of radio emission. This is similar to that found by Valentijn and Bijleveld (1983) for cD galaxies
(see also Bijleveld, this volume), but less significant.

A search has also been made for dependence of FRGs radio emission
probability on presence of X-ray luminosity excess (as in Forman and Jones, 1983). Unfortunately the too small number of clusters for which such
information is available makes any conclusion meaningless from a statistical point of view.

5) Radial Distribution of Radio Galaxies in Abell Clusters

Fig. 8 shows the distribution of Abell clusters radio galaxies, brighter than about Mpg ∿ -18.5, as a function of distance from cluster center, in Abell radii units (d/Ra). A clear trend is present as a function
of radio power, the distribution becoming less and less peaked at lower
powers. An effect of segregation also is found as function of optical
absolute luminosity of the parent galaxy, the optically brightest ones
being slightly more concentrated toward the center. These results are
tentatively explained as due to a dependence of the very bright galaxies
fraction on the radial distance. This combined to the different shapes
of the fractional RLF as a function of Mpg (see Auriemma et al. 1978;
and also the previous session), below 10exp24.5 watts/Hz, which seem to
be saturated at different powers as a function of Mpg, would explain the
observed segregation of radio luminosities. This explanation needs to
be confirmed by specific counts of ellipticals as a function of radial
distance and optical luminosity, on a representative number of clusters.

Segregation effects appear also as a function of radio morphological
types. WAT's and D's show pronounced peaks with d/Ra ∿ .1, accounting
for ∿ 40% of the class, the remaining fraction being spread over a long
uniform tail up to d/Ra = 1.0. HT sources have a broader peak (∿ 0.3

d/Ra), while small size sources (L < 40 Kpc) do have a rather broad
distribution. Again these trends can be understood as above, owing to
the radio morphological type – optical absolute luminosity correlations
seen on paragraph 2.

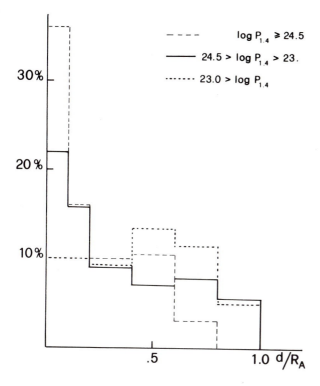

Fig. 8. Radial distribut-
of radio galaxies in Abell
clusters, as function of
radio luminosity.

It is finally noted that previous clusters surveys selecting radio
galaxies within a fraction of the Abell radius, have missed fractions
of cluster sources varying as function of radio power. For instance in
searches with d/Ra < 0.3, the fraction of cluster radio galaxies missed
at log P $\sim$ 23.0, 24.0, 25.0, are 70%, 55% and 40% respectively.

6) Conclusions

     a) There is a striking difference in radio morphology between radio
galaxies inside and outside Abell clusters, the former containing a large fraction of HT and WAT sources. These morphologies are much less
frequent, although not absent, in the latter case. It should be noted
that, since Abell clusters contain only about 20% of the radio galaxies
population, in complete sources samples, not biased in favor of cluster

members, HT and WAT sources are roughly equally found inside and outside
Abell clusters.

b) WAT's (and D's) are generally associated with the brightest
ellipticals (Mpg < -20), while HT's are equally divided between bright
and less bright ellipticals.

c) The present data show that galaxies associated with HT sources
have velocities typical of the cluster galaxies. WAT radio galaxies seem
to be moving also, but more slowly, although there are uncertainties in
the data (see comment of Burns in the discussion). The situation is
even more uncertain for D sources.

d)The orientation of HT's is random with respect to the cluster.
WAT's, on the contrary, seem to point away from the cluster center.
This last finding adds further doubts to the interpretation of the WAT's
based on the parent galaxy's motion.

e) Comparison of maps of radio galaxies with the hot intracluster
gas distribution, as deduced from X-ray observations, allows the interpretation of a number of radio source features in terms of buoyancy effects.

f) Considerations on confinement of radio galaxies both inside and
outside Abell clusters indicate similarities in the properties of the
ambient gas in the two situations. This seems also confirmed by limited
X-ray observations on group of galaxies less rich than Abell clusters
and containing radio sources.

g) The number of radio galaxies present, above a given radio luminosity, in an average cluster, depends on the richness of the cluster in
a way so as to indicate proportionally with the number of bright galaxies
contained in the cluster.

h) Sixty per cent of radio galaxies in Abell clusters are associated
with optically bright ellipticals (Mpg < -20). Accounting for the general shape of the Optical Luminosity Function of cluster galaxies, this
implies that the probability of a galaxy to become a radio source is a
function of the optical absolute magnitude in as much the same way as
shown by Auriemma et al. for non Abell cluster galaxies. Also the shapes
of the RLFs are remarkably similar to those in Auriemma et al. Absolute
agreement (namely in terms of source per galaxy) is also obtained with
reasonable assumptions on the average number of ellipticals per average
cluster.

i) The FRGs account for $\sim$ 30 - 50% of the radio galaxies present in
an average cluster above log P $\sim$ 24.4. Below this value their RLF sat-
urates more quickly than that of other less bright galaxies. Both the
shape and absolute value of the FRGs RLF is similar to that of very
bright ellipticals (Mpg < -21) given by Auriemma et al.

j) A correlation of FRGs RLF with BM type is suggested to be a combined effect of the correlation between RLF and absolute optical luminosity and of the FRG absolute magnitude with BM type. This point however
requires further investigation.

k) No correlation is found between Cluster RLF and X-ray luminosity
of the cluster. A marginally significant effect is found for the FRGs
RLF, stronger X-ray emitting clusters displaying a larger proportion of
radio emitting FRGs.

l) Points g) to k) indicate that the main parameters in determining the number of radio galaxies in a cluster are the number of elliptical galaxies contained and their optical luminosities. Furthermore the clusters RLF show remarkable similarities with that from Auriemma et al., which is largely based on non Abell clusters galaxies. It is emphasized that, while radio galaxies in Abell clusters are in space regions where the surrounding galaxy's density is of the order of 100 members Mpc^{-3}, those not in Abell clusters belong mostly to smaller aggregates with typical densities of < 10 members Mpc^{-3}. Therefore a difference of more than a factor of 10 in the surrounding galaxies density does not affect the probability of a galaxy to become a radio source.

REFERENCES

Auriemma C., Perola G.C., Ekers R.D., Fanti R., Lari C., Jaffe W.J., Ulrich M., 1977, Astron.Astrophys. 57, pp.41
Baggio R., Perola G.C. and Tarenghi M.,1978,Astron.Astrophys.70,pp.303
Burns J.O., 1979, Mon.Not.Roy.Astron.Soc., 195, pp.523
Burns J.O., Eilek, J., Owen F., 1982, IAU Symp. 97 "Extragalactic Radio Sources" e. by Heeschen and Wade, Reidel, pp.45
Burns J.O., 1983, "Astrophysical jets" ed. by Ferrari and Pacholczyk Reidel, pp.67
Burns J.O. and Owen F.N., 1977, Astrophys.J. 217, pp.34
Burns J.O., Gregory S., Holman G.D., 1981, Astrophys.J.,250,pp. 450
Colla G., Fanti C., Fanti R., Gioia I.M., Lari C., Lequeux J., Lucas R., Ulrich M.H. 1975, Astron.Astrophys. 38, pp.209
Ekers R.D., Fanti R., Lari C., Parma P.,1981,Astron.Astrophys.110,pp.169
Fanti R., Gioia I.M., Lari C., Ulrich M.H., 1979, Astron.Astrophys.Suppl. 34, pp.341
Fanti C., Fanti R., Feretti L., Ficarra A., Gioia I.M., Giovannini G., Gregorini L., Mantovani F., Marano B., Padrielli L., Parma P., Tomasi P., Vettolani P., 1982, Astron.Astrophys., 105, pp.200
Harris D.H., 1982, IAU Symp. 97 "Extragalactic Radio Sources"ed. by Heeschen and Wade, Reidel, pp.77
Hoessel J.G., Gunn J.E., and Thuan T.X., 1980,Astrophys.J.,241,pp.486
Hooley T., 1974 Mon.Not.Roy.Astr.Soc. 166, pp.259
Jaffe W.J. and Perola G.C., 1973, Astron.Astrophys., 26, pp.423
Jones C. and Forman W., 1983, preprint
Lari C. and Perola G.C., 1978, IAU Symp. 79 "The Large Scale Structure of the Universe", ed. by Longair and Einasto, Reidel, pp.137
Meier D.L., Ulrich M.H., Fanti R., Gioia I.M., Lari C., 1979, Astrophys. J., 229, 25
Miley G.K., Perola G.C., van der Kruit P.C. and van der Laan H., 1972, Nature 237, pp.269
Owen F.N. and Rudnick L., 1976, Astrophys.J., 205, pp.Li
Rudnick L. and Owen F.N., 1977, Astron.J., 82, pp.1
Sandage A. and Hardy E. 1973 Astrophys.J. 183, pp.743
Ulrich M.H., 1978, Astrophys.J., 221, pp.422
Valentijn E.A., 1978, Astron.Astrophys. 79, pp.362
Valentijn E.A. and Bijleveld W., 1983, Astron. Astrophys., 125, pp.217

DISCUSSION

BURNS: I worry about your velocity histograms, especially for the WAT sources. The centroid velocities of the clusters are poorly determined from generally < 10 galaxies. Therefore, the velocities of the WAT sources within the clusters could still be consistent with zero.

FANTI: I agree that there are uncertainties on the velocities of parent galaxies, due to the limited number of objects generally available to derive the cluster average velocity. Nevertheless at least five out of the 8 WATs here considered appear to have significant velocities with respect to the cluster.

JAFFE: What significance do you attach to the results of Stokie and collaborators showing that truly isolated galaxies have a lower RFL than the Auriemma curve predicts?

FANTI: First of all, the effect reported by Adams et al. is not inconsistent with the above conclusions, since it refers to situations where the spatial density of galaxies is a factor =10 lower than the average one of the comparison sample here used. As a second remark, there are still some uncertainties, as mentioned by the authors themselves, on the optical absolute luminosities and E/SO ratio in their galaxies sample. The result is nevertheless rather stimulating and if further confirmed would indicate the existence of another parameter, the surrounding galaxies density, relevant in the processes of radio source formation. A final remark is that in the B2 comparison sample there are a couple of objects at least, which appear totally isolated galaxies (adopting the criterium of Adam et al.). Both are of medium radio strength and have large scale radio emission and account for 1-3% of the radio galaxies space density. This is not a small fraction, if one takes into account the small, although largely uncertain space density of isolated galaxies.

REES: The evidence that wide angle tails preferentially point away from cluster centers looks rather convincing. Does this have to imply an outflowing wind or could it simply mean that most galaxies are falling inward (because the cluster is not stationary but always gaining new members)?

FANTI: If the orientation of WATs shown above would turn out to be real and if one wants to keep the galaxy motion as explanation of the radio morphologies the inevitable conclusion would be a falling inward of the more massive galaxies. However, before reaching this conclusion, it is important to gain more confidence about the evidence of motion of the WATs galaxies (see the comment by J. Burns).

WESTERBORK OBSERVATIONS OF 37 POOR CLUSTERS OF GALAXIES AT 1415 MHZ

Robert J. Hanisch
Radiosterrenwacht,
Postbus 2, 7990 AA Dwingeloo, The Netherlands

SUMMARY

Observations of a sample of 37 poor clusters of galaxies north of declination +35° have been made at 1415 MHz with the Westerbork Synthesis Radio Telescope (WSRT). The poor cluster sample has been drawn from the Zwicky catalog by locating groups of three or more galaxies brighter than $15.^{m}7$ where the surface density of galaxies exceeds the average surface density for all galaxies in the Zwicky catalog above 40° galactic latitude by at least a factor of 46.

In the 37 poor cluster fields observed, 204 radio sources with 1415 MHz flux densities greater than 10 mJy were detected. However, only 14 of these sources could be identified with galaxies that are poor cluster members. Half of the poor cluster radio sources are associated with galaxies that have close companions and, in some cases, that appear disrupted as a result of the encounter with the companion. The detection rate for galaxies with close companions was approximately five times greater than that for galaxies without close companions.

The morphologies of the identified poor cluster radio sources are not unusual - ten sources were unresolved and are probably confined to the nucleus of the associated galaxy, two are elongated at the same position angle as the associated galaxy and can be attributed to low-level disk emission, one is double, and one is a head-tail (the well known source B2 1615+35). Although several examples of head-tail radio galaxies in poor clusters are now known, the fact that further examples were not discovered in the current observations could indicate that the presence of an intracluster medium dense enough to distort the radio structure may be relatively uncommon in poor clusters of galaxies.

1. INTRODUCTION

The study of radio sources in clusters of galaxies is both reward-ing, because of the wealth of information available and its implications

F. Mardirossian et al. (eds.), Clusters and Groups of Galaxies, 203–208.
© *1984 by D. Reidel Publishing Company.*

for so many questions of astrophysical interest, and difficult owing to
the large number of interconnected factors which can affect the observed
properties of the radio sources. These studies have the potential to
tell us much about the conditions necessary for radio source formation
and how environmental characteristics influence subsequent source
evolution.

Poor clusters of galaxies are groups of a few to a few tens of
galaxies showing a significant enhancement in galaxy density over the
field, but not sufficiently populous to be counted in the rich cluster
catalog of Abell (1958). Among the reasons for studying poor clusters
are to try to isolate some of the environmental factors that influence
radio source structure and evolution through a comparison with the
sources in rich clusters and in the field, and to determine the extent
of and physical conditions in the intracluster medium.

2. THE POOR CLUSTER SAMPLE

A complete sample of poor clusters of galaxies has been selected
optically by White et al. (1983). The sample was based on the Zwicky
catalog (Zwicky et al. 1960-1968) and contains over 600 poor clusters at
distances comparable to Abell's distance classes 0 through 2 (z<0.05).
The angular extent of these clusters is generally less than 30 arcmin.
The 600 clusters have at least a factor of $100^{2/3}$ (21.5) enhancement in
the local galaxy surface density over the average surface density of
galaxies in the Zwicky catalog above 40° in galactic latitude. This
factor corresponds, on the average, to a factor of 100 enhancement in
the spatial density (Turner and Gott 1976). A subset of these clusters
was chosen for the current observations, namely those with a surface
density enhancement of 46.4 over the average (for a typical spatial
density enhancement factor of over 300) and north of declination 35°,
which makes them suitable for Westerbork observations (see Table 1). A
total of 38 poor clusters satisfies these criteria. Unfortunately,
redshifts are available for only a fraction of the galaxies in the poor
cluster sample. Thus, it is possible that some of the poor clusters are
not true physical associations, but rather only line-of-sight superpositions. The chance of foreground or background contamination is less,
however, for the clusters which meet the higher surface density enhancement criterion. Observations of approximately 150 other poor clusters
from White's sample at more southerly declinations are currently in
progress using the VLA (White, Burns, and Nelson, private communication).

3. OBSERVATIONS AND DATA REDUCTION

The observations of 37 poor clusters (one of the 38 was not observed) were done in May and June 1983 with the Westerbork Synthesis
Radio Telescope at a frequency of 1415 MHz and bandwidth of 10 MHz. In
order to observe a maximum number of clusters in a reasonable amount of

Table 1. Poor clusters observed with the WSRT

Cluster	α (1950) δ		No. members	Comments
N45-270	$08^h48^m.5$	$53°07'$	3	ZC 0855.0+05248, at edge.
N45-342	09 06.9	50 38	4	TG 2
N45-278	09 14.1	42 12	3	
N45-279	09 17.5	64 27	3	TG 6
N45-349	09 43.4	54 41	3	ZC 0943.7+5454
N56-271	10 01.7	46 56	3	
N56-276	10 17.8	43 13	3	
N56-281	10 25.8	40 04	3	ZC 1026.9+4023
N56-292	10 55.7	55 52	3	
N56-294	10 56.9	50 17	3	
N67-296	11 07.2	37 13	3	ZC 1107.7+3610, at edge
N45-371	11 17.0	73 07	5	ZC 1112.7+7259
N79-232	12 18.5	40 11	3	
N56-313	12 29.8	66 40	3	ZC 1204.8+6520, at edge
N56-314	12 33.4	64 15	3	
N67-323	13 03.5	53 53	6	
N79-248	13 22.8	36 40	3	TG 71
N79-286	13 51.4	40 32	5	ZC 1352.9+3856, HG 69
N56-318	14 05.5	55 16	3	ZC 1406.4+5513, HG 77
N67-307	14 21.0	40 36	3	ZC 1420.7+4025
N67-289	14 32.8	48 11	3	
N67-291	14 36.8	46 53	3	
N56-365	14 57.0	48 40	4	ZC 1456.2+4901
N45-313	15 15.5	69 28	3	
N56-346	15 26.4	43 08	3	ZC 1534.0+4222
N56-374	15 34.8	43 41	5	ZC 1534.0+4222, TG 101
N56-366	15 48.1	42 11	4	
N45-321	15 48.1	69 37	3	
N45-324	15 49.7	71 24	3	
N45-372	15 55.6	48 18	3	ZC 1610.3+4955
N45-334	16 10.1	52 35	3	ZC 1610.3+4955
N45-389	16 15.8	35 13	4	ZC 1615.8+3505
N45-383	16 29.4	39 57	5	ZC 1625.5+4006
N45-338	16 31.9	50 30	3	ZC 1629.7+5027
N45-339	16 36.1	36 10	3	
N45-365	16 40.7	57 56	4	ZC 1638.4+6038, at edge
N45-340	16 46.5	35 59	3	

time and yet insure adequate (u,v) coverage, each cluster was observed
with a series of 7 10-minute cuts spaced evenly over the range of hour
angles from -90° to +90°. The (u,v) data from each of the 7 cuts were
combined and transformed to make 1024 x 1024 maps and beam patterns. The
central 512 x 512 pixels in each map were CLEANed to a level of 1.5 WFU
(7.5 mJy) using a loop gain factor of 0.25. The maps were then restored
using the center of the dirty beam as a restoring beam, and the resulting clean maps were searched for sources down to the 1.5 WFU level. The
flux densities were corrected for the WSRT primary beam attenuation. The
final result for each cluster was a clean map and associated source
list. The final source lists only include sources within 36 arcmin of
the field center, where the primary beam attenuation is no worse than
0.5, and only include sources with flux densities greater than 10 mJy (a
5-sigma level detection for a source located at the primary-beam half-
power point). After a final source list was obtained for each cluster,
an attempt was made to find an optical identification for each source
using the Palomar Sky Survey "E" prints.

A total of 204 radio sources was detected in the 37 cluster fields
observed, and of these 46 were found to have reasonably good candidate
optical identifications. However, only 14 radio sources could be identified with galaxies likely to be members of the poor clusters of interest (i.e. located near the field center and being brighter than about
17^m).

4. RESULTS AND DISCUSSION

Of the 14 sources identified with poor cluster members, 10 were
unresolved and presumably are associated with the nuclei of their associated galaxies. Four sources were resolved: two are elongated at the
same position angle as the associated spiral galaxies and are attributed
to disk emission, one is a strong double source (NGC 5141) and one is
the well-known head-tail B2 1615+35 (Ekers et al. 1978, Burns and
Gregory 1982). Of these 14 sources, 7 are identified with galaxies that
have close companions and in some cases are also disrupted. Four of the
unresolved (nuclear) sources have companions, and 3 of the resolved
sources (including NGC 5141, which has NGC 5142 = Mrk 452 nearby) have
close companions. The two sources that appear to be associated with the
disks of spiral galaxies, 0917+645 and 1526+431, are very weak, and the
luminosity of 0917+645 (NGC 2820) is commensurate with the low radio
powers typically associated with disk emission ($L_{1415} = (6.9+1.4) \times 10^{27}$
erg s^{-1} Hz^{-1} for $H_0 = 75$ km s^{-1} Mpc^{-1}). Two other sources are identified
with galaxies whose velocities are known. The source 0914+422 is unre-
solved and identified with NGC 2798 and has a luminosity of $L_{1415} = (3.4+0.1) \times 10^{28}$ erg s^{-1} Hz^{-1}. NGC 2798 may not, however, be a member of
the poor cluster N45-278 - it is 1.5 magnitudes brighter than any of the
other Zwicky galaxies in the vicinity. Another unresolved source,
1351+405, is identified with NGC 5353 and has a luminosity of $L_{1415} = (4.7+0.2) \times 10^{28}$ erg s^{-1} Hz^{-1}. There is some uncertainty about cluster
membership in this case as well, however. NGC 5353 and NGC 5354 are a

close pair with magnitudes 11.8 and 12.3, respectively, whereas the other galaxies in the region are all fainter than 14^m.0.

Although a complete discussion of the general properties of radio sources in poor clusters must await the completion of all observations of the statistical sample described previously, there are several points to be considered based on the current observations. One remarkable result from the current work is the large fraction of radio sources identified with poor cluster galaxies that have (interacting) companions. Only one of these sources appears to be a typical double (NGC 5141), while the rest are primarily unresolved and presumably nuclear sources. The radio sources are associated primarily with S and SO galaxies; indeed, none of these sources are associated with giant ellipticals or cD's. Hummel (1981) has pointed out the tendency for spirals in close pairs to have nuclear radio sources with 2 to 3 times the luminosity of those in solitary spirals, and this correlation may account for the number of unresolved sources detected in galaxy pairs in the current sample.

Stocke (1978) has found that close pairs of galaxies are radio sources at about twice the frequency of galaxies not found in pairs, although the effect seems more closely related to the local galaxy density than to the presence of a companion galaxy per se. Half (7 of 14) of the poor cluster sources found here have companion galaxies, and often there is evidence of actual disruption of the galaxies as the result of this close encounter. Although our sample is too small to draw general conclusions in this regard, the data are suggestive that galaxies disrupted by a close encounter with another galaxy are more likely to be radio sources. The detection rate for all galaxies in our sample that are members of poor clusters is 11% (14 of the 127 galaxies catalogued as poor cluster members were detected), compared to the 50% detection rate (7 of 14) for galaxies with companions.

The extended sources identified with poor cluster member galaxies can be divided into two general categories: (1) disk emission from spiral galaxies, of which there are two examples, and (2) double or distorted morphologies typical of radio galaxies, of which there are also two examples. Clearly the total number of objects involved is much too small to make general conclusions, however, it might be inferred that while distorted source structures can (and do) occur in poor clusters as well as rich clusters, it is probably not the case that the majority of radio sources in poor clusters have distorted morphologies.

We wish to thank R.A. White for making the poor cluster catalog available prior to publication, as well as for useful comments throughout this work. The Westerbork Synthesis Radio Telescope is operated by the Netherlands Foundation for Radio Astronomy with the financial support of the Netherlands Organization for the Advancement of Pure Research (Z.W.O.).

REFERENCES

Abell, G.O.: 1958, Astrophys. J. Suppl. 3, 211.
Burns, J.O., and Gregory, S.A.: 1982, Astron. J. 87, 1245.
Ekers, R.D., Fanti, R., Lari, C., and Ulrich, M.-H.: 1978, Astron.
        Astrophys. 69, 253.
Hummel, E.: 1981, Astron. Astrophys. 96, 111.
Huchra, J.P., and Geller, M.J.: 1982, Astrophys. J. 258, 423.
Stocke, J.T.: 1978, Astron. J. 83, 348.
Turner, E.L., and Gott, J.R., III: 1976, Astrophys. J. Suppl. 32, 408.
White, R.A., Bhavsar, S.P., and Bornmann, P.: 1983, In preparation.
Zwicky, F., Herzog, E., Karpowicz, M., Kowal, C.T., and Wild, P.: 1960-
        1968, Catalogue of Galaxies and Cluster of Galaxies (Cali-
        fornia Institute of Technology, Pasadena, California), Vols.
        1-6.

DISCUSSION

MCGLYNN: Given that there were only 127 Zwicky galaxies in the sample does the detection of only one head-tail source really show a deficiency relative to galaxies in large clusters?

HANISCH: The sample of galaxies we looked at was strongly biased towards spiral galaxies. Since head-tail sources are usually found with ellipticals the detection of only one in our sample is not unusual.

SADLER: I have recently made a large continuum survey of early-type galaxies and found that the presence of a companion galaxy had no effect on radio emission from these galaxies, in contrast to Hummel's result for spirals. This may explain your finding that companions enhance radio emission in low-luminosity (spiral) sources but not in high luminosity (presumably elliptical) ones.

HANISCH: Your result also agrees with my observations of Abell 2243 and 2244, two binary-rich clusters (Struble and Rood 1981). None of the binary galaxies (all early-types) in these clusters were detected with WSRT at 21 cm, implying that the presence of a companion does not influence radio activity at high luminosity levels.

RADIO CONTINUUM PROPERTIES OF SPIRAL GALAXIES IN THE
COMA/A1367 SUPERCLUSTER. COMPARISON BETWEEN ISOLATED AND
CLUSTER GALAXIES

G.Gavazzi,Ist.Fisica Cosmica,Milano,Italy

W.Jaffe,Space Tel.Sci.Inst.,Baltimore,USA

E.Valentijn, E.S.O., Garching, Germany

INTRODUCTION
It has been known for several years that the morphology of
the extended components of radio galaxies is a sensitive
probe of cluster membership of their parent E galaxies .In
fact the frequency of distorted structures (head-tails, wide
angle tails) compared with that of normal doubles is much
higher within clusters than outside. As R.Fanti showed in
today's section , this evidence gained new evidence in the
last few years. This fact is currently interpreted in terms
of interaction between fast moving radio galaxies and the
dense intergalactic medium (IGM), as discussed in details by
Jaffe and Perola (1973).
On the other hand Auriemma et al (1977) showed that the
Radio Luminosity Function (RLF) of radio galaxies seems not
to differ significantly inside and outside rich clusters.
This might be related to the fact that "engines" in radio
galaxies are hidden in the galactic nuclei and shielded by
the sourronding galaxies from the intergalactic environment.
Whether the radio continuum emission in SPIRAL galaxies is
an indicator of the galactic environment is the subject of
the present discussion. The origin of the much less
spectacular radio emission from these objects is certainly
not confined to the nucleus as it is for radio galaxies. It
was shown (Hummel ,1981) that nuclear components contribute
up to 20-40 % of the total radio luminosity in late and
early-type spirals respectively, and that most of the
emissivity is due to the disk population. The emission is
mainly non-thermal in origin, due to the interaction between
cosmic ray electrons accelerated in SN explosions or
produced in bursts of star formation (Condon et al, 1982)
and the magnetic field.
To investigate whether radio properties of spiral galaxies
are somewhat more sensitive probes of their environment, as

209

F. Mardirossian et al. (eds.), Clusters and Groups of Galaxies, 209–214.

compared with radio galaxies, we compare (Section II) the
RLFs of normal galaxies in different regimes of galaxy
aggregation. The RLFs are derived in the Bivariate form
(BRLF), that is per intervals of optical luminosity to
account for the well known dependency of the radio
luminosity on the absolute magnitude as $P \propto L^{\iota}$ (see Hummel,
1981; Gavazzi and Trinchieri, 1981).
In Section III the continuum emission in spiral galaxies and
the luminosity of their Hydrogen line are compared to
investigate whether the two quantities are correlated. Since
the total HI mass provides an indirect estimate of the star
formation rate at present, HI luminosity and radio
emissivity are expected to be correlated if the latter
originates in recent star formation events.

I THE SOURCE SAMPLE
Our comparative analysis is performed on three samples of
galaxies.
a) The "Rich Cluster" sample is obtained by combining
published data on 7 rich clusters of galaxies as specified
in Table I. All clusters have been observed with the WSRT
at 0.6 GHz except for Virgo and A262 that were observed at
1.4 GHz (the radio luminosity in these two clusters were
converted to that at 0.6 GHz using $<\alpha>=-0.8$).

CLUSTER NAME	DIST Mpc	%SP	N OF S+I GAL. Surv.	N OF S+I GAL. Det.	FREQ. GHz	REFERENCE
Virgo(R<4°)	12	45	30	17	1.4	1
Cancer	46	55	23	6	0.6	2
A262	48	38	14	4	1.4	3
A1367	65	44	51	4	0.6	4
Coma	69	10	32	9	0.6	5
A2197	91	44	35	1	0.6	6
A2199	91	40	35	3	0.6	6

Table 1: The 7 clusters in Sample a. Only the galaxies with
$-18<M_p<-20$ are listed. References: 1):Kotanyi, 1980, 2):
Perola et al, 1980, 3): Fanti et al, 1982 4): Gavazzi, 1979,
5): Jaffe et al, 1976, 6): Gavazzi and Perola, 1980

b) The "Field" sample contains 319 S+I galaxies ($-18<M_p<-20$)
from the Arecibo 2.4 GHz survey (Dressel and Condon, 1978).
This subsample of the Arecibo survey was studied by Gavazzi
and Trinchieri (1981) who derived the BRLF of the S+I
galaxies with D>20 Mpc (therefore excluding the Virgo
cluster). The sample is dominated by galaxies not belonging
to Abell clusters although it contains supercluster galaxies
as well as members of Zwicky groups. The 2.4 GHz
luminosities are converted to 0.6 GHz using $<\alpha>=-.8$.

c) The "Coma/A1367 supercluster" sample contains 40 S+I
galaxies with -18<M_p<-20 belonging to four poor groups in
the Coma/A1367 supercluster (see Fig.1). The sample was
observed in 1982 with the WSRT at 0.6 GHz and 12 S+I
galaxies were detected with LogP>20.9. These observations
will be published in a forthcoming paper (Gavazzi et al, in
preparation). Here we present the preliminary results

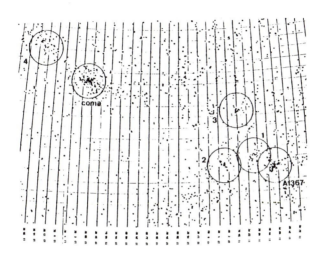

Fig.1: the distribution of the Zwicky galaxies in the
Coma/A1367 supercluster.Circles indicate the 1°.5 WSRT
fields. Sample b) is constitued of fields 1,2,3 and 4.

II RADIO LUMINOSITY FUNCTION AND GALAXY AGGREGATION
The RLFs of the three samples described in section I are
compared in Figs.2 a and b in two intervals of absolute
magnitude.
Our main conclusions are:
1) For all samples studied the dependence of the RLF on the
optical luminosity is consistent with P $\propto$ L^1 (by shifting
the RLF in Fig.1a to the right by one "radio magnitude",
(Δ LogP=0.4), one reproduces quite well the RLF in Fig.2b).
2) The RLF derived for the supercluster sample agrees,
within the errors, with that of the "Field" sample.
3) in both Figs.2 a and b it appears that the radio
luminosity of the cluster sample is identical to the other
two at high radio luminosity (LogP>21.8). Below LogP=21.8
the cluster distribution differs significantly from the
other two,i.e. it is a factor of 2-4 lower and flatter.
The difference is better understood if one takes the
integral RLF F(>P) in the luminosity interval -19<M_p<-20.We
have F(>20.9)=83.2% and F(>20.9)=47.2% for the "field" and
"rich cluster" sample respectively as seen in Fig.3. This

means that almost all of the surveyed "field" galaxies have
been detected above 20.9 and their $\langle LogP \rangle \sim 21.8$. On the
other hand only half of the "cluster" galaxies have been
detected, that is the other half must have $LogP < 20.9$.
Assuming, somewhat arbitrarily, that the remaining 50% of
the cluster sample have $LogP \sim 20.0$ it is derived
$\langle LogP \rangle \sim 21.5$ (this being probably an overestimate). One can
conclude that the average radio power of the cluster spirals
is about half of that of the field galaxies.

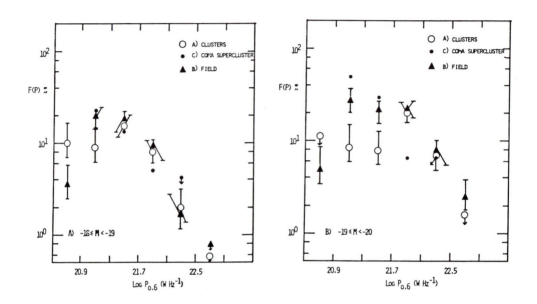

Fig.2 a and b: the RLFs for the three samples.

III RADIO CONTINUUM AND HI CONTENT
There are 34 galaxies in A1367 and in the Coma supercluster
which belong to both our continuum survey and to a 21 cm HI
line survey (Sullivan et al,1981; Chincarini et al, 1982).
9/34 are "bright" continuum galaxies ($LogP > 21.38$) and 13/34
HI "bright" ($LogMHI > 9.35$). All 9 bright continuum galaxies
are also HI bright , while none of the 21 HI "faint"is
bright in continuum. We claim this as a positive correlation
between HI content and radio continuum luminosity, although
we are aware that it can partly be spurious due to the well
known dependency of both Pcont. and MHI on the optical
luminosity. To check this point we use the integral RLF
obtained by combining the three samples to compute the
number of expected detections among the 13 HI bright
($\langle M_p \rangle = -19.7$) and among the 21 HI faint ($\langle M_p \rangle = -19.0$)

galaxies. We find F(>21.38)=63% and F(>21.38)=41% for the two optical luminosities, or

	N surveyed	N expected	N detected
HI bright	13	8.2	9
HI faint	21	8.6	0

The absence of actual detections among the HI faint galaxies suggests that our correlation is not spurious. To further check this point we include in our analysis the Virgo cluster where the percentage of detections is much higher due to the smaller distance. For each of the galaxies surveyed by Kotanyi (1980) we have searched the HI content in the catalogue by Huchtmeier et al (1983). In Fig.4 we plot LogP(cont)/LogM HI for the galaxies which were detected at both frequencies (including the detections in the Coma supercluster and 6 galaxies detected in the Cancer cluster for which the HI information is given in Schommer, 1981). All upper limits shown refer to Virgo galaxies, except for three -20<M<-21 galaxies in the Coma supercluster. From Fig. 4 a weak correlation is apparent, which an optical luminosity effect could contribute to.

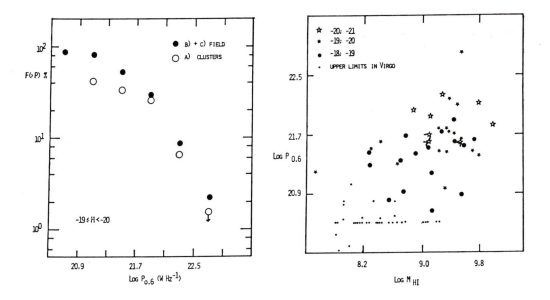

Fig.3: the integral RLF for sample a) and sample b)+c) in the interval -19<M$_P$<-20.
Fig.4: the LogP/LogM HI plot for galaxies detected at both frequencies. The data include A1367, the Coma supercluster, the Cancer cluster and Virgo.

IV CONCLUSIONS

The main conclusions of this work are:
1) Spiral galaxies in rich clusters tend to develop weaker radio sources ($<LogP>\sim21.5$) than isolated or supercluster galaxies ($<LogP>\sim21.8$) in the magnitude interval $-19<M_p<-20$.
2) There is a weak correlation between HI mass and radio luminosity indipendent of the optical luminosity. More observations are needed to establish quantitatively this relation.
The above results 1 and 2 support the idea that radio continuum emissivity is a phenomenon somewhat related to recent star formation. In fact in clusters of galaxies where the HI content of galaxies has been found deficient (with respect to isolated galaxies),as definitely shown by R. Giovannelli and C. Kotanyi at this conference, the efficiency of new star formation is expected to be lower. We have shown that these galaxies tend also to be weaker radio continuum emitters.

REFERENCES

Auriemma,C., Perola,G.C., Ekers,R., Fanti,R., Lari,C., Jaffe,W., Ulrich,M.H.:1977, A&A, 57,41
Chincarini,G., Giovannelli,R., Haynes,M.P.: 1982, ESO preprint, 227
Condon,J.J., Condon,M., Gisler,G., Purschell,J.J.: 1982, Ap.J., 252,102
Dressel,L.L., Condon,J.J.: 1978,Ap.J.Suppl., 36, 53
Fanti,C., Fanti,R., Feretti,L., Ficarra,A., Gioia,I., Giovannini,G., Gregorini,L., Mantovani,F., Marano,B., Padrielli,L., Parma,P., Tomasi,P., Vettolani,G.: 1982, A&A, 105, 200
Gavazzi,G., 1979, A&A, 72,1
Gavazzi,G., Perola,G.C.: 1980, A&A, 84, 228
Gavazzi,G., Trinchieri,G., 1981, A&A, 97,128
Huchtmeier,W.K., Richter,O.G., Bohnenstengel,H.D., Hanschildt,M., 1983, ESO preprint 250
Hummel,E.: 1981, A&A, 93, 93
Jaffe,W., Perola,G.C.: 1973, A&A, 26,423
Jaffe,W., Perola,G.C., Valentijn,E.: 1976,A&A, 49, 179
Kotanyi,C.G.: 1980, A&A Suppl., 41, 421
Perola,G.C., Tarenghi,M., Valentijn,E.: 1980, A&A, 84, 245
Schommer,R.A., Sullivan,W.T., Bothun,G.D.:1981, A.J.,86,943
Sullivan,W.T., Bothun,G.D., Bates,B., Schommer,R.A.: 1981, A.J.,86,919

THE EXTENDED RADIO SOURCE NEAR COMA A

H. Andernach
Max-Planck-Institut für Radioastronomie, Bonn, FRG

L. Feretti and G. Giovannini
Istituto di Radioastronomia, Bologna, Italy

ABSTRACT.   Maps at 2.7 and 4.75 GHz are presented for the elongated radio source SE of Coma A (3C 277.3). No radio bridge to Coma A could be found nor can any other optical object be easily identified to account for the emission. Spectral index variations across the strongly polarized source suggest a particle diffusion along the major axis from the centre to the outer ends of the source, parallel to the magnetic field direction. We argue against a galactic foreground feature, but an association with the Coma cluster or its galaxies is also uncertain.

The existence of an extended radio source SE of Coma A was first noted by Jaffe and Rudnick (1979) at 610 MHz. Later Gubanov (1982) confirmed its existence at 102.5 MHz and estimated the low frequency spectral index to be $\sim$ 1.1. The first map to give structural information on the source (HPBW 2!6 x 5') was the one obtained by Ballarati et al. (1981). The problem has since then been posed that if the source were associated with Coma A (z = 0.086) the latter would be extremely large ($\gtrsim$ 4 Mpc h^{-1} with h = H_O/75 km s^{-1} Mpc^{-1}) and essentially one-sided, whereas if the source were associated with the Coma cluster the source would reside at the SW edge of the cluster and no Coma galaxy coincides well enough in position to be a good identification.

We have mapped the source at 2.7 and 4.75 GHz with the Effelsberg 100-m telescope (HPBW 4!3 and 2!5, respectively) to look for

a)   any emission bridge to Coma A (3C 277.3),
b)   compact or unrelated features possibly blending the source,
c)   spectral variations across the source and, finally
d)   the polarization distribution over the emission region

to find a clue to the nature of the source.

A total of eight coverages were obtained at 2.7 GHz (λ11.1 cm) by scanning the area in both right ascension and declination. To minimize the atmospheric 'scanning' effects we used the 'basket weaving technique'

F. Mardirossian et al. (eds.), Clusters and Groups of Galaxies, 215–220.
© *1984 by D. Reidel Publishing Company.*

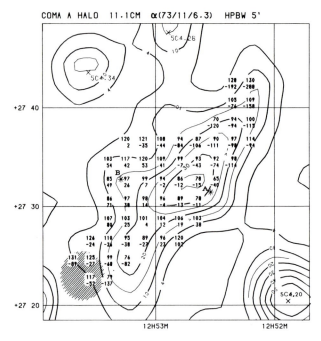

COMA A HALO 11.1CM α(73/11/6.3) HPBW 5'

Figure 1. The extended source at 2.7 GHz after subtraction of Coma A and its sidelobes, smoothed to HPBW = 5'. No other source has been subtracted from the field. Contours in mJy/beam area. A and B mark faint optical objects. Upper numbers = slope of power law fit using 0.4, 2.7 and 4.75 GHz maps; lower numbers = $\alpha(0.4/2.7) - \alpha(2.7/4.75)$ both in units of 0.01 with $S \propto \nu^{-\alpha}$.

which corrects baselevel errors determined from scan intersections. Coma A was included in the mapped field, so that its sidelobes could be effectively subtracted using the nearby strong source 3C 286 as an antenna beam probe. The final map is one of the most sensitive 2.7 GHz maps so far obtained with the 100-m dish (σ_I = 2.3 mJy/beam area; σ_{pol} = 1.1 mJy). It proves that the confusion level at this frequency is much lower than previously believed and the virtual absence of any polarization confusion. Figure 1 shows a version of this map smoothed to HPBW = 5'. There is no radio bridge to Coma A above the 2σ level. The source is strongly polarized (up to $\sim$ 30%) in its NW half with the electric vectors perpendicular to the source's main ridge.

At 4.75 GHz both double and single beam observations were carried out with scanning directions and field sizes carefully chosen to avoid limitations of the restoration technique as well as to reduce baselevel errors due to the many 5C4 sources in the field. The final map is displayed as an optical overlay in Figure 2. Sidelobes of Coma A are negligible here and the absence of any radio bridge to Coma A is confirmed. At the given HPBW of 2.5 the source breaks up into more compact features, but still none of these are point-like. The three brightest features (all near the line with DEC $\approx$ 27°32') have a deconvolved half-width of $\sim$ 1', the eastern one (5C4.31) being almost spherical in shape, the others (5C4.24 and 29) extended east-west. Even higher levels of polarization (up to $\sim$ 60%) are observed at this frequency, but smoothing is necessary to reduce the large errors due to the low signal-to-noise

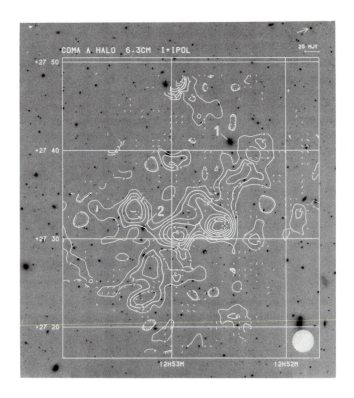

Figure 2. 4.75 GHz map of the extended source superimposed to the POSS (E) print. Noise level is σ_I = 1 mJy/beam area and contour interval 1.5 mJy/beam. E-vectors of polarized intensity shown above $2\sigma_p$ = 2 mJy/beam. Objects A and B (cf. Fig. 1) are marked by gapped plus signs. Galaxies 1 and 2 are N4798 and IC3900, respectively, the arrow at top right points to the Coma A galaxy.

ratio (see below).

The 4.75 GHz map does not shed new light on the optical identification of the source or its components. Only for the eastern- and westernmost of the three brighter condensations mentioned above are there faint red optical objects (A and B in Fig. 1) both $\sim$ 40" from the radio peaks. The two brightest Coma cluster galaxies in this region (1 and 2 in Fig. 2) are very difficult to associate with the extended source, which lies $\sim$ 70' (i.e. 0.9 Abell radii) SW of the centre of Coma (A1656) where the density of galaxies is rather low (cf. Fig. 8 in Godwin et al., 1983).

The integrated emission spectrum of the extended radio source is consistent with a single power law of slope 1.02 ± 0.14 with a slight steepening above $\sim$ 3 GHz. We find a radio luminosity of L_R = 4.3 10^{40} erg s^{-1} h^{-2} between 10^7 and 10^{10} Hz. Assuming energy equipartition of particles and fields, equal energies in electrons and protons and a filling factor of unity we derive the following physical parameters

total size (at z = 0.023)	620 x 180 kpc h^{-1}
magnetic field	5.2 10^{-7} G h$^{2/7}$
minimum energy density	2.5 10^{-14} erg cm^{-3} h$^{4/7}$
minimum total energy	8.1 10^{57} erg h$^{-17/7}$
5 GHz radiation lifetime	1.4 10^7 yr h$^{-3/7}$

The values of magnetic field and minimum energy density are comparable to those of radio halos known in the centres of some rich clusters of galaxies (A1367, A1656, A2255, A2256, A2319) and are also similar to the lowest values known in the outer lobes of complex low luminosity radio galaxies. The pronounced elongation as well as its peripheral position in the cluster is, however, in contrast to known cluster radio halos.

The spectral index distribution between 2.7 and 4.75 GHz has its minima ($\alpha \approx 0.6$) around objects A and B (cf. Fig. 2) suggesting the possibility that they are unrelated objects. The index ranges between 0.8 and 1.0 in the central part of the source. The spectral slope rises continuously along the ridge of the source due NW and SE terminating with values of $\alpha \gtrsim 2 \pm 0.5$ at the outer extremities. We used the 0.4 GHz map of Ballarati et al. (1981) to examine the spectral shape across the source and convolved all three maps to a common HPBW of 5'. The slope of a straight spectrum fitted to the three frequencies as well as the difference between low and high frequency slope is displayed in Fig. 1. This analysis marks the region around object B (or 5C4.31) as the only one with a high-frequency excess (concave spectrum), while the central part shows little spectral curvature. The above mentioned strong spectral index gradient due NW and SE manifests itself here as an increasing spectral break (convex spectrum) in these directions. We infer from this that relativistic particles are probably originating in the centre of the source and suffer increasing synchrotron radiation and inverse Compton scattering losses on their way along the main ridges to the outer ends.

The polarization results for the NW half of the source suggest that this diffusion occurs along the magnetic field direction. The intrinsic rotation measure between 2.7 and 4.75 GHz is $\lesssim 20$ rad m^{-2}, but a significant depolarization ($= m_{2.7}/m_{4.75}$) is observed, 'increasing' from ~ 0.7 near the centre to ~ 0.4 at the NW edge of the source. This requires a mixture of random and uniform magnetic fields with $(B_r/B_u)^2 \lesssim 1$. On the basis of Burn's (1966) slab model we infer that the line-of-sight component of B_u must be small, i.e. B_u near to the plane of the sky. This suggests that projection effects in this source are probably small. In the SW part of the source the polarization is much less uniform.

The nature of this source is still uncertain, though we feel that a halo-type feature associated with the Coma cluster is the most likely explanation. An interpretation as a radio-tail remnant requires fast-moving parent galaxies which have turned off their radio nuclei for a long time. There is an insufficient number of galaxies to explain a heating of the IGM by galactic wakes. An exceptionally extended IGM is however suggested to exist in the Coma cluster by HI observations (e.g. Bothun et al., 1982). We note here that in the sample observed by Chincarini et al. (1983) the only two galaxies within $\sim 45'$ from the extended radio source belong to the most HI-deficient galaxies of the Coma/A1367 supercluster. For N4798 at the NW edge of the source the present upper limit for HI (Sullivan et al., 1982) does not yet require a true deficiency. Therefore, if higher resolution radio maps confirm the diffuse halo-type appearance of the source, it could be interesting to probe the IGM of Coma by further sensitive HI observations.

REFERENCES

Ballarati, B., Feretti, L., Ficarra, A., Gavazzi, G., Giovannini, G., Nanni, M., Olori, M.C.: 1981, Astron. Astrophys. 100, 323
Bothun, G.D., Schommer, R.A., Sullivan, W.T.: 1982, Astron. J. 87, 731
Burn, B.J.: 1966, Monthly Notices Roy. Astron. Soc. 133, 67
Chincarini, G.L., Giovanelli, R., Haynes, M.P.: 1983, Astrophys. J. 269, 13
Godwin, J.G., Metcalfe, N., Peach, J.V.: 1983, Monthly Notices Roy. Astron. Soc. 202, 113
Gubanov, A.G.: 1982, Astrophysics (USSR) 18, 107
Jaffe, W.J., Rudnick, L.: 1979, Astrophys. J. 233, 453
Sullivan, W.T., Bothun, G.D., Bates, B., Schommer, R.A.: 1982, Astron. J. 86, 919

DISCUSSION

MILLS: This source reminds me very much of a steep spectrum source apparently associated with the southern cluster A85. Its origin has puzzled us for several years.

ANDERNACH: There is a similarly peculiar extended radio source at the periphery of a southern cluster described by Goss et al. (1982, MNRAS 198, 259).

REPHAELI: Could you say what is the radio luminosity of the source at $\nu > 1$ GHz if it is at the distance of Coma?

ANDERNACH: Extrapolating the spectrum with $\alpha = 1.1$ the radio luminosity from 1 to 10 GHz is $5 \ 10^{40} \ h^{-2}$ erg s^{-1} with $h = H_o/75$. It can be three times larger if one integrates up to 10^4 GHz, but it can be lower by a large factor if the spectrum steepens strongly above 5 GHz as suggested by our spectral index maps.

HANISCH: New high-dynamic range observations made with the WSRT at 50 and 21 cm indicate an S-shaped symmetry in the extended source reminiscent of an extended radio galaxy, although there is no bright galaxy located at the centre of symmetry. There is a $\sim 16^m$ galaxy within about 3 arcmin of the centre of the source, however, which might be associated with the radio source if it stopped being active $\sim 10^8$ years ago and has moved transversely at a velocity comparable to the velocity dispersion in the Coma cluster. The interpretation of the extended source near Coma A as an old radio galaxy is probably the most likely.
    Valentijn (1978) pointed out the difficulty of making polarization

measurements in the vicinity of the Coma cluster due to its proximity
to the North Polar Spur. Are you confident that the high polarization
values you have found are not contaminated by Galactic foreground emission?

ANDERNACH: Concerning your comment I agree that fast-moving parent
galaxies which turned off their radio nuclei ages ago could nicely explain the high number of unidentified radio sources in the cluster of
galaxies. On the other hand, to my knowledge the scenario you suggest
for this source has never been established for any other source yet.
Regarding polarization I rule out a sidelobe influence of the North
Polar Spur (being $\sim$ 10° away) since its effect should be quite different and extremely weak at the two frequencies. Instead, we see polarized emission at the same position coincident with the total intensity
feature at both frequencies. Assuming the extended source to be extragalactic one would expect any galactic foreground polarization features
to be unrelated in position and extent. The clear coincidence of total
and polarized emission and the absence of other extended polarized
features in our map makes such a contamination rather improbable.
Smooth variations in polarized brightness of size $\gtrsim$ 1° as observed by
Valentijn do not affect our single dish maps since these are subtracted
by the observing method. The above arguments do, of course, not rule
out a galactic nature of the source in both total and polarized intensity.

SARKAR: If the depolarization behaviour suggests that the fluctuating
component of the magnetic field is comparable to the ordered component,
then the strength of the field may be much less than the 'equipartition' value. This is, for example, the case in our own Galaxy.

HI CONTENT OF GROUPS AND CLUSTERS OF GALAXIES

W.K. Huchtmeier
Max-Planck-Institut für Radioastronomie, Bonn, F.R.G.

From early observations of extragalactic neutral hydrogen (HI) it became evident that some HI-parameters were dependent on the morphological type of galaxies. With a sample of 130 galaxies Roberts (1969, 1975) and Balkowski (1973, n=149) established the correlations of type-dependent galaxian properties. This data collection is referred to as the "standard" sample of nearby galaxies, which represents galaxies from nearby groups, the field, and a few Virgo cluster galaxies. In the following we will mainly concentrate on the relative HI-content of galaxies. This we get by normalizing the total HI-mass (M_H) by a quantity representing the size, the mass, or the luminosity of the galaxy. The distance-independent hydrogen-mass to optical luminosity ratio (M_H/L) and the distance-dependent ratio M_H to total mass (M_H/M_T) are such quantities. Fig. 1 shows the dependence on morphological type of the quantities M_H/L_B and M_T from a collection of ∿200 galaxies (Li and Liu, 1981). The trend with morphological type is obvious but the scatter within a given morphological type is considerable (much greater than the observational errors). The HI-content probably depends on more parameters than just the morphological type. Barred spirals tend to have more HI than normal spirals of the same morphological type (Li and Liu, 1981), "anemic" (van den Bergh, 1976) or smooth arm spirals seem to have less HI (Wilkerson, 1980). Low surface brightness (LSB) spiral galaxies appear to have a relative hydrogen content more than twice that of normal late-type spirals (Strom, 1982). A dependence of M_H/L on the absolute magnitude has been suggested by different authors (Strom 1980, Balkowski 1973, Bottinelli and Couguenheim 1974). From their sample of isolated galaxies Haynes and Giovanelli (1981) found such a dependence for galaxy types Sa to Sc (but not for Sd and irregular types, in agreement with Thuan and Seitzer, 1979). Fig. 2 shows this dependence, M_H/L varies by a factor of about 3.5 between magnitude −19 and −23. This has to be taken into account when comparing different samples of galaxies.

In Fig. 3 a typical HI profile of an ordinary spiral galaxy is given (NGC 7541, Shostak 1978) in order to demonstrate the derivation of some measurable quantities. As frequencies can be measured to a high degree of accuracy the limitations to velocity measurements are set by the resolu-

F. Mardirossian et al. (eds.), Clusters and Groups of Galaxies, 221–242.
© 1984 by D. Reidel Publishing Company.

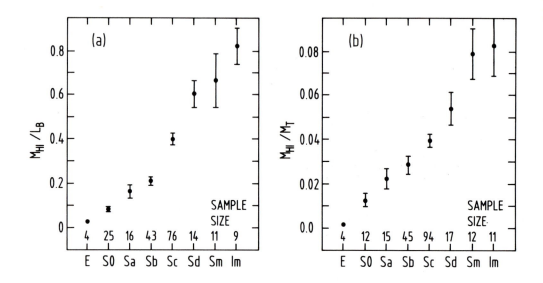

<u>Fig. 1</u>: *Relative HI-content (M_H/L_B and M_H/M_T) for various morphological types from Li and Liu (1981). Error bars correspond to twice the r.m.s. error of the mean value.*

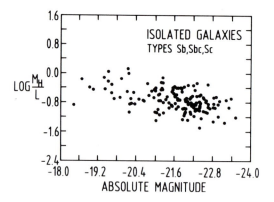

<u>Fig. 2</u>: *Log(M_H/L) versus absolute magnitude for 149 isolated galaxies of morphological types Sb to Sc (Giovanelli et al., 1981).*

tion used. The radial velocity of a galaxy and the line widths are therefore well defined quantities. Typical velocity resolutions found in the literature are around 10 km s^{-1}. The distance of a galaxy can be deduced by dividing the corrected (for motion of the galaxy and the Local Group) radial velocity by the Hubble constant. The line integral (a direct measure of the amount of HI involved in the optically thin case) is usually affected by rather high relative errors due to uncertainties involved in the flux calibration process, possible pointing errors, etc. In addition baseline problems may occur in spectroscopy

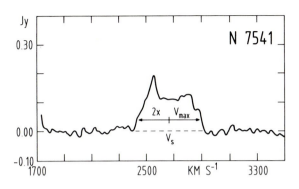

Fig. 3: HI profile of the Sbc/Sc galaxy NGC 7541 (Shostak, 1978). At the 20% level (of the peak value) the HI line width is nearly twice the maximum rotational velocity (v_{max}; Rubin et al. 1978).

which will affect primarily wide and weak profiles. Most HI fluxes have errors greater than 10% not counting uncertainties in distances. The line width obviously is a measure of the galaxy's total mass, as disk systems are dominated by rotation. Rubin et al. (1978) compared HI line widths of galaxies of different morphological types (at the 20% level: Δv_{20}^i) with optically derived rotation curves and derived

$$\Delta v_{20}^i = 2 \times v_{max}$$

where v_{max} is the maximum rotational velocity. Relevant HI-data by Bosma (1978) and van Moorsel (1982) yield similar results. Hence the total mass of a galaxy is given by its line widths (Δv_{20}^i), optical extend (D) in arc min (as the HI-extent in general is unknown), its distance (d) and a constant which is model dependent (e.g. Casertano and Shostak, 1980) and type-dependent, too (e.g. Dickel and Rood, 1978).

The extent of the distribution of neutral hydrogen seems to be related to the optical diameter. For a level of 1.8×10^{20} atoms cm^{-2} Bosma (1978) found

$$D_H/D_{Holmberg} = 1.52 \pm 0.11$$

for 20 late-type galaxies. From van Moorsel's (1982) data of binary galaxies for a similar level

$$D_H/D_{Holmberg} = 1.38 \pm 0.08$$

For our present purpose we state that these results are not very different. Hewitt et al. (1983) used the HI "isophote" D_{70} which contains 70% of the galaxy's HI derived by a model fit procedure (because of the relatively low spatial resolution). They relate D_{70} with the blue Nilson (1973) diameter a:

$$\log(D_{70}) = \log(1.2 \times a).$$

It is evident that the HI-diameter is related to the optical diameter.

For late-type galaxies the HI surface density seems to be of the same order of magnitude. Therefore it is not suprising that the hydrogen mass of a galaxy (M_H) is linked to the linear optical diameter (Shostak, 1978)

$$\log M_H = 1.85 \log A(0) + const., \quad A(0) \text{ in kpc} \qquad \text{(Fig. 4)}$$

A relation between M_H and the HI-extent (D_H) of a galaxy was given by Hewitt et al. (1983)

$$\log M_H = 1.74 \log D_H + const.$$

Different global parameters of spiral galaxies seem to be roughly proportional to each other as follows

$$M_H \sim M_T \sim L_B \sim A(0)^2 \qquad \text{(Shostak 1978, Reif 1982)}$$

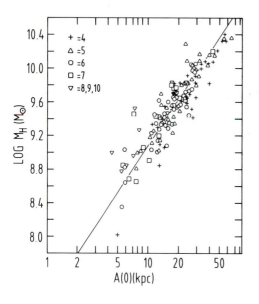

Fig. 4: HI mass M_H versus linear diameter $A(0)$. Different symbols are used for different morphological types (Shostak, 1978).

Briggs et al. (1980) and Briggs (1982) used the Arecibo radiotelescope to search for HI-emission beyond the Holmberg limit of a complete sample of galaxies. For about 90% of their sample galaxies they found only upper limits of the order of a few times 10^{18} atoms cm^{-2}. A number of galaxies with extended HI-haloes, disks, or rings have been found so far (e.g. Huchtmeier and Richter, 1982 and references therein). As an example the HI-extent of the normal galaxy M33 (Fig. 5) at a level of 2×10^{20} atoms cm^{-2} (i.e. comparable to Bosma's sample (1978)) is 1.3 times its Holmberg diameter. At a lower surface density of 4×10^{18} atoms cm^{-2} the hydrogen extends to 2.2 times the Holmberg diameter. The amount of HI beyond the Holmberg limits is only 10% of the total HI-mass. There are several galaxies known that even at a level of 10^{20} atoms cm^{-2} do extend to more than twice their Holmberg diameter. A nice example is the Sc galaxy M83 (Fig. 6) where the HI extends three times the Holmberg limits at this level. HI has been traced to 6.5 times the Holmberg diameter at a level of 10^{19} atoms cm^{-2}. In this exceptionally large HI-ring 80% of the HI is located outside the Holmberg limit. The outer HI-ring is at a different orientation than the galaxy but shows clear indication of rotation (Huchtmeier and Bohnenstengel, 1981). The

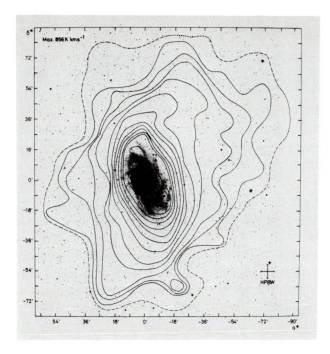

Fig. 5: The HI distri-
bution of M33 overlaid
on a deep (IIIaJ)
Palomar Schmidt plate
(courtesy of
G.A. Tammann). Contour
lines are given in steps
of 10% of the peak val-
ue (in brightness tem-
perature) down to the
10% level followed by
5%, 2%, 1% and 0.5%.
The broken line cor-
responds to the detec-
tion limit.

Fig. 6: HI distribu-
tion of M83 overlaid
on a Palomar sky
survey plate. Con-
tour lines are given
as in Fig. 5 with
the exception of
the 0.5 % level.

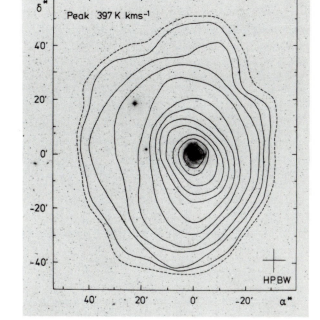

irregular galaxy IC10 as well as the Markarian galaxy NGC262 are other
prominent examples for extended HI haloes. A selection of relevant publications in Table 1 demonstrates the significant increase of HI-data
from the early standard sample. At the end of 1982 all published extragalactic HI-data referred to more than 4500 galaxies including upper
limits. In the lower part of this Table references relative to isolated
galaxies are given. Giovanelli, Haynes and Hewitt observed a reference
set of 324 isolated galaxies from the Karachentseva (1973) list of isolated galaxies within the declination range of the Arecibo telescope
($-1°$ <dec. <$38°$) including 52 extended (D $\geq 2.5'$) galaxies. The HI-mass
(M_H) is shown to be nearly proportional to the square of the linear

Table 1: Selected HI Observations of Galaxies

number of galaxies

Roberts (1969, 1975)	standard	130
Balkowski (1973)	sample	149
Li Zong-yu and Liu Ru-liang (1981)		∿200
Shostak (1978)		169
Fisher and Tully (1981)		1787
Reif et al. (1982)		359
Bottinelli et al. (1982), HI-catalogue		1189
Huchtmeier et al. (1983), HI-catalogue		∿4500

isolated galaxies:

Balkowski and Chamaraux (1981)	39
Haynes and Giovanelli (1980)	
Giovanelli and Haynes (1981)	324
Hewitt et al. (1983)	
Krumm and Shane (1982)	2
Huchtmeier and Richter (1983)	3

HI-diameter, implying that the HI surface density averaged over the disk
is almost constant. In Table 2 the distance-independent parameter M_H/L
is given for the standard sample (Sullivan et al., 1981) and for the
isolated sample (Giovanelli et al., 1981). For late-type objects differences between the M_H/L in the two samples are significantly greater
than the errors indicated. This may be due to differences between the
two samples, but we can not exclude selection effects. Studies of dwarf
galaxies (Thuan and Seitzer, 1979) indicate that there is a trend for
these galaxies to have larger M_H/L values. A few dwarfs reach M_H/L values
of 2 to 4 (e.g. NGC3109, the Sculptor dwarf, M81dwA). However, the Nilson
dwarf survey (Thuan and Seitzer, 1979) had a detection rate of only 31%,
i.e. apart from a few galaxies outside the velocity range that was
searched for HI emission many dwarfs were below the detection limit of
$10^6 M_\odot M_{pc}^{-2}$ (i.e. M_H/L <0.2 on the average). Blue compact galaxies seem

Table 2: M_H/L *values for the standard sample and the isolated sample as a function of the morphological type*

standard sample type	M_H/L_{pg}	isolated sample type	M_H/L_{pg}	(±)	n
		S0-S0a	0.137	0.14	5
Sa-Sab	0.1	Sa	0.103	0.11	10
		Sab	0.12	0.08	9
Sb	0.16	Sb	0.155	0.04	45
Sbc	0.20	Sbc	0.18	0.06	25
Sc	0.33	Sc	0.20	0.06	23
Scd	0.51	Scd-sd	0.41	0.05	13
Sd	0.74				
Sdm-Sm	0.54				
Im	0.91	Im	0.33	0.08	15

to have colours, luminosities, HI-masses, M_H/L and other parameters in the same range as normal late-type galaxies (Gordon and Gottesman, 1981; Thuan and Martin, 1981). The same seems to be the case for Seyfert galaxies (Heckman et al. 1978). Low surface brightness galaxies (LSB) have projected central disk surface brightness values approximately 3 times smaller, on average, than normal spirals in the same luminosity range. Their M_H/L is observed to be more than twice the value typical for normal late-type spirals (Romanishin et al., 1982).

ISOLATED GALAXIES

     It has been argued that the environment might have a considerable influence on galaxian evolution and hence on observable parameters (e.g. Peterson et al., 1979 and references therein). The effect of gravitational interaction has been convincingly demonstrated by Toomre and Toomre (1972). The intergalactic matter in clusters is another possible candidate for an influencing medium. A comparison between isolated and non-isolated galaxies should provide information on properties which may depend on the environment (galaxy evolution) or to different conditions at the time of galaxy formation. To establish a well-defined comparison sample is quite important. The isolated galaxy sample has been observed for this reason.

     In general galaxies in less dense environments show slight irregularities in their velocity fields and indications for warps in their HI just beyond the optical limit. Krumm and Shane (1982) studied two very isolated galaxies (NGC2712 and NGC5301) which had a very low chance that any galaxy ever passed within 100 kpc of their nuclei. These two isolated galaxies do not have more extended HI-distributions than non-isolated galaxies, so tidal stripping may not be an important factor for galaxies in low density environments. Warps and asymmetries in the gas distribution of spirals can be internally generated and sustained. A search for extended HI-emission at a lower surface density has been made with the 100-m radiotelescope at Effelsberg (Huchtmeier and Richter, in preparation) with no indication of large haloes (more than 15% of the galaxy's HI-mass).

GROUPS OF GALAXIES

About 50% of the known cases of extended HI-distributions are members of binaries or groups of galaxies. They do not seem to be common in clusters. In the NGC4631 group there are three pairs with strong indications of gravitational interaction. A very instructive example is the HI-plume extending out from NGC3628 (Fig. 7, Haynes et al. 1979). Apart from those cases where gravitational interaction is evident spirals in groups have HI properties comparable to the standard sample of field galaxies. Some recent publications on galaxies in groups are collected in Table 3. Accurate HI velocities have been used mainly to derive virial masses of groups. The IC698 group may be taken as an example (Williams, 1983). It is situated in the Coma-A1367 supercluster and yields a virial mass-to-light ratio of 5 in excellent agreement with the M_T/L ratio of spiral galaxies (e.g. Faber and Gallagher, 1977). The NGC1961 group (Shostak et al., 1982) is a very fine example of a partially displaced HI distribution demonstrating the effect of ram pressure stripping by an intergalactic gas.

INTERGALACTIC MATTER

Intergalactic gas has been postulated since early applications of the virial theorem (Zwicky, 1933) which led to the "missing mass" prob-

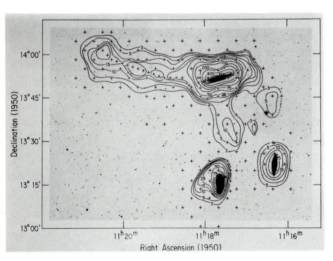

Fig. 7: Neutral hydrogen distribution in the Leo group. Contours are given in K_{T_a} levels. The long appendage extending eastwards from NGC3628 is referred to as the plume; the extension in the region between the three galaxies is the bridge (Haynes et al., 1979).

Table 3: Selected publications on groups of galaxies

	number of galaxies
Webster et al. (1979)	8
Chincarini et al. (1979)	25
Sullivan et al. (1981)	
Appleton and Davies (1982)	69
Williams (1983)	
binaries and multiple galaxies, compact groups	
Haynes et al. (1979)	3
Haynes (1981)	47
Gallagher et al. (1981)	6
Helou et al. (1982)	28
Sulentic and Arp (1983)	131
Bieging and Biermann (1983)	9

lem. Specially the discussion of the medium and high velocity cloud phenomenon motivated deep searches for intergalactic HI-clouds in groups
and clusters. No such clouds have been detected to a limit of about
10^6 $M_\odot/M_{pc}^2$ in nearby groups (Lo and Sargent, 1978, Haynes and Roberts,
1979, Materne et al., 1979, Fisher and Tully, 1981). One such cloud has
been suspected near NGC1023 (Hart et al. 1980) but recently it was suggested to be part of the NGC1023 system itself (Sancisi 1983, priv.
communication). Schneider et al. (1983) recently reported detection of
a large intergalactic HI cloud in the M96 group.

The corresponding limits to the intergalactic HI clouds in galaxy
clusters are considerably higher due to the much greater distances. The
upper limits are of the order of 4×10^{19} to 2×10^{20} atoms cm^{-2} surface
density or 8×10^{10} $M_\odot$ for the Cancer cluster and 2×10^{11} $M_\odot$ for Perseus
and Coma (Tarter and Wright 1979; Shostak et al. 1980). Many more obser-
vations should be available within the bulk of recent extragalactic HI-
observations as the general observing procedure of single dish instruments involves a reference field in the vicinity from the position of
the galaxy to be observed.

CLUSTERS OF GALAXIES

The core areas of galaxy clusters are the regions with the highest
galaxian densities. There the chance of any close passage between two
galaxies will be considerably greater than in the low density field. The
existance of an intergalactic medium in the core regions of galaxy clusters is documented by head-tail radio sources (e.g. Miley et al. 1972)
and by extended X-ray emission from these areas (e.g. Forman and Jones,
1982). The mass of this X-ray emitting gas seems to be about equal to
the total luminous mass of the galaxies and hence not sufficient for
providing most of the missing mass.

Galaxies passing through the cluster's core are expected to interact with the intergalactic medium. The mechanism would have to remove a
considerable amount of mass of the neutral gas before this can become
evident when the large scatter of relative HI-content (e.g. M_H/L) for a
given morphological type is considered. It will be difficult to call an
individual galaxy HI-deficient. However, in a statistical sense an HI-
deficiency corresponding to a factor of 1.5 to 2 should be measurable
with present observations. The HI-deficiency has been defined as the
logarithmic difference of the M_H/L of the comparison sample and the M_H/L
value of the galaxy in question:

$$DF = [\log(M_H/L)]_{cs} - \log(M_H/L)$$

(M_H/L_B values for different morphological types are given in Fig. 1).
Values of the comparison sample are given in the blue or photographic
magnitude system. The classification of small and faint galaxies is difficult and the distribution in the step function in Fig. 1 introduces
some additional uncertainty into the final results. To avoid such problems different approaches to the measurement of the HI-deficiency were
made by referring the observed HI-mass to the galaxies' linear extent
(Giovanelli et al. 1982 and references therein) and to colour indices,
which are related to morphological types themselves (Bothun et al.
1982a, b). HI-deficiency has been found in six clusters so far (Table 4)
but there is at least the same number of clusters with normal HI-content.
All HI-deficient clusters show extended X-ray emission in their core regions whereas the other clusters do not. The aggregates with normal HI-
content seem to be considerably smaller (lower velocity dispersion) and
typical for the extended components of superclusters (e.g. Coma region,
Williams and Kerr 1981).

VIRGO CLUSTER

Spiral galaxies in the Virgo cluster were the first suspected to
have a lower HI-content than galaxies of the same morphological type from
the comparison sample (see Table 4). Since 1980 all authors agree that
the Virgo cluster is HI-deficient within 2 Mpc of the centre. So far HI
has not been detected in elliptical galaxies in the Virgo cluster
(Krishna-Kumar and Thonnard, 1983). Upper limits are as low as 8×10^7 $M_\odot$
corresponding to M_H/L_B of $\sim 10^{-3}$. For S0 galaxies the detection probability is said to be lower inside the Virgo cluster than outside (Giovanardi
et al. 1983b). All spirals with blue magnitude $B_T \leq 12.80$ in an area
$15° \times 16.5°$ around the centre of Virgo have been observed so far (Helou
et al. 1983). Spirals have smaller HI-diameter within the cluster
(Giovanardi et al. 1983a, Giovanelli and Haynes 1983): $D_H/D_{opt} = 0.90\pm0.09$
inside compared to 1.23 ± 0.07 outside the cluster. The most exciting result, however, is the VLA-map (Kotanyi et al. 1984, this volume) in which
the hydrogen depletion phenomenon in Virgo is convincingly demonstrated
The amount of the deficiency corresponding to a factor of $2 - 2.7$ on the
average from all published data.

Table 4: *HI-deficiency in cluster galaxies*

			number of galaxies
1. Virgo	Davies and Lewis	(1973)	
	Huchtmeier et al.	(1976)	39
	Krumm and Salpeter	(1979)	
	Chamaraux et al.	(1980)	56
	Helou et al.	(1981)	
	Huchtmeier	(1982)	20
	Giovanardi et al.	(1983a,b)	112
	Bohnenstengel	(1983)	12
	Giovanelli and Haynes	(1983)	24
	Krishna-Kumar and Thonnard	(1983)	
	Helou et al.	(1983)	74
2. Hydra	Richter and Huchtmeier	(1983)	
3. A262	Giovanelli et al.	(1982)	67
4. Coma	Sullivan and Johnson	(1978)	11
	Sullivan et al.	(1981)	32
	Chincarini et al.	(1982)	
	Bothun et al.	(1982)	
5. A1367	Sullivan et al.	(1981)	34
	Chincarini et al.	(1983a)	51
6. Hercules (A2147)	Giovanelli et al.	(1981)	52
	Schommer et al.	(1981)	26

clusters with normal HI-content

1. Pergasus I	Schommer et al.	(1981)	22
	Richter and Huchtmeier	(1982)	75
	Bothun et al.	(1982a,b)	54
2. Cancer	Bothun et al.	(1981)	
	Schommer et al.	(1981)	19
	Bothun et al.	(1983)	
3. 774-23	Sullivan et al.	(1981)	16
4. Comar region	Williams and Kerr	(1981)	233
	Chincarini et al.	(1983b)	67
5. A2151	Giovanelli et al.	(1981)	

Independent of the classical definition of the HI-deficiency in the Virgo cluster was demonstrated by Giovanelli and Haynes (1983) plotting HI-mass against the optical diameter (log $D_{Holmberg}$) in Fig. 8. In case of the comparison sample there is a nice correlation between M_H and log D_{Ho} whereas many Virgo galaxies have less HI than would be expected for their optical extent.

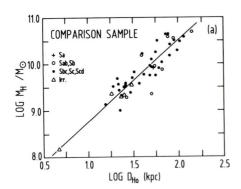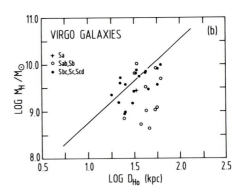

Fig. 8: *The correlation between HI mass and linear Holmberg diameter for the isolated galaxies' sample (a) and for galaxies in the Virgo cluster (b). The full line in both figures represents the best fit to the comparison sample (Giovanelli and Haynes, 1983).*

COMA CLUSTER

In contrast to the irregular spiral-rich Virgo cluster the Coma cluster is a typical case of a regular cluster dominated by elliptical galaxies (to 82%). This cluster has the highest HI-deficiency so far observed (>3). Bothun et al. (1982a,b) demonstrated a clear radial gradient in the HI-mass of Coma spirals (Fig. 9) with many upper limits within 1.5° of its centre. The other big cluster in the Coma supercluster, A1367, is irregular, spiral rich, and shows patchy X-ray emission. This cluster is HI-deficient within 1° of its centre (Fig. 10). Galaxies just outside this region are found to have normal HI-content. Recently Chincarini et al. (1983) reported individual X-ray emitting spirals in A1367 to be HI-deficient by factors 5 to 10. The outer parts of both clusters as well as groups and the intracluster population of this supercluster seem to be quite normal regarding the HI-content.

ABELL 262

In the case of the cluster A262 Giovanelli et al. (1982) demonstrated a definition of the deficiency referring to the optical diameter (D), i.e. using a distance-independent quasi HI-deficiency

$$DF' = [\log(M_H/D^2)]_{cs} - \log(M_H/D^2)$$

the comparison sample $[\log(M_H/D^2)]_{cs} = c_t + d_t \log(D)$ depends on the morphological type t. The correlation between both definitions of the HI-deficiency measures DF and DF' was checked by Giovanelli et al. (1982) and is satisfactory.

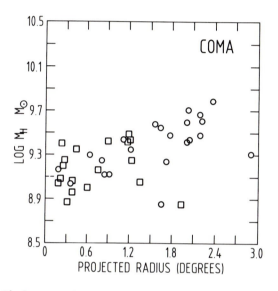

Fig. 9: *Gradient in hydrogen mass (M_H) for Coma cluster spirals. Boxes are upper limits (Bothun et al. 1982a).*

HI-deficient galaxies are strongly concentrated to the clusters' core. Further out there are some galaxies that are HI rich by at least a factor of two. Such galaxies have been observed in the Pegasus I cluster and its surrounding (Richter and Huchtmeier, 1982; Bothun et al. 1982a,b). In an attempt to define the HI-deficiency of galaxies independent of their morphological type Bothun et al. (1982a,b) measured colours for a sample of nine clusters. The colour-colour relation of their sample is largely the same as for the RC2 (de Vaucouleurs et al. 1976) sample. In particular for both samples the range of colours is similar. Comparing the distribution of colours shows

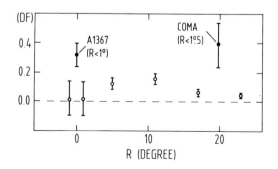

Fig. 10: *The HI-deficiency in the Coma/A1367 supercluster is limited to the core region (∿1°) (Chincarini et al. 1983). Values for the two clusters correspond to averages over 1.5° or 1° around the centre, respectively. Values relative to the low density area of the supercluster correspond to averages over 6 degrees.*

the presence of a larger percentage of red spirals than found for the
field. The gas content of disk galaxies in clusters is not well correlated with colour as a variety of gas contents can be found for any
given colour. There is a weak trend for galaxies of red colour to be
gas pour. The stellar and gaseous content of disk galaxies in clusters
has a large dispersion. However, when comparing the colour-gas relation
for the Coma and the Pegasus I cluster it is evident that blue HI rich
spirals are missing in the Coma cluster in contrast to the Pegasus I
cluster, another documentation of the HI-deficiency (see Fig. 11).

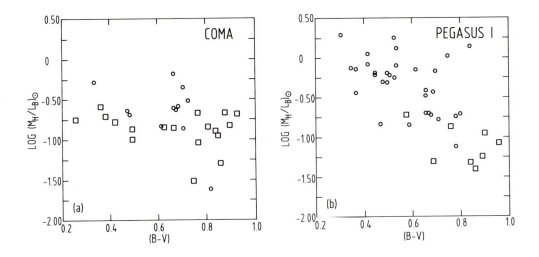

*Fig. 11: Colour-gas content relation for (a) the Coma cluster and (b)
the Pegasus I cluster. Boxes represent upper limits (Bothun et al.
1982b).*

THE TULLY-FISHER RELATION

The correlation between absolute magnitude and HI line width of
late-type galaxies known as the Tully-Fisher relation (TFR; Tully and
Fisher 1977) is an efficient tool to derive distances if this relation
is independent of the galaxies' location (environment). Indications of
such a dependence were discussed by Kraan-Korteweg (1983). Nearby galaxies with known distances and different luminosities should allow
calibration of the TFR. Such a calibration is shown in Fig. 12 (Sandage
and Tammann 1976) including error bars. In general HI line widths (Δv)
have small observational errors (see small errors in v for M31 in Fig.
Fig. 12). However, for nearly face-on galaxies where the inclination
correction is relatively large the uncertainty in v increases (see M101,
M33 in Fig. 12). On the other hand the luminosity correction for internal absorption for nearly edge-on galaxies becomes high and rather un-

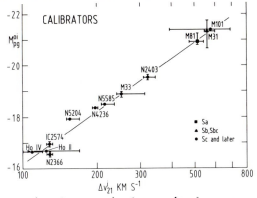

Fig. 12: The Tully-Fisher relation (absolute blue magnitude versus log of the corrected 21 cm line width) for nearby galaxies (Sandage and Tammann, 1976).

certain when optical magnitudes are used (see M31 in Fig. 12). To avoid
the problem of the correction of the blue magnitudes, Aaronson et al.
(1979) have used infrared magnitudes for the TFR.

Table 5: The slope of the Tully-Fisher relation

		author
-6.88	± 0.3	Sandage and Tammann (1976)
-6.25	0.75	Tully and Fisher (1977)
-8.6		Roberts (1978)
-5.3	0.2	Shostak (1978) local galaxies
-8.24	1.3	" late-type galaxies
-5.0	0.4	Bottinelli et al. (1980)
-13.0	1.5	Rubin et al. (1980)
-7.67	0.25	Richter (1982) Virgo galaxies
-6.84	0.19	" nearby galaxies
-10.0		Rubin et al. (1982)
-8.82	0.30	Aaronson and Mould (1983)
-5.0	0.2	Bottinelli et al. (1983)
-7.5	0.25	Meisels (1983)
-7.6	mean	
-7.27		Richter and Huchtmeier (1983)

Table 5 shows a considerable scatter in the slope of the TFR as
derived by different authors. The unweighted mean of these values is
quite close to the value derived in a rediscussion of the TFR (Richter
and Huchtmeier 1983) including new HI observations for nearby groups of
galaxies. The resultant fit is compatible with the assumption of a unique
TFR as the slopes for each of the galaxy groups and of the Virgo cluster
are the same within the errors. The fit does not show any strong dependence to the absorption correction used. Fig. 13a shows 66 Virgo spirals
with the best fit TFR; Fig. 13b shows only those spirals that were named
HI-deficient by at least two different authors (i.e. from independent
observations). The straight line is the best fit from Fig. 13a. HI-defi-

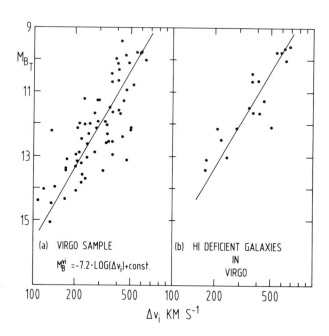

Fig. 13: The Tully-Fisher relation for Virgo cluster galaxies: a) the complete sample of 66 galaxies, b) only those 18 galaxies found to be HI-deficient by at least two different authors.

ciency does not seem to affect the TFR. This is not surprising as environmental stripping mechanisms will mostly influence the outer disk regions. The measured line width (corresponding to twice the maximum rotational velocity) is a quantity determined by the inner part of the galaxy, v_{max} is reached within a few kpc of the centre.

VELOCITY DISPERSION OF FIELD GALAXIES

The typical velocity dispersion of field galaxies is a quantity of cosmological importance. Values between 50 and 370 km s^{-1} have been published (Table 6) using different galaxy samples and different methods of reduction. Using all disk galaxies of the Kraan-Korteweg Tammann catalogue (1979) with the exception of Virgo cluster members and applying the TFR of Fig. 13 Huchtmeier and Richter (1983 in preparation) derived a value of $\sigma = 170\pm40$ km s^{-1}. Obviously this is an upper limit by sev-

Table 6: Velocity dispersion of field galaxies

velocity dispersion (km s^{-1})	number of galaxies		
<150	446	Neyman, Scott	1961
370	98	Holmberg	1964
<157		Karachentsev	1971
344±40	174	Fesenko	1973
$\lesssim$ 50	71	Sandage, Tammann	1975
152±30	42	Tsvetkov	1981
70±10	RSA	Rivolo, Yahil	1982
177±40	155	Huchtmeier, Richter	1984

eral reasons: the infall velocity of the Local group has not yet been optimized; with the catalogue common distances have been assumed for all members of a galaxy group whereas in our calculations galaxy distances were calculated from corrected redshifts directly.

DISCUSSION

Observational data accumulated in recent years provide evidence for a deficiency of spirals in clusters when compared with galaxies in low density regions of superclusters. Except for Virgo the sample of galaxies involved in HI observations per cluster is relatively small (Table 4). Groups and small irregular clusters seem to have a normal HI-content whereas the phenomenon of HI-deficiency seems to be limited to rich clusters with extended X-ray emission in their core regions. The large scatter of the hydrogen content of galaxies of a given morphological type makes it difficult to decide about the deficiency of a given galaxy. However, the effect is evident statistically. Moreover the amount of deficiency seems to vary from cluster to cluster. Mildly deficient clusters like A2151, Virgo, A1367 and more deficient clusters like Coma are observed.

Several mechanisms have been discussed as potentially effective to remove a substantial amount of the interstellar gas from galaxies in rich clusters:
a) galaxy - galaxy collisions (Spitzer and Baade 1951, Richstone 1976, Farouki and Shapiro 1980),
b) enhanced efficiency of galactic winds (Mathews and Baker 1971),
c) evaporation of the interstellar gas in conductive contact with a hot intergalactic medium (Cowie and Songeila 1977),
d) ram pressure sweeping caused by the motion of the galaxies through an intergalactic medium (Gunn and Gott 1972),
e) differences between galaxies due to different conditions for their origin.

The situation is not clear at all but some of these mechanisms are more likely than others.
a) Tidal stripping should not discriminate between X-ray and non X-ray clusters. Head-on collisions do not seem to occur frequently enough. The same argument holds for tidal interactions. Undoubtedly tides are at work (Farouki and Shapiro 1980) and they would affect both the stellar content and the gas content,
b) galactic winds should affect primarily the central parts of galaxies and not so much the outer areas contrary to the observations and would push HI and HII regions at large distances from the galactic center,
c) gas evaporation should deplete a galaxy in a rather symmetric way. There is some evidence that the HI disks are distorted. There is no strong argument that could rule out the possibility that such mechanism is at work,
d) (ram pressure) sweeping in general seems to fit the observations better and easily could account for asymmetries in the HI envelopes.

Nulsen (1982) has discussed several effective transport processes which cause stripping of galaxies moving through a hot intergalactic medium. In most rich clusters ram pressure stripping is very effective near the core, but causes negligible mass loss in the lower density regions making up the bulk of the cluster.

ACKNOWLEDGEMENTS

Numerous colleagues kindly made available data prior to publication.

REFERENCES

Aaronson, M., Huchra, J. P., Mould, J. R.: 1979, Astrophys. J. 229, 1
Aaronson, M., Mould, J. R.: 1983, Astrophys. J. 265, 1
Appleton, P. N., Davies, R. D.: 1982, Mon. Not. Roy. ast. Soc. 210, 1073
Balkowski, C.: 1973, Astron. Astrophys. 29, 43
Balkowski, C., Chamaraux, P.: 1981, Astron. Astrophys. 97, 223
Bieging, J. H., Biermann, P.: 1983, Astron. J. 88, 161
Bohnenstengel, H.-D.: 1983, Ph.D. Thesis, Univ. of Hamburg
Bosma, A.: 1978, Ph.D. Thesis, Univ. of Groningen
Bothun, G., Schommer, R. A., Sullivan, W. T.: 1982a, Astron. J. 87, 725
Bothun, G., Schommer, R. A., Sullivan, W. T.: 1982b, Astron. J. 87, 731
Bothun, G., Stauffer, J. S., Schommer, R. A.: 1981, Astrophys. J. 247, 42
Bothun, G., Geller, M. J., Beers, T. C., Huchra, J. P.: 1983, CfA,
    prepr. no. 1714
Bottinelli, L., Gouguenheim, L.: 1974, Astron. Astrophys. 36, 461
Bottinelli, L., Gouguenheim, L., Paturel, G., de Vaucouleurs, G.: 1980,
    Astrophys. J. 242, L153
Bottinelli, L., Gouguenheim, L., Paturel, G.: 1982, Astron. Astrophys.
    Suppl. 47, 171
Bottinelli, L., Gouguenheim, L., Paturel, G., de Vaucouleurs, G.: 1983,
    Astron. Astrophys. 118, 4
Briggs, F. H., Wolfe, A. M., Krumm, N., Salpeter, E. E.: 1980, Astrophys.
    J. 238, 510
Briggs, F. H.: 1982 in The Comparative HI-Content of Normal Galaxies,
    p. 50, NRAO, ed. M. Haynes, R. Giovanelli
Casertano, S. P. R., Shostak, G. S.: 1980, Astron. Astrophys. 81, 371
Chincarini, G. L.: 1982 in The Comparative HI-Content of Normal Galaxies,
    p. 93, NRAO, ed. M. Haynes, R. Giovanelli
Chincarini, G. L., Giovanelli, R., Haynes, M. P.: 1979, Astron. J. 84,
    1500
Chincarini, G. L., Giovanelli, R., Haynes, M. P., Fontanelli, P.:
    1983a, Astrophys. J. 267, 511
Chincarini, G. L., Giovanelli, R., Haynes, M. P.: 1983b, Astrophys. J.
    269, 13
Cowie, L. L., Songaila, A.: 1977, Nature 266, 501
Davies, R. D., Lewis, B.M.: 1973, Mon. Not. Roy. ast. Soc. 165, 231
de Vaucouleurs, G., de Vaucouleurs, A., Corwin, H. G., Jr.: 1976, (RC2),
    Second Reference Catalogue of Bright Galaxies, Univ. of Texas Press,
    Austin
de Vaucouleurs, G., Buta, R., Bottinelli, L., Gouguenheim, L., Paturel,
    G.: 1982, Astrophys. J. 254, 8
Dickel, J. R., Rood, H. J.: 1978, Astrophys. J. 223, 391
Faber, S. M., Gallagher, J. S.: 1979, Ann. Rev. Astron. Astrophys. 17, 135
Farouki, R., Shapiro, S. L.: 1980, Astrophys. J. 241, 928
Fesenko, B. I.: 1973, Sov. Astron. 17, 314
Fisher, R. J., Tully, R. B.: 1981, Astrophys. J. 243, L23
Forman, B., Jones, C.: 1982, Ann. Rev. Astron. Astrophys. 20, 547
Gallagher, J. S., Knapp, G. R., Faber, S. M.: 1981, Astron. J. 86, 1781
Giovanardi, C., Helou, G., Salpeter, E. E., Krumm, N.: 1983a, Astrophys.
    J. 267, 35
Giovanardi, C., Krumm, N., Salpeter, E. E.: 1983b, Astrophys. J. in press
Giovanelli, R., Haynes, M.: 1981, Astrophys. 247, 397

Giovanelli, R., Haynes, M. P., Chincarini, G. L.: 1982, Astrophys. J.
      262, 442
Giovanelli, R., Chincarini, G. L., Haynes, M. P.: 1981, Astrophys. J.
      247, 383
Giovanelli, R., Haynes, M. P.: 1983, Astron. J. 88, 881
Gordon, D., Gottesman, S. T.: 1981, Astron. J. 86, 161
Gunn, J. E., Gott, J. R.: 1972, Astrophys. J. 176, 1
Hart, L., Davies, R. D., Johnson, S. C.: 1980, Mon. Not. Roy. ast. Soc.
      191, 269
Haynes, M. P.: 1981, Astron. J. 86, 1126
Haynes, M., Roberts, M. S.: 1979, Astrophys. J. 227, 767
Haynes, M., Giovanelli, R., Roberts, M. S.: 1979, Astrophys. J. 229, 83
Haynes, M., Giovanelli, R.: 1980, Astrophys. J. 240, L87
Heckman, T. M., Balick, B., Sullivan, W. T.: 1978, Astrophys. J. 224,
      745
Helou, G., Giovanardi, G., Salpeter, E. E., Krumm, N.: 1981, Astrophys.
      J. Suppl. 46, 267
Helou, G., Salpeter, E. E., Terzian, Y.: 1982, Astron. J. 87, 1443
Helou, G., Hoffman, G. L., Salpeter, E. E.: 1983, preprint
Hewitt, J. N., Haynes, M., Giovanelli, R.: 1983, Astron. J. 88, 272
Holmberg, E.: 1964, Arkiv Astron. 3, 387
Huchtmeier, W. K.: 1982, Astron. Astrophys. 110, 121
Huchtmeier, W. K., Tammann, G. A., Wendker, H. J.: 1976, Astron. Astro-
      phys. 46, 381
Huchtmeier, W. K., Bohnenstengel, H.-D.: 1981, Astron. Astrophys. 100,
      72
Huchtmeier, W. K., Richter, O.-G.: 1982, Astron. Astrophys. 109, 331
Huchtmeier, W. K., Richter, O.-G.: 1983, Astron. Astrophys. submitted
Karachentsev, I. D.: 1971, Sov. Astron. 15, 171
Kennicutt, R. C., Jr.: 1983, preprint
Kotanyi, C.: 1981, Ph.D. Thesis, Univ. of Groningen
Kotanyi, C.: 1984, this volume, p. 261.
Kraan-Korteweg, R. C. 1983, Astron. Astrophys. 125, 109
Kraan-Korteweg, R. C., Tammann, G. A.: 1979, Astron. Nachr. 300, 181
Krishna-Kumar, C., Thonnard, N.: 1983, Astron. J. 88, 260
Krumm, N., Salpeter, E. E.: 1979, Astrophys. J. 228, 64
Krumm, N., Shane, W. W.: 1982, Astron. Astrophys. 116, 237
Li Zong-yun, Liu Ru-liang: 1981, Chin. Astr. Astrophys. 5, 205
Lo, K. Y., Sargent, W. L.: 1979, Astrophys. J. 227, 756
Materne, J., Huchtmeier, W. K., Hulsbosch, A. N. M.: 1979, Mon. Not.
      Roy. ast. Soc. 186, 563
Mathews, W. G., Baker, J. C.: 1971, Astrophys. J. 170, 241
Meisels, A.: 1983, Astron. Astrophys. 118, 21
Miley, G. K., Perola, G. C., van der Kruit, P. C., van der Laan, H.:
      1972, Nature 237, 269
Neyman, J., Scott, E. L.: 1961, Astron. J. 66, 148
Nilson, P.: 1973, Uppsala General Catalogue of Galaxies, Uppsala Astr.
      Obs. Ann. Vol. 6
Nulsen, P. E. J.: 1982, Mon. Not. Roy. ast. Soc. 198, 1007
Peterson, B. M., Strom, S. E., Strom, K. M.: 1979, Astron. J. 84, 735
Reif, K.: 1982, Ph. D. Thesis, Univ. of Bonn

Reif, K., Mebold, U., Goss, W. M., van Woerden, H., Siegman, B.: 1982, Astron. Astrophys. Suppl. 50, 451

Richstone, D. O.: 1976, Astrophys. J. 204, 642

Richter, O.-G.: 1982, Ph. D. Thesis, Univ. of Hamburg

Richter, O.-G., Huchtmeier, W. K.: 1982, Astron. Astrophys. 109, 155

Richter, O.-G., Huchtmeier, W. K.: 1983a, Astron. Astrophys. 125, 187

Richter, O.-G., Huchtmeier, W. K.: 1983b, ESO preprint no. 255

Rivolo, A. R., Yahil, A.: 1982, preprint

Roberts, M. S.: 1969, Astron. J. 74, 859

Roberts, M. S.: 1975 in Stars and Stellar Systems, Vol. IX, p. 309, ed. A. Sandage, M. Sandage, J. Kristian, Univ. of Chicago Press

Roberts, M. S.: 1978, Astron. J. 83, 1026

Romanishin, W., Krumm, N., Salpeter, E. E., Knapp, G., Strom, K. M., Strom, S. E.: 1982, Astrophys. J. 263, 94

Rubin, V. C., Ford, W. K., Thonnard N.: 1978, Astrophys. J. 225, L107

Rubin, V. C., Burstein, D., Thonnard, N.: 1980, Astrophys. J. 242, L149

Rubin, V. C., Ford, W. K., Thonnard, N., Burstein, D.: 1982, Astrophys. J. 261, 439

Sandage, A., Tammann, G. A.: 1976, Astrophys. J. 210, 7

Schneider, S. E., Helou, G., Salpeter, E. E., Terzian, Y.: 1983, Astrophys. J. 273, L1

Schommer, R. A., Sullivan, W. T., Bothun, G.: 1981, Astron. J. 86, 943

Shostak, G. S.: 1978, Astron. Astrophys. 68, 321

Shostak, G. S., Gilra, D. P., Noordam, J. E., Nieuwenhuijzen, H., de Graauw, T., Vermue, J.: 1980, Astron. Astrophys. 81, 223

Shostak, G. S., Hummel, E., Shaver, P. A., van der Hulst, J. M., van der Kruit, P. C.: 1982, Astron. Astrophys. 115, 293

Spitzer, L., Baade, W.: 1951, Astrophys. J. 113, 413

Strom, S. E.: 1980, Astrophys. J. 237, 686

Strom, S. E.: 1982 in The Comparative HI-Content of Normal Galaxies, p. 13, NRAO, ed. M. Haynes, R. Giovanelli

Sulentic, J. W., Arp, H.: 1983, Astron. J. 88, 489

Sullivan, W. T., Johnson, L.: 1978, Astrophys. J. 225, 751

Sullivan, W. T., Bothun, G., Bates, G., Schommer, R. A.: 1981, Astron. J. 86, 919

Tarter, J. C., Wright, M. C. H.: 1979, Astron. Astrophys. 76, 127

Thuan, T. X., Seitzer, P. O.: 1979, Astrophys. J. 231, 327+680

Thuan, T. X., Martin, G. E.: 1981, Astrophys. J. 247, 823

Toomre, A., Toomre, J.: 1972, Astrophys. J. 178, 623

Tully, R. B., Fisher, J. R.: 1977, Astron. Astrophys. 54, 661

Tsvetkov, D. Yu.: 1981, Sov. Astron. 25, no. 4

van den Bergh, S.: 1976, Astrophys. J. 206, 883

van Moorsel, G. A.: 1982, Ph. D. Thesis, Univ. of Groningen

Webster, B. C., Goss, W. M., Hawarden, T. G., Longmore, A. J.: 1979, Mon. Not. Roy. ast. Soc. 186, 31

Wilkerson, M. S.: 1980, Astrophys. J. 240, L115

Williams, B. A.: 1983, Astrophys. J. 271, 461

Williams, B. A., Kerr, F. J.: 1981, Astron. J. 86, 953

Zwicky, F.: 1933, Helv. Phys. Acta 6, 110

DISCUSSION

VAN WOERDEN: In discussing the diameter or extent of HI in galaxies, one should clearly define the isophote level (in radio terms: the column density, or rather: surface density, face-on) to which the diameter refers. Your HI diameter of Holmberg diameters in M83 is probably at a very low surface density.

HUCHTMEIER: At a surface density of 1.8×10^{20} atoms/cm^2 the HI extent of M83 is 3 times the Holmberg diameter. The extended HI halo reaches 6.5 times the Holmberg diameter at a level of 6×10^{18} atoms/cm^2 averaged over the telescope beam of 9 arc min.

GIOVANELLI: On the use by Hevitt et al. (1982) of an effective HI diameter that included 70% of the HI mass: with a beam which in some cases could barely resolve the disk of the galaxy, the situation is not the same as for optical photometry. We chose 70% as a compromise to stabilize the fitting procedure; at 50% the determination is too sensitive to uncertainties on the presence of a central depression in the HI distribution; at 90%, the determination is too sensitive to uncertainties on the radius at which the HI disk stops.

WARMELS: You are comparing the ratio of HI sizes to optical sizes for a sample of double galaxies (van Moorsel, thesis, 1982) and a sample of field galaxies. Did you take into account that on average the number of resolution elements van Moorsel had along the major axis is quite smaller than the average number in the field sample? this difference might influence your comparison.

HUCHTMEIER: You are completely right. Resolution effects have to be taken into account. The point is that the binary sample (van Moorsel) got HI extents not much smaller (may be 10%) than the average of the field sample.

GOUGUENHEIM: Professor de Vancouleurs mentions the interest of using effective diameters. I wish to recall that effective HI diameters have been actually used by Fouque in a work recently published in Astronomy and Astrophysics, dealing with more than one hundred galaxies. His main result is that the ratio of this effective HI diameter to optical one is tightly correlated with morphological type.

HUCHTMEIER: Thank you very much for your comment.

CLUSTER EVOLUTION AND SWEEPING: A COMPARATIVE STUDY OF NEARBY CLUSTERS

RICCARDO GIOVANELLI
NAIC, Arecibo Observatory

1. INTRODUCTION

To what an extent is the evolution of galaxies affected by the characteristics of their environment and, more specifically, can the morphological type segregation observed in clusters be in some measure the result of mechanisms made efficient by the high galaxian density? Numerous lines of evidence support the view that morphological type differentiations are inbred in the galaxian population: they must have occurred at the time of galaxy formation or forced in very early phases of evolution by an active environment, as reviewed by Dressler (1984, these proceedings). On the other hand, observations in the X-ray, optical and radio domains have accumulated that suggest that the intracluster medium (IGM) is capable of severely affecting the structure of galaxies that venture in its denser parts. In particular, 21 cm line observations indicate that in several clusters spiral galaxies appear impoverished of their HI content, a circumstance that, in absence of mechanisms that would replenish their interstellar medium, would halt star formation and lead to fading of the spiral structure. Doubts, however, have been cast on these results, on several grounds:

(i)   uncertainties in the measurement of the HI content of a galaxy depend on the operational definition of that quantity, and are in some cases very large;

(ii)   large uncertainties or biases have also been attributed to the values of optical parameters, such as luminosity, size and type, of galaxies in more distant clusters; such parameters are used to estimate the deviation from normalcy of a galaxy's HI content;

(iii)   observed samples in several clusters contain numerous non-detected galaxies with unsatisfactory upper limits for their fluxes;

(iv)   comparison samples, which are used to define the standard of normalcy, have not always been rigorously chosen.

In this paper, an attempt is made to overcome most of those difficulties, and the evidence for HI deficiency in clusters revisited. The work was carried on in collaboration with Martha Haynes.

F. Mardirossian et al. (eds.), Clusters and Groups of Galaxies, 243–249.
© *1984 by D. Reidel Publishing Company.*

2. THE SAMPLE

Table 1 lists nine clusters, and a summary of their parameters, for which a large amount of data has been collected at Arecibo by various authors (cf Huchtmeier 1983 for references). Data in the literature were combined with new observations with the 305m telescope to create a pool from which samples for each cluster were generated which include only galaxies of type Sa or later (peculiars excluded) and non detections with an inferred HI deficiency of at least a factor of 2. For Coma, A1367, A2147, A2151 and A262, types and angular sizes were obtained from high scale KPNO 4m and Las Campanas material kindly provided by G. Chincarini, A. Dressler and S. Strom. The galaxies in the final samples for each cluster, projected at the same angular scale, are plotted in figure 1 (the circle in each case identifies the points at one Abell radius, AR, which is listed in column 6 of table 1).

Table 1. Cluster Parameters

Name	RA hhmm.m	Dec ddmm	v km/s	Δv km/s	AR deg	f	$\log L_x$ erg/s
A262	0149.9	3555	5068	452	1.75	0.57	43.19
Cancer	0817.5	2114	4607	317	1.92	0.23	<42.6
A1367	1141.4	2006	6370	813	1.40	0.47	43.46
Virgo	1225.4	1240	1026	673	5.00	0.62	42.89
A1656	1257.2	2814	6950	905	1.29	0.82	44.23
Z74-23	1400.4	0934	5840	412	1.52	0.00	<42.8
A2147	1559.7	1611	10867	1189	0.83	0.64	44.05
A2151	1603.0	1756	11055	920	0.83	0.33	43.58
Pegasus	2318.0	0755	3990	616	2.20	0.07	<42.5

3. HI CONTENT

First, one must accurately define a comparison sample that sets the standard of normalcy. As such we used the sample of isolated galaxies of Haynes and Giovanelli (1983b) which, although not veritable field objects since they outline the low density regions of superclusters (Haynes and Giovanelli 1983a), represent a population as unaffected as one can find by the sweeping mechanisms whose effects we aim to investigate.

It has been common practice in the past to adopt the distance independent ratio M_H/L as the indicator of HI content. Within that scheme, the HI deficiency, Def, of a galaxy of type T is given by the difference between the average value of $\log(M_H/L)$ of all comparison sample galaxies of type T and the galaxy's $\log (M_H/L)$. This technique has shortcomings. The scatter of $\log(M_H/L)$ values around the mean is very large; moreover, a residual dependence on L exists, as first suggested by Bottinelli and Gouguenheim (1974), which, if ignored, will affect the analysis of distant clusters. We choose to measure Def by comparing the hybrid surface density M_H/D^2, where D is an optical diameter, expected for an isolated galaxy with that of the sample galaxy.

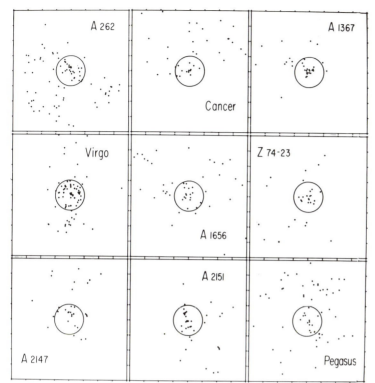

Figure 1.
Sample galaxies
for each cluster,
all projected at
the some distance.
The radius of the
circle in each
case is one Abell
radius.

Figure 2.
HI deficiency
histograms for the
galaxies in figure
1. The shaded part
of the histogram
refers to the
galaxies within
one Abell radius
from the cluster
center.

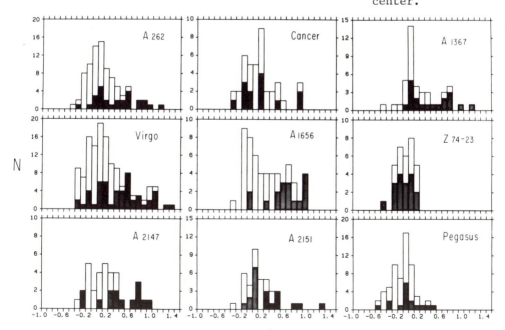

This approach has numerous advantages, the main ones being that the
dependence on morphological type, for spirals, is nearly negligible and
it allows to gauge Def with an rms accuracy of about 0.22 (Def is a
logarithmic quantity). For further details, see Haynes and Giovanelli
(1983b).

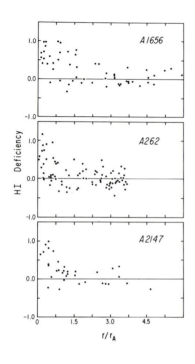

Figure 3.
Radial behavior of the HI
deficiency pattern for three
clusters: A1656 (Coma), A262 and
A2147 (in Hercules). In each case
the abscissa is scaled by the
cluster's Abell radius. Notice
that HI deficiency is a logarithmic
quantity, positive for HI deficient
galaxies. Galaxies which are not
detected are plotted at the
nominal lower limit of HI
deficiency.

4. HI DEFICIENCY

Figure 2 shows histograms of Def for the galaxies in the nine
clusters. In each case, the shaded part of the histogram represents the
subsample of galaxies projected within 1 AR from the cluster center.
The central galaxies of A1656, A2147, A1367, A262 and Virgo show a
pronounced tendency to be deficient, A2151 somewhat less markedly so.
The radial extent of the deficiency pattern is different for various
clusters. In figure 3, three of them are shown: in the case of A1656,
deficient galaxies are observed to a radial distance of 1.5 AR; in the
other clusters the effect is limited to half that far or less. In
figure 4, a "composite" cluster sample is built by combining the samples
of the six deficient clusters mentioned above, after having projected
them all at the same distance; then to each cell of this projected
distribution a Def is assigned as the distance-weighted average of Def
of the 4 nearest galaxies to that cell. Def, proportional to shade
intensity, clearly grows toward the center.

In order to compare the various clusters, we define the
"deficient fraction" f of a cluster as the ratio of the number of

galaxies with Def > 0.3 found within 1 AR, to that of all galaxies observed within the same radius, and list it in table 1. In figure 5, we plot f versus the X-ray luminosity of the cluster in the 0.5 to 3.5 keV range, L_x. HI deficiency appears to be well correlated with L_x. For its L_x, Virgo appears to have a rather high f. This circumstance may be partly the result of the following bias: because Virgo is close to us, we sample a larger number, and hence on the average, smaller galaxies than in other clusters; those galaxies may be more vulnerable to sweeping mechanisms and more likely to show high Def, thus inflating f.

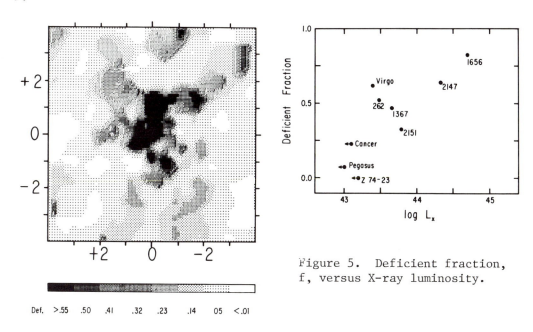

Def. >.55 .50 .41 .32 .23 .14 05 <.01

Figure 5. Deficient fraction, f, versus X-ray luminosity.

Figure 4. Average HI deficiency, as a function of projected location in the sky, for a "composite" cluster obtained combining the samples of Virgo, A262, A1367, A1656, A2147 and A2151, each scaled by its respective Abell radius. Coordinate units are Abell radii. The wealth of "structure" in the deficiency pattern is partly due to the insufficient sampling of some regions of the plane.

5. THE SO PROBLEM

In clusters that show HI deficiency, the number of deficient galaxy is usually very high, so much as to suggest that the sweeping mechanisms responsible for the observed deficiency have been active for a short time. In Virgo, Def is correlated with (B-V) (Kennicutt 1984) and (U-V) (Giovanelli and Haynes 1983). The indication is that galaxies in X-ray clusters lose gas so that star formation ceases and their spiral

structure fades. Hence one should observe, at the same local galaxian
density, an enhancement of the population of lenticular galaxies, and a
depression of that of spirals, in clusters that have undergone this gas
sweeping phase, with respect to that of all clusters. In 1980 Dressler
noticed precisely this effect (see his figures 4 and 10): the
percentage of lenticular galaxies, at any given galaxian density, is
systematically higher in clusters with log L_x > 44 than for his total
sample, while the percentage of spiral galaxies is systematically low,
by the same amount. One thus infers that the presence of a healthy IGM
does, by way of mechanisms that strip the galaxies of their interstellar
gas, affect their evolution and further enhances the already existing
pattern of morphological segregation.

    In conclusion, HI deficiency is seen clearly in several clusters of
galaxies. The magnitude and extent of the deficiency pattern are
correlated with cluster's X-ray luminosity, suggesting strong coupling
between IGM and cluster galaxy evolution. The morphological type
segregation may then be enhanced as a result of that coupling.

    The Arecibo Observatory is part of the National Astronomy and
Ionosphere Center, operated by Cornell University under contract with
the National Science Foundation.

REFERENCES

Bottinelli, L. and Gouguenheim, L. 1974, Astr. Ap. **36**, 461
Dressler, A. 1980, Ap. J. **236**,, 354.
Dressler, A. 1984, this volume, p. 117.
Giovanelli, R. and Haynes, M.P. 1983, A.J. **88**, 881
Haynes, M.P. and Giovanelli, R. 1983a, Ap. J. in press.
Haynes, M.P. and Giovanelli, R. 1983b, preprint
Huchtmeier, 1984, this volume, p. 221.
Kennicutt, R.C. 1983, preprint

DISCUSSION

FINZI: Can you make a comparison between the mass of HI that is missing
in the galaxies of a cluster, and the mass needed to produce the
observed X-ray emission?

GIOVANELLI: The HI mass missing from the galaxies is only a very
small fraction of the ICM inferred from X-ray observations.

VAN WOERDEN: Since galaxy diameters may be affected by the environment
more strongly than luminosities, is it not preferable to scale HI
content by luminosity: M_H/L_B, with a correction for L dependence,
rather than to take M_H/A^2?

GIOVANELLI: First, the HI content measure we advocated here consists in using fluxes and angular diameters. Not M_H/A^2. Answer to your question is no for various reasons: (a) optical diameters may be smaller in Virgo, but not in Hercules (according to Peterson and the Stroms) so the effect is not clearly ubiquitous; (b) luminosities may also be a function of local density. We have some data (though not conclusive) that suggest that luminosities of galaxies within a given type may be higher in denser regions (c) removing the L residual dependence of M_H/L with introduce a distance dependence, which the choice of M_H/L tried to do away with the first place; (d) the residuals of the M_H/L versus L relation are quite a bit larger than those of the fit of flux vs. angular diameter the first yielding then an inferior diagnostic tool than the second.

PERSIC: Maybe I've missed the point, but there exists any correlation between HI-deficiency and cluster size?

GIOVANELLI: If you mean between extent to which HI deficiency is observed and cluster characteristics, the answer is yes. HI deficiency is observed in Coma out to ∿2 Abell radii, while in all other cases, the extent where deficiency is observed is much smaller. Coma has a very healthy ICM.

COMPARISON OF HI SIZES OF VIRGO CLUSTER AND FIELD GALAXIES

Rein H. Warmels and Hugo van Woerden
Kapteyn Astronomical Institute, University of Groningen,
Postbox 800, 9700 AV Groningen, The Netherlands.

SUMMARY

With the Westerbork Synthesis Radio Telescope we have measured the HI distributions of 35 galaxies in the region of the Virgo Cluster. For 19 Sab to Sd-type galaxies we present the ratio of the face-on corrected HI diameter D_{HI} to the face-on corrected optical diameter D_o. Comparison with a sample of field galaxies suggests that the relative HI diameter of Virgo Cluster galaxies are smaller by about 20%. Within the present Virgo data set, we find no correlation between D_{HI}/D_o and the projected distance to M 87.

1. INTRODUCTION

Both theoretical and observational arguments indicate that environmental effects may play a major role in the formation, structure and evolution of galaxies. Hence, in studies of global properties of galaxies such as size, colour, morphology, gas content etc., it is important to compare galaxies in clusters and in the field.

During the last decade the neutral hydrogen properties of galaxies in clusters have been studied extensively. Results obtained sofar (e.g. Sullivan et al., 1981) suggest that the ratio of HI content to luminosity tends to be smaller in regions of higher galaxy density. Giovanelli and Haynes (1983) note that spiral galaxies in clusters containing an extended X-ray source show a more pronounced HI deficiency than spirals in clusters without such a source. This correlation may suggest that the X-ray source is fuelled by the missing HI gas. It is however not at all clear what processes reduce the gas content: collisions (including tidal interactions), mergers or sweeping mechanisms. Detailed comparison of the gas distributions and motions in cluster and field galaxies may help to answer this question.

The Virgo cluster is especially suitable for a study of the HI distributions: it is nearby, contains galaxies of a variety of

251

F. Mardirossian et al. (eds.), Clusters and Groups of Galaxies, 251–259.

morphological types, has been partly mapped in X-rays by the Einstein
satellite (Forman et al., 1981), and additional data (e.g. velocities,
magnitudes, colours etc.) are available for a large number of cluster
members.

In the last few years several studies of the HI content and size
of Virgo Cluster galaxies and based on single dish instruments have
been published. Recently Giovanelli and Haynes (1983) showed that the
HI disk size of galaxies in the Virgo Cluster core with a given optical
diameter is on average a factor 1.6 smaller than the corresponding disk
size of a field galaxy. They also have measured an average HI
deficiency of a factor 2.7 for the core galaxies with respect to their
field sample. These findings confirm the results by Chamaraux et al.
(1980) who found a deficiency factor of 2.2, and those of Giovanardi et
al. (1983) who found a 26% decrease in HI disk size for Virgo Cluster
galaxies inside a 6° radius compared with galaxies outside.

The observations at Arecibo and those at Nançay (Chamaraux et
al.), having angular resolutions of 3-4 arcmin in two dimensions or in
only one, barely resolve the HI distributions in Virgo Cluster
galaxies. Also, the disk size determinations are based on the
assumption of gaussian gas distributions inside both cluster and
comparison galaxies. Since it is unclear how the (yet unknown) removal
mechanisms may have affected the HI distributions, the sizes derived
should be treated with caution.

With the Westerbork Synthesis Radio Telescope (WSRT) we have
measured the HI distributions in 35 galaxies in the Virgo Cluster with
~25 arcsec resolution in one dimension. For 19 galaxies we have
compared the HI diameters at a specific (face-on) surface density
level, σ_{HI}, with a sample of field galaxies.

2. OBSERVATION AND REDUCTION

Full synthesis mapping with the Westerbork Synthesis Radio
Telescope (WSRT, Bos et al., 1981) in the Virgo Cluster area would
provide a synthesized beam of 0.2 arcmin in right ascension and about 1
arcmin in declination using the 3 km array. Because of the coarse
resolution in declination the observing time was limited to 2 × 2 hours
per galaxy rather then 12 hours. The Digital Line Backend (DLB) was
used to observe 31 frequency channels over bandwidth ranging from 0.625
to 5.0 MHz, and selected per galaxy, resulting in a channel spacing of
8.3 to 66 km/s. Each of the short observations then provides 31 monochromatic brightness distributions ("channel maps"), $T_b(p,q)$, having
high angular resolution along one axis (the resolution or p-axis) and a
low resolution perpendicular to that (q-axis); the position angle of
the resolution axis and the resolution obtained depend on the time of
observation.

After subtraction of the continuum emission we integrated each of the channel maps in the poorly resolved direction. This procedure results in a position-velocity (p-V) map showing the projection of the HI distribution upon the resolution axis (p) for each observed velocity channel (V). To remove sidelobe effects in this map due to the limited observing time and to correct the baseline level we applied the procedure clean (Schwarz, 1978) to the map. Occasionally we smoothed in velocity and/or position to increase the signal-to-noise ratio.

In order to derive the radial distribution of HI surface density, $\sigma_{HI}(r)$, we integrate the p-V map in velocity, resulting in the integrated HI distribution projected upon the resolution axis, to be called the strip integral, $S_{HI}(p)$. Given the position angle of the resolution axis, the orientation of the galaxy (i.e. inclination and position angle) and $S_{HI}(p)$, we compute $\sigma_{HI}(r)$, using the iterative method described by Lucy (1974) under the assumptions of azimuthal symmetry and planar distribution of the HI. Since this method can be applied to both sides of the strip distribution with respect to the galaxy centre, asymmetries in $\sigma_{HI}(r)$ between two halves of the galaxy can be calculated.

3. RESULTS

In Figures 1, 2 and 3 we present the results of two observations of NGC 4254 obtained with different orientations of the resolution axis. Figure 1 shows the orientations of the resolution axis super-imposed on a size IIIa-J plate. In Figure 2 the position-velocity maps together with the strip integrals are given. Finally Figure 3 shows the average radial distribution, $\sigma_{HI}(r)$, obtained by combining the results of the two separate observations. Note the asymmetry of the HI distribution in the outer parts of the galaxy; at the level of 10^{20} atoms/cm^2 the western part of the galaxy extends about 0.5 arcmin farther out than the eastern part.

At present radial distributions have been obtained for a total of 19 galaxies in the Virgo Cluster. The morphological types of these galaxies range from Sab to Sd; their absolute magnitudes ranges from −19 to −22. We restricted ourselves to galaxies for which our spatial resolution is such that the ratio optical diameter/beamsize is greater than about 4. For these galaxies we have determined the HI diameter, D_{HI}, defined as the diameter of the surface density distribution at a level of $\sigma_{HI} = 1.82 \times 10^{20}$ atoms/cm^2, and compared this to the face-on, extinction-corrected diameter D_0 from the Second Reference Catalogue (de Vaucouleurs et al., 1976).

Table 1 lists the results of the 19 objects; in column (1) the NGC number; in column (2) the morphological type (Sandage and Tammann, 1981); in column (3) the extinction corrected face-on optical diameter, D_0, (de Vaucouleurs et al., 1976); in column (4) the face-on HI dia-meter just defined, D_{HI}, and in column (5) the distance to the galaxy from M 87 on the sky.

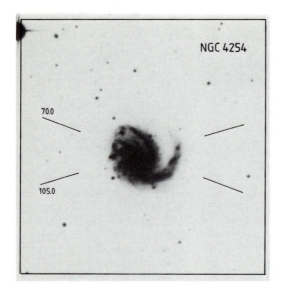

Figure 1 – Orientations of the resolution axis of two observations of NGC 4254 superimposed on a IIIa-J print taken by Dr. J. Jones.

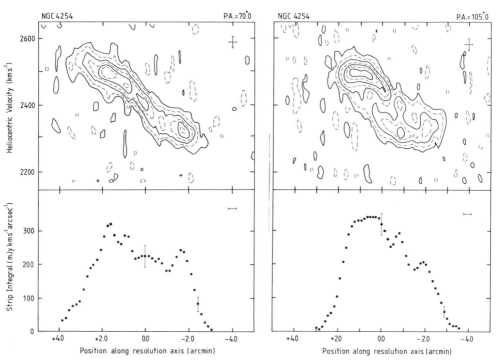

Figure 2 – Left top: Position-velocity (p-V) map of the observation along p.a. = 70°. The resolution both in velocity and in position is indicated; east is at the left; west at the right. Left bottom: Strip integral $S_{HI}(p)$ of the HI obtained from the p-V map by integration in velocity. Right top: p-V map of the observation along p.a. = 105°. Right bottom: Strip integral obtained from the p-V map.

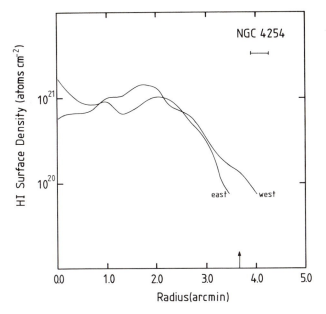

Figure 3 - Surface density distribution of the HI, $\sigma_{HI}(r)$, in NGC 4254 obtained by combining the results of the two observations presented in Figure 2.

TABLE 1

OPTICAL AND HYDROGEN DIAMETERS OF VIRGO CLUSTER GALAXIES

NGC (1)	Type (2)	D_o (arcmin) (3)	D_{HI} (arcmin) (4)	R_{M87} (degrees) (5)
4116	SBc	3.5	5.2	11.3
4152	Sc	2.3	3.4	6.2
4178	SBc	4.2	7.0	4.8
4189	SBc	2.5	3.3	4.4
4254	Sc	5.5	7.3	3.6
4294	SBc	2.6	3.8	2.5
4299	Sd	1.7	2.4	2.5
4303	Sc	6.2	7.8	8.2
4321	Sc	7.1	6.8	4.0
4519	SBc	3.0	5.2	3.8
4535	SBc	6.6	8.0	4.3
4579	Sab	5.4	3.6	1.8
4639	SBb	2.8	3.7	3.1
4647	Sc	3.0	2.5	3.3
4651	Sc	3.6	3.7	5.1
4654	SBc	4.4	5.6	3.4
4689	Sc	4.1	2.3	4.5
4713	SBc	2.6	3.4	8.5
4900	Sc	2.4	2.2	12.4

Figure 4a shows the frequency distribution of D_{HI}/D_o for these galaxies. For this ratio we find an average value of 1.22 with a 1σ scatter of 0.31.

From published HI synthesis data we have constructed a comparison sample of 25 galaxies, using the same selection criteria we applied for the Virgo Cluster galaxies. For most galaxies in this sample D_{HI} was given in some form by the authors; in some cases however, we had to derive D_{HI} from the published HI maps. The frequency distribution statistics of D_{HI}/D_o for the comparison sample is given in Figure 4b. For the ratio D_{HI}/D_o we here find an average of 1.50 with a 1σ scatter of 0.57. If we restrict the comparison sample to Westerbork observations only, this ratio drops to 1.47±0.55 (non-Westerbork observations are indicated by the hatched boxes).

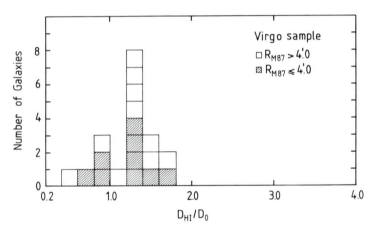

Figure 4a – Frequency distribution of D_{HI}/D_o for 19 galaxies in the Virgo Cluster. Galaxies in the core are represented by the hatched boxes.

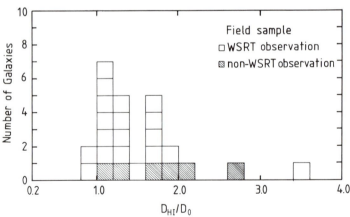

Figure 4b – Frequency distribution of D_{HI}/D_o for 25 field galaxies. Non-Westerbork observation are indicated by the hatched boxes.

4. DISCUSSION

High-resolution HI observations have shown that the ratio of HI diameter and optical diameter does not strongly depend on morphological type (Bosma, 1981). For this reason and because of the small number of objects we have in the present analysis not taken account of possible type dependence. We also have neglected the small differences in luminosity between the Virgo and field samples. We discuss the comparison of the D_{HI}/D_o ratios for Virgo Cluster galaxies with those of the field sample; thereafter we comment on the radial gradient of D_{HI}/D_o within the Virgo Cluster itself.

4.1. Comparison between Virgo Cluster and field

The comparison between the frequency distributions of the Virgo Cluster and field sample (Figures 4a, 4b and Table 2) reveals the following differences:
a) The HI diameter D_{HI} (the level of $1.82 * 10^{20}$ atoms/cm^2) for the Virgo Cluster galaxies with a given optical diameter D_o, is on average 22% smaller than for galaxies in the field with the same D_o. This difference drops to 17% if we exclude field galaxies not observed with the WSRT.
b) The D_{HI}/D_o distribution for the sample of field galaxies shows a number of galaxies with high D_{HI}/D_o values (>1.8); we do not find such high values in our sample of Virgo galaxies.
c) The D_{HI}/D_o distribution of the Virgo sample shows some galaxies with low D_{HI}/D_o values (<1.0). Although the statistics is clearly not strong on this point, we find that this part of the distribution is less prominent in the field sample.

Since the bulk of the two distributions looks rather similar, we might conclude that on average the environmental effect on the HI sizes of galaxies in our cluster sample are not drastically different from those on galaxies in the field. The field galaxies with high D_{HI}/D_o values may then be explained as systems which have suffered no significant environmental effects, while the galaxies with low D_{HI}/D_o values in the Virgo Cluster might indicate recent stripping processes.

Our comparison of the ratio D_{HI}/D_o between the Virgo Cluster sample and the field sample is in good agreement with the results of Giovanelli and Haynes (1983). Excluding their upper limits for 5 galaxies, they measured an average decrease of 20% in the HI size of Virgo Cluster galaxies with respect to a field sample. In an earlier, preliminary paper (Warmels and van Woerden, 1983) we announced a 35% difference between Virgo Cluster and field galaxy sizes. The difference is due to a) the larger size of the present sample and b) and improved method of diameter determination.

4.2. Radial Gradients in the Virgo Cluster

    In Figure 4a we have indicated galaxies which are located in the
cluster core, i.e. within four degrees of M 87. In the present Virgo
sample of 19 galaxies we do not find a difference between the HI sizes
of galaxies in the cluster core and those in the outer regions: D_{HI}/D_o
= 1.22±0.30 for the cluster core galaxies, D_{HI}/D_o = 1.23±0.32 for those
outside the 4° radius (cf. Table 2). This result seems to be in contra-
dication with the radial dependence of d_{HI}/d_{Ho} which was found by
Giovanelli and Haynes (1983) and by Giovanardi et al. (1983). However
this difference might be the result of the different methods of
determining the HI sizes (effective diameter for the Arecibo results;
isophotal diameter for our measurements).

TABLE 2

HI SIZES IN THE VIRGO CLUSTER AND IN THE FIELD

Sample	Distance range to M 87	Average distance to M 87	Number of galaxies	Average D_{HI}/D_o and dispersion
Virgo	1.8-12.8°	–	19	1.22±0.31
	1.8- 4.0	2.8	9	1.22±0.30
	4.0-12.8	7.0	10	1.23±0.32
Field			25	1.56±0.57
Field (WSRT observation only)				1.47±0.55

5. CONCLUSIONS

    We have sofar analysed the HI sizes of 19 Virgo Cluster galaxies
and compared the results with HI sizes of field galaxies. We find on
average Virgo Cluster galaxies have 20% smaller HI sizes than field
galaxies; the difference shows no radial dependence within the cluster
itself. However it is clear that the remaining 16 galaxies of our
observing program and the existing data on 16 additional field galaxies
will help to obtain a more accurate result. Another item to be
considered is the possible difference in the radial distributions of
the neutral hydrogen between cluster and field galaxies. Analyses of
this kind of the WSRT data are in progress.

6. ACKNOWLEDGEMENTS

    We wish to thank Renzo Sancisi and Seth Shostak for many
stimulating discussions. We also thank Bart Wevers for providing us
with HI data of the Palomar-Westerbork survey prior to publication. RHW
acknowledges support by the Netherlands Foundation for Astronomical

Research (ASTRON). The Westerbork Radio Observatory is operated by the
Netherlands Foundation for Radio Astronomy, with financial support from
ZWO.

7. REFERENCES

Bos, A., Raimond, E., van Someren Greve, H.W. 1981, Astron. Astrophys.
    98, 251
Bosma, A. 1981, Astron. J. 86, 1825
Chamaraux, P., Poalkowski, C., Gerard, E. 1980, Astron. Astrophys. 83,
    38
Forman, W., Jones, C. 1982, Ann. Rev. Astron. Astrophys. 20, 547
Giovanelli, R., Haynes, M.P. 1983, Astron. J. 88, 881
Giovanardi, C., Helou, G., Salpeter, E.E., Krumm, N. 1983, Astrophys.
    J. 267, 35
Lucy, L.B. 1974, Astron. J. 79, 745
Sandage, A., Tammann, G.A. 1981, Revised Shapley-Ames Catalogue of
    Bright Galaxies, Carnegie Institute, Washington
Sullivan, W.T., Bothun, G.D., Bates, B., Schommer, R.A. 1981, Astron.
    J. 80, 919
de Vaucouleurs, G., de Vaucouleurs, A., Corwin, H.G. 1976, Second
    Reference Catalogue of Bright Galaxies, University of Texas Press,
    Austin
Warmels, R.H., van Woerden, H. 1983, IAU Symposium No. 106, Early
    Evolution of the Universe and Its Present Structure, eds. G.O.
    Abell, G. Chincani, Reidel Publishing Company, p. 261

HI DEFICIENCY IN THE VIRGO CLUSTER: VLA OBSERVATIONS

J.H. Van Gorkom, NRAO, Socorro, U.S.A.
C. Balkowski, Observatoire de Paris, Meudon, France
C. Kotanyi, ESO, Garching b. München, F.R.G.

SUMMARY

    HI line observations of 10 bright spiral galaxies within 6° of the centre of the Virgo Cluster were made at a resolution of about 45" using the VLA. HI disks within 2° of the cluster centre are HI deficient and much smaller than the optical disks, whereas disks outside 3° are normal. NGC 4254 and NGC 4654 situated at about 2° show asymmetric HI and optical structures suggesting the action of ram-pressure stripping. It is concluded that stripping is nearly complete and instantaneous and may account for the Hubble type segregation of spiral galaxies within rich clusters.

    The neutral hydrogen deficiency of spiral galaxies in rich clusters is a prime evidence for environmental effects on galaxies. The deficiency was first discovered in the Virgo cluster (Lewis and Davies, 1973; Chamaraux et al., 1980) which is the nearest spiral-rich cluster. Observations at Arecibo indicate that the neutral hydrogen disks have a smaller size in the cluster centre, in addition to being strongly deficient (Giovanardi et al., 1983). Accurate sizes and the detailed structure of the emission can however not be measured through single-dish observations. We have therefore used the nearly completed spectral line system at the Very Large Array (VLA) to map the HI line in spiral galaxies in the Virgo Cluster. We present here the data on the brightest galaxies situated within a 6° radius of M87 (the cluster core).

    The VLA in the D configuration (base lines ranging from 40 to 700 m) was used to map 10 of the brightest galaxies in the cluster core. The galaxies are listed in Table 1, in order of increasing radial distance R from M87. The integration time was between one and two hours on each galaxy, depending on the expected signal. The number of 25 m antennas available was typically 20. The total frequency bandwidth was divided in 32 frequency channels separated by 20 km/s. The system temperature was 60°K. The resulting r.m.s. noise was about 2.5 mJy. The continuum

F. Mardirossian et al. (eds.), Clusters and Groups of Galaxies, 261–265.
© *1984 by D. Reidel Publishing Company.*

Table 1: HI observations of bright spiral galaxies

NGC	R (degrees)	D_H/D_0	Δ	Type	B_T	V_0 (km/s)
4438	1.0	0.2	1.28	S0apec	10.85	182
4388	1.3	0.3	1.06	Sb	11.83	2535
4402	1.4	0.6	0.61	Sb	12.50	-77
4569	1.7	0.3	0.99	Sab	10.23	-382
4579	1.8	0.2	1.00	Sb	10.61	1730
4501	2.0	0.9	0.47	Sb	10.27	1989
4654	3.4	1.3	0.00	Scd	11.10	970
4254	3.6	1.5	0.02	Sc	10.42	2324
4535	4.3	1.1	0.17	Sc	10.66	1853
4192	4.9	1.3	0.09	Sab	10.86	-206

emission was subtracted from each channel to give pure line emission maps which were cleaned and restored with a nearly circular gaussian beam with a HPW of typically 45".

The channel maps were added together applying a cutoff of 2 times the r.m.s. noise, giving a map of the total distribution of neutral hydrogen near each galaxy. These maps are collected in Fig. 1, in a composite showing the whole cluster core area. In this composite map the positions of the galaxy centres are marked by crosses. The position of M87, the approximate cluster centre, is also indicated. In each single galaxy map, the angular scale has been magnified with a factor of 11 relative to the global coordinate frame, in order to show clearly the individual structures. The lowest contour is at a level of 5×10^{19} atoms/ cm^2 throughout the composite. The contour increments are about 5×10^{20} atoms/cm^2 for maps falling outside 2° of M87, and about 10 times less for galaxies inside 2°. The peak fluxes within 2° are therefore nearly 10 times lower than those outside 2°, which is a reflection of the gas deficiency of the central galaxies.

Fig. 1 shows at a first glance that the neutral hydrogen disks close to the cluster centre are all much smaller than those further out. The HI sizes D_H are given in Table 1, normalized to the optical diameters D_0. D_H gives the major axis size of the lowest HI contours, and D_0 is the 25th magnitude isophote diameter from the RCBG2 (de Vaucouleurs et al., 1976). Galaxies within R = 2° have all $D_H/D_0 \ll 1$. Between R = 2° and $R \simeq 3°$, D_H/D_0 increases sharply to a normal value > 1. The deficiency factor Δ (taken from Chamaraux et al., 1980) is consistently high within R < 2° and close to 0 outside, i.e., outside $R \simeq 3°$ galaxies have essentially a normal HI content.

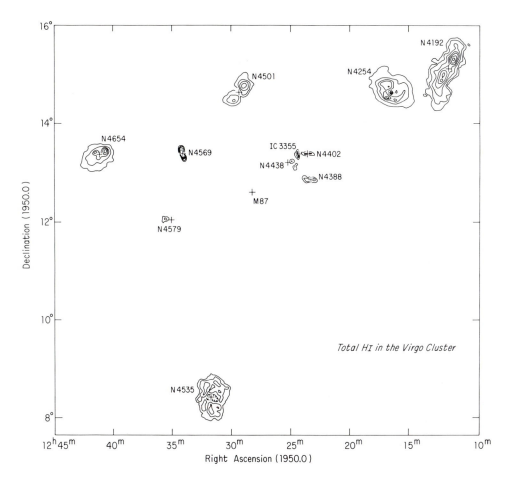

Figure 1:   VLA map of the HI in the core of the Virgo cluster

Whereas the HI content and size seem to vary with R, a variation of the Hubble type, and some variation of the magnitude are also apparent in Table 1. The well-known type and luminosity radial segregations in the cluster (e.g., Chamaraux et al., 1980) introduce selection effects which should be taken into account since the HI content M_H/L and size in spiral galaxies are known to be dependent on the type and luminosity (see Huchtmeier, this volume). We do not discuss these selection effects here, referring to Chamaraux et al., 1980, for an extensive discussion. We comment instead on the individual structures.

The observations indicate the existence of two regions sharply separated at R ≃ 2° to 3°. Near the separation R ≃ 3° two of the three galaxies, NGC 4254 and NGC 4654, have strongly asymmetric HI distributions. In these galaxies the HI maps show a sharp gradient on one side, with a smoother tail on the opposite side, pointing in a direction roughly opposite to the cluster centre in both cases. The optical structures show asymmetry similar to that of the neutral hydrogen. If R = 3° is the boundary of the hot intergalactic gas halo in the cluster, it might be that these two galaxies are in the process of being stripped of their interstellar gas content as a result of ram-pressure caused by their radial motion towards the cluster centre, as they meet the dense hot halo centred on M87 (see Forman and Jones, this volume). From these two cases it would seem that ram-pressure stripping affects not only the HI distribution, but the optical structure as well.

This hypothesis is supported by the case of NGC 4438, which may be an extreme case of stripping. This galaxy is the bright spiral galaxy nearest to M87, apparently undergoing heavy stripping, which should account for its strongly asymmetric optical appearance. An estimate of the ram-pressure indicates a stripping time scale $\tau \sim 10^8$ y (Kotanyi and Ekers, 1983; Kotanyi et al., 1983), which is short compared to the time t to cross the cluster, suggesting that the galaxy approaches the cluster centre on a nearly radial orbit and is rapidly stripped. The cases of NGC 4254 and NGC 4654 may be similar.

We note that both NGC 4438 and NGC 4654 have optical companions, and may be therefore affected tidally. However, ram-pressure as a cause of the asymmetry is strongly indicated in at least the case of NGC 4254, which has no obvious optical companion.

In summary, the observations presented here indicate that ram-pressure stripping is nearly complete within $R \simeq 3°$ and absent outside. We might conclude that the stripping process is complete and instantaneous (as compared with the cluster crossing time), as is suggested by the case of NGC 4438. Moreover the asymmetric optical morphologies suggest that the stripping does modify the Hubble type of spiral galaxies. Indeed, it is perhaps the one single cause for type segregation within a cluster.

Detailed analysis of the present data and synthesis HI observations of a larger sample of Virgo Cluster spirals, presently under way, will be presented elsewhere.

The National Radio Astronomy Observatory is operated by Associated Universities, Inc., under contract with the National Science Foundation.

REFERENCES

Chamaraux, P., Balkowski, C., Gerard, E.: 1980, Astron. Astrophys. 83, 38.
Giovanardi, C., Helou, R.D., Salpeter, E.E., Krumm, N.: 1983, Astrophys. J. (in press).
Kotanyi, C.G., Ekers, R.D.: 1983, Astron. Astrophys. 122, 267.
Kotanyi, C.G., van Gorkom, J.H., Ekers, R.D.: 1983, Astrophys. J. 273, L7

Lewis, M., Davies, R.P.: 1973, Monthly Notices Roy. Astron. Soc. <u>165</u>,
    231.
de Vaucouleurs, G., de Vaucouleurs, A., Corwin, H.G.: 1976, The Second
    Reference Catalogue of Bright Galaxies.

DISCUSSION

MCGLYNN: Can your claim that stripping begins suddendly at 2° from
the Virgo cluster be supported from your data, since your sample
includes no galaxies between 2° and 3°.4 from the center?

KOTANYI: There may be a gradient between 2° and 3°.4, indeed. We
have observed more galaxies which will maybe fill that gap.

PEEBLES: Brent Tully has a cloud of spirals that in his model are
falling toward the Virgo Cluster for the first time. Have you observed
these spirals for stripping, or is there a possibility of doing
it?

KOTANYI: These spirals are outside the 6° core and therefore we have
not observed them. They can be of course observed by the same technique
we used.

GAVAZZI: Did you say that in galaxies that you find more HI deficient
(and smaller) the HI extension is confined well within the optical
disks or, in other words, that stripping is effective not only on
the extended HI haloes, but also well inside the optical disks?

KOTANYI: Yes, but one must be careful that also the optical disks
might be smaller.

MICROWAVE DECREMENT MEASUREMENTS IN GALAXY CLUSTERS

R.D. Davies and A.N. Lasenby
University of Manchester
Nuffield Radio Astronomy Laboratories
Jodrell Bank, Macclesfield, Cheshire SK11 9DL

1.   INTRODUCTION

Photons from the 3K cosmic microwave background originating at $z \sim 1500$ are inverse Compton scattered on passing through the hot (10^7 to 10^8K) gas in galaxy clusters. Photons from the Rayleigh-Jeans ($\lambda \gg 1mm$) part of the spectrum are scattered to $\lambda \ll 1mm$, thereby producing a reduction in radiation intensity at $\lambda > 1mm$. This microwave decrement or "cooling" in clusters has been actively pursued observationally in the 10 years since its prediction by Sunyaev and Zeldovich (1970,1972). In this paper we will describe the significance of such observations, the present observational status and the prospects for the future.

The physical conditions of the intracluster gas in clusters of galaxies can be established by combining measurements of the microwave decrement with data on the X-ray emission from the gas. For a cluster with single values for its electron density (n_e and temperature T_e, the X-ray luminosity L_X
        (integrated over its volume) $\propto n_e^2 \, T_e^{-\frac{1}{2}} \, r^3$ (1)
and the decrement dI/I
        (integrated through the cluster) $\propto n_e \, T_e r$ (2)
where r is the linear radius of the cluster. T_e can be obtained from the observed X-ray spectrum. In principle, observations of either L_X or dI/I will give n_e directly if the electron distribution is uniform. Given the X-ray, the microwave decrement and T_e distributions across a cluster the following astronomical insights will be gained:-

(a)   the establishment of the clumpiness factor $\langle n_e^2 \rangle / \langle n_e \rangle^2$ which is necessary for deriving the gas mass of the cluster. This is an important determination because estimates derived from X-ray data suggest that $\sim$10 percent of the mass of some clusters is in the form of hot intracluster gas (Gursky & Schwartz 1977).

(b)   the determination of the ratio of gas to luminous matter (galaxies) as a function of radius through the cluster.

267

F. Mardirossian et al. (eds.), Clusters and Groups of Galaxies, 267–272.

(c) the detection of a microwave decrement in large redshift clusters will confirm the cosmological nature of the 3K background.

(d) equations (1) and (2) can be used to determine r which, when combined with the observed angular radius of the cluster, will provide a direct estimate of the distance of the cluster and hence of the Hubble constant H_0. This method is independent of the chain of distance indicators and of any evolution effects (see for example Silk & White 1978).

2.   THE OBSERVATIONAL RECORD

The first to attempt a measurement of the microwave decrement was Parijskii (1973) who claimed a decrement of 1.2 mK in the Coma cluster at an observing wavelength of 4 cm. Since that time a number of groups of observers have made deep surveys of rich clusters at a range of wavelengths in the Rayleigh-Jeans part of the 3K blackbody spectrum. Since the predicted decrements are measured in fractions of milliKelvins this is a technically demanding observational programme. Plots of the

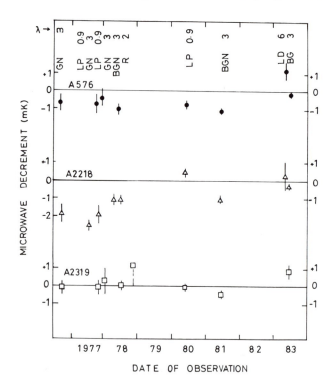

Figure 1 Microwave decrement as a function of epoch for the galaxy clusters A576, A2218 and A2319. The authors and the wavelengths of observation are shown at the top. Errors (vertical bars) are the errors claimed by the authors (see references).

decrements claimed in 3 clusters as a function of time are shown in Fig. 1. A576 and A2218 are two clusters for which a claimed decrement persisted for a number of years on repeated measurements by two independent groups. No significant decrement is claimed in the most recent measurements (BG 1983) however. A2319 has a decrement not significantly different from zero, except in the most recent measurements, where it changed sign. There is of course no question that the actual decrement has changed with time; the variation reflects the difficulty of this type of measurement and the fact that statistical errors do not reflect the true uncertainty in these experiments. The experience gained is used in what follows to evaluate the prospects for improving the sensitivity of this work.

3.   DESIDERATA FOR MEASUREMENTS OF THE MICROWAVE DECREMENT

Based on the X-ray emission from clusters which indicate $n_e \sim 10^{-3} cm^{-3}$ and $T_e = 10^7$ to $10^8 K$, estimates of the microwave decrement are in the range 0.1 to 1.0 mK for different clusters. As an example Silk & White (1978) predict a decrement ~ 0.15 mK for A576 for which the observations are illustrated in Fig.1. If the gas has significant clumping these estimates will be somewhat reduced. Clearly, observations should be made as sensitive as possible by reducing the system noise of the receiver and increasing its bandwidth. Typical receivers with a system noise of 50K and a bandwidth of 500 MHz give an rms noise ~ 0.05 mK for 1 hr of observation. Attaining these sensitivities is the least of the problems. The major concerns are the systematic effects encountered in this work, the most important of which are the following.

(a) The atmosphere. The main constituents of the atmosphere which produce emission in the Rayleigh-Jeans part of 3K spectrum ($\sim 10K$ total and increasing with frequency) are H_2O and O_2. The most serious effects arise from H_2O because of its variability in time and angle as observed from the ground. Even in good observing conditions at ground level the fluctuations on angular scales of degrees may be ~ 10 mK. These effects may be reduced at high sites where the humidity is reduced to a few mm of precipitable water. They are further reduced by beam switching to a nearby reference field.

(b) The telescope environment. Substantial systematic effects are encountered as the radio telescope tracks a cluster in azimuth and elevation due to far sidelobes and spillover lobes picking up radiation at $\sim 300K$ from local obstacles and the changing horizon. Where observations to identify these effects have been performed, systematic effects of a few to a few tens of mK have been measured. We believe this is the cause of most of the discrepancies between observers in the earlier experiments. These problems can only be overcome by making observations of suitable comparison fields closely adjacent to the cluster centre and at the same declination so that the sidelobes and spillover trace out the same systematic effects (e.g. Lasenby & Davies 1983, Birkinshaw & Gull 1983). In principle, aperture synthensis observations should obviate this problem; however because of the lack of

filling of the aperture, very long observations are required and the
problem then may be the reluctance of the time assignment committees
to do their duty. The VLA has responded positively; several long
integration programmes have been completed and are in the process of
analysis.

(c) Source confusion. The search for any weak signal at the sub-
milliKelvin level will be plagued by the presence of confusing sources
(quasars and radio galaxies) in the cluster field and in the adjacent
field. Since most, but certainly not all, sources have spectra which
fall with increasing frequency, most observations have been made at
high frequency ($\nu \gtrsim 10$ GHz) and the effects of confusion have been
ignored. This can give misleading results, particularly as clusters
often contain a dominant cD radio galaxy as well as a number of head-tail
sources shaped by the intracluster gas. The flat spectrum sources in
the reference field always remain an embarrassment. The only effective
remedy is to survey the field with a high resolution instrument at
several frequencies so that confusing sources can be identified and
their spectral indices determined. This is the approach which we have
taken in our observations of clusters at $\lambda 6$ cm (Lasenby & Davies 1983).
The basic decrement observations were made with the 25 x 35m MK II
telescope which has a 10' x 8' beam. This deep survey was supplemented

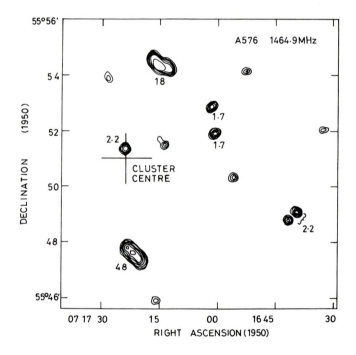

Figure 2 $\lambda 20$ cm (1464.9 MHz) observations of A576 with the C array of
the VLA. The resolution is 12 arcsec. Source fluxes are given in mJy.
The cluster centre is marked by a cross.

by a modest depth λ6 cm survey with the Bonn 100m telescope which was
used to identify the sources in the field with diameters ≲3'. The procedure with the VLA observations was to make the deep observations of
the decrement at λ6 cm and to make an observation at λ20 cm to identify
the confusing sources at a longer wavelength where they are stronger;
spectral indices are obtained for all the sources in the field. Fig.2
shows the dominant confusing sources in the field of A576. Such sources
are responsible for the λ6 cm emission shown in A576 and A2218 in Fig.1.

4.   THE PROSPECTS FOR MICROWAVE DECREMENT OBSERVATIONS

The solution for the physical conditions in the cluster gas and the
achievement of the objectives outlined in Section 1 require (i) detection
of the microwave decrement and its radial distribution (ii) X-ray observations of sufficient sensitivity and resolution to determine the radial
distribution of the X-ray brightness and (iii) the gas temperature distribution. At both radio and X-ray wavelengths the effect of confusing
background objects or cluster galaxies must be removed before a solution
is made for the n_e, T_e and $<n_e^2>/<n_e>^2$ distributions across the cluster.
An analysis could follow along the lines set out by Boynton et al. (1982)
for the cluster A2218; however the situation has changed from that
paper since the decrement is now believed (see Fig.1) to be significantly
less than the ∼1.0 mK used.

We consider that it should be possible with current technology to detect
microwave decrements as low as 0.1 to 0.2 mK and obtain an indication of
the radial distribution of this decrement. Clearly, it would be
necessary to identify and measure the systematic effects inherent in
the observing system to a level ≲0.1 mK. Because of the systematic
effects, observations on adequate reference regions will significantly
increase the required observing time. The choice of observing frequency
is also important. Observations at the highest frequencies (∼100-300
GHz) are most effected by atmospheric water and are confused by emission
from galactic dust. Irregularities in galactic dust on the scale of
arcminutes have not yet been investigated, and if low enough in intensity may not effect the decrement measurements. Observations at
longer wavelengths (say ∼6 cm) are not degraded by these effects but
are subject to source confusion; such confusion can be eliminated by
the techniques described in Section 3. Since the decrement varies in a
predictable manner as a function of frequency, an unambiguous detection
of a microwave decrement requires confirmatory observations at several
frequencies. Likewise, because of the presence of difficult-to-identify
systematic effects which are different for every observing system,
observations are needed with a range of systems including both single
antennas and aperture synthesis instruments. Experiments in progress
at the present and planned for the near future hold out the exciting
prospect of significant detections of the Sunyaev-Zeldovich effect in
galaxy clusters.

REFERENCES

Birkinshaw, M., Gull, S.F. & Northover, K.J.E., 1978. Nature, 275, 40.
Birkinshaw, M., Gull, S.F. & Northover, K.J.E., 1978. Mon.Not.R.astr.
    Soc., 185, 245.
Birkinshaw, M., Gull, S.F. & Northover, K.J.E., 1981. Mon.Not.R.astr.
    Soc., 197, 571.
Birkinshaw, M. & Gull, S.F., 1983. Mon.Not.R.astr.Soc., in press.
Boynton, P.E., Radford, S.J.E., Schommer, R.A. & Murray, S.S., 1982.
    Astrophys.J., 257, 473.
Gull, S.F. & Northover, K.J.E., 1976. Nature 263, 572.
Gull, S.F. & Northover, K.J.E., 1977. IAU Symposium No.74, p.313,
    Reidel, Holland.
Gursky, H. & Schwartz, D.W., 1977. Ann.Rev.Astr.Astrophys., 15, 541.
Lake, G. & Partridge, R.B., 1977. Nature, 270, 502.
Lake, G. & Partridge, R.B., 1977. IAU Symposium No.74, p313, Reidel,
    Holland.
Lake, G. & Partridge, R.B., 1980. Astrophys.J., 237, 378.
Lasenby, A.N. & Davies, R.D., 1983. Mon.Not.R.astr.Soc., 203, 1137.
Parijskii, Y.N., 1973. Sov.Astr. 16, 1048.
Rudnick, L., 1978. Astrophys.J., 233, 37.
Silk, J. & White, S.D.M., 1978. Astrophys.J., 226, L103.
Sunyaev, R.A. & Zeldovich, Y.B., 1970. Astrophys.Space Sci., 7, 1.
Sunyaev, R.A. & Zeldovich, Y.B., 1972. Comm.Astr.Space Sci., 4, 173.

THE FORMATION OF HI RINGS AROUND SO GALAXIES

P. N. Appleton
Department of Astronomy
University of Manchester

INTRODUCTION

Recent studies of the HI emission from a number of SO galaxies (Shostak et al., 1979; van Woerden et al., 1983) have often shown the gas to be distributed in rings outside the optical image of the galaxies. In some cases the rotation axis of the rings appears to differ from that of the stellar component of the galaxy. Can such rings of HI be the result of the capture of gas tidally torn from extended low-surface brightness hydrogen rich companions? Of the SO galaxies observed in HI to date, two systems show definite signs of fitting into this picture. The first possible candidate is NGC 1023 (Hart, Davies and Johnson, 1980; Sancisi et al., in preparation). In this case a clumpy HI ring is observed surrounding the optical image and in addition a long plume of HI is seen extending from one end of the major axis of the galaxy. Near the root of the plume a peculiar blue companion galaxy is observed. The velocities at the tip of the plume are almost 300 km s^{-1} different from that of NGC 1023. In another example, the SO galaxy NGC 4026 may also be interacting with a hydrogen rich dwarf (Appleton, 1983; van Driel, Appleton and Davies, in preparation). Here a 17th mag. HI rich dwarf is observed only 37 kpc in projected distance from NGC 4026. A 55 kpc long clumpy filament of gas is observed pointing away from the dwarf and this gas shows some of the kinematic properties of a tidal tail (Toomre and Toomre, 1972).

TIDAL MODELS

This is a progress report on some 3-body computer models of the NGC 1023 system. A sequence of models have been run assuming that the blue dwarf object near the Eastern tip of NGC 1023 is a remnant of a once more extensive and more massive dwarf galaxy. Various prograde encounters were explored. A final model is not yet available which fits all the observational data. The best results have come from slow elliptical encounters with a quite high mass dwarf of 5 x 10^{10} M$_\odot$ (assumed mass ratio 1:8) and large initial disc radius of 15 kpc. The plume geometry

F. Mardirossian et al. (eds.), Clusters and Groups of Galaxies, 273–274.
© *1984 by D. Reidel Publishing Company.*

is easily produced in most of the models and the ring is the result of
material captured from the disc of the dwarf. The high initial mass of
the dwarf (almost a factor of 10 times larger than that inferred from
the blue object) is needed to produce the high velocities in the plume.
In the course of the rather violent interaction with NGC 1023 much of
this mass is expected to be stripped along with the gas in a fully
self-consistent model.

The 3-body results suggest that tidal stripping of the outer envelopes of LSB galaxies can provide a mechanism for forming HI rings around
SO galaxies. In all such cases a remnant nucleus and associated HI plume
would also be expected to be found somewhere close to the SO. The plume
may however be of lower surface brightness than the ring and would be
expected to be more extensive.

REFERENCES

Appleton, P. N., 1983, Mon. Not. R. astr. Soc., 203, 533.
Hart, L., Davies, R. D. and Johnson, S. C., 1980, Mon. Not. R. astr. Soc.
     191, 269.
Shostak, G. S., van Woerden, H. and Schwarz, U. J., 1979, Photometry,
     Kinematics and Dynamics of Galaxies, ed. D. S. Euchs, University
     of Texas, p.213.
Toomre, A. and Toomre, J., 1972, Astrophys. J., 178, 623.
Woerden, H. van, Driel, W. van, and Schwarz, U. J., 1983, IAU Symposium
     100, D. Reidel Pub. Co., Dordrecht, Holland, p. 99.

INVESTIGATIONS ON THE HI-CONTENT OF FIELD GALAXIES, VIRGO CLUSTER GALAXIES AND UMA-I GROUP GALAXIES

H.-D.Bohnenstengel
Hamburger Sternwarte, D2050 Hamburg 80
Gojenbergsweg 112
present address: Hochschule der Bundeswehr Hamburg
2000 Hamburg 70 Holstenhofweg 85

In this poster paper we will present some important results of mapping 28 Sc I galaxies at λ = 21 cm. The observations were made between 1975 and 1977 (Bohnenstengel 1983 and 1984 a) with the 100 m radio telescop in Effelsberg. The goal of these observations is to find characteristic differences in the HI-distribution and in the HI-content of field cluster and group galaxies as already suggested in several papers (e.g. Lewis et al. 1973, Huchtmeier et al. 1976, Sullivan et al. 1978, Krumm et al. 1979 and Chamaraux et al. 1980). Therefore half of the observed objects are field galaxies and the other half are Virgo cluster galaxies or UMa I group galaxies and model calculations for the HI-distributions were made for this purpose (Bohnenstengel 1983 and 1984 b).

In all cases the observed HI-profiles can be matched by ringshaped HI-distributions where the radial density distribution of these rings is gaussian. We find that the radii of these rings, which are measured from the optical centre of the galaxy to the maximum density, are small for field galaxies and large for group galaxies, whereas the values for the Virgo cluster galaxies lie between these two categories, as shown in table 1.

Table 1: Mean values of radii an thickness of HI-rings

	R_{HI}^{+}	σ_{HI}^{+}
field	0.51 ± 0.26	0.26 ± 0.08
UMa I	0.69 ± 0.29	0.25 ± 0.15
Virgo	0.55 ± 0.23	0.25 ± 0.13

$R_{HI}^{+} = R_{HI}/R_{Holm}$; $\sigma_{HI}^{+} = \sigma_{HI}/R_{Holm}$; R_{HI} : radius of the ring in ['] ; R_{Holm} : Holmberg radius of the galaxy in [']

F. Mardirossian et al. (eds.), Clusters and Groups of Galaxies, 275–276.
© 1984 by D. Reidel Publishing Company.

and σ_{HI} : thickness of the Gaussian distribution in $[']$

On the other hand we find for the total extend in HI – measured at 10^{20} atoms cm^{-3} kpc – the highest values for the field galaxies and smaller values for the observed group and cluster galaxies. The same applies to the HI-mass and we can give the following relation between these two parameters

$$\log M_{HI} = 1.67 \log d_{HI} + 7.00 \tag{1}$$

where M_{HI} is given in $[10^6 M_\odot]$ and d_{HI} in $[kpc]$, while we find

$$\overline{M_{HI}}(\text{field}) \approx 1.7 \; \overline{M_{HI}}(\text{cluster}) \quad \text{and}$$

$$\overline{M_{HI}}(\text{field}) \approx 3.2 \; \overline{M_{HI}}(\text{group}) \tag{2}$$

where the $\overline{M_{HI}}$ are the mean values of the HI-mass of the corresponding categories. This indicates group and cluster galaxies to have lost most of the outer part of their HI-envelope, e.g. by interactions with intergalactic gas. Finally there are significant differences for the maximum rotational velocities v_{max} at comparable inclinations i. We find that the field galaxies rotate faster than the group galaxies as shown in table 2. This last effect indicates

Table 2: Approximate bounderies of $\frac{1}{2}\Delta v_{50}$

	min	max	
field	130	200	km/s
UMa I	50	100	"
Virgo	80	160	"

Δv_{50} : linewidth of the central profile at 50% of maximum intensity

that the rotation of the group galaxies in particular has been retarded by interactions with galaxies in the neighbourhood. The differences in the HI-ring radii support this conclusion.

References:
Bohnenstengel,H.-D., 1983, Dissertation, Universität Hamburg
Bohnenstengel,H.-D., 1984a, in preparation
Bohnenstengel,H.-D., 1984b, in preparation
Chamaraux,P., Balkowski,C., Gérard,E.,1980, Astron.Astro-
        phys. 83, 38
Krumm,N., Salpeter,E.E., 1979, Astrophys.J. 227, 776
Huchtmeier,W.K., Tammann,G.A.,Wendker,H.J, 1976, Astron.
        Astrophys. 46, 381
Lewis,B.M., Davies,R.D., 1973, Mon.Not.R.astr.Soc. 165, 213
Sullivan,W.T.III, Johnson,P.E., 1978, Astrophys.J. 225, 751

NEW METHODS OF INVESTIGATION FOR EXTRAGALACTIC BACKGROUNDS IN THE INFRARED

C.Ceccarelli,G.Dall'Oglio[+],P.de Bernardis,S.Masi,B.Melchiorri[+],
F.Melchiorri,G.Moreno,L.Pietranera.

+ Istituto di Fisica dell'Atmosfera del C.N.R.,Roma.
Dipartimento di Fisica,Università di Roma,Roma.

If we look at the infrared sky between 1 and 1000 μm we observe several
contributions to the background coming from different parts and/or epochs
of the universe. We consider the problem of distinguishing between 'local'
contributions (planetary and galactic dust,atmospheric emission,dashed
lines) and the extragalactic backgrounds. In the figure are shown the
computed emissions for rich clusters of galaxies (RC) and integrated
infrared galaxies (IRG). CMB is the cosmic microwave background.

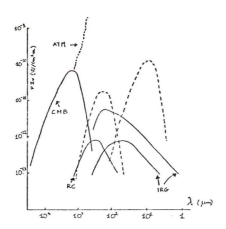

We have performed two methods to observe 'indirectly' the presence of
such extragalactic backgrounds. In the first one we have overcome the
problem of local contamination by performing differential measurements
looking for the dipole anisotropy. In fact such kind of signature is
mainly due to the motion of our galaxy inside the local supercluster; in
such a case emission coming from our galaxy should not be affected by
such a signature. Let us recall some useful formulas:
the angular distribution of a radiation spectrum $I(\nu)$ seen from an
object moving with velocity v is :

$$I_{obs}(\nu_{obs}) = \gamma I(\nu)(1 - \beta\cos\vartheta)^3$$

F. Mardirossian et al. (eds.), Clusters and Groups of Galaxies, 277–278.

where
$$\nu_{obs} = \gamma \nu (1 - \beta \cos \vartheta)$$

and ϑ is the angle between the line of sightand the direction of v.
For the dipole anisotropy ($\vartheta = 180°$) and for $\gamma \cong 1$ we get:

$$\Delta I_{obs} = \int I(\nu)(3-\alpha)\beta T(\nu)d\nu$$

where $T(\nu)$ is the spectral response of the instrument considered and α is the spectral index: $\alpha = d \ln I(\nu)$

In order to test this method we used data from three experiment designed
to observe the CMB. For details see Ref.1; subtracting dipole anisotropy
of CMB found by Weiss (ref.4) we got upper limits for dipole anysotropies

ν eff. (cm^{-1})	upper limit (W/cm^{2}sr cm^{-1})
13 - 18	2×10^{-12}
15 - 21	5×10^{-12}

Although spectral responses are inadequate for our porposes we get results which are consistently below the atmospheric emission.We are
performing a proper designed esperiment to reach the level of extragalactic backgrounds. The second method is conceived to test the existence
of the strong extragalactic background found by Matsumoto et al. (ref.3).
Such a radiation should be absorbed by the intergalactic dust and then
reemitted at longer wavelenght.
Using the data from the esperiment described above and considering a
model for dust emission (Ref.2) we obtained an upper limit for the epoch
of dust formation and for its density:

z form.	n_{dust} (cm^{-3})
20	3×10^{-20}
10	1×10^{-19}
5	3×10^{-19}

REFERENCES

1 Ceccarelli,Dall'Oglio,de Bernardis,Masi,Melchiorri B.,Melchiorri F.,
  Moreno,Pietranera, 1983,Ap.J.lett.,in press.

2 Ceccarelli,Dall'Oglio,de Bernardis,Masi,Melchiorri B.,Melchiorri F.,
  Moreno ,Pietranera,1983,submitted to Astr.Astrophys..

3 Matsumoto T.,Akiba M.,Murakami H.,1983,preprint.

4 Weiss R.,Ann.Rev.Astr.& Astrophys.,1980,18,489.

DUST FAR INFRARED EMISSION AT HIGH GALACTIC LATITUDES

C.Ceccarelli,G.Dall'Oglio[+],P.de Bernardis,S.Masi,B.Melchiorri[+],
F.Melchiorri,G.Moreno,L.Pietranera
+Istituto di Fisica dell'Atmosfera del C.N.R. Roma
Dipartimento di Fisica G.Marconi,Università di Roma

In order to confirm the cosmological origin of the quadrupole anisotropy
detected in the cosmic background radiation by Fabbri et.al.(1),we
performed a balloon borne experiment devoted to the study of diffuse
radiation in the far infrared.We used two germanium bolometers,in the
wavelenght bandwidths 150-400 µm and 350-3000µm.The sky modulation was
achieved by wobbling a metallic mirror . The modulation amplitude was 6
degrees.The sensitivity for far infrared gradients in the sky was
2×10^{-14}W cm^{-2}sr^{-1}deg^{-1}Hz$^{-1/2}$.For further details see ref.(2).
The instrument was flown by balloon at 45 Km of altitude from the italian
base of Trapani-Milo of C.N.R. Piano Spaziale Nazionale,on July 19,1980.
A real time analysis of the data shows the presence of a large scale
signal which is negative when the field of view approaches the galactic
plane and is positive when it leaves it.In order to explain the signal
we formulated three different hypothesis : 1)Atmospheric temperature
increase from north to south.This gives a dipole-like anisotropy; 2)Inter
planetary dust emission following a cosec-law in ecliptical latitude; 3)
Interstellar dust emission following a cosec-b law in galactic latitude
as suggested by optical data;(3),(4).The comparison between data and fits
immediately shows a good agreement between the cosec-b law and the data.
This is confirmed by the χ^2 analysis for both NIR and FIR channels.We can
conclude we have detected for the first time the galactic dust emission
at high galactic latitudes($10^{\circ} < b < 60^{\circ}$).
The slope of the best fit line in fig.1
provides information about the extrapolated dust emission at the N.G.P..For
the two channels we obtained :

$$\int_0^\infty I(\lambda) t_{NIR}(\lambda) d\lambda = (1.0 \pm 0.3) \cdot 10^{-11} \, W cm^{-2} sr^{-1}$$

$$\int_0^\infty I(\lambda) t_{FIR}(\lambda) d\lambda = (5.1 \pm 1.5) \cdot 10^{-13} \, W cm^{-2} sr^{-1}$$

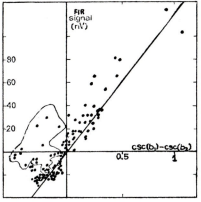

279

on the detectors i.e. the spectral transmittances t_{NIR} and t_{FIR} are not corrected for.We can rewrite the dust emission spectrum

$$I(\lambda) = kN\lambda^{-\alpha}BB(\lambda,T_d)$$

where N,T_d are the column density and the temperature of the dust,while α is the spectral index for dust emissivity.For $T_d = 13$ K we obtain $\alpha = 2.4+0.4$; for $T_d = 20$ K we have $\alpha = 1.3+0.4$;for $\alpha = 2$ we obtain $T_d = 15+2$ K. By combining our data (a) with those of Lubin(1983)(c)(5) and with those of the 1978 flight(6,1)(b)we get:

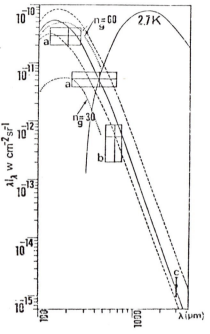

$$I(\lambda) = \frac{(6.4 \pm 1.3) \cdot 10^4 \ W cm^{-2} sr^{-1} \mu m^{-1}}{\lambda^{(6.8 \pm 0.2)} \{exp[hc/k\lambda(17\pm2)°k] - 1\}}$$

It must be stressed that our estimate of the N.G.P. emission is an extrapola tion from measurements carried out at b ranging from 10 to 70 deg.The real N.G.P.emission could be substantially different.Assuming the values listed in the last formula for α and T we get

$$\lambda I_\lambda (200\mu m) = (2.9 \pm 0.9) \cdot 10^{-11} \ W cm^{-2} sr^{-1}$$
$$\lambda I_\lambda (440\mu m) = (5.4 \pm 1.6) \cdot 10^{-12} \ W cm^{-2} sr^{-1}$$

The dust column density has then been derived:

$$N_g = (3 \pm 2) \cdot 10^9 \ cm^{-2}$$

For the(extrapolated column density at the N.G.P.).This value corresponds to an extinction m=(0.3+0.2)mag at NGP, in good agreement both with the old Shane and Wirtanen(3)data and with the critical reanalysis by de Vaocouleurs (4).The dust emission is so high that it must be detected in submillimetric obrervations of CBR.This seems to be the case in Gush's experiment(7) where the spectrometer went as close as 7 deg.to the galactic plane. Part,if not all of the submillimetric excess he observed at high frequency may be due to galactic emission.On the other hand the excess around 15 cm^{-1} cannot be explained in this way.If it is genuine,it must be extra-galactic.However,if it is so,one would expect a contribution to the di-pole anisotropy due to the motion of the observer with respect to the extragalactic matter responsible for the radiation.A rough estimate suggests that the dipole anisotropy should be as large as 5mK,in contrast with current experimental values(2).

1)Fabbri R.,Guidi I.,Melchiorri F.,Natale V.,Phys.Rev.Lett.44,1563,(1980)
2)deBernardisP.,Masi S.,Melchiorri B+F,Moreno G.,Ap.J.in press(1984)
3)Shane C.D. Wirtanen C.A. Publ.Lick.Obs. 22,1,(1967)
4)de Vaocouleurs (1983) preprint
5)Lubin P.M.,Epstein G.L.,Smoot G.F., Phys.Rev.Lett. 50,616,(1983)
6)Bussoletti E.,Guidi I.,Melchiorri F.,Natale V., A.&A. 105,184,(1982)
7)Gush H.P. Phys.Rev.Lett.47,745,(1981)

DUST DISTRIBUTION NEARBY THE SOLAR SYSTEM AND ITS POSSIBLE IMPLICATIONS ON COSMIC RADIATION BACKGROUND ANISOTROPY

C.Ceccarelli, G.Dall'Oglio[+], P.De Bernardis, S.Masi,
B.Melchiorri[+], F.Melchiorri, G.Moreno, L.Pietranera

Dipartimento di Fisica,Università di Roma, Roma
[+]Istituto di Fisica dell'Atmosfera, C.N.R., Roma

ABSTRACT

It is well known that the two darkest clouds in the sky are located in the regions of ρ-Ophiuchus ($l"=354°$; $b"=16°$) and Taurus ($l"=174°$; $b"=-13°$). The approximate simmetry of these two regions around the Sun has suggested that they may be linked by a very elongated cylindrical cloud. Indirect evidence supporting this view has been obtained by studying the colour excess of nearby stars and the relation between galaxy counts and HI column density (Turon and Menessier, 1975; Lebrun, 1979). Observations of the diffuse sky radiation in the far infrared (150 μm $< \lambda <$ 3mm), performed during a balloon flight on July 19, 1980, appear to be consistent with the above model. It turns out, in fact, that signals detected when the field of view of the experiment crossed the hypothezed dust cloud differ significantly from the general trend of the data,which reflects the global dust distribution in the Galaxy.
Fitting the data in terms of the physical parameters of the cloud leads to values of the column density ranging from 2×10^9 cm^{-2} (for a dust temperature T=20 K) to 1.5×10^{10} cm^{-2} (for T=10 K).
Emission of the local dust cloud may have relevant implications in studying the anisotropy of the Cosmic Background Radiation (C.B.R.).
In figures 1 and 2 the trend of dipole and quadrupole anisotropies are sketched vs. α and δ (plots are based on results obtained by Fabbri et al.,1980, analyzing far infrared observations, 500 μm$< \lambda <$ 3mm,performed during a balloon flight in 1978). In figure 3 the expected signals from the dust cloud are plotted in the same format. It is seen that maxima of functions plotted in figure 2 and 3 are approssimately located at the same directions, thus suggesting that the local dust cloud emission may simulate a quadrupole anisotropy of the C.B.R.. A re-analysis of 1978 data is now in progress to establish if the contribution of the dust emission to the apparent quadrupole anisotropy is in fact significant.

F. Mardirossian et al. (eds.), Clusters and Groups of Galaxies, 281–282.
© *1984 by D. Reidel Publishing Company.*

REFERENCES

Fabbri R., Melchiorri B., Melchiorri F., Natale V., Caderni V.,
Shivanandan K.: Phys. Rev. D 21, 2095 (1980)
Lebrun F.: Astron. Astrophys. 79, 153 (1979)
Turon P., Mennessier M.O.: Astron. Astrophys. 44, 209 (1975)

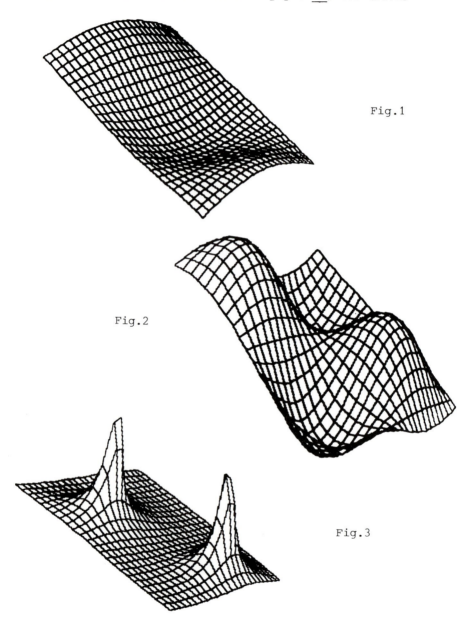

Fig.1

Fig.2

Fig.3

THERMAL RADIATION NOISE AND THE 3 K BACKGROUND.

C.Ceccarelli,G.Dall'Oglio[+],P.de Bernardis,S.Masi,B.Melchiorri[+]
F.Melchiorri,G.Moreno,L.Pietranera .

Dipartimento di Fisica G.Marconi,Università di Roma,Italy
+ Istituto di Fisica dell'Atmosfera del C.N.R.,Roma,Italy

All the measurements of the Cosmic Background (CBR) spectrum in the submillimetric region are consistent with a Planckian decrease at short wavelenghts ; this agreement is,however,only qualitative (1) .
New,independent experimental methods are required,in order to clarify the situation and to detect indisputably the presence of distortions which could have important cosmological consequences.
For this reason we designed a balloon borne far infrared correlator, operating in the interval 800-2000 μm,in order to measure the quantum fluctuations of the CBR.Let us recall some formulas :

$$\bar{n} = \frac{g}{e^{(E-\mu)/kT} + \xi}$$

$$\xi \begin{cases} = -1 & \text{Bose Einstein} \\ = 0 & \text{Maxwell} \\ = +1 & \text{Fermi-Dirac} \end{cases}$$

μ = chemical potential
g = degeneration

is the distribution function for an ideal gas in thermal equilibrium and the mean square fluctuation is given by :

$$\overline{\Delta n^2} = \overline{n^2} - \bar{n}^2 = kT\left(\frac{\partial \bar{n}}{\partial \mu}\right)_{T,V} = \bar{n} - \frac{\xi}{g}\bar{n}^2$$

In the case of blackbody radiation

$$\overline{\Delta n_\nu^2} = \bar{n}_\nu\left[1 + \frac{1}{e^{h\nu/kT} - 1}\right]$$

For power fluctuations in a detector with acceptance area A,solid angle Ω and electrical bandwidth Δf we found

$$\frac{\sqrt{\overline{\Delta W^2}}}{W} = 2.77\cdot10^{-18}(A\Omega)^{1/2}\left[T^5\int_{x_1}^{x_2}\frac{x^4 e^x dx}{(e^x - 1)^2}\right]^{1/2}\Delta f^{1/2} \qquad x = \frac{h\nu}{kT}$$

For a blackbody between 800 and 2000 m we have :

$$\sqrt{\overline{\Delta W_{BB}^2}} = 3.6\cdot10^{-16} \quad W/\sqrt{Hz\,cm^2\,sr}$$

283

F. Mardirossian et al. (eds.), Clusters and Groups of Galaxies, 283–284.
© *1984 by D. Reidel Publishing Company.*

For a grey-body we have :

$$\sqrt{\overline{\Delta W^2}_w} = 2.77 \cdot 10^{-18} (A\Omega)^{1/2} \left[T^5 \int \frac{\mathcal{E} (e^x - 1 + \mathcal{E}) x^4 dx}{(e^x - 1)^2} \right]^{1/2} \Delta f^{1/2}$$

($\mathcal{E}$ = emissivity).We assumed an atmospheric millimetric emission at balloon altitude like that of a grey-body with T=250 K and $\mathcal{E}$=5x10^{-4}.We get :

$$\sqrt{\overline{\Delta W^2}_{ATM}} = 1.6 \cdot 10^{-16} W / \sqrt{Hz cm^2 sr} \qquad \text{at the zenith}$$

We can separate the atmospheric effect from the CBR using the cosec law for atmospheric emission and extrapolating it to 0.The instrument consisted of two Ge bolometers operating at 0.3 K,a mylar beamsplitter and a Cassegrain optics at 1.8 K.The electrical NEP was 8x10^{-16}W/$\sqrt{hz}$,the throughput ($A\Omega$) was 0.05 cm^{2}sr,with a field of view of two degrees. For further details see Dall'Oglio and de Bernardis (1982). The correlated noise found during the flight is plotted vs elevation in fig.1.We get an upper limit of 3.1 K for the thermodinamic temperature of the CBR in our wavelenght range (800-2000 microns).

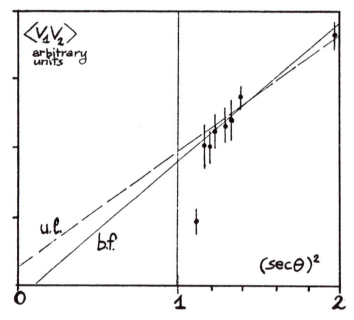

Fig.1.Correlated noise at different tipping angles is plotted.The extrapolation to sec(θ)=0 is shown in the right side. bf is the best fit secant law and ul is our upper limit at one standard deviation.

References :(1)Gush H.P. Phys.rev.Lett.47,745,(1981)
              (2)Dall'Oglio G.Proc of the International School of Physics
                 E.Fermi on Gamow Cosmology (Varenna) (1982).

THE VOIDS AND THE COSMIC MICROWAVE PHOTONS

C.Ceccarelli,G.Dall'Oglio[+],P.deBernardis,S.Masi,B.Melchiorri[+],
F.Melchiorri,G.Moreno,L.Pietranera.

Dipartimento di Fisica,Università di Roma,Roma.
[+]Istituto di Fisica dell'Atmosfera,C.N.R.,Roma.

Recently large voids in galaxy distribution have been found by means of
red shift surveys. We consider the possible effect of the presence of
one of these voids (the one found by Kirshner et al. (1982) in Bootis)
on the cosmic microwave background photons.
We computed the distortion of the CMB photons due to the inverse Compton
interaction with the hot Intergalactic Medium (IGM).
By using this method Field and Perrenod (1979) derived an upper limit
on the IGM density of:

$$\Omega_{IGM} \leq 1$$

from the limits on distortion of the CMB spectrum.
During the far-infrared balloon-borne experiment performed in 1978
(see also Fabbri et al.,1979) the instrument scanned several times the
region in Bootes where the void has been found. The average of the scans
is shown in the figure as a function of Right Ascension.

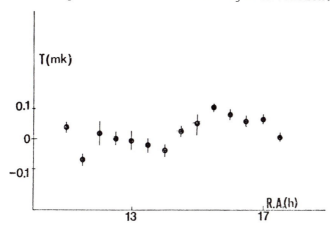

The signals are compatible with the absence of a source within a 2 σ
error of 0.1 mK.
The theoretical interpretation of this limit depends strongly on the

285

F. Mardirossian et al. (eds.), Clusters and Groups of Galaxies, 285–286.
© 1984 by D. Reidel Publishing Company.

model assumed for the physics of the void.
If we assume, following Zeldovich (1982),that the void is completely
empty,the distortion should be due to the difference of mean density
with respect to the surroundings. In such a case we get anupper limit

$$\Omega_{IGM} \lesssim 0.5$$

On the other hand,if we assume, in accordance with Rees (1982) that the
'void' is filled with hot gas which prevented galaxy formation we can
put some constraints on the gas excess inside the void. Assuming a mean
density Ω_{IGM} = 1 our result prove that the amount of gas inside the
void does not differs by more than 10% from that of the surroundings.

This work has been supported by Servizio Attività Spaziali del C.N.R. di
Roma.

REFERENCES.

Fabbri R.,Guidi I.,Melchiorri F.,Natale V.,1979 in Proceeding of the 2nd
M.Grossman Meeting on general Relativity,ed.R.Ruffini (Amsterdam N.Holl.)
p.889.
Field G.B.,Perrenod S.C.,1977,Ap.J.,215,717.
Kirshner R.P.,Oemler A.,Schecter P.L.,Schectman S.A.,1981,Ap.J.lett.,
248,L57.
Rees M.,1983,IAU Symposium 104,ed.G.Abell and G.Chincarini(Dordrecht
Reidel).
Zeldovich Ya B.,Einasto J.,Shandarin S.F.,1982,Nature,300,407.

OBSERVATION OF A LARGE HALO AROUND M87

C.Ceccarelli[1], G.Dall'Oglio[2], P.De Bernardis[1], N.Mandolesi[3],
S.Masi[1], B.Melchiorri[2], F.Melchiorri[1], G.Moreno[1], G.Morigi[3],
L.Pietranera[1]

[1]Dipartimento di Fisica,Università di Roma, Roma
[2]Isituto di Fisica dell'Atmosfera,C.N.R., Roma
[3]Istituto T.E.S.R.E.,C.N.R.,Bologna

During 1978 balloon borne infrared experiment we observed a signal coming
from Virgo Cluster (see Ceccarelli et al.,1982).
The instrument was working at a wavelenght of 1mm;the amplitude of the
signal was of (0.25±0.1)mK over a field of view of 6 degrees.
In order to check the spectrum of such a signal we performed radio measurement with the Medicina isotropometer (Ceccarelli et al.,1983) at
2.8 wavelenght and with a field of view of 2 degrees.
We found the signal shown in figure 1 (the best fit of the experimental
data is represented by the solid line), which is about three times larger than the expected one by measurements of Virgo A (dashed line, see
for example Andernach et al.,1981).

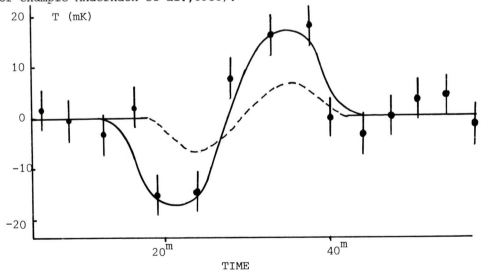

The best fit curve gives an excess amplitude of 16 mK for a source exten
ded 2 degrees . Such a signal could have escaped detection with small
beam radiometers because of its large estension.

287

F. Mardirossian et al. (eds.), Clusters and Groups of Galaxies, 287–288.
© 1984 by D. Reidel Publishing Company.

Both infrared and radio signals could be explained by assuming thermal
emission from a tepid (T= 10^5-10^6 K) halo whose mass, within a 350 Kpc.
diameter is of the order of 10^{13} solar masses, that is the mass reque-
sted to close gravitationally M87.
This hypothesis could be confirmed or rejected by comparison with measurements in the U.V. spectral region using large beam instruments.
This work has been completely supported by Servizio Attività Spaziali
del C.N.R. in Rome.

REFERENCES.

Andernach H.,Baker J.R.,Vonkaperr A.,Wielebinsky R.,1979,Astr.Ap.,74,93.
Ceccarelli C.,Dall'Oglio G.,Melchiorri B.,Melchiorri F.,Pietranera L.,
1982,Ap.J.,260,484.
Ceccarelli C.,Mandolesi N.,Melchiorri F.,Morigi G.,Pietranera L.,1983,
Atti della Scuola E.Fermi di Varenna,Italy,in press,ed.F.Melchiorri and
R.Ruffini.

HIGH RESOLUTION   RADIO AND X-RAY OBSERVATIONS OF 3C28   IN A115

Feretti, L., Gioia, I.M.[1], Giovannini, G.,
Gregorini, L., Padrielli, L.
Istituto di Radioastronomia, 40100 Bologna, Italy
1) also Center for Astrophysics, Cambridge, MA 02138

The cluster of galaxies A115 (D=6, R=3, BM-Type=III) is one of the most distant clusters (z=0.1971) for which a considerable number of redshifts are available in the literature and a study of the system dynamics has been attempted (Noonan, 1981; Beers et al., 1983). The latter authors have measured redshift for 29 galaxies in the region, 19 of which are cluster members, and mapped the galaxy distribution finding three major clumps of galaxies. The cluster is classified as "double" on the basis of IPC X-Ray data (Forman et al., 1981), which show two peaks of emission coincident with the two main subcondensations of galaxies. The brightest galaxy of the cluster lies in the northern X-Ray clump and is identified with the strong radio source 3C28.

The present observation of this source was performed at 1.4 GHz with the VLA, with a resolution of $\sim$3". No emission is detected in the position of the optical galaxy, at a level of 3 mJy/beam area; the two components, located on both sides of the galaxy, show low brightness tails in the western direction (Fig. 1a). The total linear extent of the structure is about 100 kpc (H_o=100 km s^{-1} Mpc^{-1}; q_o=0). There is a slight misalignement of the radio components with respect to the galaxy position, still in the western direction. The radio components have a steep spectral index, $\alpha \simeq 1$.

An X-Ray observation of this region of sky was obtained with the HRI on board the Einstein Observatory, with a resolution of $\sim$3" (FWHM): strong X-Ray emission is detected in the position of 3C28 (Fig. 1b). We interpret this X-Ray emission to originate from hot gas, which either is accreting onto the galaxy or is in hydrostatic equilibrium and is part of a massive dark halo around it (as in M87).

The radio components are strongly confined by the surrounding gas, which has a pressure $\sim$10 times greater than that of internal magnetic field plus relativistic particles. External confinement counteracts adiabatic losses in the source and enhances the lifetime of the radio source, which loses energy mainly by synchrotron emission. The lack of

F. Mardirossian et al. (eds.), Clusters and Groups of Galaxies, 289–290.
© *1984 by D. Reidel Publishing Company.*

detection of a strong nuclear source, together with the steep spectral
index of the components, indicates that little or no activity is present
in the core and radiation losses are dominant; injection of fresh
relativistic particles should have stopped $\sim 2 \times 10^7$ years ago. In order
for the electrons to diffuse in this time from the nucleus to the
extreme parts of the radio source, they must travel at a speed ~ 1000
km/s, which is not unreasonable.

The complex radio structure of 3C28 is understood by invoking a
combination of external ram pressure plus buoyancy effects of light
radio plasmoids. It seems reasonable to assume that radially directed
buoyancy effects play an important role in producing the observed radio
tails. Infact, from a superposition of the radio and X-Ray contour maps,
we can see a tendency of the radio tails to diffuse in the directions of
lower X-Ray emission. The misalignment of the radio components,
instead, could be due to ram pressure originated from the motion of the
galaxy in the subcluster potential well and/or from a large scale motion
which could reasonably derive from the merger of the subclusters.

References

Beers, T.C. Huchra, J.P. Geller, M.G.: 1983, Ap. J. <u>264</u>, 356.
Forman, W., Bechtold, J., Blair, W., Giacconi, R., van Speybroeck, L.,
       Jones, C.: 1981, Ap. J. (Letters) <u>243</u>, L133.
Noonan, T.W.: 1981, Ap. J. Suppl. <u>45</u>, 613 .

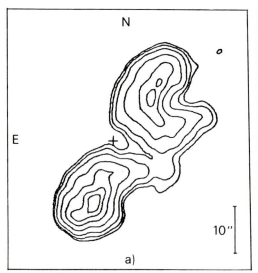
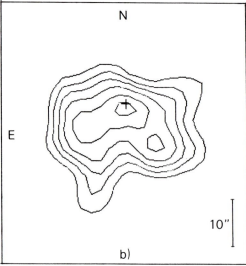

Fig. 1 - a) VLA radio map at 1.4 GHz of 3C28; contours are 2, 3.5, 7,
17, 35, 50, 62 mJy/beam area; b) HRI X-Ray map of 3C28; contours
correspond to 0.12, 0.16, 0.19, 0.23, 0.27, 0.30 net counts. The cross
marks the optical galaxy (RA= $00^h 53^m 09.1^s$, DEC= 26° 08' 24").

THE RADIO LUMINOSITY FUNCTION OF FIRST RANKED GALAXIES IN ABELL CLUSTERS

L. Feretti, G. Giovannini, L. Gregorini
Dipartimento di Astronomia dell'Università di Bologna, Italia

Using a recent WSRT survey of Abell clusters (Fanti et al. 1982, 1983) and the data previously published, we have good radio data for a very large number of Abell clusters with distance class (D) $\leqslant$ 3 and $\delta \geqslant$ 15° for which optical data are also available. This sample of 49 clusters, includes all the Abell clusters of D 1, 3 (75%) of D 2 and 29 (58%) of D 3 with $\delta \geqslant$ 15°. We note that this sample is homogeneous and useful for statistical considerations, since no selection biases, from optical or radio point of view, are present.

We have derived the RLF of the first ranked galaxies (FRGs) of these clusters, according to the method of Avni et al., 1980. The result is shown in Fig. 1 where the galaxies have been divided in two classes of magnitude. The absolute magnitudes have been derived from Zwicky and Herzog, 1963, after correction for the galactic absorption (H_0 = 100 km/s Mpc). In agreement with Auriemma et al. 1977, we find that the more luminous galaxies have a greater probability to show radio emission.

We have selected the FRGs with $-21 < M_{pg} \leqslant -20$ for a comparison with other galaxies of the same optical range. A comparison in the brighter class is not possible because of the poor statistics. The comparison samples are:
1) The optically selected cD galaxies in Abell clusters and in poor groups (Valentijn and Bijleveld, 1983).
2) The E+SO galaxies in Abell clusters of D 1, 2 (Fanti et al. 1982), after subtracting the contribution of FRGs.
3) The E+SO galaxies outside rich clusters (Auriemma et al. 1977)

For FRGs and for galaxies of samples 2 and 3 we have used the photographic magnitude (Zwicky and Herzog, 1963). The cD galaxies of sample 1 have a photometric magnitude within a galaxy diameter of 19 kpc (Valentijn and Bijleveld, 1983). To be consistent with the Zwicky magnitude we have added 0.4 m to have an absolute magnitude within a galaxy diameter of 43 kpc.

From a first comparison, we find that the RLF of cDs and FRGs in Abell clusters (not cD) are in agreement, indicating that cDs are not different from FRGs with respect to the probability of radio emission. Therefore, in order to have a better statistic, we have added the cD galaxies to the FRGs. The RLF computed in this way has been compared

F. Mardirossian et al. (eds.), Clusters and Groups of Galaxies, 291–292.

with that of galaxies of samples 2 and 3, which have been added together, since their RLFs are in agreement (Fanti, present volume). The RLF of these two final samples is shown in Fig. 2.

     The RLFs of the two classes of galaxies are clearly separated: the FRGs have a larger probability to be radio sources than the other E and SO galaxies of the same luminosity. This probably reflects the import- ance of their location at the gravitational centers of the clusters. Dominant galaxies, in fact, can collect gas more easily than normal gal- axies also in presence of removing mechanisms like supernova winds or ram pressure sweeping. A fraction of this gas could be the fuel of the radio source.

     An alternative interpretation of the difference between the two RLFs is suggested by noting that the RLF of FRGs is in good agreement with the RLF of galaxies not in Abell clusters of the higher optical ra- nge ($M_{pg} \lesssim -21$) given by Auriemma et al. 1977. This may indicate that while the Zwicky or the photometric magnitude in a size of 43 kpc is the total magnitude of an elliptical galaxy, it can be seriously underestim- ated for the cDs and the FRGs, due to their large halos (see also Fanti, present volume). In this case the higher probability of FRGs and cDs to become radio sources would be the consequence of their higher optical luminosity only.

REFERENCES
Auriemma et al.: 1977, Astron. Astrophys. 57, 41

Avni et al.: 1980, Astrophys. J. 238, 800
Fanti et al.: 1982, Astron. Astrophys. 105, 200
Fanti et al.: 1983, Astron. Astrophys. Suppl. Ser. 51, 179
Valentijn and Bijleveld. : 1983, Astron. Astrophys. 125, 223
Zwicky and Herzog : 1963, Catalogue of Galaxies and Clusters of
     Galaxies, Calif. Inst. of Tech., Pasadena

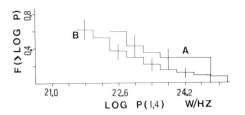

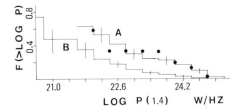

Fig. 1: The integral RLF of FRGs at 1.4 GHz. (A) Mpg $\lesssim -21$, (B) $-21 <$ Mpg $\lesssim -20$. Here and in Fig. 2, error bars represent 1 r.m.s.

Fig. 2: The integral RLF for $-21 <$ Mpg $\lesssim -20$ of: (A) FRGs+cDs, (B) E+SO galaxies. Dots repre- sent the RLF of Auriemma et al., 1977 for gala- xies with Mpg $\lesssim -21$.

EXTENDED RADIO SOURCES AND THE DIFFUSE MEDIUM IN THE
COMA–A1367 SUPERCLUSTER

W. J. Jaffe
Space Telescope Science Institute, Baltimore
G. Gavazzi
Istituto di Fisica Cosmica, Milan
E. A. Valentijn
European Southern Observatory, Garching bei München

Studies of extended, thermally confined radio sources in rich clusters provided estimates of the pressure and temperature in the cluster medium that confirmed the values obtained from x-ray studies (Jaffe & Perola, 1973). We have now found some extended sources in the Coma–A1367 supercluster that may yield similar information on the supercluster medium.

The source 1253+276 (Jaffe & Rudnick 1979, Andernach et al., this meeting) is a very extended source (10'x22') seen in the direction of the supercluster. It has no confirmed identification and could be connected with the background source 3C277.3. However, recent Westerbork observations (Hanisch, Strom & Jaffe, in preparation) indicate no connection to 3C277.3 and that it is most likely a remnant source in the supercluster. If so, the standard equipartition arguments imply a confining pressure of 2×10^{-13} dyne cm^{-2} or a density-temperature product of $nT \sim 600$ $cm^{-3}K$. This compares to a value of some $5\times10^{4}cm^{-3}K$ in the coma cluster core. The source appears to be some 2Mpc ($H_{o} = 100$) from the cluster center. This corresponds to some six times the core radius of the cluster gas distribution. Thus the above pressure may represent the outer envelope of the cluster distribtuion or an intermediate region between the cluster and the supercluster.

The wide angle tailed source 1201+205 was found in a survey of small groups within the supercluster (Gavazzi et al. this meeting). It is identified with NGC 4065 in the NGC 4061 group of galaxies. Its relaxed components have an internal pressure of at least 1×10^{-12} dyne cm^{-2}, implying a confining gas with nT of 3000 $cm^{-3}K$, a factor of 16 below the Coma center. The central galaxy density N_{o} in the group is a factor of 4 below the Coma center and the velocity dispersion in the group, σ, is about half that of the rich cluster. Hence a simple

F. Mardirossian et al. (eds.), Clusters and Groups of Galaxies, 293–294.
© *1984 by D. Reidel Publishing Company.*

model, in which the gas in groups and clusters is shed at low velocity by the member galaxies and then thermalizes, agrees with our data, since it predicts $nT \sim N_o \sigma^2$.

The gas in rich clusters, however, show signs of additional heating, since the spatial distribution of the gas is usually broader than that of the galaxies in the clusters. This heating is usually attributed to supernovae that also provide the iron content of the gas. A commensurate heating in a small group like that containing NGC 4065 would expell the gas from the group. It seems likely then that either the heating comes from an external medium around the group (as suggested for example by C. Jones at this meeting), or that an external medium exerts a pressure to contain the gas in the group.

In conclusion, the two radio sources we have studied in the Coma-A1367 supercluster suggest the presence of an gaseous medium with a pressure equivalent to $nT \sim 500$ cm^{-3}K, either in large volumes around the individual groups and clusters, or filling the supercluster.

References

Jaffe, W., Perola, G.C. 1973, Astr & Astrophys. 26 423

Jaffe, W.J., Rudnick. L. 1979, Astrophys. J. 233, 453

PRELIMINARY RESULTS OF THE 20-CM VLA SURVEY OF ABELL CLUSTERS

F.N. Owen, National Radio Astronomy Observatory
J.O. Burns, University of New Mexico
R.A. White, Goddard Space Flight Center

During the past three years, we have been conducting a 20-cm radio frequency continuum survey of Abell clusters using the NRAO VLA. The survey consists of clusters which meet the following criteria:

(1) All clusters with m_{10} <$17^m_.0$, and integrated flux densities S_{20} >100 mJy within 0.5 of an Abell cluster radius as measured with the NRAO 91-m Green Bank telescope (Owen et al. 1982);

(2) All Abell clusters with richness class ≥ 3 that were detected above 100 mJy with the 91-m telescope;

(3) All sources in the Abell cluster fields observed with the 91-m telescope ($\sim 30'$ in extent) which have $S_{20} \geq 400$ mJy; and

(4) All distance class ≤ 3 clusters with $S_{20} \geq 30$ mJy.

The goal of the survey is to produce a large (~ 300 cluster sources) sample of radio galaxies in rich clusters for an indepth statistical study.

Sources were first observed with the VLA in the C-configuration with a resolution of $\sim 17''$ to map the extended structure. Follow-up observations in the B-configuration with a resolution of $\sim 5''$ were performed to resolve structures such as tails and jets, and to allow accurate optical identifications. Examples of some of the more interesting sources and source structures discovered in our survey are shown in Figure 1.

Although we have not fully completed the reductions of all the Abell cluster fields, we have sufficient data to report on five interesting trends. First, the survey has resulted in the discovery of many new tailed sources, thereby greatly increasing the number of known examples of the head-tail phenomenon (by a factor of 5-10). Second, there are numerous examples of multiple sources (including several head-tail sources) in a single cluster (see e.g., A119, A194, and A400 in Figure 1). Third, a preliminary comparison with the Einstein x-ray observations of Jones and Forman (1983) and Mushotzky (1984) suggests that clusters with cooling cores tend to have compact radio emission

F. Mardirossian et al. (eds.), Clusters and Groups of Galaxies, 295–296.
© 1984 by D. Reidel Publishing Company.

associated with the dominant galaxy. However, these radio sources tend to be less powerful than those associated with central galaxies in other clusters. Fourth, the more extended, powerful, wide-angle tailed sources do not appear to be associated with cooling core clusters. Fifth, there is a tendency for the largest radio sources to be in relatively poor clusters (i.e., richness class 0), and these clusters sometimes have no detectable x-ray emission.

A more detailed comparison of the Einstein x-ray and VLA radio results is being performed in collaboration with groups at CFA and Columbia University.

J.O.B. wishes to gratefully acknowledge partial support of this research and travel by NASA, and additional travel funds from the American Astronomical Society and Conference organizers.

Jones, C. and Forman, W., 1983, Ap. J., in press.
Mushotzky, R.F., 1984, this volume.
Owen, F.N. White, R.A. Hilldrop, K.C., and Hanisch, R.J., 1982, A.J., 87, 1083.

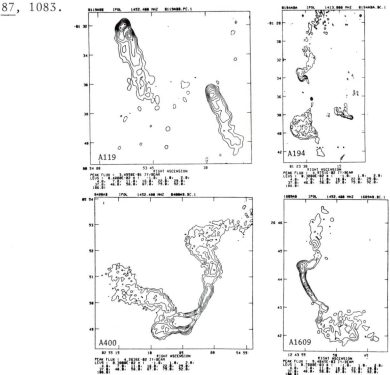

Figure 1. Contour maps of 4 interesting Abell cluster fields. A separate galaxy is coincident with the north, leading edge of each of the tailed sources in A119. Similarly, a galaxy is identified with the compact core at the center of the eastern source in A194 and a separate cluster galaxy lies at the "head" of the western source. A most unusual multiple-tailed morphology is associated with a dumb-bell galaxy near the center of A400. Both nuclei appear to be active, coincident with compact radio cores near the apex of the emission.

HOT GASEOUS CORONAE AROUND EARLY-TYPE GALAXIES

W. Forman, C. Jones, and W. Tucker
Harvard/Smithsonian Center for Astrophysics

ABSTRACT

We discuss the nature of the extended x-ray emission around three elliptical galaxies in the Virgo cluster. Based on energy spectra obtained with the Einstein Observatory, we argue that this emission is hot (~1 keV) gas. Interactions of these coronae with their environment of hot intracluster gas allows their detection at x-ray wavelengths.

INTRODUCTION

One of the important problems in astronomy has been the fate of the gas shed by evolving stars in early-type galaxies. Faber and Gallagher (1976) reviewed this problem in detail and concluded that the increasingly stringent limits on the mass of gas at various temperature regimes required that the gas be expelled from the galaxies in winds like those modeled by Mathews and Baker (1971). In general, these winds are very difficult to detect, having luminosities at soft x-ray energies of 10^{36} ergs sec^{-1} (Mathews and Baker 1971). However, using Einstein imaging observations we have detected hot gaseous coronae around early-type galaxies in clusters. We suggest that in a cluster environment, i.e., in the presence of an intracluster medium, the gas shed by stars becomes detectable in x-rays.

The first systematic study of x-ray coronae around galaxies was carried out by Bechtold et al. (1983) for the cluster A1367. Using high resolution x-ray observations of the cluster, they found nine x-ray sources of characteristic size ~1 arcminute (30 kpc at the distance of A1367, assuming H_0 = 50 km sec^{-1} Mpc^{-1}) associated with galaxies in the cluster. Eight of these galaxies were early types (3 ellipticals, 4 S0s, 1 Sa). Within the ninth x-ray corona, two galaxies were found of types Sa and Sc. Clearly there is a preponderance of early-type galaxies and it is possible that the coronae are associated with only early-type galaxies if the last mentioned corona is associated with the Sa galaxy.

F. Mardirossian et al. (eds.), Clusters and Groups of Galaxies, 297–305.
© *1984 by D. Reidel Publishing Company.*

Bechtold et al. discussed the difficulty of maintaining these coronae against ram pressure and evaporation by the diffuse intracluster gas observed in the cluster. One possibility noted by the authors is that these galaxies may not be permanent residents of the cluster core where the mass loss is most severe. Thus, if these galaxies are only transiting the core, they could have accumulated their coronae in the outer regions of the cluster. The galactic winds, which normally expel gas from the galaxy, would have been halted by the intracluster gas but the ram pressure and evaporation losses might not have been excessive.

In order to study the corona phenomenon in more detail, we have analyzed x-ray images of the Virgo cluster where the proximity of the cluster allows a much more detailed study. Figure 1 shows an image of the Virgo cluster in x-rays made by combining about 20 imaging proportional counter (IPC) fields. The x-ray emission is dominated by that around M87 but several other sources, which we discuss below, stand out; notably NGC4472 (M49) to the south and NGC4406 (M86) to the west of M87. In addition, we will present observations of NGC4374 (M84) which is just visible to the west of NGC4406.

I.   NGC4472 — A PRESSURE CONFINED WIND

We begin with a discussion of the x-ray emission around the elliptical galaxy NGC4472 which optically is comparable in brightness to M87 although it does not have the extended and extensive globular cluster populations of M87 (Harris and Racine 1979 and Van den Bergh 1983). The high resolution x-ray image of NGC4472 is shown in Figure 2 and the basic source parameters are given in Table 1. The surface brightness profile using both the IPC and HRI observations is detectable to a distance of 6.5 arcminutes or 38 kpc (assuming a distance of 20 Mpc to the Virgo cluster). The radial profile (at large radii) can be characterized by a power law of slope -1.5. In addition to studying the spatial distribution of the gas, we have also analyzed the integrated spectral data. We find that the best fit thermal spectrum has kT = 1.1 ± 0.2 keV for a hydrogen column density appropriate to the direction of NGC4472 (Heiles 1975). A power law fit gives a χ^2 of about 150 for 8 degrees of freedom and, therefore, emission mechanisms giving power law spectra are not viable. Although NGC4472 is a weak radio source, the small source extent and low luminosity of the radio source (Kotanyi 1980) argue against its being directly related to the x-ray emission which is ten times larger in extent and 10^5 times more luminous. One other emission mechanism, the integrated luminosity of the globular cluster population, also seems unlikely. First, the typical x-ray luminous globular cluster has an exponential spectrum characterized by kT greater than 5 keV (Hertz and Grindlay 1983). Second, the integrated emission of the 2500 globular clusters in NGC4472 is at most 10^{40} ergs sec^{-1} which is only 5% of the required luminosity (this assumes the globulars in NGC4472 have the same x-ray luminosity function as those in M31 as derived by Van Speybroeck 1983).

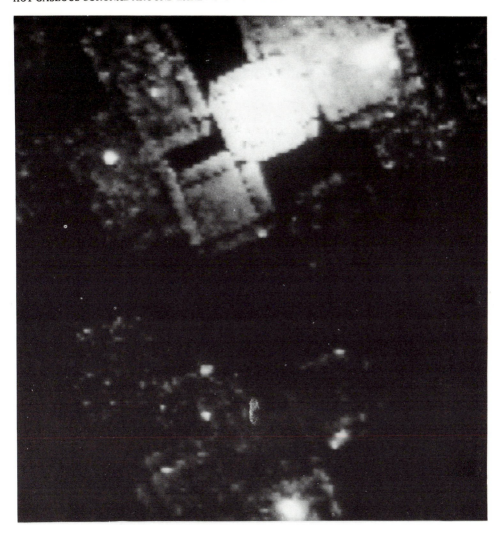

Figure 1. About 20 Imaging Proportional Counter fields in the
Virgo cluster have been merged into a single image. Increasing
brightness corresponds to increasing x-ray surface brightness.
The tic-tac-toe patterns in each individual field are produced by
the detector window supports. The cluster x-ray emission is
dominated by that surrounding M87 (top center) which can be seen
to extend more than one degree from the galaxy which is located
at the center of the very bright field. To the south of M87
(bottom of the figure) extended emission is seen around NGC4472.
To the north-west (upper right) of M87 an extended region of
emission is seen and corresponds to M86. The x-ray emission
around M84 is just barely visible to the west of M86. The scale
of the figure is given by the dimensions of each IPC field of one
degree on a side.

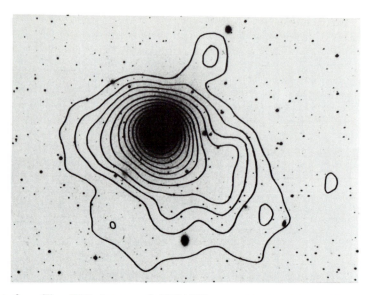

Figure 2. The HRI image of NGC4472 smoothed with a 30 arcsecond Gaussian. Contour levels begin 3σ above background and one spaced in steps at 1σ. The intrinsic resolution of the detector and telescope is about 4 arcseconds. Caution should be used when looking for structure since an individual contour level is usually not a significant indicator.

TABLE 1

	Luminosity (0.5-3.0 keV)	Extent (arc minutes)
NGC 4472 (M49)	2×10^{41}	8
NGC 4406 (M86)	3×10^{41}	10
NGC 4374 (M84)	5×10^{40}	0.5

The most likely emission mechanism is thermal radiation from a hot diffuse gas. However this mechanism also would appear to have some difficulties. First, the gas around NGC4472 is probably not bound to the galaxy. The critical temperature needed for generating a wind is given by Mathews and Bregman (1980) as

$$T_{crit} \sim 10^7 \, M_{12}/R_{15} \qquad (1)$$

where T_{crit} is in degrees, M_{12} is in units of $10^{12} M_\odot$, and R_{15} is in

units of 15 kpc. With a temperature of $\sim 10^{7}\,^{\circ}$K one expects the outer
regions of the hot corona to be unbound (even if the galaxy mass is a
few times 10^{12}M$_{\odot}$). Furthermore, the mass loss rates of ~ 100 M$_{\odot}$
year^{-1} would deplete the entire gas supply in only 10^{8} years. While we
can not detect cluster gas at the distance of NGC4472 away from the
cluster center, we would expect that the gas around M87 continues to
decrease in density to distances of 2-3 Mpc from the cluster center (as
it does in other clusters). Then if we require that the cluster gas and
the NGC4472 galaxy gas be in pressure equilibrium, this equilibrium
occurs at about 35 kpc from the center of NGC4472. If NGC4472 lies out
of the plane of the sky (and thus further from M87) or if the cluster
gas density has begun to fall off faster than r^{-2} or the cluster gas is
cooler at large radii, then the equilibrium radius would be larger.
Given these uncertainties, the calculated value is in good agreement
with the observed size of the corona around NGC4472. Thus, the external
gas from the cluster apparently serves to confine the galactic corona
and prevent its dispersal in a wind.

II. M86 — RAM PRESSURE STRIPPING AT WORK

The second x-ray corona we studied is that around M86 (Forman et
al. 1979, Fabian et al. 1980, Forman and Jones 1982). This system is
more complicated than NGC4472 and shows interesting structure as
discussed by the aforementioned authors and as seen in Figure 3. In
addition to a core of x-ray emission around the galaxy, we also see a

Figure 3. The HRI image of M86 is shown smoothed with a 20
arcsecond Gaussian. The levels are as described in Figure 2. In
addition to emission centered on the galaxy, emission is seen to
the north. Both sections are embedded in a common envelop (see
Figure 1).

plume of emission to the north. At lower surface brightness levels, which are detected in the IPC, both features are surrounded by a common envelope of emission. Table 1 gives the parameters of the source determined from the x-ray observations.

We used the IPC pulse height data to study the energy spectrum for this source. We obtained similar results for the core, the plume, and the entire region including the lower surface brightness emission. For a thermal spectrum, with abundances varied from 50% to 200% of solar, we find kT = 1.0±0.2 keV. As for NGC4472, a power law fit to the energy spectrum gives an unacceptably high value of χ^2.

The arguments in favor of thermal emission are stronger for M86 than for NGC4472 since M86 shows no evidence for unusual activity at radio or optical wavelengths. Furthermore, the peculiar distribution of the emission is difficult to explain with globular clusters. Thus thermal emission from a hot gas appears to be the probable source of the x-rays.

The plume of gas to the north of M86 is most likely gas which has been ram pressure stripped by the intracluster gas which fills the cluster core. In addition, we see in Figure 1 a lower surface brightness region which envelops both the core of gas around the galaxy itself and the detached plume. This gas is probably stripped material in the process of merging with the diffuse cluster gas. Thus, we observe at least one mechanism for the injection of gas into the cluster medium. This gas was originally shed by stars and therefore must contain significant abundances of heavy elements.

III. M84 — PERPETUAL ABLATION

Figure 4 shows a high resolution x-ray image of M84 superposed on the optical galaxy. The x-ray envelop around this galaxy is physically smaller and less luminous than those observed around M86 and NGC4472 (see Table 1). With its velocity of 1011 km sec^{-1} (Tonry and Davis 1981) compared to the mean of the Virgo cluster (1172 km sec^{-1}; Huchra 1983), M84 is probably a resident of the Virgo cluster core. Unlike the other two elliptical galaxies, M86 and NGC4472, M84 is a strong radio source (3C272.1) which has been studied by DeYoung (1980). Both the radio map and the x-ray contours (Figure 4) suggest a small velocity (in a westerly direction) with respect to the cluster gas.

The gas we observe in M84 could be a stable feature since it is tightly bound in the galaxy core and can withstand the ram pressure of the cluster gas. The evaporative losses are also modest with a mass loss rate of 2.5 $(T/3x10^7)^{5/2}$(R/10 kpc) solar masses yr^{-1} where T is the cluster gas temperature in degrees and R is the radius of the corona in kpc. This loss could be replenished by the normal mass loss from stars with a specific mass loss rate of a few 10^{-12} yr^{-1} and a galactic mass of $10^{12} M_\odot$.

Figure 4. The HRI image of M84 smoothed with a 60 arcsecond Gaussian. The levels are as described in Figure 2. A point source is seen to the lower left of the galaxy. The contours are compressed to the west (right) compared to the east. This suggests motion through the intracluster gas in a westerly direction.

One difficulty for the corona around M84 is its short cooling time. In the core, the cooling time is $10^8(T/10^7)^{1.6}/(n/10^{-2})$ years where T is the gas temperature in degrees and n is the central density in units of cm^{-3}. This energy loss from radiation could be replenished either by conduction from the hot external cluster gas or by supernova heating.

IV. CONCLUSION

We have discussed three examples of extended x-ray sources around individual elliptical galaxies in clusters. We have presented arguments showing that these sources are produced by thermal emission from a hot gas. Thus, this gas is most likely that produced by the evolving stars in the galaxies and made detectable by its interactions with the intracluster gas. The detection of this gas opens a variety of new avenues of research to study the gas itself, its interaction with the intracluster gas, and the importance of galactic winds for the formation of the intracluster medium.

ACKNOWLEDGMENTS

We thank K. Modestino for help in preparing this manuscript. This work was supported by NASA Contract NAS8-30751.

REFERENCES

Bechtold, J., Forman, W., Giacconi, R., Jones, C., Schwarz, J., Tucker, W., and Van Speybroeck, L. 1983, Ap.J., 265, 26.

DeYoung, D., Condon, J., and Butcher, H. 1980, Ap.J., 241, 511.

Faber, S. and Gallagher, J. 1976, Ap.J., 204, 365.

Fabian, A., Schwarz, J., and Forman, W. 1980, MNRAS, 192, 135.

Forman, W., Schwarz, J., Jones, C., Liller, W., and Fabian, A. 1979, Ap.J.(Lett.), 334, L27.

Forman, W. and Jones, C. 1982, Ann. Rev. of Astron. and Astrophys., 20, 547.

Harris, W.E. and Racine, R. 1979, Ann. Rev. of Astron. and Astrophys., 17, 241.

Heiles, C. 1975, Astr. and Astrophys. Suppl., 20, 37.

Hertz, P. and Grindlay, J. 1983, preprint.

Huchra, J. 1983, Conference Proceedings of the Trieste Meeting on the Large Scale Structure of the Universe.

Kotanyi, C. 1980, Astron. and Astrophys. (Supp), 41, 421.

Mathews, W.G. and Baker, J.C. 1971, Ap.J., 170, 241.

Tonry, J. and Davis, M. 1981, Ap.J., 246, 666.

Van den Bergh, S. 1983, Conference Proceedings of the Trieste Meeting on the Large Scale Structure of the Universe.

Van Speybroeck, L. 1983, private communication.

DISCUSSION

G. Illingworth: Can you estimate how often M86 has passed through the core of Virgo?

W. Forman: If you assume M86 is presently at the cluster center and has its maximum velocity, then using a rough mass for the cluster we find the time between core crossings to be about 5×10^9 years. Since Virgo is a young system it is possible that this is M86's first pass through the cluster.

G. Illingworth: Then do you expect that M86 will lose all its x-ray gas in his first pass?

W. Forman: It is likely that M86 will lose the bulk of its mass, although it may retain a small amount of gas tightly bound at the galaxy center as we see in M84.

G. Burbidge: Why do you not consider the possibility that the x-ray coronae are due to Compton scattering, or even globular clusters not radiating at the same level as that in our own Galaxy or M31? I suspect that you are going much too far in interpreting everything as due to hot gas.

W. Forman: There are several arguments why the emission we see is mostly likely due to hot gas. Even if you postulate that the clusters around the elliptical galaxies are more luminous, the energy spectra that we know from globular cluster x-ray sources in our own Galaxy are different (much harder) than those of the x-ray coronae. Thus, you need to invent a different process (or modify the process) which produces x-ray emission from individual compact objects in globular clusters.

Second, for two of the three coronae I have discussed, the x-ray spectrum is consistent with thermal emission and inconsistent with a power law slope which you might expect from a non-thermal process.

Y. Rephaeli: In connection to Prof. Burbidge's remark we can, presumably, tell whether there is non-thermal emission by comparison with radio observations from these galaxies.

W. Forman: M84 is a radio source but the structure of the radio source is very different from that in the x-rays. M86, on the other hand, shows no radio emission. Optically, neither galaxy shows emission lines suggesting nuclear activity. Thus, there is no evidence from the optical or the radio of any non-thermal activity in M86 and that in M84 has a different morphology.

IMPLICATIONS OF HEAO A-1 AND EINSTEIN SATELLITE OBSERVATIONS
OF CLUSTERS OF GALAXIES

M. P. Ulmer
Northwestern University

M. P. Kowalski, R. G. Cruddace, K. S. Wood, and G. Fritz
Naval Research Laboratory

Here we present a review of X-ray observations we have made over
the past several years. This work is a survey of all rich clusters with
the HEAO A-1 experiment, plus some follow-up work with the Einstein
Observatory. For brevity, we touch on a few salient points. A full
account of the HEAO A-1 work can be found in Johnson et al. 1983,

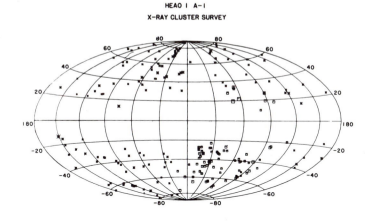

HEAO I A-I
X-RAY CLUSTER SURVEY

Figure 1.
Map of X-ray bright
clusters detected
by HEAO A-1.

Kowalski et al. 1983a, Kowalski et al. 1983b. In Figure 1, we show the
clusters that were detected as X-ray sources. These were found by
explicitly examining the X-ray data associated with each cluster in the
Abell catalog (Abell 1958) and the southern sky catalog of Duus and
Newell (1977). From this, we have derived two complete samples--one
from Abell's "statistical" sample (richness ≥ 1 clusters in a well
defined area of the sky) and that from the southern clusters plus the
richness = 0 Abell clusters in the same portion of the sky as the statistical sample. The resulting luminosity functions are shown in Figures
2a and 2b. There are several points we would like to make based on
these figures. First, it is clear that a simple exponential or power
law function will not fit the data (2a). Second, we see that the data

307

F. Mardirossian et al. (eds.), Clusters and Groups of Galaxies, 307–312.
© *1984 by D. Reidel Publishing Company.*

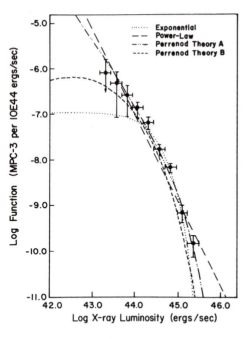

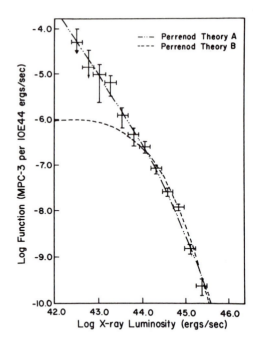

Figure 2a   Figure 2b

favour (at the ∿2σ level), the Perrnod mode A over model B. These
models are based on different hypotheses of cluster evolution, "A" being
based on the hypothesis that galaxies formed before clusters and vice
versa for "B". Figure 2b will, we hope, encourage theoreticians to
explore how sensitive the predicted luminosity function is to the
assumed initial conditions--particularly in the case of cluster
formation prior to galaxy formation. Third, with a well defined
luminosity function, one can search for evolution of the average X-ray
luminosity of clusters. We find no evidence for evolution in the mean
volume emissivity out to a redshift of $z = 0.2$. The sensitivity of our
result is that a change of ± ∿50% could have been detected between
$z = 0.1$ and $z = 0.2$.
     Next, we show three correlations with optically determined prop-
erties of clusters. In Figure 3, we have the X-ray luminosity versus
richness. This is based on the hypothesis that there should be a
correlation between the number of galaxies in a cluster and the amount
of (hot) intra-cluster gas. We have used a low redshift sample only.
This correlation is significant at the 99% level for the sample shown
($z \leq 0.1$). Since selection effects could dominate the high redshift
sample (see Kowalski et al. 1983b), we do not discuss that sample here.
We note that there is a correlation, but not a relation; i.e., there is
a wide range of luminosities within a richness class. Although there
may be some misclassification of richness in the sample, we do not
believe it is enough to account for this dispersion. Next, in Figure 4,
we show the velocity dispersion X-ray luminosity relation. The solid

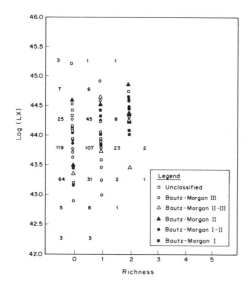

Figure 3

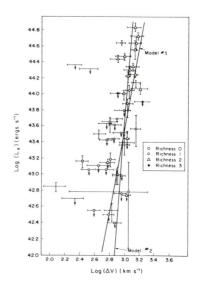

Figure 4

lines are from models by Silk (1976).
Neither model fits the data well. Again,
there is a correlation, but there are
counter examples, as well. Finally, we
show the spiral fraction versus X-ray
luminosity relation in Figure 5. This
correlation holds for the individual
richness classes. This is reassuring,
because there could be a correlation
between richness and spiral fraction
which could have led to an apparent
correlation between spiral fraction and
X-ray luminosity. We note that yet again
there are exceptions to the rule. From
this study of the above correlations, we
suggest that the true effects are due to
a combination of factors; e.g., spiral
fraction, richness, and velocity
dispersion. Thus, clusters cannot be
simply characterized by a single attribute
and they must be measured in detail before
the origin of the X-ray emission can be
understood.

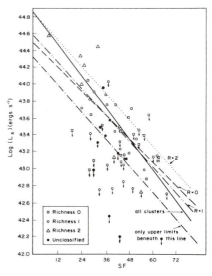

Figure 5

One area of further research is X-ray morphology. In Figure 6, we
see a sample shown from our HEAO-2 follow-up survey. All these clusters
have the same X-ray luminosity within a factor of ∿2; yet, as can be
seen, the structures are all quite different (all these sources are well
resolved, all redshifts of ∿.15 ± .05). We also remark that ∿10% of
all clusters seem to be clumpy, and that these clumpy clusters have been

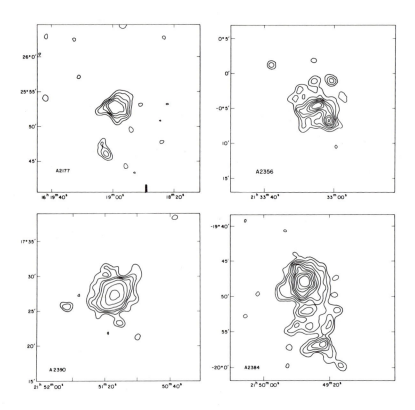

Figure 6.   0.7 - 3.5 keV X-ray brightness contour plots
            from 2000 sec HEAO-2 observations with the IPC

hypothesized to evolve in the time scales as short as 10^9 years. That we
don't see any evidence for evolution of the luminosity function is
suggestive that these clusters evolve more slowly.

A second line of research is to make energy dependent maps of these
clusters to determine temperature profiles and element abundances. We
have such an experiment planned for August 1984 on the Space Shuttle
(STS-16). This experiment will have about 1000 cm^2 collecting area and
a passive collimator with a 3^0 x 5' (full width half maximum) field of
view. We expect to have about 40 hours of observing time and we plan to
map several bright clusters and supernova remnants in the ∿1 - 15 keV
energy range. The experiment is called SPARTAN-1, and it is shown in
Figure 7.

In conclusion, the origin of the X-ray emission from clusters of
galaxies is complex. Further detailed studies of X-ray, optical, and
radio properties are needed in order to understand the origin and
evolution of the X-ray emission.

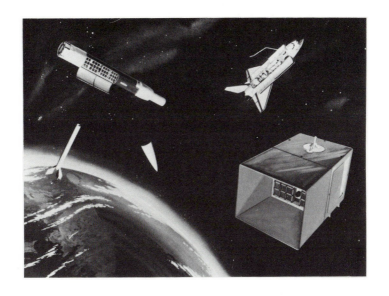

Figure 7.   Artist's conception of the SPARTAN-1
            project-sounding rocket technology
            flown on the space shuttle.

REFERENCES

Abell, G. O., 1958, Ap. J. Supp., 3, pp. 211-288.
Duss, A. and Newell, B., 1977, Ap. J. Supp., 35, pp. 209-219.
Johnson, M.W., Cruddace, R.G., Ulmer, M. P., Kowalski, M. P., and
      Wood, K. S., 1983, Ap. J., 266, pp. 425-445.
Kowalski, M. P., Ulmer, M. P., and Cruddace, R. G., 1983a. Ap. J., 268,
      pp. 540-551.
Kowalski, M. P., Ulmer, M. P., Cruddace, R. G., and Wood, K. S., 1983b,
      submitted to Ap. J. Supp.
Silk, J., 1976, Ap. J., 208, p. 646.

DISCUSSION

DRESSLER: I understood you to say that the spiral fraction is correlated with cluster richness. The fundamental correlation of spiral fraction is with the galaxy density in a region, and any derivative correlation with cluster richness may be influenced by selection effects in choosing cluster samples.

ULMER: I did say that I thought richness and spiral fraction could be correlated. Probably galaxy density is a much better parameter to use, but we have used richness because there are so many more values available.

BURBIDGE: What is the luminosity of the brightest X-ray cluster? Is there anything morphologically peculiar about the high luminosity systems?

ULMER: I'm really not sure, but I would guess the brightest ones are centrally condensed smooth clusters with a cD galaxy in the center. Perhaps Christine Jones will tell us more in her talk.

DARK MATTER IN POOR CLUSTERS OF GALAXIES

Gerard A. Kriss
Department of Astronomy
University of Michigan
Ann Arbor, MI 48109-1090

X-ray observations of 16 poor clusters containing central dominant galaxies have been performed with the Imaging Proportional Counter of the Einstein Observatory. Twelve clusters were detected, and in each case the X-ray emission is centered on the dominant galaxy. For the six brightest clusters we find extended X-ray emission that is smooth, centrally peaked, and reasonably symmetric. The X-ray surface brightness implies density profiles that are roughly inversely proportional to radius. We find total binding masses of $\sim 10^{13}$ $M_\odot$ within 0.5 Mpc and mean mass to light ratios of 100 M/L_V. Comparison of the deduced distribution of binding mass with the light distribution of the central galaxies of four clusters shows that M/L_V rises to over 200 in the galaxy halos.

1. INTRODUCTION

The clusters studied here were catalogued by Morgan, Kayser, and White (1975, MKW) and by Albert, White, and Morgan (1977) as possible examples of cD galaxies in poor clusters. Since cD galaxies are usually found at the centers of rich clusters, these poor clusters comprise an interesting sample. The environments of these poor clusters are different from those of rich clusters, and so they provide fertile ground for investigations of the formation and evolution of the central galaxies. This may proceed through the merging of smaller galaxies (e.g., Hausman and Ostriker 1978; Roos and Norman 1979) and the tidal stripping of other cluster galaxies. The accretion of gas onto the dominant galaxy may also influence its evolution (Fabian, Nulsen, and Canizares 1982). The review talk by A. Dressler covers each of these points in more detail.

Although they are poor, these clusters are a smooth continuation from the richer Abell systems. Many of their optical morphological characteristics have been studied by Bahcall (1980). They typically are at or below richness class 0 on the Abell system, have low velocity dispersions, low to moderate central densities, and fairly high spiral fractions.

F. Mardirossian et al. (eds.), Clusters and Groups of Galaxies, 313–318.

2. OBSERVATIONS AND PHYSICAL PARAMETERS

Details of the observations and the data analysis are given by Kriss, Cioffi, and Canizares (1983). The six resolved images show the same centrally peaked, symmetrical structure which Jones et al. (1979) and Forman and Jones find for relaxed, rich clusters with cD galaxies (their XD clusters). This morphology could describe the unresolved clusters as well - their extent would be masked by poor photon statistics. The X-ray contours of AWM 7 and AWM 4 show clear ellipticity that extends over most of the image. In each case the major axis of the X-ray distribution is approximately aligned with the major axis of the central galaxy.

Spectra were obtained for the bright, extended sources as described by Fabricant and Gorenstein (1983). The best fit temperatures for each cluster are given in Table 1. The error bars are 90% confidence limits for two interesting parameters (kT and N_H). Using our knowledge of the temperature, we can derive radial density profiles from the surface brightness profiles of the X-ray emission. This involves iteratively deprojecting the X-ray image via a method similar to Fabian et al. (1981). The radial density profiles are illustrated in Figure 1. Each profile is well represented by a power law as a function of radius R of the form $n_0 R^\alpha$. The constants n_0 and α are listed in Table 1. The density of the central bin n_c is also given for each cluster along with the cooling time for that bin. For all six of the resolved clusters the central gas densities at the observed temperatures are sufficiently high that the cooling time t_c of the gas is shorter than a Hubble time ($H_0=50$). This implies that a radiative accretion flow onto the central galaxy can have been established.

TABLE 1.   PROPERTIES OF EXTENDED SOURCES

Cluster	kT	n_0 (cm^{-3})	α	n_0 (10^{-3}cm^{-3})	t_c (10^9 yr)
MKW 4	1.7 ± 0.5	0.21	-1.22	3.8	8
MKW 9	>1.0	0.87	-1.43	2.1	>6
MKW 3s	3.2 ± 2	2.17	-1.35	1.2	8
MKW 4s	>1.0	0.12	-1.06	1.3	>8
AWM 4	2.4 ± 1	0.58	-1.28	2.2	9
AWM 7	$1.0^{+4.0}_{-0.5}$	0.32	-1.04	5.3	2

3. BINDING MASSES AND MASS-TO-LIGHT RATIOS

The temperatures and the radial density profiles from the preceding section enable us to estimate the mass required to bind the X-ray emitting gas to the cluster. If we assume that the hot gas is in hydrostatic equilibrium, then the gradient of the density profile will give the binding mass M_b within a radius R:

$$M_b = \frac{-kT}{G\mu m_p} \left(\frac{d \ln \rho}{d \ln R} + \frac{d \ln T}{d \ln R} \right) R \qquad (1)$$

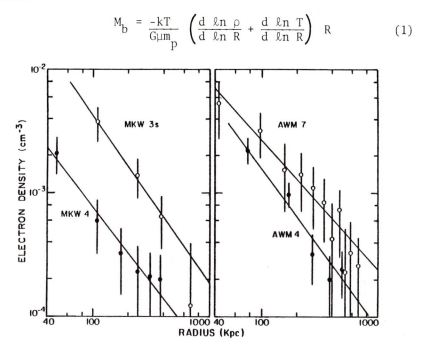

Figure 1. Electron density vs. radius for the X-ray emitting gas. The lines are power law fits, $n_e = n_0 R^\alpha$. Values of n_0 and α are in Table 1. Reproduced with permission from Kriss, Cioffi, and Canizares (1983).

Here μ is the mean mass per particle and M_p the proton mass. We assume that the gas is nearly isothermal, so the derivative of T is ~0. It is unlikely that T changes by more than a factor of 2-3 over a decade in R, so this assumption contributes uncertainties of ~30-50%. Our poor determination of average temperatures can contribute uncertainties of a factor of 2. For uniformity we tabulate the mass of the gas M_{gas} and the binding mass M_{bind} within R=0.5 Mpc in Table 2. The visual luminosities within 0.5 Mpc listed in Table 2 include the luminosity of the first brightest galaxy from Thuan and Romanishin (1981), Bahcall's (1980) galaxy counts, and a slight correction for fainter galaxies using a Schechter luminosity function.

TABLE 2.   MASSES, VISUAL LUMINOSITIES, AND M/L_V

Cluster	M_{gas} $(10^{12} M_\odot)$	M_{bind} $(10^{12} M_\odot)$	L_V $(10^{11} L_\odot)$	M/L_V $(M_\odot/L_\odot)$
MKW 4	2.8	35	5.0	70
MKW 9	3.5	49	6.3	78
MKW 3s	14	73	7.8	94
MKW 4s	3.9	36	-	-
AWM 4	5.5	52	5.1	100
AWM 7	12	71	8.8	81

From 30% to 70% of the light in each cluster is coming from the central dominant galaxy. This light is much more highly concentrated than the binding mass derived from the X-ray surface brightness profiles. We infer that M/L_V must rise with radius in the central galaxies much as Dressler (1979) found for the cD galaxy in the rich cluster A2029. In more quantitative terms we can compute M/L_V as a function of radius for those galaxies for which both optical and X-ray surface brightness profiles are available. Figure 2 shows the binding mass density ρ_b, the luminosity density ρ_V, and M/L_V all versus radius (the moderate ellipticities have been neglected). For all four clusters the derived M/L_V distribution rises by nearly an order of magnitude within the galaxy.

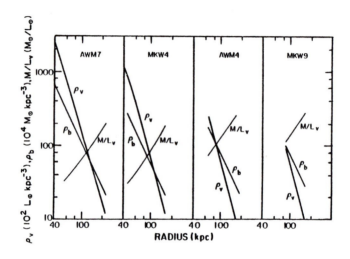

Figure 2. Radial distributions of binding mass density ρ_b, visual luminosity density ρ_V, and M/L_V. Reproduced with permission from Kriss, Cioffi, and Canizares (1983).

4.   DISCUSSION

As described above, the X-ray data on poor clusters with dominant galaxies reveal that these systems bear a striking similarity to rich clusters with cD galaxies. The optical data show additional similarities between the rich and poor clusters. The dominant galaxies of the MKW and AWM clusters have larger core radii and lower central surface brightnesses than normal ellipticals and often show multiple nuclei (Thuan and Romanishin 1981). In several cases the dominant galaxy of the cluster is more luminous for its central velocity dispersion than is predicted by the relation derived for normal ellipticals (Malumuth and Kirshner 1981, 1983). Malumuth and Kirshner (1981) argue that this is typical of cD galaxies that have formed in mergers.

The only real apparent difference, then, between the dominant gala-xies of the MKW and AWM clusters and the cD galaxies in rich clusters is the lack of a luminous envelope surrounding the dominant galaxies in poor clusters. This point has been emphasized by Thuan and Romanishin (1981), for if it is true, the environmental difference between poor and rich clusters may be directly responsible for the formation of the enve-lope (see also Oemler 1976). Malumuth and Kirshner (1981) and Thuan and Romanishin (1981) propose that both merging and stripping are at work in the formation of the central galaxy. Mergers form the main body of the galaxy while stripping forms the envelopes. If poor clusters are smaller and have lower velocity dispersions than rich clusters, then mergers pro-ceed just as quickly in poor clusters, but stripping is more probable in rich clusters. Our X-ray data indicate that the dominant galaxies of poor clusters satisfy one major requirement of the merger hypothesis - they lie at the centers of their cluster potentials.

REFERENCES

Albert, C.E., White, R.A., and Morgan, W.W. 1977, Ap.J., 211, 309 (AWM).
Dressler, A. 1978, Ap.J., 226, 55.
Dressler, A. 1979, Ap.J., 231, 659.
Fabian, A.C., Hu, E.M., Cowie, L.L., and Grindlay, J. 1981, Ap.J.,248,47.
Fabian, A.C., Nulsen, P.E.J., and Canizares, C.R. 1982, M.N.R.A.S.,
     201, 933.
Fabricant, D., and Gorenstein, P. 1983, Ap.J., 267, 535.
Hausman, M.A., and Ostriker, J.P. 1978, Ap.J., 224, 320.
Jones, C., Mandel, E., Schwartz, J., Forman, W., Murray, S.S., and
     Harnden, F.R., J. 1979, Ap.J.(Letters), 234, L21.
Kriss, G.A., Canizares, C.R., McClintock, J.E., and Feigelson, E.D. 1980,
     Ap.J.(Letters), 235, L61.
Kriss, G.A., Cioffi, D.F., and Canizares,C.R. 1983, Ap.J., 272, 439.
Malumuth, E.M., and Kirshner, R.P. 1981, Ap.J., 251, 508.
_____. 1983, preprint.
Morgan, W.W., Kayser, S., and White, R.A. 1975, Ap.J., 199, 545(MKW).
Oemler, A.,Jr. 1976, Ap.J., 209. 693.
Roos, N., and Norman, C.A. 1979, Astr.Ap., 76, 75.
Schwartz, D.A., et al. 1980, Ap.J.(Letters), 238, L53.
Schwartz, D., Schwarz, J., and Tucker, W., 1980, Ap.J.(Letters),238,L59.
Thuan, T.X., and Romanishin, W. 1981, Ap.J., 248, 439.

DISCUSSION

PORTER: I don't think that all "poor cluster cDs" lack envelopes, even if you define an envelope to be an excess at larger r above a fit to the galaxy surface brightness at small r. Chris Morbey thinks that some of the AWM and MKW galaxies have such envelopes, and their profiles are often very shallow even if they don't. One example I can rouch for is WP23: optically, it is a poor group, but its X-ray emission is unusually bright and extended, and my surface photometry shows that it does have an envelope. The point is that a poor group is not necessarily a low density environment.

KRISS: Shallow profiles probably are the best defining characteristic of a cD. The envelopes Morbey sees for some AWM and MKW galaxies are much lower in surface brightness and are a much smaller fraction of the total luminosity than similar structures in rich clusters. True, a poor group is not necessarily a low density environment, but the central densities of the AWM and MKW clusters are low to moderate at best. The densest clusters are ~ 2 times less dense than say, A2029 or A2256.

BURBIDGE: How do you know there are "cooling flows"? How do you know the matter is falling in rather than being ejected?

KRISS: Cooling flows are only inferred for these clusters since their central cooling times are shorter than a Hubble time. We have no direct measure of cooler gas at the centers of these clusters as exists from M87 and NGC1275. If temperatures at the center are lower, however, for reasonable profiles of the gravitational potential, only infalling material will satisfy the hydrodynamic equations.

HUCHRA: Be careful when you say taht cD's depart significantly from the Faber-Jackson $L \propto \sigma^{\alpha} v$ relation. The original paper (preprint) of Malumuth & Kirshner did not correct the luminosities of their comparison galaxies – most of which are at low velocities – for the infall into Virgo. When the correction was made, the cD's fell very close to the normal relationship.

KRISS: Further work by Malumuth in his thesis supports the original result. I believe the Virgo infall has been corrected, but I am not sure.

EINSTEIN IMAGES OF CLUSTERS OF GALAXIES:
GALAXY HALOES, THE INTRACLUSTER MEDIUM, AND THE INTERCLUSTER GAS

C. Jones and W. Forman
Harvard/Smithsonian Center for Astrophysics

ABSTRACT

Analyses of the x-ray images obtained with the Einstein Observatory have provided a new and different view of clusters of galaxies. The primary topics discussed in this review are the role of the central dominant galaxy, the origin of SO galaxies, cluster classification and the dynamically young state of most clusters, the large scale height of the gas compared to the galaxy distribution, and the effects of a hot intercluster gas on the intracluster gas.

INTRODUCTION

The x-ray imaging capability of the Einstein Observatory has provided a wealth of material for the study of clusters of galaxies. Since the gas traces the cluster gravitational potential, the x-ray images of the intracluster gas are a powerful tool in the study of cluster structure. These images also provide essential information for analysis of the interaction of galaxies with the intracluster gas as well as information about the environment outside clusters. Analysis of an x-ray observation of the gas in a cluster of galaxies allows the determination of

> X-ray luminosity
> Cluster structure (mass distribution)
> Core radius and gas scale height
> Gas temperature
> Gas density (mass)
> Cooling times and inflow rates
> Cluster virial mass.

With our knowledge of these cluster parameters, one can address many problems of current interest in cluster research. We have chosen a few of those topics and ordered their discussion by scale.

F. Mardirossian et al. (eds.), Clusters and Groups of Galaxies, 319–340.
© 1984 by D. Reidel Publishing Company.

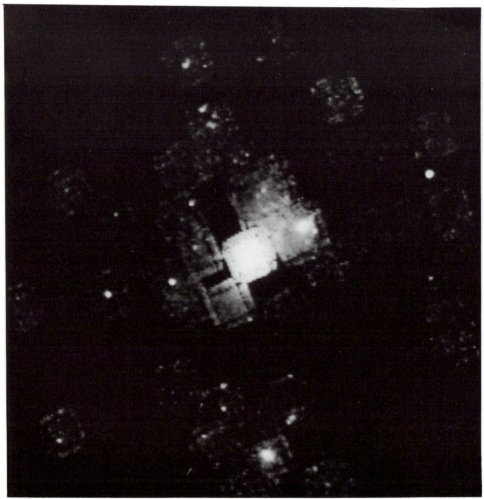

Figure 1. The 0.5–3.0 keV image of the Virgo cluster. X-ray emission from M87's halo dominates the central region. Each field is a single $1^\circ N^\circ$ IPC image.

I. GALAXIES AND THE INTRACLUSTER MEDIUM

A.1. The central dominant galaxy

In nearly all cluster classification schemes, the presence of a large, central galaxy has played a fundamental role. In the Bautz–Morgan system (Bautz and Morgan 1970), the classification is based on the difference in magnitude between the first ranked galaxy and the second brightest member. In the Rood–Sastry system (Rood and Sastry 1971) the cD class contains those clusters with the most spectacular central galaxies. In x-rays, we have used the term XD for clusters with small core radii and central dominant galaxies.

It was primarily through the x-ray observations that the dominant role played by the central galaxy in clusters at early stages of their dynamical evolution came to be fully realized. The Virgo cluster where M87 plays this role, is one of the best examples.

Figure 1 shows the x-ray emission from the central ~$10 \times 10^{\circ}$ region of the Virgo cluster. The dominance of the emission from M87 is apparent. Fabricant and Gorenstein (1982) have shown that the x-ray halo around M87 requires a large unseen mass to bind the hot gas. Figure 2 shows the integrated mass as a function of radius. The x-ray observations require a mass for the halo of M87 in excess of 10^{13} $M_{\odot}$. The mass to light ratio increases with radius such that at a $20^{\prime}$ radius, it is

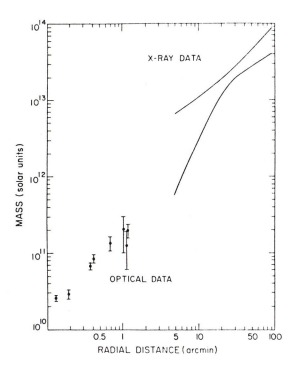

Figure 2. From Fabricant and Gorenstein (1983); the integrated mass of M87's halo is shown as a function of radius. The allowed mass range derived from the Einstein observation is bounded by the two solid lines.

~200, a value approaching that for clusters of galaxies. The presence
of this dark halo is not unique to M87 but is observed around other
central dominant galaxies including those in the Centaurus and Hydra
clusters.

The existence of dynamically unevolved clusters (clusters with low x-ray
gas temperatures and low velocity dispersions, high spiral fractions,
irregular galaxy distributions) with massive centrally located galaxies
has important implications for the formation of these galaxies, for the
subsequent dynamical evolution of clusters with and without such
galaxies, and for the general classification of clusters.

The presence of these massive aggregations in dynamically young clusters
would at first seem to conflict with ideas of dynamical evolution where
for example Hausman and Ostriker (1978) suggested that the increasing
dominance of the brightest galaxy in the Bautz-Morgan "sequence of
cluster types is essentially one of increasing dynamical
evolution....with denser systems (those with shorter internal relaxation
times) progressing further in a given time". In fact, however, as more
realistic and detailed work has been carried out on numerical modeling,
we find that the theory does agree with the observations. In the
hierarchical clustering scenario for the formation of clusters, small
groups form first. Subsequent hierarchical clustering ultimately
results in a single relaxed system. Cavaliere and colleagues have been
exploring the consequences of such a scenario including mergers and
stripping of individual galaxy halos. In particular Carnevali et
al. (1981) find that merging is most important in the small groups.
They describe the phenomenon as a "merging instability" with extensive
galaxy merging and the stripping of halos. This is not surprising since
the timescales for two-body interactions depend on the velocity
dispersion of the system and the galaxy cross sections. In addition
Merritt (1983) has argued that little further merging occurs subsequent
to the violent relaxation of the cluster. Thus, the presence of a
dominant central galaxy is largely determined at the early stages of a
cluster's dynamical evolution.

Galaxies like M87 probably owe their dominance to their unique position
at the cluster center where their sizes are not subject to tidal
limitations by the cluster's gravitational field. The x-ray
observations of dynamically young clusters like Virgo show that central
dominant galaxies with massive extended halos are often present.

2. Stripping and evaporative effects - Origin of SO Galaxies

The intracluster medium can effect the evolution of galaxies by the
removal of interstellar material through the processes of ram pressure
stripping and evaporation by the hot medium. These mechanisms had been
suggested by Gunn and Gott (1972) and Cowie and Songalia (1977) to be
responsible for the removal of interstellar material from spiral
galaxies to produce SO systems. Dressler (1980a) and Burstein
(1979a,b,c) have demonstrated structural differences between SO and

spiral galaxies and cast considerable doubt on the possibility that SO galaxies are stripped spirals. The x-ray observations confirm that not all SO galaxies are stripped spirals but that some are formed as a distinct Hubble class.

In particular by combining the parameters of gas density and temperature with galaxy velocities, we can calculate the ram pressure stripping force and the evaporative rates. The ram pressure depends on the galaxy velocity which is measured by the three dimensional cluster velocity dispersion and the density of the intracluster medium which we determine from the x-ray observations. Figure 3 shows the velocity plotted against the central gas density and the velocity and density required if a spiral galaxy passing through the cluster center is to be stripped of its interstellar material and the velocity and density required if a galaxy at 3 core radii is to be stripped. Clusters that are spiral rich are marked with open symbols and those that are spiral poor with solid symbols. This graph shows that the spiral rich clusters tend to be located in regions where the ram pressure force is at best only effective in the core region. Although generally spiral poor clusters are in regions of high ram pressure, two spiral poor systems (A194 and A1314) fall in the region where stripping should not be effective. In A194, about half the galaxies were classified by Oemler (1974) and Dressler (1980b) as SO's. This large population of SO galaxies could not be produced by the ram pressure stripping of gas from normal spirals.

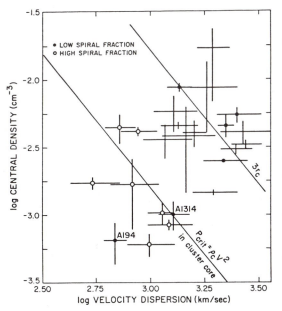

Figure 3. Central gas density is ploted against cluster velocity dispersion. Clusters with high spiral fractions (>30%) are shown as open symbols which those with low spiral fractions (<30%) are identified with solid symbols. The remaining clusters have no published spiral fractions. The solid line labeled "in cluster core" divides the diagram such that galaxies in clusters to the lower left of the line experience ram pressures less than the critical stripping value at the cluster center.

The alternative mechanism for spiral to S0 conversion is evaporation of interstellar material by the hot intracluster gas. Figure 4 shows a plot of the intracluster gas temperature against the gas density. Since the calculated rates are probably capable of depleting the interstellar medium in all galaxies in the cores of all rich clusters, the argument that evaporation is not effective is based on the comparison of spiral rich systems with A194 and A1314. Since the density and temperature of the intracluster medium in A194 and A1314 is comparable to that in other spiral-rich systems, Jones and Forman (1983) argued that evaporation would not preferentially act in one type of cluster more than in another. Therefore, evaporation would be suppressed and could not be used to explain the presence of S0's in some clusters but not others.

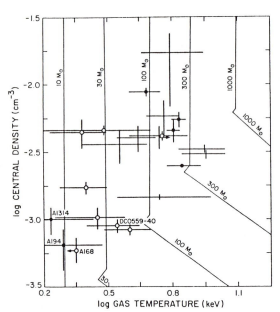

Figure 4. The central gas density is plotted against the gas temperature. The solid lines show the evaporation rate of interstellar material in solar masses per year from a galaxy with a radius of 15 kpc. The break in the lines indicate the transition from the classical (upper left) to saturated regime (lower right). Solid symbols are used for clusters with low spiral fractions (<30%) and open symbols for clusters with high spiral fractions (>30%).

B. Galactic Coronae

Although evaporation and ram pressure stripping probably do not play a substantial role in the formation of S0 galaxies, these processes do substantially effect galaxies and the x-ray emitting or HI coronae observed around galaxies in clusters.

In addition to the central dominant galaxy, other early type galaxies are observed to have extended x-ray halos (see Forman 1984, this volume for a discussion of the galactic coronae in the Virgo cluster). A few examples are M84, M86, and NGC4472 in Virgo. In A1367, Bechtold et al. (1983) observed early type galaxies with extended x-ray emission.

In clusters like A194 (Figure 5), it is not possible to distinguish whether the extended x-ray emission around the galaxy pair NGC545/547

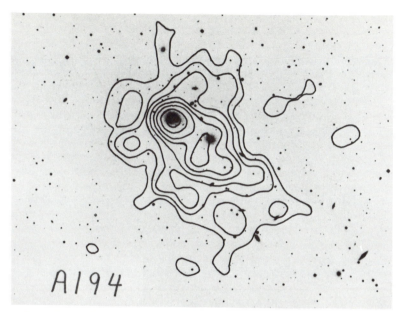

Figure 5. Iso-sigma contours of the 0.5-3.0 keV image of A194 are shown superposed on an optical PSS photograph. The image was first smoothed with a Gaussian of FWHM of 60 arcseconds. The contour levels start at 3 standard deviations above background and are separated by 1σ.

should be classed as a small central M87 type halo or as a corona. A morphological continuum of extents and masses of galactic corona probably exists. In the southern cluster DC0622-64 (Figure 6), there are two extended coronae, one of which may be centered on a central stationary galaxy.

For the coronae for which we have suitable observations to determine the parameters, we measure cool gas temperatures of ~1 keV, x-ray luminosities up to ~10^{42}ergs/sec and typical extents of ~50 kpc. Many of the galaxies with corona are located well outside the cluster cores. Observations are consistent with observing hot x-ray emitting coronae only around early type galaxies.

II. STRUCTURE OF THE INTRACLUSTER MEDIUM

The x-ray observations provide a different view of clusters of galaxies than has been obtained in optical light. Although we sometimes observe x-ray emission from individual galaxies, the primary source of the emission is generally the hot, diffuse, low density medium that permeates the entire cluster. This gas is a tracer of the gravitational potential and, therefore, we can use the gas distribution to study the dynamical evolution of clusters.

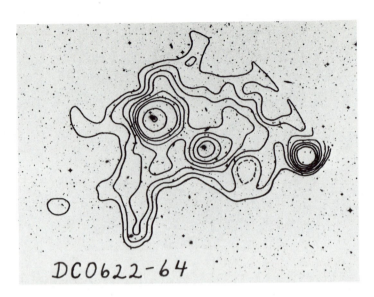

Figure 6. Iso-sigma contours of the 0.5-3.0 keV image of the southern cluster DC0622-64 are shown superposed on an optical photograph. The image was smoothed with a 60 arcsecond FWHM Gaussian. In addition to the cluster x-ray emission, two extended x-ray sources are observed around to two brightest D and E galaxies. The x-ray source on the extreme right is an unidentified point source.

A. Cluster Structure and Dynamical Evolution

Table 1 lists the dynamical indicators for clusters of galaxies and the relevant physical processes.

TABLE 1
The Relation of Optical Classification Schemes to Dynamical Evolution

	Open		Regular
Zwicky Type	Open	⎫	Compact
Symmetry	Irregular	⎬ Violent	Spherical
Central		⎪	
Concentration	Very little	⎬ Relaxation	High
Substructure	Significant	⎭	Very little
Bautz-Morgan	II-III, III	⎫ Mergers and	I, I-II, II
Rood-Sastry	Flat, Irregular	⎬ Dynamical Friction	cD, B
Content	Spiral-rich	Ram pressure, evaporation or formation in dense regions	Elliptical-rich

Figure 7. The 0.5-3.0 keV image composed of a composite of IPC
fields around the double cluster SC0627-54. The extended source
north of SC0627-54 is another cluster in this supercluster as is
the weaker source at the upper right (just west of Canopus).

The evolutionary indicators can be tied directly to the dynamical
processes that occur during the cluster collapse. The regular clusters
are often considered to be dynamically more evolved and certainly have
shorter dynamical timescales.   The irregular clusters would be less
dynamically evolved and would have longer timescales.   The cluster
symmetry, central density and central concentration would all be
expected to increase during the cluster collapse and violent relaxation.
One would expect subclustering to progressively disappear and the galaxy
velocity dispersion to increase as the central potential deepens.   With
regard to galactic content as an indicator, Dressler (1980a) argued that

SO's are preferentially formed in locally denser regions rather than being transformed spirals. However, since denser regions have shorter dynamical timescales, population should remain, locally, an indicator although somewhat poorer of the dynamical age of a cluster.

One of the most striking demonstrations of the dynamical evolution of clusters was the discovery in the Einstein images of clusters that had a bimodal mass distribution (Forman et al. 1981). These had previously been selected optically as single systems. Figure 7 shows an x-ray composite of several imaging proportional counter fields, including the southern double cluster SC0627-54. This double cluster is part of a supercluster which includes the x-ray bright cluster to the north, and the fainter cluster located west of the bright star Canopus.

The interpretation of these double systems was facilitated by a single numerical simulation that White (1976) had analyzed and which is shown in Figure 8. Beginning with an initially expanding uniform distribution one sees subclustering on larger and larger scales until only two large aggregations remain. Their separation and sizes are comparable to the double clusters discovered in the x-ray observations.

Ten years ago Gunn and Gott (1972) remarked in a paper that "The present is very much the epoch of cluster formation". While the richest, most regular clusters, like Coma, have been dynamically quiescent for about five billion years, clusters with much lower densities are not yet

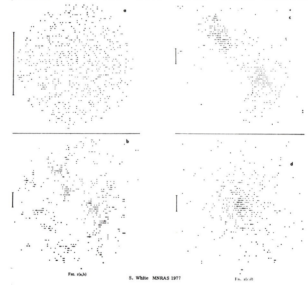

Fig. 1(a,b)   S. White MNRAS 1977   Fig. 1(c,d)

Figure 8. The numerical simulation of White (1976) shows an initially expanding set of galaxies (upper left), followed by the formation of small groups (lower left). A bimodel subclustering follows (upper right) which may correspond to the evolutionary state of the double clusters. At the lower right is the final relaxed cluster.

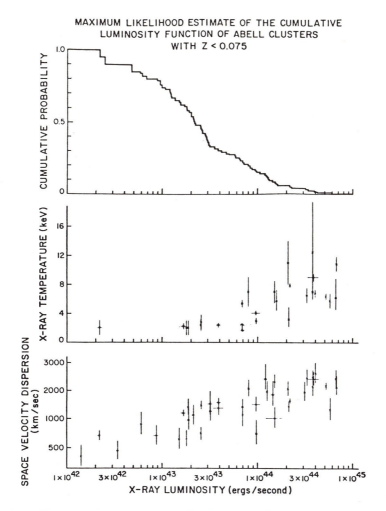

Figure 9. The cumulative x-ray luminosity function is shown for a complete sample of Abell clusters. Below it are plotted measured cluster velocity dispersions and gas temperatures.

virialized. The dynamical timescale for the Coma cluster (R ~3 Mpc, M ~2 x 10^{15} $M_\odot$) is roughly 4 x 10^9 years. Since the dynamical timescale is given by T_D ~$(G\rho)^{-1/2}$ then clusters with densities 1/10 that recombination and would not be virialized systems. Thus, all but richest (more properly densest) clusters are not yet virialized systems of galaxies.

For a complete sample of Abell clusters, we have estimated the percentage of dynamically evolved and unevolved systems. Of the many dynamical indicators, the one which is available for a large sample of clusters is the x-ray luminosity.

Using x-ray observations of all Abell clusters with z <0.075 in the part of the sky contained in Abell's (1958) complete survey, we have generated an x-ray luminosity function for rich clusters. For this luminosity function we need to select a luminosity threshold which can be used to separate virialized from non-virialized systems. We do this based on the disappearance of substructure and x-ray coronae around galaxies and the change in x-ray gas temperature from cool (~2 keV) to hot (~7 keV) which corresponds to a transformation from a condition in which gas can be associated with galaxies to one in which gas can only be contained by the entire cluster potential.

Figure 9 shows this cumulative luminosity function computed from the complete sample of nearby Abell clusters. Beneath it, against luminosity, are plotted x-ray gas temperatures and cluster velocity dispersions. Although there is not a precise break, but a smooth transition from unevolved to evolved systems, if we select L_x ~1 x 10^{44} ergs s^{-1} as a cutoff, then about 30% of clusters would be virialized systems. This is consistent with Dressler's (1980b) comment that based on the low central concentrations of galaxies in the 55 clusters he studied, only 20-30% had completed the violent relaxation phase and with the substructure analysis of these same clusters by Geller and Beers (1983). The x-ray images show us observationally what Gunn and Gott (1972) noted 10 years ago based on theoretical grounds - most clusters are _not_ relaxed virialized systems.

B.   Cluster Structure and Classification

Using Einstein imaging observations we have studied a sample of about 50 nearby clusters to explore their properties (Jones and Forman 1984). The clusters are described analytically with a simple two parameter model - the hydrostatic-isothermal model - as suggested by Cavaliere and Fusco-Femiano (1976,1981), and Bahcall and Sarazin (1977). Two parameters which we measured are the core radius of the virial mass distribution as traced by the intracluster gas and the ratio of energy per unit mass in galaxies to that in gas. In this formulation the x-ray surface brightness distribution can be written as

$$S(r) = S(0) \ [1+(r/a)^2]^{-3\beta+1/2}$$

where a is the core radius and $\beta = \mu m v^2/3kT$ is the ratio of the energy per unit mass in galaxies to that in gas. In addition, we allowed for cooling in the central core regions of clusters by omitting the central few bins from our fits to the hydrostatic-isothermal models. Abramopoulus and Ku (1983) have completed a similar analysis of clusters but have kept β fixed at 1.0.

All the clusters in our sample have comparable limiting surface brightness distributions at "large" radii with β ~ 0.6-0.8 and most interesting for half the clusters, β is constrained by the surface brightness profile to be less than 1.0. We will return to this point in section II.C.

The cluster core radii vary from <100–800 kpc with all of the small core radii associated with clusters having bright galaxies at their centers. We performed the following test to see if there was a significant difference between small and large core radii systems. We divided clusters into two groups – those in which the brightest galaxy in the core (within 2 core radii or at least 0.5 Mpc) lay at the cluster center as determined from the x-ray observations and those in which it did not. We then formed the cumulative distributions in core radius for these two classes which are shown in Figure 10. The probability that these two distributions are drawn from the same parent population is about 10^{-3}. Therefore, we conclude that the presence of a bright centrally located galaxy is related to the small core radius of the parent cluster. It is at present difficult to know which property is the more fundamental but it seems that both are well established by the end of the violent relaxation phase.

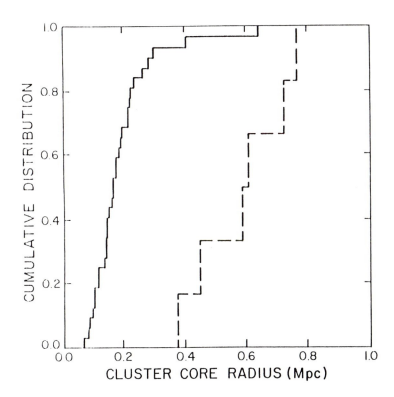

Figure 10. The solid line shows the cumulative distribution versus cluster core radius for clusters with an optically dominant galaxy at the centroid of the x-ray emission. The dashed line is the cumulative distribution for clusters which do not have a central bright galaxy. The K-S test shows that the probability is $<10^{-3}$ that both distributions are drawn from the same parent population.

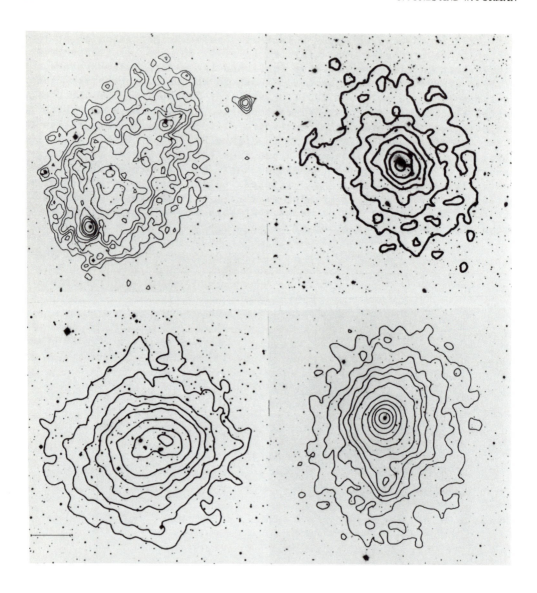

Figure 11;   the 0.5–3.0 keV isointensity contours of A1367 (upper
left),   A2256   (lower   left).   A262 (upper right), and A85 (lower
right) are shown superposed on optical prints.   A1367   and   A2256
represent   the   dynamically   unevolved and evolved states for the
class of clusters with no central dominant galaxies (nXD systems)
while A262 and A85 represent these dynamical states for the class
of clusters with central dominant galaxies (XD systems).

Table 2

nXD	XD
~400–800 kpc core radius (no central dominany galaxy)	$\leq$300 kpc core radius (central dominant galaxy)
Dynamically Young clusters A1367	**Dynamically Young Clusters** A262, Virgo
Low L_x($<10^{44}$ergs/sec)	Low L_x
Cool gas (1–4 keV)	Cool gas
Coronae around galaxies	X-ray emission dominated by a single galaxy
High spiral fraction (>40%)	High Spiral fraction
Low Central galaxy density	Low Central galaxy density
Asymmetric	Irregular Optical Structure
No cooling flows	May have cooling flows
Evolved Systems – Coma, A2256	**Evolved Systems – A85, A1795**
High L_x(>3x10^{44}ergs/sec)	High L_x
Hot Gas ($\geq$6 keV)	Hot Gas
High Velocity Dispersion	High Velocity Dispersion
Low Spiral Fraction (<20%)	Low Spiral Fraction
High Central Galaxy Density	High Central Galaxy Density
Regular structure	Symmetric x-ray emission centered on dominant galaxy
No cooling flows	Cooling flows

The x-ray observations also suggest that clusters with a dominant galaxy are more common than previously supposed. From optical studies about 20% of clusters are early B-M types or cD systems. From the x-ray observations we estimate nearly twice this fraction are XD clusters containing central dominant galaxies. Cooling flows are found to be present in a large fraction of these XD systems (Stewart et al. 1984, Jones and Forman 1984).

The observations suggest that clusters can best be described in a two dimensional system with dynamical relaxation and the dominance of a central galaxy as the two coordinates. We have described a zero order system in which we simply separated clusters into two subfamilies. Undoubtedly as our sophistication grows, we will want to take better advantage of the second dimension to allow us to more precisely classify clusters and thereby understand the relation of one parameter to the other.

Figure 11 shows four clusters which we have taken to be typical of their respective types - A1367 and A2256 for the less and more evolved clusters having no single central galaxy and A262 and A85 as the less and more evolved clusters with a single central galaxy. The properties of these clusters are listed in Table 2.

C. Scale Height of the Intracluster Gas

The second parameter, β, in the hydrostatic-isothermal model for clusters provides a measure of the energy per unit mass in the galaxies to that in the intracluster gas.

From the x-ray surface brightness profiles, we find β values consistent with ~0.6-0.8 and from our first sample of 46 clusters, 27 have β values constrained to be less than 1.0, each at the 90% confidence level. The β value also can be computed from the cluster velocity dispersion and the gas temperature. Table 3 shows a comparison of some of the measured β values and those computed. In all of these, the assumption of isotropic galaxy velocities has been made. If instead the galaxy velocities are anisotropic as Abell (1977) suggested and Kent and Sargent (1983) have observed in Perseus, the calculated value of β would be reduced. In general we find from the profile of the x-ray gas that the gas distribution is flatter than that expected for an isothermal sphere. In other words, the x-ray gas has a larger scale height than the galaxies and thus contains more energy per unit mass.

Table 3

Cluster	β_{fit}	β_{CAL}
A85	.62±.03	1.84(+1.05,−.70)
A119	----	.67(+.31,−.22)
A194	.7±.3	.47(+.27,−.26)
A262	.55±.05	.43(+.20,−.19)
A399	.52±.05	1.36(+.58,−.48)
A400	.57±.08	.43(+.22,−.17)
A401	.67±.03	1.54(+.75,−.55)
A426	.57±.03	1.55(+.28,−.25)
A576	.49±.03	1.15(+.40,−.36)
A1060	.67±.08	.91(+.47,−.37)
A1367	.53±.13	.93(+.49,−.43)
A1656	.76±.10	.62(+.08,−.07)
A1775	.66±.14	<2.6
A1795	.72±.08	.54(+.31,−.22)
A2029	.73±.10	1.20(+.61,−.46)
A2063	.62±.05	.23(+.13,−.10)
A2142	----	.85(+.54,−.28)
A2199	.68±.05	1.12(+.36,−.31)
A2256	.73±.05	1.35(+.82,−.58)

III. THE CLUSTER ENVIRONMENT

A. Effects on the Intracluster Gas by a hot IGM

One of the interesting possibilities is that the excess energy in the intracluster gas as compared to the galaxies is provided by heat conduction from a hot intercluster medium (see Forman, Jones, and Tucker 1984 for a detailed discussion). If a small amount of gas were leftover from galaxy and/or cluster formation, then this gas could be heated at an early epoch (z = 2-3) by an evolving population of active galaxies as in the Field and Perrenod model (1977). This gas would cool adiabatically with the expansion of the Universe to the present epoch temperature. Subsequent heat conduction from such a hot ($>10^8$ °K) intercluster medium could heat the intracluster gas above its equivalent virial temperature.

A cluster embedded in a hot IGM can dispose of the additional heat from conduction in two ways. The cluster gas temperature can rise or the gas can be expelled from the cluster in a wind. The criterion for a wind (Bregman 1978) is

$$T_{critical} > 6.9 \times 10^7 \, V^2_{1500}$$

where V is the cluster velocity dispersion in units of 1500 km/sec. Thus, when the temperature exceeds this critical value, the gas is no longer bound to the cluster and a wind is established. For a typical luminous (massive) cluster like Coma, V_{Coma} = 1570 km/sec and T_{Coma} = 9 $\times 10^7$ °K. Thus, one might begin to expect the existence of a wind in the outer portions of the Coma cluster.

B. Distant Clusters and the Limit on the IGM

What is the effect of such a wind? We would expect dynamically evolved clusters in the past to be more x-ray luminous than present epoch clusters since they would have contained more gas. If we can set a limit on the luminosity evolution of clusters from high redshifts to the present, then we can limit the amount of hot gas in the intercluster medium.

In selecting a sample of distant clusters to compare to nearby clusters, we must restrict ourselves to relaxed systems. Otherwise we will be confused by the dynamical evolution of the clusters themselves. Therefore, we restrict ourselves to high x-ray luminosity clusters since high L_x implies a large virial mass and therefore a high mass density and a short relaxation time. Table 4 shows a comparison of luminous high redshift clusters to low redshift systems. A detailed analysis carried out by Henry et al. (1982) found no evidence for a change in luminosity.

Table 4

L (0.5–4.5 keV)
x 10^{45} ergs/sec

Low Redshift

A401	0.8
Perseus	1.7
A644	0.5
A754	0.6
A854	0.7
Coma	0.8
A1689	1.0
A2065	0.5
A2142	1.5
A2319	0.7

High Redshift

3C295	(0.46)	0.9
CL0016+16	(0.55)	2.5
PKS0116+08	(0.59)	0.5
3C343.1	(0.75)	0.3
3C318	(0.75)	0.3
3C6.1	(0.84)	0.4
3C184	(0.99)	<0.6

To set a limit on the amount of intracluster gas that could have been removed from the cluster, we used the present epoch (z <0.075) cluster luminosity function and "evolved" the luminosity of each cluster by adding additional gas. When 1.4×10^{14} $M_\odot$ of gas (half that observed in A85) is added to each nearby cluster, the luminosity function of these "evolved" nearby clusters exceeds that of the distant cluster sample (which consists of only six z >0.5 clusters). This corresponds to a limit on a hot IGM of $0.6 \times 10^{-6} cm^{-3}$ (which could produce 36% of the x-ray background).

We can make the following prediction. If there is a small amount of gas ($\sim 10^{-8} cm^{-3}$) remaining from galaxy and cluster formation which was heated at early epochs, then clusters could have been heated to their critical wind temperatures and could have modest winds. By combining the definition of β with the equation of the critical wind temperature we would predict that β = constant = 5/6. This value is roughly independent of the details of each cluster.

In summary clusters can be used to probe the intercluster medium and thus limit the contribution of a hot IGM to the x-ray background and to the mass density of the Universe. In addition a modest density, hot IGM can explain the larger scale height of the intracluster gas compared to the galaxy distribution.

To pursue this problem in more detail, we need extensive observations of distant clusters to determine:

1) The epoch of cluster formation and gas injection.
2) The limit on any change in cluster x-ray luminosity and thereby gas mass.
3) The detailed temperature structure in the outer regions of clusters.

ACKNOWLEDGMENTS

We thank K. Modestino for help in the preparation of this manuscript. This work was supported by NAS8-30751.

REFERENCES

Abell, G.O. 1958, Ap.J.Suppl., 3, 211.
Abell, G.O. 1977, Ap.J., 213, 327.
Abramopoulus, F. and Ku, W. 1983, Ap.J., 271, 446.
Bahcall, J.N. and Sarazin, C.J. 1977, Ap.J.(Lett.) 213, L99.
Bautz, L.P. and Morgan, W.W. 1970, Ap.J.(Lett.), 162, L149.
Bechtold, J., Forman, W., Giacconi, R., Jones, C., Schwarz, J., Tucker, W., and Van Speybroeck, L. 1983, Ap.J., 265, 26.
Bregman, J. 1978, Ap.J., 224, 768.
Burstein, D. 1979a, Ap.J.Suppl., 41, 435.
Burstein, D. 1979b, Ap.J., 234, 435.
Burstein, D. 1979c, Ap.J., 234, 829.
Carnevali, P., Cavaliere, A., and Santangelo, P. 1981, Ap.J., 249, 449.
Cavaliere, A. and Fusco-Femiano, R. 1976, Astron.Astrophys., 49, 137.
Cavaliere, A. and Fusco-Femiano, R. 1981, Astron.Astrophys., 100, 194.
Cowie, L.L. and Sangalia, A. 1977, Nature, 266, 501.
Dressler, A. 1980a, Ap.J., 236, 351.
Dressler, A. 1980b, Ap.J.Suppl., 42, 565.
Fabricant, D. and Gorenstein, P. 1983, Ap.J., 267, 535.
Field, G. and Perrenod, S. 1977, Ap.J., 215, 717.
Forman, W., Bechtold, J., Blair, W., Giacconi, R., Van Speybroeck, L., and Jones, C. 1981, Ap.J.(Lett.), 243, L133.
Forman, W., Jones, C., and Tucker, W. 1984, Ap.J. (Feb. 1).
Geller, M. and Beers, T. 198 , PASP
Gunn, J.E. and Gott, J.R. 1972, Ap.J., 176, 1.
Hausman, M. and Ostriker, J. 1978, Ap.J., 224, 320.

Henry, J.P., Branduardi, G., Briel, U., Fabricant, D., Feigelson, E.,
     Murray, S., Soltan, A., and Tananbaum, H. 1979, Ap.J.(Lett.), 234,
     L15.
Jones, C. and Forman, W. 1983, Submitted to Ap.J.
Jones, C. and Forman, W. 1984, Ap.J. (Jan. 1).
Kent, S.M. and Sargent, W.L.W. 1983, Ap.J.
Merritt, D. 1983, Ap.J., 264, 24.
Oemler, A. 1974, Ap.J., 194, 1.
Rood, H.J. and Sastry, G.N. 1971, PASP, 83, 313.
Stewart, G.C., Fabian, A.C., Jones, C., and Forman, W. 1984, Ap.J.,
     submitted.
White, A. 1976, MNRAS, 177, 717.

DISCUSSION

R.A. Shafer: Though the model where a hot ICM provides the bulk of the
x-ray 2-40 keV background has acknowledged problems, your limit on its
contribution may have a significant dependence on the assumed thermal
history, in particular that the heating is at $z \gtrsim 3$. There are
independent arguments (made by Fabian and Kembhavi among others) that
the CXB spectrum requires a much earlier epoch of heating ($z \sim 5$), where
the increased efficiency of emission, longer period of adiabatic
cooling imply lower values of both the current epoch temperature and
density ($T_0 \sim 10$ keV). In this case the presence of such gas, only
slightly hotter than typical cluster temperatures, is difficult to
constrain.
C. Jones: As long as the intercluster gas is hotter than the cluster
gas, the cluster gas can still be heated. As Figure D shows, this
heating is characterized as the product of the IGM temperature and
density so that one actually places a limit on that product. Then for a
specific temperature such as that measured for the x-ray background, one
can get a limit on the IGM density.
Y. Rephaeli: I take it that the nucleosynthesis limit that the baryonic
contribution to mean mass density in the Universe is much smaller than
the critical density, is ignored in taking seriously the possibility of
the x-ray background being thermal bremsstrahlung emission. Also, I
thought that the conclusion used to be that quasars cannot, probably,
heat up closure density IGM to a temperature $\sim(10^8)$ K.
C. Jones: One of the problems with a dense hot IGM as the source of the
x-ray background is the enormous heating imput required from quasars.
However as Figure D shows, the density required to heat the clusters from
an IGM with a temperature of 3×10^8 °K (appropriate to the x-ray
background) is only $\sim 10^{-8}/cm^3$ so that the heat input required from
quasars is reduced to reasonable levels.
W. Jaffe: Radio observations of an extended source some 1 Mpc from the
Coma cluster show a containment pressure similar to that predicted by
your supercluster gas model.
S. van den Bergh: In NGC4438 both the gas and the stars seem to exhibit
a warped distribution. Since ram pressure does not act on stars one
might perhaps prefer to interpret this as a case of tidal damage rather

than ram pressure.

<u>C. Jones</u>: In Kotanyi et al.'s (1983, Ap.J., 273, L7) analysis of the Einstein high resolution observation of NGC4438, they show that the extended x-ray emission is offset from the galaxy. If their interpretation of this phenomenon as due to ram pressure stripping is correct, the stars would have been formed in this medium after it was stripped out of the galaxy.

<u>H. van Woerden</u>: Tidal damage to NGC4438 may have been caused not by the low-mass nearby companiona NGC4435, but rather by one of the (fairly) nearby, massive ellipticals: M86 or M87.

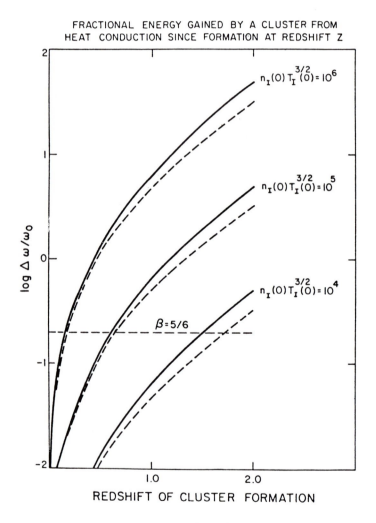

FRACTIONAL ENERGY GAINED BY A CLUSTER FROM
HEAT CONDUCTION SINCE FORMATION AT REDSHIFT Z

Figure D. From Forman et al.(1984); The fractional energy gained by a cluster from heat conduction is shown as a function of its redshift of formation for several values of $n_I(0)T_I(0)^{3/2}$.

EXOSAT OBSERVATIONS

G. Branduardi-Raymont
Mullard Space Science Laboratory, University College London
Holmbury St.Mary, Dorking, Surrey, UK

ABSTRACT

A short status report of the EXOSAT mission is given: the in-flight performance of the four experiments onboard is described and a brief comment is made on the problems encountered in orbit and on their resolution. At the time of the meeting only CMA image data of one cluster, Virgo, were available and these are presented, as well as, for demonstration purposes, the only PSD image of an extended X-ray source existing at the time, that of the supernova remnant Cassiopeia A.

THE EXOSAT MISSION

The European X-ray Observatory Satellite EXOSAT was launched on the 26th May 1983 and has been making celestial observations for the last 3 months. During this short period the work of the scientists involved with the mission has been devoted mainly to calibrate and verify the performance of the instruments, so a detailed analysis of the data from the astrophysical point of view is starting only now. I will give a short status report of the EXOSAT mission and will show some of the data obtained so far.

The EXOSAT Observatory carries two identical Low Energy (LE) telescopes; either a microchannel plate detector (the Channel Multiplier Array or CMA) or an imaging proportional counter (the Position Sensitive Detector or PSD) can be positioned at the focus of each of the mirrors. Parameters of the LE experiment are given in the following table (the spatial resolution results were obtained from the in-flight data of Cyg X-1):

Telescopes

Geometric area	90 cm^2 (each)
Energy range	.05 - 2 keV

341

F. Mardirossian et al. (eds.), Clusters and Groups of Galaxies, 341–346.
© *1984 by D. Reidel Publishing Company.*

	PSD	CMA
Field of view	$\overline{1.5}^\circ$ (diameter)	$\overline{2.2}^\circ$
Energy range	.1-2 keV	.05-2 keV
Spatial resolution	1.3´ at 3´	15"-20" at 3´
	1.8´ at 30´	40" at 30´
Energy resolution	$\Delta E/E = .44/\sqrt{E}$	--
	(36% at 1.5 keV)	

The main difference between CMA and PSD detectors, apart from the much better spatial resolution of the former, is the lack of intrinsic energy resolution of the CMAs. However, broadband spectroscopy can be performed with the use of filters in conjunction with the CMAs. There are 6 filters available: Boron, 4000 and 3000 Å Lexan, Aluminium-Parylene, Polypropylene and Magnesium fluoride, in order from the least to the most trasmitting. Some of these filters are transparent to UV radiation in various degrees. To exclude the possibility of UV contamination the Boron filter must be selected. High resolution spectroscopy at the expense of high efficiency is available through the use of 2 trasmission gratings (500 and 1000 lines/mm) which can be positioned in the focussed X-ray beam. They have a wavelength resolution of 2-4 and 1-2 Å respectively above 40 Å and 1 Å below 40 Å.

The major problem encountered in flight by the LE experiment has concerned the PSDs. One of them, a month after launch and after many hours of successful operations, showed evidences of anomalous behaviour and was turned off, while investigations of the problem were carried out. Since then the PSD has been turned on again at a lower gain than nominal and has proved to be working properly. The other PSD showed the onset of some form of high-voltage breakdown at the first turn ON and has never been used for observations yet. It is believed that the problem resides outside the detector and that it could be cured by further outgassing. More trials of this PSD will be attempted later.

EXOSAT also carries two non-imaging experiments, the Large Area Proportional Counter Array (or ME, for Medium Energy, experiment) and the Gas Scintillation Proportional Counter (or GSPC), all coaligned with the LE telescopes.

There are 8 ME detectors organized in 4 quadrants; half of the array can be offset from the other to permit collection of background data simultaneously with a source observation. Each detector consists of two chambers, the upper one filled with Argon, the lower with Xenon. Particle background rejection is achieved by anticoincidence and pulse rise time discrimination. The following table lists some of the parameters of the ME experiment:

Total geometric area	1800 cm^2	
Energy range: Argon chamber	1-15 keV	
Xenon chamber	5-50 keV	
Collimator field of view	45´ FWHM	
Energy resolution: Argon	$\Delta E/E = .49/\sqrt{E}$	(20% at 6 keV)
Xenon	$\Delta E/E = .89/\sqrt{E}$	

The GSPC is a 95% Xenon, 5% Helium filled counter which comprises a photo-absorption and a scintillation region. The burst of light subsequent to the absorption of an X-ray photon is viewed by a photomultiplier and the amplitude of the signal in this is proportional to the energy of the X-ray photon. Particle background rejection is achieved by discrimination on the basis of the burst length. Parameters of the GSPC are as follows:

Total geometric area	160 cm^2
Energy range	2-40 keV
Collimator field of view	45´ FWHM
Energy resolution	$\Delta E/E = .27/\sqrt{E}$ (10% at 6 keV)

Figure 1 gives an exploded view of the EXOSAT Observatory.

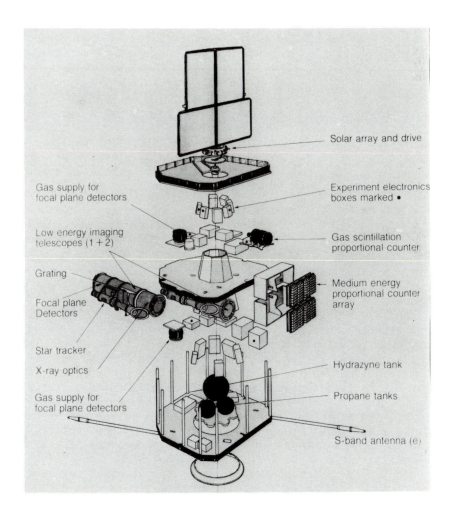

Figure 1. An exploded view of the EXOSAT Observatory.

A remarkable difference from all previous X-ray astronomy satellite missions is the very eccentric orbit in which EXOSAT operates. This was originally chosen being the most suited to the performance of lunar occultations, which was expected to be the principal operational mode of the satellite when first concieved, more than 10 years ago. Since then the EXOSAT payload has changed substantially. In particular, the introduction of imaging detectors has eliminated the need of occultations to achieve goood spatial resolution. However, the EXOSAT orbit has remained the same. Here is a list of the orbit parameters at the time of launch:

Perigee	355 km
Apogee	~190,000 km
Inclination	72.5$^\delta$
Period	90.5 hrs
Lifetime	estimated at 2.9 yrs

During the 78 hrs of experiment ON-time every orbit there is continuous data acquisition at the ESA receiving station of Villafranca, near Madrid. Due to EXOSAT unusual orbit, only a rough estimate of the level of particle induced background in the experiments could be made before launch. The in-orbit measurements are ~1.5 counts/cm^2/s in the CMAs (somewhat higher than predicted), .9 counts/cm^2/s in the PSD (roughly double than expected), ~6 counts/s/keV for the full ME array and ~1 count/s/keV between 2 and 10 keV in the GSPC; for the last two experiments these values correspond to a reduction by a factor of 2 and ~10 respectively on that calculated before flight, with a substantial increase in sensitivity on what predicted.

EXOSAT OBSERVATIONS

During the two-month calibration and performance verification phase of the EXOSAT mission observations of a variety of objects were performed. With regard to clusters of galaxies, regions in Virgo and Coma were observed, and so was the core of the Perseus cluster, with the active galaxy NGC1275. Figure 2 shows a composite image of the region of the Virgo cluster containing M87. The 4 exposures, all of about an hour duration, were taken with the CMA and various filters. The image with the polypropylene filter shows enhanced emission with respect to the others: interesting results on the distribution of energy and mass in the cluster may be produced by a detailed analysis of these data. In view of demonstrating the performance of the PSD, Figures 3 a) and b) show images of the supernova remnant Cassiopeia A in the energy bands .6-1.1 and 1.1-2.5 keV taken soon after launch: note the qualitative difference in appearance with X-ray energy. The manoeuvrability of the EXOSAT spacecraft has been exploited on several occasions to point at 'targets of opportunity' quickly after observations at other wavelengths showed the occurrance of something unusual in them. This was the case of a supernova explosion in M83 (no X-ray detection was made then) and the optical outburst of the old nova GK Per (a 6 min periodical modulation in the X-ray flux was discovered by the ME experiment).

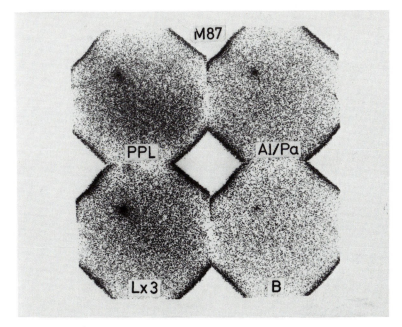

Figure 2. The M87 region in the Virgo cluster.

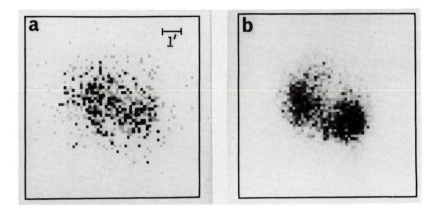

Figure 3. PSD images of Cas A: a) .6-1.1keV, b) 1.1-2.5 keV.

ACKNOWLEDGEMENTS

EXOSAT is an ESA project to which several European institutes and industries have contributed. The LE experiment is responsability of a Consortium constituted by the Mullard Space Science Laboratory of University College London, the Cosmic Ray Working Group, Leiden and the Space Science Laboratory, Utrecht, The Netherlands.

DISCUSSION

PORTER: What time resolution do the detectors have?

BRANDUARDI-RAYMONT: Various time resolutions, up to 7.6 µsec.

DIFFUSE Ly_α EMISSION AROUND THE cD GALAXY IN A1795[+]

H.U. Nørgaard-Nielsen, H.E. Jørgensen, L. Hansen
Copenhagen University Observatory
Øster Voldgade 3, DK-1350 Copenhagen K, Denmark

<u>Summary</u> We report briefly the discovery of diffuse Ly_α emission coinciding with the optical filaments around the cD galaxy in the X-ray cluster A1795. The derived Ly_α/H_β ratio is close to the recombination value. From this ratio we find that the content of <u>smoothly</u> distributed HI is less than 100 solar masses. The non-detection of a stellar UV continuum implies that the formation rate of stars more massive than $3M_\odot$ is less than 1 per cent of the accretion rate. Combining UV, optical and radio data for the non-thermal nuclear source we find $\alpha \sim .8$.

1. Introduction

Because of the high density of the cool X-ray component in the central parts of clusters of galaxies the gas may be cooling in a time scale less than a Hubble time giving rise to pressure-driven inflow of gas into the central galaxy (Cowie and Binney, 1977, Fabian and Nulsen, 1977). Thermal instabilities in the cooling flow will lead to formation of blobs of cooled material around the galaxy. The cooled material is identified with optical filaments around several dominant galaxies: NGC1275 in the Perseus Cluster (Kent and Sargent, 1979), M87 in Virgo (Ford and Butcher, 1979), NGC4696 in the Centaurus Cluster (Fabian et al., 1983, Nørgaard-Nielsen and Jørgensen, 1983). Here we present the discovery of extended Ly_α emission around the dominant galaxy in A1795 from ultraviolet spectra obtained by IUE. Beforehand extended Ly_α emission has only been detected in the Perseus Cluster (Fabian et al., 1984, hereafter FNA).

F. Mardirossian et al. (eds.), Clusters and Groups of Galaxies, 347–349.

2. The Observations

We have obtained two 3 hours exposures with the SWP camera
on IUE in the low resolution mode (~ 6Å). During the
observations the major axis of the large aperture was
north - south within 1^0. Extended emission (~ 10") at a
position corresponding to redshifted Ly_α (z = .0631) is
clearly seen on both exposures. The optical emission is
coming from the same region of the galaxy (van Breugel et
al., 1984, hereafter BHM).

From our estimate of the total flux in Ly_α (~ 2.8 x
10^{-13} erg/cm^2/sec) and the total flux in H_β given by BHM we
find a Ly_α/H_β ratio ~ 23. Since a small amount of galactic
reddening could be present the ratio is close to the value
33 from classical recombination theory. A weak (F1500Å ~
1.5 x 10^{-15} erg/cm^2/sec/Å) central continuum source is also
present in both SWP exposures. For the extended component,
just like in NGC1275, the CIV 1550Å emission is absent and
a UV continuum is missing. We estimate that $F_{CIV1550Å} \leqslant$
3 x 10^{-14} erg/cm^2/sec and $F^{cont}_{1500Å} \leqslant 10^{-15}$ erg/cm^2/sec/Å (2
sigma level) for an extended component.

3. Discussion

The X-ray gas around A1795 is according to Mushotsky (1983)
cooling at a rate of $560 M_\odot/y$. This means that about $10^{13} M_\odot$
could have been accreted in a Hubble time. From limits on
the extended UV flux from NGC1275 FNA place important
constraints on the star formation rate assuming 'normal'
IMF. FNA consider three different types of IMF used for our
Galaxy: a) power-law, b) log-Gaussian, c) broken power-law.
By scaling the results of FNA to A1795 we find limits to the
star formation rate of 2-10 $M_\odot/y$ for normal IMF. In order to
accommodate a star formation rate of 560 $M_\odot/y$ we must impose
an upper mass cut off $\leqslant 2.5 M_\odot$ for all three types of IMF.
Also by scaling the results for NGC1275 we estimate that the
star formation rate for stars with masses larger than 3 $M_\odot$
is less than 1 $M_\odot/y$. Fabian et al. (1983) has argued from
considerations about the Jeans mass in a high pressure
environment without dust that only low mass stars should be
formed in the cooling flow.

The basic assumption in the determination of the cooling
rate is that there is no extra energy supply to the cool
gas. Wirth et al. (1983) and Canizares et al. (1982) have
pointed out that supernovae produced by massive stars could
disrupt a cooling flow. Tucker and Rosner (1983) have
discussed the X-ray data of M87 for which Mushotsky (1983)
predicts a cooling rate of ~ 10 $M_\odot$. In their hydrostatic
model Tucker and Rosner include thermal conduction and non-
thermal heating by relativistic electrons in the radio lobes.

Such a model may fit the available X-ray data without predicting any radiative accretion flow.

The optical emission lines seen in A1795 are typical of low ionization nuclear emission regions (LINERs). The ionization and excitation source of LINERs has been widely discussed. Both photoionization models with a low ionization parameter and low velocity shock models have been suggested. The Ly$_\alpha$/H$_\beta$ ratio found in A1795 is close to the value calculated from the two types of models implying that the absorption in Ly$_\alpha$ must be less than a factor of 2. From the figures given by Sargent et al. (1980) we estimate that $N_H \lesssim$ 5×10^{14} cm^{-2} and that the total amount of <u>smoothly</u> distributed HI cannot exceed a few hundred solar masses. This is in great contrast to the considerable amount of general reddening found in the Centaurus Cluster (Jørgensen et al., 1983) and A496 (Nørgaard-Nielsen et al., 1984) implying a mass M(HI) $\sim 10^6 M_\odot$ for a normal gas to dust ratio. Of course several orders of magnitude more HI could be present in A1795 if it is sufficiently clumpy.

For the nuclear source we have combined our UV data with optical and radio data. We find that the spectrum can be represented by a power-law ($f_\nu \sim \nu^{-\alpha}$) with $\alpha \sim .8$, close to the mean value 1.1 found for Seyfert 2 galaxies (Halpern and Steiner, 1983).

<u>Literature</u>
Canizares, C.R., Clark, G.W., Jernigan, J.G., Markert, T.H., Astrophys. J. <u>262</u> p33 (1982)
Cowie, L.L., Binney, J., Astrophys. J. <u>215</u> p723 (1977)
Fabian, A.C., Nulsen, P.E.J., Mon. Not. Roy. Ast. Soc., <u>180</u> p479 (1977)
Fabian, A.C., Nulsen, P.E.J., Arnand, K.A., preprint
Ford, H.D., Butcher, H., Astrophys. J. Suppl. Ser. <u>41</u> p147 (1979)
Jørgensen, H.E., Nørgaard-Nielsen, H.U., Astron. Astrophys. <u>122</u>, p301 (1983)
Kent, S.M., Sargent, W.L.W., Astrophys. J. <u>230</u> p667 (1979)
Mushotzky, R.F., Proceedings from 'Workshop on Very Hot Astrophysical Plasma', Nice, November 1982
Nørgaard-Nielsen, H.U., Jørgensen, H.E., Proceedings from 'Workshop on Very Hot Astrophysical Plasma', Nice, November 1982
Tucker, W.H., Rosner, R. Astrophys. J. <u>267</u> p547 (1983)
Wirth, A. Kenyon, S.J., Hunter, D.A., Astrophys. J. <u>269</u> p102 (1983)

[+]Based on observations by the International Ultraviolet Explorer collected at the Villafranca Satellite Tracking Station of the European Space Agency.

CLUSTER X-RAY LUMINOSITY AND FIRST RANKED GALAXY SIZE

Alain Porter[1]
Harvard-Smithsonian Center for Astrophysics

ABSTRACT.   We examine the relation between the metric X-ray luminosity of a cluster of galaxies and the optical core radius of its brightest member. Implications for theories of galaxy growth are briefly discussed.

1.   OBSERVATIONS

The X-ray luminosities of a large sample of clusters of galaxies have been measured with the Einstein Observatory (Jones and Forman 1984). These luminosities are for the energy range 0.5-3.0 keV within 0.5 Mpc of the cluster center. Optical surface photometry of the dominant galaxies in many of these clusters has been done by Hoessel, Gunn, and Thuan (1980, HGT). They measured the seeing-corrected core radius a in the approximation to a King model

$$I(r) = \frac{1}{1 + \left[\frac{r}{a}\right]^2} \tag{1}$$

We have obtained photometry of 15 bright cluster galaxies with the CCD camera on the 61 cm telescope at Whipple Observatory (described in Gursky et al 1981), confirming HGT's core radii and extending the sample.

In Figure 1, we have plotted the X-ray luminosity against the galaxy core radius. The correlation is clearly strong: the Spearman nonparametric rank correlation coefficient is 0.99959. Fitting a simple power law to the data shows

$$L_x \propto a^{1.9}, \tag{2}$$

and the 90% confidence region for the exponent runs from 1.5 to 2.2.   To test for the possibility of Malmquist bias, we repeated the analysis on the subset of clusters with z   0.06, eliminating 41% of the sample.   The results were not significantly affected.

F. Mardirossian et al. (eds.), Clusters and Groups of Galaxies, 351–352.
© *1984 by D. Reidel Publishing Company.*

352   A. PORTER

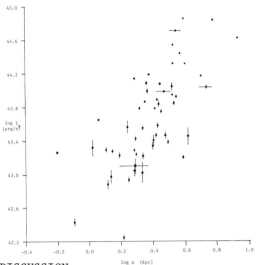

Figure 1. Cluster X-ray
luminosity plotted against
galaxy optical core radius.

2. DISCUSSION

These observations provide insight into the physical processes by
which a cluster and its brightest member evolve. One such process is
galaxy mergers, of which the most frequently discussed theory is the
homologous merger model of Ostriker and collaborators (e.g. Ostriker and
Hausman 1977, Hausman and Ostriker 1978). A simple dimensional argument
based on this model and the $L_x \propto \sigma^4$ relation leads to

$$L_x \propto a^{8/3},$$ (3)

which is too steep to account for the observations. If mergers were the
only means of growth available to a cD galaxy, this could be taken as
evidence that galaxies are disproportionately bloated by mergers. However,
tidal stripping (Richstone 1976) and cooling gas inflow (Fabian et al 1982,
Jones and Forman 1984)also operate in many clusters, and mergers may not
dominate the $L_x(a)$ relation. In particular, tidal stripping, by preferentially adding matter to the outer regions of a dominant galaxy (Richstone
1976, Dressler 1979), might increase its size faster than homologously with
respect to L_x. Detailed work will be presented elsewhere.

1. Present address: Robinson Laboratory, California Institute of Technolog

REFERENCES

1. Dressler, A. (1979) ApJ 231 pp 659-70
2. Duncan, M., Farouki, R., and Shapiro, S. (1983) ApJ 271 pp 22-31
3. Fabian, A., Nulsen, P., and Canizares, C. (1982) MNRAS 201 pp 933-8
4. Forman, W., and Jones, C. (1982) Ann.Rev.Ast.Ast. 20 pp 547-96
5. Gursky, H., Geary, J., Schild, R., Stephenson, T., and Weekes, T. (1980)
    Proceedings of the SPIE 264 p. 14
6. Jones, C., and Forman, W. (1984) ApJ in press (January 1)
7. Hausman, M., and Ostriker, J. (1978) ApJ 224 pp 320-36
8. Hoessel, J., Gunn, J., and Thuan, T. (1980) ApJ 241 pp 486-92
9. Ostriker, J., and Hausman, M. (1977) ApJ 217 pp L125-9
10. Richstone, D. (1976) ApJ 204 pp 642-8

GROUPS OF GALAXIES

Margaret J. Geller
Harvard-Smithsonian Center for Astrophysics
60 Garden Street
Cambridge, MA 02138

I. INTRODUCTION

If galaxies in the universe were uniformly distributed we would
have had to find another excuse to enjoy ourselves in Trieste.
Fortunately groups of galaxies are a sufficiently messy subject to
require at least one more meeting.

The study of groups of galaxies gives limits on the mean mass
density of the universe (Rood, Rothman, and Turnrose 1970; Gott and
Turner 1977; Huchra and Geller 1982; see Faber and Gallagher 1979 for
additional references). Groups are important for understanding the
development of galaxy clustering and they provide laboratories for
studying the interactions between individual galaxies and their
environment (Dressler 1980; McGlynn and Ostriker 1980: Geller and
Postman 1983, Mardirossian, Giuricin, and Mezzetti 1983; Mezzetti
et al. 1983). The study of groups of galaxies thus has bearing upon a
number of deep questions in the study of large-scale structure in the
universe. However, the impact of these studies is moderate by several
unpleasant statistical problems in the identification and analysis of
systems with small numbers of members.

In this paper I will concentrate on the analysis of statistically
complete group catalogs rather than on detailed studies of individual
systems. I will focus on four aspects of these analyses: (1) the
definition and construction of group catalogs, (2) the extraction of
physical results (particularly mass-to-light ratios) from the study of
groups of galaxies, (3) the statistical problems involved in the study
of groups, and (4) the overlap or "continuum" of properties of "groups"
and "clusters" of galaxies. I hope that emphasis on the third of
these topics and an attempt to define some of the problems will lead
to new methods of analysis.

F. Mardirossian et al. (eds.), Clusters and Groups of Galaxies, 353–366.
© *1984 by D. Reidel Publishing Company.*

II. APPROACHES TO THE STUDY OF GROUPS OF GALAXIES

Conventionally a group is a collection of galaxies which are likely to be physically (and dynamically) associated. The usual operational translation of this definition is that groups are density enhancements - either in surface density or redshift-space volume density. Groups identified this way may be neither bound nor virialized. According to the general wisdom, the greater the density contract, the more likely the associations (or parts of a system) are to be real physical systems.

I will consider only groups containing three or more galaxies brighter than the relevant survey limit. When the number of members of a group is small (<10), erroneous group membership assignments caused by accidental superpositions on the sky are particularly problematic.

One method of "solving" the membership assignment problem is to study the "ensemble average" of groups in a galaxy catalog without assigning any particular galaxy to any particular group (Geller and Peebles 1974; Fall 1975; Davis and Peebles 1983). These "statistical virial theorems" compare the potential energy in fluctuations (measured by the two-point correlation function for the galaxy distribution) with the kinetic energy (measured by the pairwise velocity dispersion) to obtain an estimate for the mean mass density, Ω. The most recent application of these methods to the CfA redshift survey (Davis and Peebles 1983) gives $\Omega = 0.2 \ e^{\pm 0.4}$. The methods are reviewed in detail by Peebles (1980).

More traditional approaches face the group assignment problem head on. Before the availability of complete redshift surveys, identifications of group members were based either on subjective data (de Vaucouleurs 1975; Holmberg 1969; Karachentseva 1973) or on two-dimensional criteria (Turner and Gott 1976; Materne 1978, 1979; Hickson 1982). De Vaucouleurs, for example, assigns group membership on the basis of similarity in redshift, apparent magnitude, and morphology as well as on positional coincidence. Morgan, Kaiser, and White (1975; MKW) and Albert, White and Morgan (1977; AWM) select groups by the "cD-like" appearance of a member. Although most of the subjective group identifications are confirmed by more objective "observer blind" procedures, it is difficult to determine selection biases and incompleteness effects in catalogs based on subjective choices.

Turner and Gott (1976) pioneered in the application of automatic "observer blind" group finding algorithms. Their two-dimensional computer algorithm (applied to a magnitude limited galaxy catalog before the availability of complete redshift information) identifies group members on the basis of angular separation. "Groups" are collections of galaxies in regions where the galaxy surface number density is enhanced by a specified factor. One difficulty with this selection criterion is that the corresponding projected spatial

separation varies with distance - relatively nearby groups of large
angular, but small spatial scale cannot be identified. The method will
not yield the same groups when applied to samples which cover the same
region of the sky to different limiting apparent magnitudes. In
another contribution, Hickson et al. (this volume) discuss the physical
properties of compact groups selected as surface density enhancements
on the sky (Hickson 1982).

The recent completion of redshift surveys for regions of large
angular scale now makes "three-dimensional" group selection possible.
So far there have been two analyses. Press and Davis (1982; PD)
purport to identify virialized systems of galaxies which have crossing
times short compared with the Hubble time. Huchra and Geller (1982;
HG) and Geller and Huchra (1983; GH) find density enhancements in
"redshift space". Many statistical problems remain, but some
difficulties are circumvented by using redshift information.

III. GROUP SELECTION FROM REDSHIFT SURVEYS

Two complete redshift surveys are now available for the extraction
of group catalogs: (1) a whole-sky survey complete to $m_{B(0)}$ = 13.2
(1312 galaxies; Huchra et al. 1983 - denoted NB) and (2) the CfA
redshift survey complete to $m_B(0)$ = 14.5 for about 20% of the sky
(2396 galaxies; Huchra et al. 1983 - denoted CfA). We define groups
as number density enhancements in three-dimensional redshift-position
space (other definitions are discussed by HG, GH, and PD).

One of the difficulties in group selection is the treatment of
magnitude limited samples. For groups with large mean redshift, only
the intrinsically brightest members are seen. In order to treat the
magnitude limit we assume the galaxy luminosity function, ϕ (M), is
independent of position. We take the Schechter form with $M_{B(0)}^*$ = -19.4
and α = -1.3 for the CfA survey and $M_{B(0)}^*$ = -19.1 and α = -1.0 for
the NB Survey (H_o = 100 kms^{-1} Mpc^{-1}). The computer algorithm searches
for "friends of friends", and identifies all pairs with projected
separation

$$D_{12} = \sin(\theta/2)(v_1 + v_2)/H_o \leq D_L(v_1, v_2, m_1, m_2)$$

and line-of-sight velocity separation

$$V_{12} \leq V_L(v_1, v_2, m_1, m_2)$$

where v_1 and v_2 refer to the redshifts of the galaxy and its companion,
m_1 and m_2 are their magnitudes, and θ is their angular separation.
Pairs with a member in common are linked into a single group. The
selection parameters D_L and V_L are scaled to account for the magnitude
limit of the galaxy catalog. The ratio D_L/V_L is kept constant. The
number density contour surrounding each group corresponds, then, to a
fixed number density enhancement relative to the mean:

$$\delta\rho/\rho = \frac{3}{4\Pi D_L^{\,3}} \left[\int_{-\infty}^{M_{12}} \phi(M)\,dM \right]^{-1} - 1 \,.$$

M_{12} is the limiting absolute magnitude for each galaxy pair. All the groups have at least this density contrast. Further details are given in HG and GH.

The selection parameters D_o and V_o can obviously be varied to produce a set of group catalogs. If D_o and V_o are small few groups are found: in the extreme case each galaxy in the catalog is a "group" by itself. If D_o and V_o are large enough virtually all the galaxies in the catalog will be dumped into a single "group". Some intermediate range is evidently desirable. We choose values of D_o and V_o which give $\delta\rho/\rho = 20$. The values lie in a range where the derived dynamical parameters of the groups are only weakly, if at all, dependent on the selection parameters. In these group catalogs, the positions of group centers are particularly robust. There are 92 groups in the NB catalog; 176 in the CfA catalog. About 60 percent of the galaxies in the catalog are assigned to groups.

Once a group catalog is selected, it is important to check that the physically interesting derived parameters of the groups are distance independent. Figure 1 shows, for example, that the mass-to-light ratio for groups in the NB catalog is distance independent.

Automatic selection of group members from redshift surveys reduces but does not eliminate the problems caused by "interlopers" i.e. galaxies which are not physical members of a group but which are superposed on the region of the sky covered by the group. Sometimes obvious non-members are included in a group; sometimes two groups are merged. However, given the selection criteria, it is possible to calculate _a priori_ the expected number of interlopers in any group and to assign confidence levels. Rather than culling the groups in this

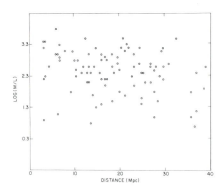

Figure 1. Mass-to-light ratios for groups in the NB catalog as a function of distance.

way, we use the median parameters of the unculled groups to measure
the typical properties of systems in the sample. Gott and Turner
(1977) discuss the merits of this approach.

There are several tests which the three-dimensional search
algorithm should and does pass. The procedure identifies well-known
systems. Virtually all the original de Vaucouleurs groups are
identified: in every case, the de Vaucouleurs groups not found automatically are accidental superpositions of galaxies on the sky. Some
feeling for the power of the search algorithm may be obtained from
Figure 2, a map of section of the plane of the Local Supercluster.
The solid contour marks a density enhancement $\delta\rho/\rho = 100$, the dashed
is $\delta\rho/\rho = 20$ and the dashed and dotted is $\delta\rho/\rho = 2$. Both the core
of the Virgo Cluster and Leo association are clearly defined. The
plane of the Local Supercluster is obvious.

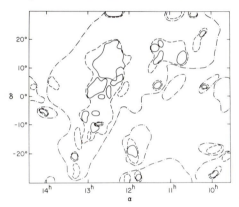

Figure 2. A contour plot of the region around 12^h and 0°. The
central region of the Virgo cluster lies near 12^h20^m $+10^o$ and the Leo
association at 11^h $+15^o$. In one region near 12^h20^m $+7^o$ the dotted
line indicates a background group -- this plot is a two-dimensional
representation of a three-dimensional contour plot.

Another test is the application of the finding algorithm to
galaxy surveys which cover the same region of the sky to different
limiting apparent magnitudes: the method should yield very nearly
the same groups. Comparison of the NB and CfA surveys is a check of
this expectation. Every NB group in the CfA survey region is identified
once again as a CfA group (see GH for details). In general the NB and
CfA realizations of the group differ only by the addition of fainter
members from the deeper galaxy survey.

IV. DYNAMICAL QUANTITIES

Once the group catalog is defined, a set of standard dynamical
parameters are easily calculated. In particular, the virial mass is

$$M_{vt} = \frac{3\Pi}{G} \sigma_r^2 r_H$$

and the virial crossing time (in units of the Hubble time) is:

$$t_v = \frac{6}{5^{3/2}} r_H/\sigma_r$$

where σ_r is in line-of-sight velocity dispersion and r_H is the mean harmonic radius. (For the definition of r_H given in HG, there is an error of a factor of 2 in the expressions for M_{vt} and t_v. All virial masses and crossing times there are underestimated by a factor of 2.) Groups in the catalog with $\delta\rho/\rho > 20$ have crossing times $< 0.8\ H_o^{-1}$. Short crossing time is often taken as evidence of the physical reality of a group. However, there is no guarantee because velocity interlopers produce some bias toward short crossing time. Figure 3 shows the distribution of crossing times for the NB catalog. The few groups with long crossing times are small systems (usually triples) with small velocity dispersion and large angular scale. The distribution of crossing times for the CfA groups is similar: only 11 of the 176 groups have crossing times $> 0.8\ H_o^{-1}$ (see Geller and Huchra 1984).

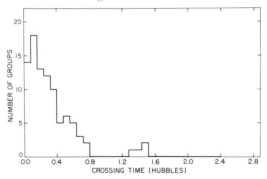

Figure 3.   Crossing times for NB groups in units of the Hubble time.

The distribution of mass-to-light ratios is not nearly so well-behaved as the distribution in crossing time. Figure 4 shows the histogram of mass-to-light ratio, $M/L_{B(0)}$, for groups in the CfA catalog. The B(0) luminosity for each group includes a correction for incomplete sampling of the galaxy luminosity function. The histogram has a long tail at high mass-to-light ratio. Many of the groups in the tail contain obvious interlopers of lie near enough to the galaxy survey limits that large (and uncertain) corrections to the group luminosity are required. The median mass-to-light ratio is $M/L_{B(0)} = 440\ M_\odot/L_\odot$. The median one-dimensional velocity dispersion is $220\ kms^{-1}$ at a scale of 0.9 Mpc. Table 1 is a comparison of median virial mass-to-light ratios obtained from various group catalogs. If the groups are bound but not virialized, the mass-to-light ratios will be smaller by a factor of two.

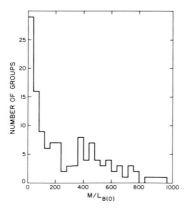

Figure 4. Histogram of mass-to-light ratios for groups in the CfA catalog. 55 of the 176 groups in the catalog have mass-to-light ratios greater than 1000 $M_\odot/L_\odot$.

TABLE 1

COMPARISON OF MEDIAN MASS-TO-LIGHT RATIOS

Catalog	Number of Groups	$M/L_{B(0)}$
CfA	176	440 $M_\odot/L_\odot$
NB	92	340
CfA(>10 members)	27	600
NB(>10 members)	12	430
GT[o]	39	280
DV[*]	45	400

[*]Membership determined from complete redshift data.
[o]Estimates based on incomplete velocity data (see GT).

Within the errors ($\sigma(M/L_{B(0)}) \stackrel{\sim}{=} 100\ M_\odot/L_\odot$) the estimates in Table 1 are consistent with one another.

   In principle, correlations among the physical parameters of the groups can give insights into the physics of galaxy clustering and the nature of the "missing mass" (see, for example, PD; Mardirossian et al. 1983). However the task of disentangling physical effects from selection biases is formidable. It is important that the group-finding algorithm be simple enough for an a priori, internally consistent, estimation of interloper probabilities. N-body simulations can provide valuable tests, but these can be misleading because of subtle differences between the simulations and the actual data. Press and Davis (1982; PD), for example, concluded that M/L ∝ R for groups they identified in the CfA survey. This dependence on scales > 1 Mpc is now known to be a result of a flaw in the finding algorithm. The discriminant used for group membership, crossing time, is inadequate.

In a number of the larger scale groups, the galaxies are superposed
on the sky but are well-separated in velocity. The crossing time of
such a "system" is short, but has no physical meaning. N-body
simulations are useful for testing group selection algorithms, but
there are limitations. PD use simulations by Efstathiou and Eastwood
(1981) to search for systematic biases in their selection procedure.
The simulations were run at critical density ($\Omega = 1$) procedure and
then artificially scaled only in the velocity coordinate to make $\Omega > 1$.
In these simulations, regardless of the quoted Ω, the "fingers of God"
(the elongation of systems of galaxies along the velocity coordinate
in redshift space) are much better defined than they are in the actual
data (see, for example Davis et al. 1982). The simulated groups are
generally well separated on the sky and there are not enough accidental
superpositions (interlopers) to produce a trend of M/L with R on large
scales; the simulations do not provide a sufficient test for this
systematic problem. HG and GH find no detectable tendency for the
mass-to-light ratios of groups to increase with scale.

V. MASS ESTIMATORS

     Except for a recent paper by Bahcall and Tremaine (1981), there
has been little incisive work on the statistical properties of the
estimators used to derive the physical parameters of groups (or, for
that matter, larger systems) of galaxies. As a result, both the large
uncertainties and the biases in parameter estimation are often ignored.
Furthermore, there have been few searches for better estimators.

     The distribution of mass-to-light ratios for groups of galaxies
is broad with a long tail extending to high mass-to-light ratio
(Figure 4). Much of the tail is due to systematic errors (e.g.
interlopers, large corrections for the unseen contribution to the group
luminosity), but, even ignoring the tail, the distibution is broad.
Is this width a measure of intrinsic variations in the mass-to-light
ratios of galaxies in groups or is it merely comparable with the
expected variance in the virial theorem estimator?

     Turner et al. (1979) show that the spread in mass-to-light ratio
obtained for their group catalog is the same as the spread obtained
for an artificial catalog generated from n-body simulations. The
spread of mass-to-light ratio in the simulated catalog is entirely
statistical - there is no intrinsic spread. Turner and Gott therefore
conclude that the observed scatter in the mass-to-light ratios of
groups is probably not intrinsic: it is a reflection of the variance
in the virial theorem estimator.

     The same conclusion can be reached by a comparison of group
catalogs derived from the two available redshift surveys. (See Geller
and Huchra 1984 for details). Because the CfA and NB catalogs overlap,
a comparison of groups found in both catalogs gives an indication of
the dominant statistical errors in the analysis. Figure 5 is a plot

of the virial mass-to-light ratio for individual CfA groups as a
function of the virial mass-to-light ratio for the same group as
identified in the NB catalog. In general, only faint members are
added in the CfA realization of the group. There are 29 groups in
the plot. Groups which have obvious interlopers are denoted x; groups
without obvious interlopers and with more than ten members in at
least one catalog are denoted o. Note that the relatively well-sampled
groups lie close to the line $M/L_{B(0)}$ (CfA) = $M/L_{B(0)}$ (NB). The groups
containing interlopers contribute substantially to the remarkably
large scatter. The scatter for the uncontaminated group is still
large, but it is comparable with the variance in the virial theorem
estimator for groups with small numbers (∼5) of members (see Bahcall
and Tremaine, Figure 4). In addition to the scatter, there is a bias:
There is a tendency for greater underestimation of the mass when fewer
group members are included in the sample. In agreement with the
conclusions of Turner and Gott (1979), the scatter in observed mass-
to-light ratio does appear to come primarily from variance in the
virial theorem estimator.

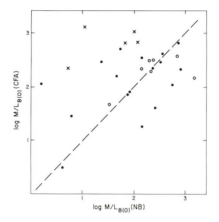

Figure 5. A comparison of mass-to-light ratio estimates for groups
identified in both the CfA and NB catalogs. The dashed line is
$M/L_{B(0)}$ (CfA) = $M/L_{B(0)}$ (NB).

Because of the unpleasant statistical properties of the virial
theorem estimator, Bahcall and Tremaine suggest an alternative,
the projected mass estimator:

$$M_{pm} = \frac{24}{\pi GN} \sum_i V_i^2 R_i$$

where R_i is the projected distance and V_i the velocity of a test
particle from the central mass. In this form the estimator applies
to systems in which N test particles orbit a central mass. The
coefficient $24/\pi = 1/2(16/\pi + 32/\pi)$ is an average of the coefficient
$16/\pi$ appropriate for isotropic orbits and the coefficient $32/\pi$
appropriate for linear orbits. In the case where the mass in the

system is distributed (e.g. a system with N galaxies of roughly equal mass) the coefficient for isotropic orbits is $32/\Pi$ and the one for linear orbits is $128/\Pi$ (Tremaine 1983). The velocities and separations are then taken relative to the center of mass of the group. In the distributed mass case, the coefficients differ by a factor of 4, a large contribution to the uncertainty in the mass estimation.

Bahcall and Tremaine derive a lower limit to the fractional standard deviation of the virial theorem estimator

$$\sigma_{vt} \gtrsim \Pi^{-1} (2 \ln N)^{1/2} N^{-1/2}$$

They emphasize that the estimator is both inefficient and biased. For the projected mass estimator, the fractional standard deviation is:

$$\sigma_{pm} \simeq 1.3 N^{-1/2}$$

In the absence of interlopers, the projected mass estimator is well-behaved. However, the estimator is sensitive to velocity interlopers which tend to have large R_i. Heisler and Tremaine (1984) are developing a method for treating this problem.

For the loose and generally poor sampled groups in the CfA and NB surveys, the inefficiency of the virial mass estimator and the effect of interlopers on the projected mass estimator makes a detailed comparison of mass estimates for individual groups uninformative. For a coefficient between the linear and isotropic orbit case, the median projected mass and virial mass estimates agree for both the CfA and NB catalogs (Geller and Huchra 1984).

A set of better sampled groups originally selected by Morgan, Kayser, and White (1975; MKW) and Albert, White and Morgan (1977; AWM) provide another testing ground for these mass estimators. We have measured and tabulated redshifts for 6 of the MKW-AWM groups (Beers et al. 1984; for a discussion of other properties of these groups, see the article by Kriss, this volume). Table 2 lists the number of group members (interlopers have been culled from the samples) along with the virial theorem and projected mass estimates of the mass-to-light ratios.

TABLE 2

MASS-TO-LIGHT RATIOS FOR MKW-AWM GROUPS

Group	Number of Redshifts	M_{vt}/L	M_{pm}/L
AWM1	12	$120 \pm 60\ M_\odot/L_\odot$	$250 \pm 100\ M_\odot/L_\odot$
AWM3	8	430 ± 270	1550 ± 930
AWM7	20	660 ± 180	680 ± 180
MKW1	8	150 ± 90	100 ± 60
MKW4	13	440 ± 150	500 ± 190
MKW12	11	70 ± 30	60 ± 20

We estimate errors in the dynamical parameters with the "jacknife"
procedure (Diaconis and Efron 1983). We calculate dynamical parameters
for all possible subsets of N-1 galaxies (where N is the total number
of galaxies in the system). For any particular parameter, the error
is the standard deviation about the mean value for the N subsets.
The error estimates are in excellent agreement with the analytic
predictions of Bahcall and Tremaine (1981) in equations (1) and (2)
above. The jacknife is a straightforward way of obtaining the error
in the virial theorem estimate. For all the groups in Table 2, the
two estimates of the mass-to-light ratio agree within better than 2
standard deviations. The significance of the spread in mass-to-light
ratio among these systems cannot be evaluated without better photometric data. Lack of deep magnitude limited samples with reliable
photometry introduces uncertainties of at least a factor of 2 in
the mass-to-light ratios. Comparably large uncertainties in the
mass-to-light ratios for rich clusters also stem from inadequate
photometric data.

VI. THE CONTINUUM OF "GROUP" AND "CLUSTER" PROPERTIES

The distinction between "groups" and "clusters" is more a matter
of semantics than science. In the vernacular, "groups" are systems
containing at most a few tens of bright galaxies: "clusters" are richer
systems. There is no obvious dividing line and there is a substantial
overlap of physical characteristics. The mass-to-light ratios of
groups are comparable with those obtained for well-studied rich
clusters (Kent and Gunn 1982; Geller et al. 1984; Shectman 1983).
Many of the groups identified in the NB and CfA surveys are loose,
but the compact groups discussed by Hickson et al. (this volume)
often have galaxy densities comparable with or sometimes even greater
than the densities typical of the cores of rich clusters.

Because the groups in a catalog span a wide range in density,
they are useful for assessing the relationship between the morphology
of galaxies and their environment. One of the observational constraints
on models for the morphological differentiation of galaxies is the
decrease in spiral population with increasing local galaxy density and
the accompanying increase with density of the fraction of S0 and
elliptical galaxies. This relationship was originally derived by
Dressler (1980b) for a sample of rich clusters (see also the review by
Dressler in this volume). The morphology-density relationships for
compact groups and for groups in the CfA and NB catalogs are consistent
with Dressler's relation. Combining the results for groups in the
general field with those for clusters extends the morphology-density
relationship over six orders of magnitude in space density. Figure 6
shows the relationship for the CfA catalog (solid histogram) and for
Dressler's (1980b) cluster sample (dashed histogram and solid curve).
Some features of this apparently universal relation suggest a
qualitative physical interpretation. There is no dependence of
morphology on density in regions where the dynamical timescale is

comparable with or greater than the Hubble time. At densities greater
than ∿ 600 galaxies Mpc^{-3}, SO's dominate the galaxy population. At
these densities, stripping mechanisms are likely to operate. At
densities greater than ∿ 3000 galaxies Mpc^{-3}, the fraction of ellipticals
rises steeply. In these regions, the collapse time is short compared
with the typical timescales for the formation of disks.

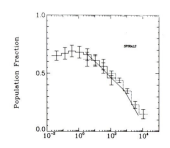 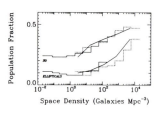

Figure 6. The morphology-density relation for the CfA redshift survey
catalog. Note the agreement between the relation for the redshift
survey (solid histogram) and Dressler's cluster sample (dashed
histogram).

VII.   CONCLUSIONS

     The properties of groups of galaxies provide constraints on
models for galaxy clustering and for the morphological differentiation
and dynamical evolution of individual galaxies. Galaxy morphology
and the dynamical timescale in a region appear to be related.

     Large, statistically complete samples of groups of galaxies
derived from complete redshift surveys yield a median mass-to-light
ratio M/L = 400 e$^{\pm0.3}$. For the luminosity density 1.1 x 10^8 L$_\odot$
determined from the CfA redshift survey (Davis and Huchra 1982),
this mass-to-light ratio yields an estimate for the mean mass density,
Ω = 0.2 e$^{\pm0.3}$, in excellent agreement with the estimates obtained
with correlation function techniques (Davis and Peebles 1983, Bean
et al. 1983). The estimate can be improved by (1) reliable photometric data, (2) larger redshift surveys, and (3) careful analysis of
methods of group selection and mass estimation.

REFERENCES

Albert, C.E., White, R.A., and Morgan, W.W., 1977, Ap.J., 211, 309 (AWM).
Bahcall, J. and Tremaine, S., 1981, Ap.J., 244, 805.
Bean, A.J., Efstathiou, G., Ellis, R.S., Peterson, B.A., and Shanks,
     T., 1983, M.N.R.A.S., 205, 605.
Beers, T.C., Geller, M.J., Huchra, J.P., Latham, D.W., and Davis,
     R.J., 1984, Ap.J., in press.
Bothun, D.G., Geller, M.J., Beers, T.C., and Huchra, J.P., 1983,
     Ap.J., 268, 47.

Cavaliere, A., Santangelo, P., Tarquini, G., and Vittorio, N., 1982,
    Proc. Paris Symposium on Clusters of Galaxies, ed. D. Gerbal and
    A. Mazure.
de Vaucouleurs, G. 1975, In Galaxies and the Universe, ed. A. Sandage,
    M. Sandage, and J. Kristian (Chicago: University of Chicago Press),
    p. 557.
Davis, M. and Huchra, J. P., 1982, 254, 437.
Davis, M., Huchra, J., Latham, D., and Tonry, J., 1982, Ap.J., 253, 423.
Davis, M. and Peebles, P.J.E., 1983, Ap.J., 267, 465.
Diaconis, P. and Efron, B. 1983, Sci. Am. 248, 116.
Dressler, A., 1980a, Ap. J. Suppl., 42, 565.
------------, 1980b, Ap. J., 236, 351.
Efstathiou, G. and Eastwood, J.W., 1981, M.N.R.A.S., 194, 503.
Faber, S.M. and Gallagher, J.S., 1979, Ann. Rev. Astr. and Ap., 17, 135.
Fall, S.M., 1975, M.N.R.A.S., 172, 23p.
Geller, M., Beers, T.C., Bothun, G.D., and Huchra, J.P., 1984,
    Ap.J., in press.
Geller, M.J. and Huchra, J.P., 1983, Ap.J. Suppl., 52, 61 (GH).
--------------------------, 1984, in preparation.
Geller, M.J. and Peebles, P.J.E., 1973, Ap.J., 184, 329.
Gott, J.R. and Turner, E.L., 1977, Ap.J., 213, 309 (GT).
Heisler, J. and Tremaine, S.D., 1984, in preparation.
Hickson, P., 1982, Ap.J., 255, 382.
Holmberg, E., 1969, Arkiv fur Astr., 5, 305.
Huchra, J.P., Davis, M., Latham, D.W., and Tonry, J., 1983, Ap.J.
    Suppl., 52, 89.
Huchra, J.P. and Geller, M.J., 1982, Ap.J., 257, 423 (HG).
Huchra, J.P. et al., 1984, in preparation.
Karachentseva, V., 1973, Soob. Spetr. Astrofiz. Obs., No. 8.
Kent, S.M., and Gunn, J.E., 1982, A.J., 87, 945.
Mardirossian, F., Giuricin, G., and Mezzetti, M., 1983, Astr. and
    Ap., 126, 86.
Materne, J., 1978, Astr. and Ap., 63, 401.
----------, 1979, Astr. and Ap., 74, 235.
Materne, J. and Tammann, G., 1974, Astr. and Ap., 37, 383.
Mezzetti, M., Giuricin, G., Malagnini, M.L., and Mardirossian, F.,
    1983, Astr. and Ap., 125, 368.
Morgan, W.W., Kayser, S., and White, R.A., 1975, Ap.J., 199, 545 (MKW).
Peebles, P.J.E., 1980, The Large-Scale Structure of the Universe
    (Princeton University Press: Princeton).
Postman, M. and Geller, M.J., 1984, Ap.J., in press.
Press, W.H. and Davis, M., 1982, Ap.J., 259, 449 (PD).
Rood, H., Rothman, V., and Turnrose, B., 1970, Ap.J., 162, 411.
Sandage, A. and Tammann, G., 1975, Ap.J., 196, 313.
Schechter, P., 1976, Ap.J., 203, 557.
Shectman, S., 1983, private communication.
Tremaine, S., 1983, private communication.
Turner, E.L. and Bott, J.R., 1976, Ap.J. Suppl., 32, 409.
Turner, E.L., Aarseth, S.J., Gott, J.R., and Mathieu, R.D., 1979,
    Ap.J., 228, 684.

DISCUSSION

APPLETON: Is there any physical justification for the building of groups by the "friends of friends" approach?

GELLER: The friends of friends approach identifies redshift space volume density enhancement – it is just a way of realizing the definition of groups.

DE VAUCOULEURS: The density enhancement rule is not the only one to use to identify group of galaxies; similarity of morphology, color, resolution ... is also a useful guide. A multiparameter search for clumping in phase space should be a natural extension of your work. Also note that a simple density enhancement rule might miss the galactic equivalent of expanding stellar associations; this was noted in the chapter on groups, volume 9 of "Stars and Stellar Systems".

GELLER: It would certainly be possible to do a multi-parameter search.

THE STRUCTURE OF COMPACT GROUPS OF GALAXIES

Paul Hickson and Zoran Ninkov
 University of British Columbia, Vancouver, B.C., Canada
John P. Huchra
 Smithsonian Astrophysical Obs., Cambridge, Ma., U.S.A.
Gary A. Mamon
 Princeton University, Princeton, N.J., U.S.A.

1. INTRODUCTION

Compact groups are small close isolated associations of galaxies. High space densities ($\lesssim 10^4/Mpc^3$) and short crossing times (~0.1 H^{-1}) make them particularly susceptible to dynamical evolution. Several well known groups contain galaxies with very discrepant redshifts, casting doubt upon the physical reality of compact groups in general or on the cosmological redshift hypothesis. The small population of these groups, typically less than ten galaxies, makes the interpretation of the structure and dynamics of an individual group difficult at best. It is often more informative to investigate the statistical properties of a homogeneous sample of groups. Galaxies in such groups are generally too faint to appear in standard galaxy catalogs, however, so they cannot be found by automated techniques used to study loose groups (Turner and Gott, 1976, Geller, this volume). For this reason a survey was undertaken of the Palomar Observatory Sky Survey (POSS) prints for groups satisfying three selection criteria of population, isolation, and mean surface brightness. The resulting catalog of 100 groups (Hickson, 1982) forms our study sample.

2. GALAXY MORPHOLOGY

Images for all group members were inspected on the red and blue POSS prints in order to classify galaxies as Spiral (S), elliptical (E), or "lenticular" (L). L types include obvious S0s and red lens-shaped galaxies. This approach clearly suffers from uncertainties, particularly for the fainter galaxies, and we are now obtaining two-colour CCD images for all galaxies. The present results are based on the POSS. The percentage distribution of types in the sample is shown in figure 1 for all galaxies, and for first-ranked galaxies (the brightest galaxy in each group) only. Spirals comprise about 43% of the total sample, considerably less than the approximately 75% spiral fraction of field galaxies (Gisler 1980). Surprisingly, the first-ranked galaxies are spiral 52% of the time. Since numerical simulations predict merger products to be elliptical-like (White 1982), it seems unlikely that many

367

F. Mardirossian et al. (eds.), Clusters and Groups of Galaxies, 367–373.
© *1984 by D. Reidel Publishing Company.*

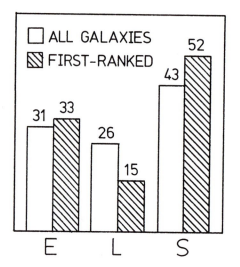

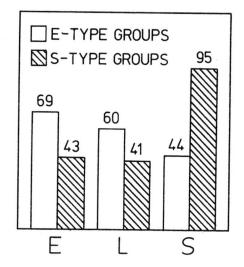

Fig. 1. Distribution of galaxy types   Fig. 2. Types vs group type

first-ranked galaxies in these groups are formed by mergers; indeed, most of these first-ranked spirals appear to be relatively normal. Figure 2 shows the distribution of fainter (not first-ranked) galaxies in groups whose brightest member is classified E or L (E-type) or S (S-type) respectively. It appears that if the brightest galaxy is spiral the fainter galaxies are more likely to be spiral also, ie. like types tend to be found together. There is a strong morphology density relation in the groups (figure 3). The mean spiral fraction decreases from about 75% in the looser groups to 30% in the most compact. This is similar to the relation found for rich clusters (Dressler 1980).

3. THE LUMINOSITY FUNCTION

Spectroscopic observations have been made with the Whipple

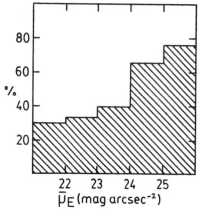

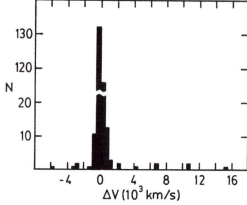

Fig. 3. Spiral fraction vs density   Fig. 4. Plot of velocity residuals

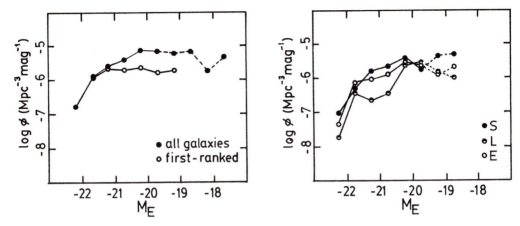

Fig. 5. Galaxy luminosity function   Fig. 6. Variation with galaxy type

Observatory 60-inch telescope of some 281 galaxies in these groups so
far. The distribution of velocity residuals (from the median) is shown
in figure 4. Of the 72 groups with more than one velocity, 9 contain
one or more "discrepant" redshifts (13%). Of the 281 total redshifts 13
are discrepant (5%). We have obtained redshifts for a total of 94 groups
including all in the "complete" subsample of 67 groups with total red
magnitude of 13.0 or brighter. This subsample was used to determine a
preliminary luminosity function for compact group galaxies (figures 5
and 6). Because present data extends only 3 mag down from the brightest
galaxy in each group the dashed portions of the curves indicate LOWER
BOUNDS only. The luminosity function for all galaxies appears to be
similar to that of field galaxies (Felton 1977) and corresponds roughly
to a Schechter (1976) curve with $\alpha \lesssim -1$. There is no evidence for significant deviations with galaxy type (figure 6). This is in contrast with
the result of Heiligman and Turner (1980) who found a turnover at
fainter magnitudes in 10 looser groups. However, the distribution of
luminosity ratios with respect to the first-ranked galaxies (Hickson
1982) is inconsistent with all galaxies in a group being drawn equally
from a normal luminosity function. Rather, it appears that there are
groups of faint galaxies and groups of bright galaxies.

4. SPATIAL STRUCTURE

4.1 Group ellipticities

    Axial ratios of the compact groups were determined by the method
applied by Rood (1979) of fitting a rectangular box to each group. The
resulting distribution of axial ratios are shown in figures 7 and 8 for
all groups of 4 (n=4) and more than 4 galaxies respectively. They are
compared with solid curves indicating the results of static Monte-Carlo
models of 5000 groups; "s" denotes a model in which galaxies are placed
at random within a randomly oriented spheroid then projected on the sky.
For n=4 groups the best fit spheroid is triaxial with axial ratio
1.0:0.5:0.2.; a sphere is incompatible with the data. The curve labelled

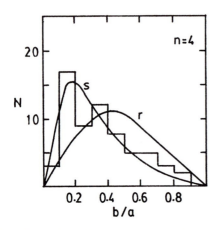
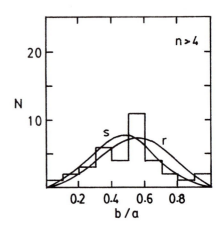

Fig. 7. Distribution of axial ratios   Fig. 8. Same for n>4 groups

"r" is the result of randomly scattering galaxies on the plane of the
sky; it is intended to test the hypothesis that groups may be transient
configurations of high apparent density in loose groups caused by "chance
crossings" (Rose, 1977). This model is strongly excluded by these data
(KS probability <0.1%). n>4 groups are considerably less elongated on
average. Their b/a distribution fits either a spherical model (s) or a
2-dim. random projection (r).

4.2 Radial profile

        A mean radial surface density profile was computed by scaling
groups to the same apparent size and superimposing them. (figures 9 and
10). The groups are more centrally concentrated than either the homogeneous spheroid or the 2-dim. random models, the latter being excluded at
the 0.1% and 3.8% levels for n=4 and n>4 groups.

4.3 Dynamic models

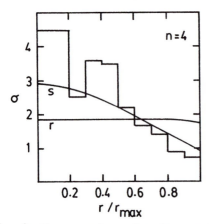
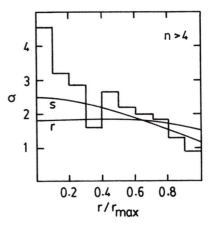

Fig. 9. Mean surface density profile   Fig. 10. Same for n>4 groups

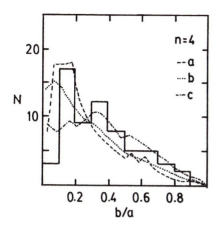 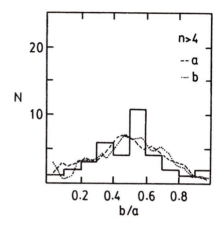

Fig. 11. Comparison with dynamical models   Fig. 12. Same for n>4 groups

    Dynamic models of groups of four or more galaxies were evolved from
an initial random spherical configuration. The axial ratios b/a fluctuate
strongly on a crossing timescale while the intrinsic shape fluctuates
between prolate and oblate forms. A mean b/a distribution was obtained
by sampling at equal time steps (figures 11 and 12). The models are not
only consistent with the data but also reproduce well the apparent
disparity between n=4 and n>4 groups. For n=4, the curves a, b, and c
indicate models of 4 galaxies, and subgroups of 4 galaxies (satisfying
the selection criteria) in groups of 5 and 10 galaxies respectively. For
n>4 the models a and b are n=5 groups and n=5 subgroups in groups of 10
galaxies respectively. All are statistically consistent with the data.

5. CONCLUSIONS

    The results presented here suggest that most compact groups are true
dynamical systems and are not transitory. The absence of much dynamical
evolution of the brightest galaxies suggests then that these groups have
only recently attained high densities and must have a short lifetime
after the onset of merging. The morphology density relation then is more
likely to be a result of "initial conditions" than of evolution.

REFERENCES

Dressler, A.: 1980, Ap. J. 236, 351.
Felton, J.E.: 1977, A.J. 82, 861.
Gisler, G.R.: 1980, A.J. 85, 623.
Heiligman, G.M., and Turner, E.L.: 1980, Ap. J. 236, 745.
Hickson, P.: 1982, Ap. J. 255, 382.
Rood, H.J.: 1979, Ap. J. 233, 21.
Rose, J.A.: 1977, Ap. J. 211, 311.
Schechter, P.L.: 1976, Ap. J. 203, 297.
Turner, E.L., and Gott, J.R.: 1976, Ap. J. Suppl. 32, 409.
White, S.D.M.: 1982, "Internal Kinematics and Dynamics of Galaxies",
    IAU Symposium No. 100 (ed. E. Athanassoula), 337, D. Reidel Pub. Co.

DISCUSSION

DRESSLER: Jim Rose argued that the lack of intergalactic light in compact groups indicated that little tidal stripping had occured, and therefore that compact groups are transient phase of looser associations. How do you data and conclusions compare with his?

HICKSON: We find that the morphology of galaxies in our groups indicates that little dynamical evolution has generally occured, in agreement with Rose. However, this cannot be explained by groups being a transient phase of looser stable associations; there are far too many highly elongated groups. The distribution of axial ratios for the quartets studied by Rose is essentially identical to our own even though only two groups are common to both samples.

GIOVANELLI: Could you expand on your last question (why so many spirals?)? It looks to me that if clusters form continuously, as you suggest they may, then one should expect to find even more spirals than you actually do (i.e. about like the field population).

HICKSON: The looser groups have the same spiral fraction as the field. Denser groups have fewer, but even the most compact groups generally contain one or more spirals. This may be understood in the continuous formation scenario provided that the timescale for destruction of spirals by ram pressure stripping is at least as long as the timescale for destruction of the groups through galaxy mergers.

TURNER: You mentioned that you had discovered a number of new examples of discrepant redshift galaxies in your compact groups. Do these new discrepant redshift galaxies have systematically different properties from the other group memebers?

HICKSON: That's an interesting point. Galaxies that have positive discrepant redshifts (higher velocity) are consistently smaller and fainter than the average group member, and those with negative discrepant redshifts are consistenty larger and brighter.

GIOVANELLI: If you assume some sort of $L \propto D^{1.8}$ law, do spirals in compact groups fit with galaxies found in the field or do they look smaller?

HICKSON: I have measured sizes and luminosities for all galaxies int the catalog. There is a trend of decreasing size and luminosity with increasing density for all galaxies including spirals in the groups. They tend to be less luminous and have smaller diameters than typical field galaxies.

A DYNAMICAL STUDY OF TWO GROUPS IN THE COMA/A1367 SUPERCLUSTER

B. A. Williams
National Radio Astronomy Observatory
Charlottesville, Virginia USA

Located ~12° southwest of A1367, the IC 698 group and the NGC 3825
group are two systems of galaxies within the Coma/A1367 supercluster.
All of the spiral galaxies in both groups were observed and detected at
21 cm with the 305-m Arecibo radio telescope. The dynamical stability
of each group was analyzed by using the accurate radial velocities and
the direct estimates of the masses obtained from these 21-cm line
observations. Application of the virial theorem to the IC 698 group
gives a mass which is comparable to the sum of the individual masses of
its galaxies. When the virial theorem is applied to the NGC 3825
group, its virial mass is found to be ~6 times larger than the sum
of the masses derived for the individual members.

1. INTRODUCTION

The existence of small-scale clustering, i.e., groups, has been
confirmed from the surface distribution of nearby galaxies (Geller and
Peebles 1973). Most of the groups identified in various lists
(Karachentsev 1966; de Vaucouleurs 1975; Turner and Gott 1976; Huchra
and Geller 1982) are gravitationally unbound and appear to have masses
much larger than the values inferred within the luminous parts of the
individual members. In this paper, HI line observations are used to
investigate the dynamical stability of two groups of galaxies in the
southern extension of the Coma/A1367 supercluster, the IC 698 group and
the NGC 3825 group.

2. THE IC 698 GROUP AND THE NGC 3825 GROUP

The IC 698 group and the NGC 3825 group have mean velocities near
6000 km s^{-1} and contain mostly spiral galaxies. The IC 698 group has
an angular size of ~17' and is one of the smallest systems that have
been observed in this supercluster. The NGC 3825 group, located ~10°
from the IC 698 group, has an angular size of 40', is found within
Zw 68-21 (11^{h}38^{m}3 + 10°24'), a medium-compact cluster (Zwicky *et al.*

375

1961), and consists of 9 galaxies, eight of which are spirals.

2.1. Optical Properties

The IC 698 group is well isolated from other galaxies having
$m_p < 15.7$ and contains the following members: IC 2850, 2853, 2857,
696, 698 and 699, all of which are spirals. The optical properties of
these galaxies are described in an earlier article by Williams (1983).
Five of the galaxies in the NGC 3825 group are concentrated toward the
center of the group in what appears to be a subsystem, while the other
four galaxies lie along the outer boundary in a "halo" surrounding this
subsystem. The galaxies that form the subsystem have already been
identified as a poor cluster (Morgan, Kayser and White 1975) and also
as a compact group (Hickson 1982). The outer boundary of the NGC 3825
group is not as well-defined as that of the IC 698 group. There are
five galaxies that lie within 1/2 degree of the nearest member of the
NGC 3825 group. These galaxies have radial velocities very close to
the mean velocity of the group and probably should be included in its
dynamical analysis.

2.2 HI Properties

All of the spiral galaxies in the IC 698 and the NGC 3825 groups
were detected in neutral hydrogen. In the field surrounding the
NGC 3825 group, the detection rate of spiral galaxies was also very
high (> 85%). The HI properties of the galaxies in the IC 698 group
have been discussed in Williams (1983). The M_H/L ratios measured for
the spirals in the compact subsystem of the NGC 3825 group are between
0.06 and 0.15 solar units while the four spirals in the "halo" have
M_H/L ratios between 0.30 and 1.03 solar units. The average uncertainty
in this ratio due to errors in the determination of the hydrogen mass
is 0.03 for the galaxies in the core and 0.05 for the galaxies in the
"halo". Unfortunately, the morphological types are known for only four
of the eight spiral galaxies in the group. Two early-type spirals are
part of the compact subsystem and the other two late-type spirals are
located in the "halo". There would be a distinct segregation between
early- and late-type spirals in the high and low density regions in the
NGC 3825 group if the other four galaxies had morphological types
consistent with their M_H/L values.

3. GROUP DYNAMICS AND VIRIAL ANALYSIS

The radial velocity and width of the HI profile of each galaxy in
the group were used to determine the potential and kinetic energies
defined as

$$\Omega = -2/\pi \ G \sum_{pairs} M_i M_j / r_{ij}, \tag{1}$$

$$T = 3/2 \sum_i M_i \Delta v_i{}^2, \tag{2}$$

respectively, where r_{ij} is the projected linear distance between galaxies with masses M_i and M_j and Δv_i is the line-of-sight velocity (corrected for the observational errors in the radial velocity) of the galaxy relative to that of the group's center of mass. (In cases where the inclination angle of the galaxy is less than 30°, the mass is estimated by assuming a mass-to-light ratio of 7 solar units). The crossing time of each group was determined from the relation

$$t_c = 2/\pi \; (\langle r\rangle/\langle v\rangle), \tag{3}$$

where $\langle r\rangle$ is the mean projected distance of members from the center of mass and $\langle v\rangle$ is the average absolute value of the velocities relative to the center of mass. Assuming that both groups are relaxed systems, the virial theorem can be applied, where the virial mass is given by

$$M_v = -2MT/\Omega, \tag{4}$$

and M is the sum of the individual masses. The dynamical parameters of the IC 698 group are summarized in Table 1. The velocity dispersion

TABLE 1.   DYNAMICAL PARAMETERS OF THE IC 698 GROUP

$\langle v_g\rangle$	6170 ± 46 km s^{-1}
Ω	-5.8×10^{58} ergs
T	3.6×10^{58} ergs
t_c	0.17 H^{-1}
M_v	8.2×10^{11} M$_\odot$
M	6.4×10^{11} M$_\odot$
M_v/L	5 solar units

about the mean velocity of the IC 698 group is surprisingly small, only 46 km s^{-1}. Since the velocity dispersion of nearby groups containing 4 to 10 members is typically 170 km s^{-1} (Huchra and Geller 1982), the value measured for the IC 698 group is unusual. Within the observational errors of the line-of-sight velocity, the total energy of the group is negative and the system is most probably stable against disruption. The short crossing time given in Table 1 is also an indication that the group is probably bound. The mass of the group derived from the virial theorem is comparable to the sum of the individual masses within the photometric radii of the galaxies in the IC 698 group. There is no evidence for any hidden mass which is apparently needed to bind most other groups of galaxies.

Because the boundary of the NGC 3825 group is not well defined, the dynamical analysis was done on the compact subsystem as if it were an

independent system, on the NGC 3825 group, and on the NGC 3825 group
including the five outlying galaxies that may be possible members. The
dynamical parameters of these three groupings are listed in Table 2.

TABLE 2. DYNAMICAL PARAMETERS

	COMPACT SUBSYSTEM	NGC 3825 GROUP	NGC 3825 GROUP + FIVE OUTLYING GALAXIES
$\langle v_g \rangle$	6054 ± 116 km s^{-1}	6013 ± 120	6001 ± 120
Ω	-1.0×10^{59} ergs	-1.3×10^{59}	-2.2×10^{59}
T	2.2×10^{59} ergs	3.8×10^{59}	5.3×10^{59}
t_c	0.05 H^{-1}	0.12	0.24
M_v	3.5×10^{12} M$_\odot$	6.2×10^{12}	8.8×10^{12}
M	0.75×10^{12} M$_\odot$	1.1×10^{12}	1.9×10^{12}
M_v/L	31 M$_\odot$/L$_\odot$	40	29

The velocity dispersion of the NGC 3825 group is typical of that
measured for compact groups of galaxies (Faber and Gallagher 1979,
Burbidge and Sargent 1971). The total energy of the group is positive
so that the system appears to be gravitational unstable. Because the
crossing time is considerably smaller than the Hubble time, this
suggests that the system must be bound, otherwise it would have
dispersed by now. If the group were unbound, it would have had to begin
its expansion less than 1.6×10^9 years ago (H = 75 km s^{-1} Mpc^{-1}). The
virial mass of the group is ~6 times larger than the sum of the
individual masses within the luminous disk of the galaxies. Within the
uncertainties of the virial mass, there is no difference in the virial
mass-to-light ratio of the compact subsystem and that of the NGC 3825
group with or without the five outlying members. The value of the
virial mass-to-light ratio given in Table 2 agrees very well with the
average mass-to-light ratio of small loose groups (Turner and Gott 1976;
Tammann and Kraan 1978; Tully and Fisher 1978) and that of compact
groups (Faber and Gallagher 1979, Burbidge and Sargent 1971).

4. CONCLUSIONS

     If most of the true velocity dispersion of the IC 698 group lies
along the line of sight, then this system is gravitationally stable.
The velocity dispersion measured for the IC 698 group is 4 times
smaller than that observed for typical groups in the Local
Supercluster. The excellent agreement between the virial mass and the
sum of the individual galactic masses implies that there are no large
amounts of hidden mass within the photometric radii of the galaxies in

the group. Unlike the IC 698 group, the NGC 3825 group is more similar dynamically to loose groups of galaxies that have been studied in the Local Supercluster. Because the virial mass is larger than the sum of the individual masses by a factor of six, there is strong evidence for hidden mass within the NGC 3825 group.

REFERENCES

Burbidge, E. M., Sargent, W.L.W.: 1971, in *Nuclei of Galaxies* ed. D.J.K. O'Connell, (Amsterdam: North-Holland), pp. 351-386.

de Vaucouleurs, G.: 1975, in *Stars and Stellar Systems*, Vol. 9, *Galaxies and the Universe*, eds. A. Sandage, M. Sandage, J. Kristian (Chicago: University of Chicago Press), pp. 557-600.

Faber, S. M., Gallagher, J. S.: 1979, Ann. Rev. Astron. Astrophys. 17, pp. 135-187.

Geller, M., Peebles, P. J.: 1973, Astrophys. J. 184, pp. 329-341.

Hickson, P.: 1982, Astrophys. J. 255, pp. 382-391.

Huchra, J., Geller, M.: 1982, Astrophys. J., 257, pp. 423-437.

Karachentsev, I. D.: 1966, Astrofizika 2, pp. 81-99. (English trans., 1966, Astrophysics 2, pp. 39-49).

Morgan, W. W., Kayser, S., White, R. A.: 1975, Astrophys. J. 199, pp. 545-548.

Tammann, G. A., Kraan, R.: 1978, IAU Symp. No. 79, *The Large Scale Structure of the Universe*, eds. M. S. Longair, J. Einasto (Dordrecht: Reidel), pp. 71-91.

Turner, E. L., Gott, J. R.: 1976, Astrophys. J. Suppl. 32, pp. 409-427.

Tully, R. B., Fisher, J. R.: 1978, IAU Symp. No. 79, *The Large Scale Structure of the Universe*, eds. M. S. Longair, J. Einasto (Dordrecht: Reidel), pp. 31-47.

Williams, B. A.: 1983, Astrophys. J. 271, pp. 461-470.

Williams, B. A., Kerr, F. J.: 1981, Astron. J. 86, pp. 953-980.

Zwicky, F., Herzog, E., Wild, P.: 1961, *Catalogue of Galaxies and Clusters of Galaxies*, Vol. 1, (Pasadena: Calif. Inst. Technol.).

DISCUSSION

GIOVANELLI: I would like to know the sizes of spiral galaxies in Hickson's compact groups.

HICKSON: Most galaxies in compact groups tend to be smaller than normal field galaxies.

HANISCH: It is possible that the low velocity dispersion you have obtained might have resulted from the fact that the galaxies are as close together as the Arecibo beamwidth at 21 cm 3'.3? Could this cause confusion in the interpretation of the observed HI profiles?

WILLIAMS: The two closest galaxies in the subsystem are separated by 3'.2 along an E-W direction while their major axes lie nearly N-S. There is little reason to worry about confusion that might produce large errors in the determination of the radial velocities.

JAFFE: What radius did you use to determine masses of galaxies from HI velocity widths?

WILLIAMS: The radius used to determine the masses of the galaxies is the Holmberg radius.

COMPUTER MODELLING OF THE GAS DYNAMICS IN SMALL GROUPS OF INTERACTING GALAXIES

P. A. Foster
Department of Astronomy, University of Manchester, England.

Previous attempts to model interacting galaxies using computers have typically used particles either moving under the influence of external potentials and/or the potential of the particles themselves. The range of approaches to this problem is large; from the simple Toomre three body models (Toomre and Toomre, 1972) to the complex, self-consistent models of Miller and Smith (1980). Although these models represent the dynamics of the stellar component of galaxies there is good reason to believe that the gas component will have a very different behaviour. The dynamics of the gas will differ from the dynamics of the particles because of the effect of pressure forces; this means, for instance, that colliding gas clouds will interact more strongly than colliding particle clouds. Observationally also there is evidence for this different behaviour; Appleton (c.f. this conference) has described the interacting galaxy NGC4747 which has a gaseous plume and a stellar plume which are offset from one-another. Observations of the M81/M82 system (Appleton et al, 1981) show it to have very extended HI emission, suggesting the gas has been affected by the interaction more strongly than the stars.

The objective of the current work is to model the gas dynamics in interacting systems. To do this a programme is currently being developed towards the point where it can follow the dynamics of two rotating discs of cold gas moving through a hot intergalactic medium. The whole system is to move in an external potential, the potential due to two softened point potentials for instance (compare the Toomre three body models). The programme has been developed from a code obtained from Alistair Nelson and Tim Johns of University College Cardiff. It uses a finite difference scheme and applies the Flux Corrected Transpent (F.C.T.) technique as described by Boris and Book (1973). Previous applications of the programme have been to quasi steady state models of gas flows in galaxies and so it is untested on the fully time dependent problems proposed here.

The programme currently suffers from certain limitations. Firstly, it can only handle problems in two dimensions, because of the necessary computer time to run three dimensional models. This problem is not a

381

F. Mardirossian et al. (eds.), Clusters and Groups of Galaxies, 381–384.
© *1984 by D. Reidel Publishing Company.*

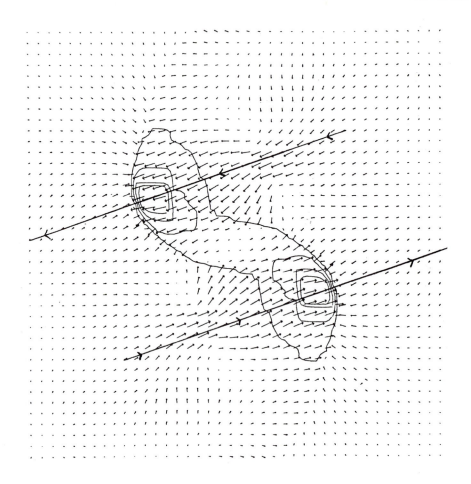

Figure 1

severe one and it is possible to run interesting models in two dimensions. More problematic are the effects of the Cartesian grid, in particular, a rotating disc of gas will develop four spiral arms, reflecting
the geometry of the grid. A comparison of the same rotating disc on a
20 x 20 grid and a 40 x 40 grid suggests that this problem can be substantially reduced by using a finer grid. The programme is currently
being transferred to a CRAY (from a VAX 11/780) which will increase the
available computer power a hundred-fold and allow both these problems to
be overcome.

A further limitation of the programme is that it is isothermal and
hence the hot, intergalactic gas cannot be included in the model. It may
be necessary to include an energy equation in the programme and this is
currently being investigated. A simpler solution may be to model isothermally a two phase gas consisting of a hot component and a cold component. Both these approaches are being investigated and some initial
one dimensional results are optimistic.

It now remains to show what can be done with the current programme.
Some results from a run are shown in figure 1, the arrows show velocity
and the contours show the gas density. In this run two identical, non-
rotating gas clouds have started from the top-right hand and bottom-left
hand corners and are pictured in the process of passing. The clouds sit
in soften point potentials which move with the same velocity as the gas
clouds, following the straight lines shown in the figure. The gas clouds
are passing through a cold intercloud medium with a velocity of one and
a half times the sound velocity. This medium has approximately 1/30th
the density of the clouds, too dense for a realistic intergalactic medium.
This is due to the isothermal nature of the programme and is necessary
to stop the clouds expanding into empty space.

These results show that a bridge has formed between the two clouds,
an interesting result, but, for the reasons given above, this cannot be
regarded as a realistic interaction. What this does show is the ability
of the programme to handle this fully time dependent problem and serves
as a useful test of the programme.

Work is continuing on the problems discussed here and it is hoped
that a study of the gas dynamics in interacting systems can be performed
soon. Likely problems to be tackled are the passage of a single gas
cloud through an intercloud medium and the effects of tidal interaction.
Also under development is a quasi particle, gas dynamics code; this may
be a more suitable approach to the problem.

REFERENCES

Appleton, P. N., Davies, R. D. and Stephenson, R. J.: 1981, M.N.R.A.S.,
          195, p.327.
Boris, J. P. and Book, D. L.: 1973, J. Comp. Phys., 11, p.38.
Miller, R. H. and Smith, B. F.: 1980, Ap. J., 235, p.421.
Toomre, A. and Toomre, J.: 1972, A. J., 178, p.623.

DISCUSSION

SHAPIRO: Just a cautionary comment: an isothermal equation of state will not be appropriate for describing shock waves, which result from cloud-cloud collisions which are at great enough relative velocities that the post-shock temperatures are high. If the post-shock gas is not enough, that is, radiative cooling will be unable to cool the gas on a timescale, which is short compared to the dynamical time scales of your problem. In that case, an assumption of isothermality would be incorrect and would, for example, prevent the numerical simulations from showing cloud ablation and dispersal.

FOSTER: Yes, there are problems with an isothermal code. Work is underway to include an energy equation into the programme.

CAN GAS AND STARS BE SEPARATED IN A TIDAL INTERACTION?

P. N. Appleton
Department of Astronomy
University of Manchester

INTRODUCTION

A great deal of attention has been given in recent years to the dynamical behaviour of stars in tidal interactions between galaxies as inferred from computer models (e.g., Toomre and Toomre, 1972). However very little work has been done on the response of the gas component of a galaxy to strong tidal forces. Much of our observational data on the kinematics of galaxies are based on measurements of either neutral or ionised gas rather than stars. If meaningful dynamical information is to be extracted from the stellar models it is important to question whether the kinematics of the gas always approximately follows that of the stars in tidally interacting galaxies. Foster (in a companion paper in this volume) describes some preliminary attempts by the Manchester group to model gas in interacting galaxies. In this present paper I shall briefly report on some observations and a particle simulation which supports the view that gas dynamical effects may be important in some tidal encounters.

Single dish and Aperture Synthesis HI observations have recently been made by Wevers, Appleton, Davies and Hart (in preparation) on the interacting galaxies NGC 4725/47 = Arp 159. The galaxies lie at a distance of 16 Mpc (Ho = 75 km s^{-1} Mpc^{-1}). NGC 4725 is a barred Sab-pec galaxy and lies 24 arcmins S.W. of NGC 4747 which is a peculiar barred Scd galaxy. The latter galaxy has a very distorted appearance and new deep plate material presented by Wevers et al. shows it to have two faint plumes of optical emission extending from it. The brightest plume extends 5 arcmins to the N.E. of the object and a second plume of shorter length points towards NGC 4725 in the S.W. The HI observations show a wealth of detail beyond the scope of this report. Of interest here, however, is the detection of HI associated with the optical plumes. Hayes (1979) has already shown that large amounts of HI are associated with the N.E. plume. Our observations reveal, however, a surprising offset between the optical and HI emission in the plumes. The HI plume is found to lie at a position angle 11° different from that of the optical plume. Although the two components appear initially overlapping in the centre of NGC 4747, the

385

F. Mardirossian et al. (eds.), Clusters and Groups of Galaxies, 385–387.
© *1984 by D. Reidel Publishing Company.*

386   P. N. APPLETON

difference in position-angle leads to a marked offset further from the
centre.

    The offset in HI may have a very straight forward explanation. It
is possible that the neutral hydrogen which now appears in the N.E. plume
of NGC 4747 was not distributed in the same way as the stars before the
tidal interaction took place. For instance, some of the HI gas could
have been distributed in an outer HI ring around the stellar component.
During the course of the interaction, tidal forces would preferentially
distort and warp the outer parts of the galaxy more than the inner parts.
I have performed a number of computer simulations of tidal interactions
between two discs of 'test particles'. The models were performed using
a constant time-step 3-body numerical scheme on the VAX 11/780 Starlink
Computer in Manchester. A mass ratio of 20:1 was used for the relative
masses of NGC 4725 and NGC 4747 respectively. The general features of
the distribution and kinematics of the N.E. HI plume was well reproduced.
However, an offset between different bands of particles corresponding to
different initial radial distribution of particles in NGC 4747 was not
observed. Similar conclusions can be drawn from Toomre and Toomre (1972)
for more equal-mass galaxy-galaxy interactions. I have therefore concluded that except for a contrived initial distribution of gas and stars
in NGC 4747, the offset between the stars and gas is not easy to explain.
Further modelling of this system is underway.

    A second possibility is that the dynamics of the stars and gas could
be markedly different. There are at least two possible gas dynamical
processes which could produce an offset; (i) the interaction of NGC 4747
with a dense, possibly ionised, intra-cluster gas or an extended ionised
halo around NGC 4725; or (ii) the redistribution of angular momentum of
the gas in NGC 4747 due to large scale cloud-cloud collisions occuring
in the disc of NGC 4747 during the strong tidal interaction. Work is
underway in Manchester to investigate both these effects (c.f. Foster,
this conference).

REFERENCES

Haynes, M. P., 1979, Astron. J., 84, 1830.
Toomre, A. and Toomre, J., 1972, Astrophys. J., 178, 623.

DISCUSSION

GIOVANELLI: Is there a bridge between the two galaxies in your model? There did not seem to be one in the map of the observed HI distribution, is that right?

APPLETON: The WSRT observations do not show a connecting bridge from NGC4747 to NGC4725 to a level of $\sim 5 \times 10^{19}$ atom cm^{-2}. There is however a short plume which points back towards NGC4725. The model I showed was not a serious attempt to model the details of the interactions but meanly to show that an explanation for the separate NE HI plume offset could not be explained by a segregation of the HI/stellar populations before the interaction.

A Neutral Hydrogen Study of Seyfert's Very Compact Group

John R. Dickel
Astronomy Department, University of Illinois

Herbert J. Rood
Box 1330, Princeton, NJ 08542

B.A. Williams
National Radio Astronomy Observatory[1], Charlottesville, Virginia 22901

Abstract

    A cloud of neutral hydrogen with an extent greater than that of any individual galaxy is present in the direction of Seyfert's very compact group of galaxies. At each position within this cloud, the velocity width is about $160\ km\ s^{-1}$ and the radial velocity lies in the range $4530\ km\ s^{-1}$ to more than $4600\ km\ s^{-1}$. The largest radial velocity of any galaxy in the group is $4503\ km\ s^{-1}$.

I. Introduction

    Seyfert's group (commonly called 'Seyfert's sextet'), No. 79 in Hickson's (1982) homogeneous catalog of 100 compact groups, is an extremely compact group of galaxies with an angular diameter of 1.3 arcmin (Hickson 1982). The members are touching or nearly touching one another; the luminosity density of the group exceeds that averaged over the core of a rich cluster. The four brightest members of the group have a mean heliocentric radial (line-of-sight) velocity of $4350\ km\ s^{-1}$ and a radial velocity dispersion of $160\ km\ s^{-1}$ (P. Hickson and J. P. Huchra, private communication 1983, Burbidge and Sargent 1971, Chincarini and Rood 1972). For calculation purposes, we adopt a Hubble constant of $100\ km\ s^{-1}\ Mpc^{-1}$ giving a distance to the group of $48\ Mpc$. Thus the linear diameter of Seyfert's group is only $18\ kpc$, comparable to that of the Milky Way! The brightest member, NGC 6027, is an S0p and has an attached plume the size of a galaxy but with very low surface brightness. A fifth galaxy has a radial velocity of $19,900\ km\ s^{-1}$ and we shall not consider it further. The only late-type spiral galaxy is NGC 6027C, the faint elongated galaxy on the southern edge of the group (de Vaucouleurs, de Vaucouleurs, and Corwin, 1976). A sketch of the group is shown in Figure 1.

    The group has been observed in the 21-line of neutral hydrogen with the 1000-foot Arecibo telescope by Gallagher, Faber, and Knapp (1981) and by Bjermann, Clark, and Fricke (1979), and observations by us with the Arecibo telescope[2] and the 300-foot telescope of NRAO show a similar integrated profile for the group (Dickel,

1. National Radio Astronomy Observatory is operated by Associated Universities, Inc., under contract with the National Science Foundation.
2. The Arecibo Observatory is operated by Cornell University under contract with the National Science Foundation.

F. Mardirossian et al. (eds.), Clusters and Groups of Galaxies, 389–394.
© *1984 by D. Reidel Publishing Company.*

Rood, and Williams 1983). It is similar to that of a single galaxy with a central radial velocity near 4560 km s^{-1} and a width at the 20% level of 225 km s^{-1}. Some emission may extend toward lower velocities and there is an indication that the hydrogen emission may extend northward beyond the optical outline of the group (Gallagher, Faber, and Knapp, 1981). To better understand the distribution and kinematics of the neutral hydrogen in this group we have undertaken new observations with the VLA telescopes of the NRAO. The observations and data will be presented in section II; the results and conclusions will be discussed in section III.

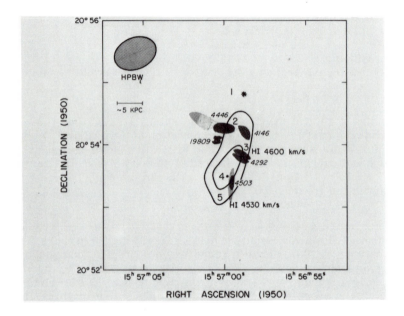

Figure 1. Schematic diagrams of Seyfert's Sextet. The tilted numbers are optically determined radial velocities of the galaxies represented by the cross hatching. The only spiral is the southernmost galaxy, NGC 6027C, with a radial velocity of 4503 km s^{-1}. The larger erect numbers are reference labels for the positions of the profiles shown in Figure 2. The asterisks mark the positions of small-diameter continuum sources. The contours represent the surface brightness of neutral hydrogen in units of 3 $mJy/beam$ at the peak of the profile. Because the profiles remain about the same width throughout the whole region containing hydrogen, these contours are approximately proportional to the total number of neutral hydrogen atoms present . The central velocities of the hydrogen at positions 3 and 5 are indicated.

II. Observations

Thirteen antennas of the VLA were used in its most compact D configuration during November 1982. The synthesized beam had half-power widths of 38 x 35 *arcsec* at a position angle of 110 degrees. Sixty-four spectral channels covered a total bandwidth of 6250 *kHz* with a channel separation of 97.7 *kHz* or 20.6 *km s*$^{-1}$. The rms noise per channel was ∼ 2 *mJy/beam*. A few channels at each end of the spectrum were left out because of the difficulty in establishing the bandpass calibration near the edges so the usable data covered a velocity range from 3680 *km s*$^{-1}$ to 4760 *km s*$^{-1}$. Also, the channel precisely at 1400 MHz (∼ 4360 *km s*$^{-1}$) was not usable because of interference.

Individual maps of each channel were made and CLEANed in the standard manner (Clark, 1980). Line profiles at chosen positions were constructed from the channel maps. These are shown in Figure 2; the location of each is designated by the appropriate number in Figure 1. No baselines other than an initial bandpass correction to the raw data have been fitted to these results.

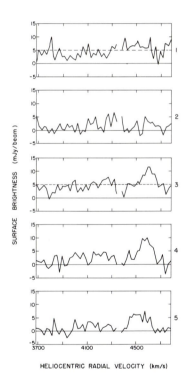

Figure 2. Profiles of the 21-cm hydrogen line radiation at the positions designated in Fig. 1. The velocity resolution was 20 *km s*$^{-1}$. The dashed lines on profiles 1 and 3 represent the adopted flux densities of the continuum sources at those positions. The velocities are heliocentric using the optical convention.

Two continuum sources with small angular diameters are present in the vicinity of the group. One lies at the position of the Sap galaxy NGC 6027A (position 3) and the other is apparently outside the group to the north (position 1). They each have peak brightnesses of about 6 $mJy/beam$. Burke and Miley (1973) found an integrated flux density of 27 mJy for the group, and evaluations of the total power recorded from Seyfert's group during the Arecibo observations gives a flux density of 20 ± 8 mJy. Thus, there could also be some faint extended emission present in the area.

The HI line maps and profiles show hydrogen extending from just below the southern-most galaxy in the group, the SBcp galaxy NGC 6027C, northward at least to the edge of the group defined by the galaxies. A schematic map of the distribution of surface brightness for the extended HI cloud is shown by the contours in Figure 1. The integrated flux in neutral hydrogen measured by us with the 300-ft telescope is 3.5 $Jy\ km\ s^{-1}$ corresponding to a total mass of 2 x $10^9 M_\odot$. The integrated value from the VLA observations is about the same which indicates that there is essentially no hydrogen distributed more diffusely over a larger area.

The velocity centers of the hydrogen profiles display a gradient of ~ 100 $km\ s^{-1}$ per arc-minute from south to north along the cloud. We note that the full-widths of the line profiles remain about 160 $km\ s^{-1}$ everywhere. There must be a large amount of internal turbulence or other mechanism for line broadening throughout the volume occupied by the possibly rotating hydrogen cloud.

III. Discussion

Seyfert's group is characterized by a very large number density of galaxies [4 galaxies per $4\pi/3(9\ kpc)^3$] and small radial velocity dispersion $(\sigma_V/\sqrt{3} = 160\ km\ s^{-1})$. These properties translate into very fast rates for most dynamical processes (relative to the corresponding rates for loose groups and rich clusters). The average time between successive collisions of a given galaxy in Seyfert's group with any of its other members, i.e., the galaxy-galaxy collision time, is $t_{gg} \sim 5 \times 10^7\ yrs$ (in comparison, $t_{gg} > 10^{10}\ yrs$ for loose groups). In a tidal collision the material (both stars and gas) in the outer parts of the galaxies can be removed. In an interpenetrating collision of two spiral galaxies, the stars in the colliding galaxies generally interact with each other only weakly (cf., Gerhard and Fall 1983) but the gas can be significantly affected (Spitzer and Baade 1951).

In an interpenetrating collision, the translational kinetic energy of the relative motion of the interstellar media of the two galaxies prior to the collision becomes thermalized; for a relative velocity of the colliding galaxies ~ $\sqrt{2}\sigma_V = \sqrt{2}\ 277\ km\ s^{-1}$, the ionized gas cloud initially has a temperature T = 3 x $10^6 K$, but appreciable cooling can occur as the colliding galaxies pass through each other, so that the expansion velocity of the resulting nearly freely expanding gas cloud is determined by its temperature and collective motions at the instant it finds itself gravitationally free of the colliding galaxies. If we assume an initial number density of ionized hydrogen ~ 1 $cm\ s^{-1}$, calculate the *square-root of the density-squared averaged over the estimated time that has elapsed since the collision event*, and apply the cooling curves plotted by Turnrose and Rood (1970), we find that the entire store of thermal energy in the cloud is radiated away by thermal Bremstrahlung, radiative recombination, and line-emission in only 2 x $10^7\ yrs$, i.e., an HI cloud is formed 2 x $10^7\ yrs$ after the collision event (for comparison, the cooling time of a Spitzer-Baade cloud formed in a rich cluster is longer than a Hubble time so the cloud remains ionized).

The phenomenon found here may be common with that in the well-studied compact group *Stephan's* quintet (Allen and Sullivan 1980, Peterson and Shostak 1980, Sullivan 1980, and van der Hulst and Rots 1981) which contains at least two large isolated HI clouds.

The evidence for the existence of the intergalactic cloud in *Seyfert's* group is twofold: (a) the HI is distributed over an asymmetrical region of length-scale ~ 18 *kpc* which is large compared to the size of any of the galaxies and is not centered on any of them, and (b) the radial velocities in the cloud range from about 4530 *km* s^{-1} to over 4600 *km* s^{-1} which are larger than the radial velocity of any of the galaxies. The neutral hydrogen with the lowest radial velocity in this range (4530 *km* s^{-1} is located near the galaxy NGC 6027C which has a radial velocity of 4503 +/− 43 *km* s^{-1} (Huchra and Hickson, private communication 1983). Therefore some of the gas in that region may lie within NGC 6027C, and this galaxy might be considered the cloud's prime HI doner candidate.

We thank Drs. P. Hickson, J. P. Huchra, R. A. Schommer, C. N. Caldwell, E. W. Greisen, R. Giovanelli, and R. Sancisi for their help and counsel. JRD gratefully acknowledges a partial travel award from the American Astronomical Society. This study was done during a 1982-1983 visit by HJR at the Institute for Advanced Study; the hospitality of the IAS faculty and members is gratefully acknowledged. This material is based partly upon work supported by the National Science Foundation under Grant No. AST-8215303.

References

Allen, R. J., and Sullivan, W. T. 1980, A. Ap., *84* ,181.
Biermann, P., Clarke, J. H., and Fricke, K. J. 1979, A. Ap., *75*,, 19.
Burbidge, E. M., and Sargent, W. L. W. 1971, *Nuclei of Galaxies* ed. D. J. K. O'Connell (American Elsevier Publication Co., Inc., New York), 351.
Burke, B. F., and Miley, G. K. 1973, A. Ap., *28* , 379.
Chincarini, G., and Rood, H. J. 1972, A. J., *77*, 448.
Clark, B. G. 1980, A. Ap., *89* , 377.
de Vaucouleurs, G., de Vaucouleurs, A., and Corwin, H. G. 1976, *Second Reference Catalogue of Bright Galaxies* (Austin, Texas: University of Texas Press).
Dickel, J. R., Rood, H. J., and Williams, B. A. 1983, A. J., in preparation.
Gallagher, J. S., Knapp, G. R., and Faber, S. M. 1981, A. J., *86* , 1781.
Gerhard, O. E., and Fall, S. M. 1983, M.N.R.A.S., *203* , 1253.
Hickson, P. 1982, Ap. J., *255* , 382.
Peterson, S. D., and Shostak, G. S. 1980, Ap. J., *241* , L1.
Spitzer, L., and Baade, W. 1951, Ap. J., *113* , 413.
Sullivan, W. T. 1980, A. Ap., *89* , L3.
Turnrose, B. E., and Rood, H. J. 1970, Ap. J., *159* , 773.
van der Hulst, J. M., and Rots, A. H. 1981, A. J., *86* , 1775.

DISCUSSION

HICKSON: Is the mass of the hydrogen cloud that you detect consistent with it being produced by stripping from the group galaxies?

DICKEL: The cloud has a mass of $\sim 2 \times 10^9 \, M_\odot$ which is much less than the virial mass of the group or the total of a typical galaxy.

GIOVANELLI: If the radio continuum of the northern source is only 6.5 mJy, wouldn't you need some enormous optical depth to get a detectable absorption?

DICKEL: This is in reference to a possible dip in the profile at position 1 in figure 2. The answer is yes, it is probably noise.

HOT GAS IN GROUPS OF GALAXIES

P. Biermann[1] and P.P. Kronberg[2]
[1]Max-Planck-Institut für Radioastronomie, Bonn, West Germany
[2]University of Toronto, Toronto, Canada

ABSTRACT
    The properties of sparse groups of galaxies dominated by early Hubble type galaxies are a natural extension of those of larger clusters. The evolutionary model by Tinsley and Larson for the formation of elliptical galaxies is strongly supported by the detection of $\sim 10^{11}$ $M_{\odot}$ of hot gas around the elliptical galaxy NGC5846 which sits in only a sparse group. The virial mass required to contain the gas is $\sim 10^{13}$ $M_{\odot}$ which is about the same mass as required to hold the entire group.

Contributed paper at the meeting on clusters and groups of galaxies held at Trieste, September 1983.

I. INTRODUCTION

    Most attempts to understand the formation and evolution of early Hubble type galaxies have concentrated on galaxies in clusters. It is now known that clusters of galaxies contain large amounts of hot gas which must be accounted for. In addition, there are some sparse groups of galaxies, also mostly populated by early Hubble type galaxies.

    To what degree are the properties of sparse groups an extrapolation from larger agglomerations of galaxies like poor to large clusters? What can we learn from the properties of early Hubble type galaxies in sparse groups about their formation and evolution?

    To answer these questions we have undertaken an extensive observing program of two sparse groups of galaxies, first searching for cool gas in the 21 cm HI line at Arecibo (Bieging and Biermann, 1977; Biermann, Clarke, Fricke, 1979) and then searching for hot gas in X-rays with the EINSTEIN satellite (Biermann, Kronberg, Madore, 1982; Biermann, Kronberg, 1983).

    Here we review our findings and report on our newest results.

F. Mardirossian et al. (eds.), Clusters and Groups of Galaxies, 395–397.
© *1984 by D. Reidel Publishing Company.*

2. RESULTS

HI observations of the galaxies in the two groups around the S0-galaxy NGC3607 (Biermann, Clarke, Fricke, 1979) and the gE-galaxy NGC5846 (the complete HI work is referenced in Biermann and Kronberg, 1983) have shown that there is no measurable neutral hydrogen in most galaxies close to the dominant galaxy. Since the red giants in a galaxy keep emitting fairly cool gas, a process is required to either remove the gas from the galaxy and/or to heat it as to become invisible in HI. A third alternative would be to consume the gas in a process of star formation as yet undetected.

The X-ray observations of NGC3607 (Biermann, Kronberg, Madore, 1982) suggest by comparison with the calculations of Lea and De Young (1976), that stripping is the process responsible for keeping NGC3607 and the surrounding galaxies clean. In this interpretation the gas ejected by the stars is removed from the galaxy by ram-pressure from intergalactic hot gas. Once stripped the cool gas is heated to ambient temperatures. This interpretation requires that NGC3607, the dominant galaxy of the group, moves at about sonic speed with respect to the gas, which is contained in a larger potential well encompassing the entire group.

The X-ray observations of NGC5846 (Biermann, Kronberg, 1983) show, on the other hand, that this gE galaxy is surrounded by extended hot gas. In the region just of the size of the optical galaxy, this gas amounts to already 10^{10} $M_\odot$. Carefully subtracting background using another comparable field, yields more information: The emission continues to about 100 kpc radius, is very well described by complete symmetry around NGC5846, and comprises a total mass of about 10^{11} $M_\odot$. Thus the $M(gas)/L_B$ ratio is about unity, just the same as the upper limit of the range of $M(gas)/L_B$ in giant Sc spiral galaxies (Huchtmeier and Bohnenstengel, 1981), measured in the HI line. The virial mass required to hold the gas is $\sim 10^{13}$ $M_\odot$ similar to that of M87 (Fabricant and Gorenstein, 1983), one of the locally dominant galaxy in the Virgo cluster. Using the (M/L) values derived for the group around NGC5846 by Gott and Turner (1977), we also arrive at a virial mass of $\sim 10^{13}$ $M_\odot$ suggesting that NGC5846 sits right at the center of the potential well.

3. DISCUSSION

Comparing the properties of the NGC3607 and NGC5846 group with those of poor clusters (Kriss, Cioffi, Canizares, 1983) we find that even these poor groups of just about a dozen members fit well into the sequence of groups of clusters; their X-ray luminosities span five orders of magnitude.

The total amount of hot gas detected around NGC5846 strongly supports the evolutionary model by Tinsley and Larson (1979) of mergers of subsystems with associated star formation explosions: Here for the first time the hot gas around an elliptical galaxy can be uniquely associated with it.

A detailed discussion of these results is being prepared for publication.

ACKNOWLEDGEMENTS

Discussions with Drs. D. Fabricant, P. Gorenstein and M. Lecar are gratefully acknowledged.

REFERENCES

Bieging, J.H., Biermann, P.: 1977, Astron. Astrophys. 60, pp. 361-368
Biermann, P., Clarke, J.N., Fricke, K.J.: 1979, Astron. Astrophys. 75, pp. 7-13
Biermann, P., Kronberg, P.P., Madore, B.F.: 1982, Astrophys. J. 256, pp. L37-L40
Biermann, P., Kronberg, P.P.: 1983, Astrophys. J. 268, pp. L69-L73
Fabricant, D., Gorenstein, P.: 1983, Astrophys. J. 267, 535-546
Gott III, J.R., Turner, E.L.: 1977, Astrophys. J. 213, pp. 309-322
Huchtmeier, W.K., Bohnenstengel, H.-D.: 1981, Astron. Astrophys. 100, pp. 72-78
Kriss, G.A., Cioffi, D.F., Canizares, C.R.: 1983, Astrophys. J. 272, pp. 439-448
Lea, S.M., De Young, D.S.: 1976, Astrophys. J. 210, pp. 647-665
Tinsley, B.M., Larson, R.B.: 1979, Monthly Notices Roy. Astron. Soc. 186, pp. 503-517

ON MERGING IN GALAXY GROUPS

M. Mezzetti, G. Giuricin, F. Mardirossian
Osservatorio Astronomico di Trieste

INTRODUCTION

In the present paper we have examined a subsample of
the catalogue of groups of galaxies compiled by Huchra and Geller
(1982, HG). Our aim was to test whether galaxy merging and cannibalism
are active in loose groups, as suggested by McGlynn and Ostriker
(1980) who studied a sample drawn from the Turner and Gott's (1976)
catalogue. Our sample is formed of 49 of HG groups (that is all
of them having $|b^{II}| \gtrsim 40°$), to avoid serious reddening problems.

In order to avoid biases induced by possible selection
and contamination effects, the properties and the correlations
studied have been tested first on the nearer groups and then on
the complete sample (49 groups).

MORPHOLOGY

The population of loose groups seems to be intermediate
between the field and more compact systems (i.e. compact groups
and clusters).

Seventy-percent of all first ranked galaxies in our
sample are Spirals. Since, on the basis of numerical experiments,
it seems hard to produce spiral galaxies from mergers, there is
no stringent morphological evidence that a large fraction of the
first-rank members of loose groups may be products of galaxy mergers.

DYNAMICAL TIMES AND MAGNITUDE DIFFERENCE

The existence of strong, real correlations of the dynamical
relaxation time t_R with the magnitude difference ΔM between the
two brightest members of each group may be evidence of the existence

F. Mardirossian et al. (eds.), Clusters and Groups of Galaxies, 399–400.

of efficient cannibalism or merging processes in galaxy groups. McGlynn and Ostriker (1980) found such correlations in a subsample of the groups gathered in Turner and Gott's (1976) catalogue. The same correlations were found to be less significant for de Vaucouleurs' (1975) groups. We found only spurious correlations both for Turner and Gott's revised groups (Giuricin et al., 1982) and for the HG selected groups of this paper. In fact, the correlation is very likely to be spurious, simply induced by the definition of $t_R = t_c L_g/(4\pi L_1)$, where t_c is the crossing time, L_g is the total luminosity of the group, and L_1 is the brightest member luminosity.

CONCLUSIONS

        The examination of morphological properties and correlations between some relevant parameters of loose groups of galaxies points toward poor evidence of merging processes. This suggests either that loose groups are young objects or that galactic cannibalism is quite inefficient inside them.

        A more detailed paper is in progress.

REFERENCES

de Vaucouleurs, G: 1975, in Stars and Stellar Systems, Vol. 9, ed. A. Sandage, M. Sandage and J. Kristian, University of Chicago Press, Chicago.
Giuricin, G., Mardirossian, F., Mezzetti, M.: 1982, Astrophys. J. 255, 361.
Huchra, J.P., Geller, M.J.: 1982, Astrophys. J. 257, 423.
McGlynn, T.A., Ostriker, J.P.: 1980, Astrophys. J. 241, 915.
Turner, E.W., Gott, J.R.: 1976, Astrophys. J. Suppl. 32, 409.

THE BLUE TULLY-FISHER RELATION FOR THE LOCAL GROUP OF GALAXIES AND THE VIRGO CLUSTER OF GALAXIES

Otto-Georg Richter
European Southern Observatory, Garching bei München
Federal Republic of Germany

INTRODUCTION

Numerous attempts have been made to calibrate the Tully-Fisher relation (between absolute magnitude and total HI linewidth) and to apply it
in different wavebands since it is one of the most powerful extragalactic
distance indicators. For a recent review and a recalibration using blue
magnitudes see Richter and Huchtmeier (1984).

THE TULLY-FISHER RELATION

In this contribution all spiral and irregular galaxies in the Local
Group of galaxies listed by van den Bergh (1981) are used for an even
better calibration. [IC 5152 has been excluded because no reliable distance estimate is available.] Thanks to the work by Lo et al. (1984) high
quality HI observations exist now also for the faintest irregulars in the
Local Group. All modern distance determinations (except the new result
for M33 by Sandage, 1983) have been taken into account, especially those
from the new studies of Cepheids in nearby galaxies in the infrared. The
correction for internal absorption inside the galaxies has been applied
following Sandage and Tammann (1981), because their procedure maximizes
the correlation coefficient as compared to other models. Then (for internal consistency) also the model for galactic absorption was taken from
Sandage and Tammann (1981). The corrections of the HI linewidth for turbulence and inclination are discussed by Richter and Huchtmeier (1984).

Despite some large uncertainties in the data (mainly in distances
and inclinations) and obvious peculiarities (e.g. Magellanic system) one
finds a surprisingly tight correlation for the Local Group galaxies. The
total residual scatter is only $0^{m}\!.39$ which for equal errors in both re-
lated quantities translates to $\Delta M_B^{o,i} = 0^{m}\!.28$ and $\Delta \log \Delta v_i = 0.037$. Note
that these values are - although they include distance errors - smaller
than all other ones for previously published versions of the Tully-Fisher
relation, even those in the infrared. This new relation has been used
together with the sample of Virgo cluster galaxies to derive a revised
calibration valid over a range of more than 11 magnitudes:

F. Mardirossian et al. (eds.), Clusters and Groups of Galaxies, 401–403.
© *1984 by D. Reidel Publishing Company.*

$$M_B^{o,i} = (-7\overset{m}{.}36 \pm 0\overset{m}{.}26) \cdot \log \Delta v_i - (1\overset{m}{.}40 \pm 0\overset{m}{.}54).$$

This gives now a distance modulus for the Virgo cluster of $(m-M)_o = 31\overset{m}{.}68 \pm 0\overset{m}{.}29$ corresponding to a linear distance D = 21.65 ± 2.85 Mpc. With a corrected mean radial velocity of 1076 ± 46 km/s (based on 248 radial velocities of galaxies in the inner 6° radius region; Richter, in preparation) one derives a local Hubble ratio of 50 ± 7 km/s/Mpc. The independently derived distance modulus of the Coma cluster relative to the Virgo cluster $\Delta(m-M)_o = 3\overset{m}{.}87 \pm 0\overset{m}{.}15$ (Dressler, 1983) leads to its distance being 128.8 ± 18.4 Mpc. With a mean radial velocity of 7010 km/s this gives a (global) value of the Hubble ratio of 54 ± 8 km/s/Mpc!

Now consider new evidence from work by Williams (1983). She observed a (physically stable) group in the Coma/A1367 supercluster, the IC 698 group. For this one finds via the same Tully-Fisher relation a distance of 80.2 ± 16.6 Mpc which, with a radial velocity of 6130 km/s, gives a local Hubble ratio of 76 ± 16 km/s/Mpc!

We know now that the Local Group moves towards the Virgo cluster with ~250 km/s. This component has to be added to the velocities of all the aggregates considered here (angular distance to Virgo < 20°) to get a better estimate for their cosmological redshifts. From the Virgo and Coma clusters we find then H_o = 59 ± 7 km/s/Mpc. Applying this value to the IC 698 group gives a peculiar velocity of 1650 km/s for that aggregate! It is hard to believe that such an extreme value can be caused by the gravitational attraction of the whole Coma/A1367 supercluster, its high mass notwithstanding. Rather there may exist systematic errors in the blue magnitudes for the galaxies in the IC 698 group of about $0\overset{m}{.}6$ (they were transformed from the photographic magnitudes given in the Zwicky et al. catalogue). However, an alternative suggestion is that the IC 698 group (1) was formed in isolation, (2) quickly settled into a relaxed state, and (3) accelerated over a long time towards the Coma/A1367 supercluster.

Anyhow, the global value of the Hubble constant is most probably less than 75 km/s/Mpc and perhaps somewhat larger than 50 km/s/Mpc. So if one were to choose between the two competing extragalactic distance scales: the (blue) Tully-Fisher relation is voting for the Sandage-Tammann scale.

ACKNOWLEDGEMENT

I thank Dr. W.L.W. Sargent for putting the new VLA HI data on extreme dwarf irregulars at my disposal as they were of crucial importance for this work.

REFERENCES

Dressler, A.: 1983, preprint.
Lo, K.-Y., Sargent, W.L.W., and Young, K.: 1984, in preparation.
Richter, O.-G., and Huchtmeier, W.K.: 1984, Astron, Astrophys., in press.
Sandage, A.: 1983, Astron. J. 88, 1108.

Sandage, A., and Tammann, G.A.: 1981, A revised Shapley-Ames catalogue of
    bright galaxies, Carnegie Institution of Washington Publ. 635.
van den Bergh, S.: 1981, in The structure and evolution of normal
    galaxies, S.M. Fall and D. Lynden-Bell (Eds.), Cambridge University
    press, p. 201.
Williams, B.A.: 1983, Astrophys. J. <u>271</u>, 461.

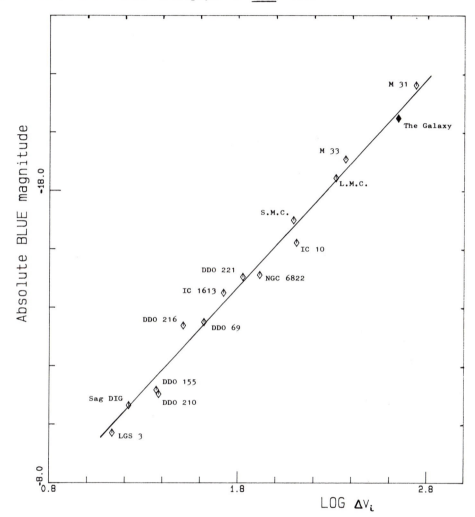

Figure 1:   The blue Tully-Fisher relation for the spiral and
            irregular galaxies in the Local Group. The shown
            regression line is the new calibration derived in
            the text.

HIERARCHICAL CLUSTERING

P. J. E. Peebles
Joseph Henry Laboratories
Princeton University

Studies of how galaxies and clusters of galaxies have formed have reached an interesting stage where good arguments can be adduced in favor of quite different scenarios. I will explain here why I think the evidence for the "top-down" picture, where galaxies are fragments of pre-existing protoclusters, seems a little questionable, and then present my assessment of the evidence on the "bottom-up" picture where galaxies form as bound objects before clusters. As some of my comments may seem overly skeptical I do hope it is understood that they are meant in a spirit of friendly debate from which we may hope will emerge a deeper understanding.

1.   THE TOP-DOWN PICTURE

The space distribution of galaxies is a complex pattern (1) that has been variously described as cloudy, noisy, frothy and filamentary. The last is particularly interesting because in the most popular versions of the top-down scenario protoclusters are pancakes (2). Of course, long thin filaments could form in other ways - G. Burbidge mentioned alignment by jets, P. Corbyn alignment by massive cosmic strings - any of which would be of great interest. Connections between clusters might also develop through gravitational interactions in a clumpy distribution of galaxies that started out with no preference for lines, but we might hope to distinguish these as broader and more chaotic. And there is unfortunately a third class of filaments, those generated strictly by chance in a clumpy random process that has no preference for lines.

It is difficult to take account of this last class of filaments because the eye is so adept at finding patterns even in noise, and also because statistical accidents do produce curious-looking arrangements across statistically independent regions. (Indeed if such accidents were not seen the regions could not have been independent!) To test how important these effects might be in contributing to the impression of filaments in the Lick galaxy map (3,4) Ray Soneria and I devised a

F. Mardirossian et al. (eds.), Clusters and Groups of Galaxies, 405–414.
© *1984 by D. Reidel Publishing Company.*

random statistical process that matches pretty well the observed low
order galaxy correlation functions and made maps of sky projections of
realizations of this process (5). Redshift maps based on this prescription are shown in reference (6). My impression is that these realizations lack some of the crispness of texture of the true galaxy
distribution and perhaps also the marked connectedness of the real
pattern. But still one can pick out strikingly long sinuous filaments
that we know are accidents because the prescription has no preference
for lines.

The problem in devising a measure of the incidence of filaments in
such models or in the data is not so much in finding statistics that are
sensitive to linear structures as in finding measures that are insensitive to (or can be calibrated for) irrelevant details. An example is
the percolation statistic introduced by Zel'dovich, Einasto and
Shandarin (2). A sphere of radius r is centered on each galaxy and one
examines the lengths of chains of connected spheres as a function of r.
There is a marked difference in the behavior of this statistic in the
CfA redshift sample (7) and a realization of the statistical prescription mentioned above. That could mean there are more filaments in the
data, but unfortunately other interpretations are possible. In the
prescription galaxies are placed in bounded clustering hierarchies, or
clumps, and the clumps are placed at random. Because the clumps have
rather sharply defined edges percolation between clumps is suppressed.
It would have been more realistic to have added a few galaxies in a
loose halo around each clump, for we surely expect to find that in a
truly dynamical model relaxation processes within and between clumps
cause them to spread out. Such halos would have little effect on the
low order correlation functions the prescription was designed to
reproduce, or on the appearance of maps of angular distributions, but
they could greatly encourage percolation between clumps. Thus it may
be that the percolation statistic is pointing to a defect in the
prescription, but one that has nothing to do with large-scale structure.
Dekel, West and Aarseth arrive at a similar conclusion from a study of
the percolation statistic in N-body model distributions (8).

Striking indirect evidence of a linear structure comes from
Binggeli's study of the orientations of clusters of galaxies (9).
Binggeli found that if a cluster has a neighboring Abell cluster at a
distance $\lesssim$ 15 h^{-1} Mpc (H = 100 h km s^{-1} Mpc^{-1}) then the long axis of the
cluster shows a marked tendency to point toward the neighbor. There
also is some tendency for a similar alignment among neighbors as far as
50 h^{-1} Mpc away. As Binggeli notes it would be very hard to account
for the latter effect by tides; it seems to require something like
pancakes. Of course, control of systematic errors in such a measurement
is difficult, so further studies of Binggeli's effect will be followed
with great interest.

Let us turn now to another question: is it reasonable to assume
that protoclusters formed before galaxies? The problem is that galaxies
appear to be old, clusters young (10,11).

The Local Group provides an instructive example. The Andromeda nebula M31 is approaching us with a crossing time r/v comparable to the Hubble time H^{-1}, as if the group were only now forming, M31 approaching us for the first time. Consistent with this is the fact that the crossing time for the outlying members of the Local Group all are within a factor of 2 of the Hubble time (12), a coincidence that would not be expected if the group had formed long ago and the group members had already completed several orbits. The nearby groups of galaxies outside the Local Group are drifting away from us roughly in accordance with Hubble's law (to perhaps 30% accuracy) so it seems reasonable to presume they were closer by the factor $1 + z_f$ at the epoch z_f of galaxy formation. As quasars with high element abundances are seen at $z \sim 3.5$ it is reasonable to take $z_f \gtrsim 4$. That would mean that at z_f the five nearest groups listed by de Vaucouleurs (12) were at distances $\lesssim 1$ Mpc, within the present Local Group, which is difficult to reconcile with the idea that these groups existed at z_f. But if the Local Group and its neighbors did not exist at z_f what is the site of formation of the galaxies now in these groups? Could the galaxies have been ejected from the pancake that made the Virgo cluster? That seems unreasonable because the nearby galaxies share a common steaming motion toward the Virgo cluster (relative to the Hubble flow) rather than away from it as might be expected if these galaxies were ejecta (13). Could the nearby galaxies have been produced in the collapse of a protosupercluster? I have in mind an object that produced the whole Local Supercluster. Again, the problem is that we want to form the galaxies at $z_f \gtrsim 4$, when the mean density of the universe was $(1 + z_f)^3 \gtrsim 100$ times the present value. The collapse of the protosupercluster presumably made its density higher than the mean at z_f. Why then is the local density of galaxies, within say 7 Mpc of the Sun, within a factor of 2 or so of the present large-scale mean density? If on the other hand the Local Supercluster is young, only now forming, as in the "bottom-up" picture, this seems more understandable, as does the local peculiar motion toward the Virgo cluster.

One way around this argument is to imagine that the galaxies in the nearby groups formed singly while galaxies in the denser regions formed in protoclusters or pancakes. The problem with this, as discussed in the next section, is that the continuity of properties of galaxies seems to argue for rather similar conditions of formation of galaxies found now in very different environments.

Another possibility is that while pancakes were collapsing in one dimension they were expanding in the other two so as to keep the mean density averaged over spheres equal to the cosmological mean. Zel'dovich (14) showed how one can analyze this in a sensible approximation. In the classical pancake scenario protoclusters form where the peculiar velocity field $v(x)$ has intersecting orbits. The velocity field is produced by the gravitational field g of the mass density fluctuations $\delta\rho = \rho(x) - <\rho>$ left over after decoupling. This means $v \propto g$ so we can write $v = - \nabla\phi$ with $\nabla^2\phi \propto \delta\rho(x)$. In the Zel'dovich approximation pancakes form where an eigenvalue λ_1 of the tensor

$$T_{\alpha\beta} = - \partial v^{\alpha}/\partial x^{\beta} = \phi_{,\alpha\beta} \tag{1}$$

is positive and unusually large. The plane of the pancake is normal to
the corresponding eigenvector q of T. Another interesting quantity is
the divergence of the peculiar velocity field, which is the sum of the
three eigenvalues,

$$- \nabla \cdot v = \lambda_1 + \lambda_2 + \lambda_3 \equiv \lambda = \partial \rho / \partial t \quad . \tag{2}$$

If $\delta\rho$ is a homogeneous and isotropic random process then

$$< \lambda_1 \lambda > = < (\lambda_1)^2 > \quad . \tag{3}$$

The fact that λ_1 is correlated with λ means that where λ_1 is positive
and unusually large $\lambda = - \nabla \cdot v \propto \delta\rho$ tends to be positive and large. As
one would expect, where gravity causes a pancake to form the mass
density tends to be high and so also tends to make the density averaged
over spherical shells increase relative to the background.

We can be more explicit if we add the assumption that $\delta\rho$ is a
random Gaussian process. Then I find that the joint distribution in the
three eigenvalues λ_i of $T_{\alpha\beta}$ is

$$dP \propto e^{-A} |(\lambda_1 - \lambda_2)(\lambda_2 - \lambda_3)(\lambda_3 - \lambda_1)| d\lambda_1 d\lambda_2 d\lambda_3 \quad ,$$
$$A = \frac{3}{4\sigma^2}[\lambda_1^2 + \lambda_2^2 + \lambda_3^2 - \lambda^2/5] \quad . \tag{4}$$

The velocity is normalized to $<T_{11}^2> = \sigma^2$. The resulting distribution
in a single one λ_1 of the eigenvalue and the sum λ is shown in Figure
1. We see again that when λ_1 is unusually large $- \nabla \cdot v$ almost certainly
is large too. More definitely, the distribution says that for fixed
and large λ_1 the mean and standard deviation of the sum of the other two
eigenvalues are

$$\lambda_2 + \lambda_3 = \frac{2}{3}\lambda_1 \pm 2\sigma(\lambda_1)/3^{1/2} \quad , \tag{5}$$

where $\sigma(\lambda_1) = <\lambda_1^2>$ is the standard deviation of λ_1 averaged over all
points in space. So if λ_1 is positive and unusually large, $\lambda_1 << \sigma(\lambda_1)$,
$\lambda_2 + \lambda_3$ is comparably large. This says the pancake tends to collapse
along its two long axes on a time-scale of the same order as the time
for the formation of the pancake, and so it tends to produce a bound
system with density comparable to the cosmological mean when the pancake

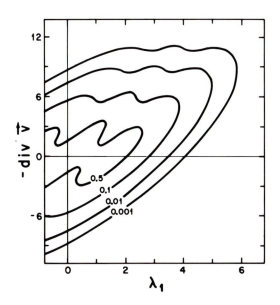

Fig. 1.   The joint distribution in – V·v and an eigenvalue λ_1 of $v^{\alpha}{}_{,\beta}$ for a random Gaussian process. The distribution is normalized to unity at the peak value.

formed. This would say that if galaxies were produced in clusters they would tend to stay in them, which certainly is not observed.

A fascinating way out of this bind has appeared with the recognition that a hole in an otherwise homogeneous mass distribution in an expanding universe grows with time, the mass piling up in a relatively thin ridge (15-17). This is because the hole acts as a negative source for the peculiar gravitational field. The mass in the ridge may wholly or partly cancel this negative mass, but as on the average a particle in the ridge "sees" only half the mass of the ridge the ridge is repelled by the hole. The ridge could fragment into galaxies while the repulsion of the hole scatters the galaxies, preventing overproduction of dense clusters. However, this picture does require special initial conditions: too many seeds would make holes cross at high redshift and so destroy the nice balance that discourages cluster formation; too few seeds would leave unacceptably large voids and an overly neat distribution of galaxies on shells. I am not aware of any natural way to arrange this.

2.   THE BOTTOM-UP PICTURE

There are indications that the mass distribution measured on scales from galaxies to superclusters is approximated by a single power law, $\rho \propto r^{-\gamma}$, $\gamma \sim 2$. If this were so the continuity between galaxies and superclusters would suggest that these objects were assembled by similar processes. Since galaxies are old and, as argued in the last section, superclusters seem to be young, only now forming, we could interpret this to mean that creation proceeded in a sequence of increasing mass, decreasing density.

The continuity is by no means convincingly established, and there is even a counter-indication in the segregation of galaxy types in clusters and the field. This could be interpreted to mean that the denser parts of pancakes preferentially produced elliptical galaxies, the less dense parts, lenticular types. On the other hand, as Dressler has described at this conference, there is striking evidence of continuity among masses of galaxies: the properties of ellipticals are rather well fitted by a simple sequence and, while spirals require more parameters, there is a close relation between luminosity and rotation velocity and hence mass at a fixed radius (18,19,20). Since these relations show no very pronounced correlation with present galaxy environment they suggest that galaxies formed when large-scale clustering was considerably less pronounced than it is now (20,21), that conditions at the regions of incipient clusters and incipient voids must have been similar enough at galaxy formation to have ensured that roughly similar amounts of luminous and dark matter were assembled in the brighter galaxies. At the same time, we do require conditions to have been different enough to have produced a different abundance ratio of elliptical and lenticular forms at the future sites of voids and clusters.

The evidence of continuity in the hierarchy of mass clustering comes from two sets of observations. The space distribution of galaxies is a close approximation to a scale-invariant clustering hierarchy, the cluster density scaling with length as $n \sim r^{-\gamma}$, $\gamma = 1.77 \pm 0.04$ (5,22). When observed by starlight galaxies stand out above this hierarchy as distinct islands. However, the mass distribution in at least some galaxies is a good deal more spread out than the light. A statistical measure of the mean mass distribution around a galaxy is the rms value of the relative line-of-sight velocity $w = <(v_1 - v_2)^2>^{1/2}$ of galaxy pairs as a function of projected separation r_p. Estimates of w from the CfA (7,23) and Durham-Australia (24) redshift samples and from the Turner (23,25) and Karachentsev pairs (11,26) suggest that $w(r_p)$ is quite flat, ranging from $w = 220 \pm 20$ km s^{-1} at $r_p \sim 10$ h^{-1} kpc to $w = 300 \pm 50$ km s^{-1} at $r_p \sim 1$ h^{-1} Mpc. The former value is about what would be expected from typical circular velocities ~ 200 km s^{-1} within bright galaxies. The flatness of $w(r_p)$ is just an extension of the now familiar observation that isolated spirals tend to have nearly flat rotation curves (19,27), and the interpretation is the same: the mean mass density found at distance r from a bright galaxy scales as $\rho \sim r^{-\gamma}$ with $\gamma \sim 2$. This applies within galaxies and on out to at least $hr_p \sim 1$ Mpc, where the estimates of $w(r_p)$ become unreliable. But on this scale we are in the regime of galaxy clustering, where, as we have just noticed, the galaxy distribution approximates a clustering hierarchy with about the same power law index, $\gamma \sim 1.8$. If mass clusters like galaxies on scales $\gtrsim$ 1 Mpc we conclude that the mean value of the mass density at distance r from a galaxy, $\rho(r)$, approximates a power law with index $\gamma \sim 2$ or slightly less from the interiors of galaxies out to hr $\sim$ 10 Mpc.

This is the continuity phenomenon mentioned above. It is not established with high precision, and indeed we would be surprised if $\rho(r)$ were an accurate power law from galaxies to clusters because the lack of continuity in the distribution of starlight suggests there has been appreciable settling of baryons relative to dark matter on the scale of galaxies. On the other hand, the continuity is close enough that I think it would be a rather surprising coincidence if galaxies and superclusters were assembled by different physical processes.

3. DISCUSSION

The controversy over the sequence of creation of the large-scale structure of the universe centers on three questions. Identification of distinct long linear structures in the galaxy distribution would point to pancakes or something equally interesting, but I think caution is indicated because of the difficulty of correcting for accidental effects. Broader connections between neighboring clusters could be relics of pancakes, but might also be produced by gravitational inter-actions in a "bottom-up" scenario. Results of N-body simulations will be useful here. Second, the mass clustering hierarchy is a pretty close approximation to a power law, which would seem to me to be a surprising coincidence if galaxies and superclusters formed by different processes. But again caution is indicated because it has not been demonstrated that this scale-invariant behavior really would arise in a natural way in a "bottom-up" scenario, and really would not arise in a "top-down" model. Third, we must consider whether portions of the galaxy clustering pattern really could be as old as galaxies themselves. There is no problem with the galaxies in rich clusters, but I think it is difficult to see where the galaxies in loose groups such as the Local Group came from on this picture.

REFERENCES

1. Oort, J. H.: 1983, Ann. Rev. Astron. Astrophys. 21, p. 373.
2. Zel'dovich, Ya. B., Einasto, J. and Shandarin, S. F.: 1982, Nature 300, p. 407.
3. Shane, C. D. and Wirtanen, C. A.: 1967, Publ. Lick Obs. 22, part 1.
4. Seldner, M., Siebers, B., Groth, E. J. and Peebles, P. J. E.: 1977, A.J. 82, p. 249.
5. Soneira, R. M. and Peebles, P. J. E.: 1978, A.J. 83, p. 845.
6. Peebles, P. J. E.: 1982, in Scripta Varia 48, "Astrophysical Cosmology," ed. H. A. Brück, G. V. Coyne and M. S. Longair, p. 165.
7. Huchra, J., Davis, M., Latham, D. and Tonry, J.: 1982, Ap. J. Suppl. 52.

8. Dekel, A., West, M. J. and Aarseth, S. J.: 1983, Ap. J. in the press.
9. Binggeli, B.: 1982, Astron. Astrophys. 107, p. 338.
10. Peebles, P. J. E.: 1983, Ap. J. 274.
11. Peebles, P. J. E.: 1983, in the Proceedings of the Eleventh Texas Symposium, Austin, Dec. 1982.
12. de Vaucouleurs, G.: 1975, in Stars and Stellar Systems, Vol. 9, "Galaxies and the Universe," eds. A. Sandage, M. Sandage and J. Kristian (Chicago: University of Chicago Press) Chapt. 14.
13. Davis, M. and Peebles, P. J. E.: 1983, Ann. Rev. Astron. Astrophys. 21, p. 109.
14. Zel'dovich, Ya. B.: 1970, Astron. Astrophys. 5, p. 84.
15. Peebles, P. J. E.: 1982, Ap. J. 257, p. 438.
16. Sato, H.: 1983, in Proceedings of the Workshop on "GUT and Cosmology," Tsukuba, January, 1983.
17. Ostriker, J. P. and McKee, C. F.: 1983, Ap. J. in the press.
18. Terlevich, R., Davies, R. L., Fisher, S. M. and Berstein, D.: 1981, M.N.R.A.S. 196, p. 381.
19. Rubin, V. C.: 1983, Science 220, p. 1339.
20. Dressler, A.: 1983, Ap. J. in the press.
21. Ostriker, J. P.: 1982, in Scripta Varia 48, "Astrophysical Cosmology," ed. H. A. Brück, G. V. Coyne and M. S. Longair, p. 473.
22. Groth, E. J. and Peebles, P. J. E.: 1977, Ap. J. 217, p. 385.
23. Davis, M. and Peebles, P. J. E.: 1983, Ap. J. 267, p. 465.
24. Bean, A. J., Efstathiou, G., Ellis, R. S., Peterson, B. A. and Shanks, T.: 1983, M.N.R.A.S. 205.
25. White, S. D. M., Huchra, J., Latham, D. and Davis, M.: 1983, M.N.R.A.S.
26. Tifft, W. G.: 1982, Ap. J. Suppl. 50, p. 319.
27. Bosma, A.: 1978, Ph.D. dissertation, University of Groningen.
28. This research was supported in part by the National Science Foundation of the U.S.A.

DISCUSSION

SHAPIRO: I will be responding to some of your criticisms of the pancake picture in my talk tomorrow morning. However, one point, in particular, bears some repetition, so I will make it here, as well. You have argued that galaxy formation in the pancake picture (if galaxies are assumed to have formed at $z \gtrsim 3.5$) necessarily implies that galaxies would now commonly be found in very high density environments, which is only rarely observed to be the case. You have argued that the density in galaxy forming pancakes can only increase with time, as a result of the nature of the gravitational force. The flaw in this argument is as follows: As the detailed 3 D collisionless particle simulations of the growth of perturbations in the pancake scenario show (e.g. Centrella and Melott 1983; Shapiro, Struck-Marcell, and Melott 1983), one must consider the interactions of neighbouring pancakes in order to describe the evolution of density contrasts inside pancakes. The pancakes which form in these simulations do not collapse in the direction parallel to their plane of symmetry (relative to the Hubble expansion). Instead, the density at the centers of the sheets actually decreases relative to the mean density, since matter tends to flow toward the edges of pancakes, towards the intersections of pancakes where there are filaments and knots. If galaxies form preferentially in the sheets, away from the intersections (where the density contrasts greater), therefore, the density of galaxies across most of the sheet can actually decrease at least as fast as the mean density does thru Hubble expansion. Continuity, of course, requires that a minority of the galaxies be found in high density contrast regions. (There may also be nongravitational forces with compete with gravity in moving matter around, such as would accompany the explosive release of energy when for example, the first bound, star-forming objects form, producing supernovae, as in the Ostriker model for galaxy formation by amplification of intergalactic explosions).

PEEBLES: What you describe as the flow of matter toward the intersection of pancakes is what I had in mind when I talked of the collapse along the long axes of the pancake. It is my impression that in the N-body simulations that show a nice filamentary structure the matter in the filaments was placed there only recently. If galaxies form after gas pancakes then that would be acceptable only if we could imagine that galaxies in filaments formed relatively recently, at redshifts less than one or two, say, and then you would have to explain why we do not see young galaxies forming.

MACGILLIVRAY: I would like to add my support to your comments on the distribution of galaxies in the Lick Catalogue. I have recently examined objectively-obtained galaxy samples, using techniques similar to that of Kuhn, and from these I find no statistical evidence for filamentary structure.

OORT: Is the skepticism which you empressed regarding the reality of supercluster justified? While your approach is statistical, my interest is more in specific objects. I disagree with your implicated conclusion that the supercluster features that have been discussed may be just only chance configurations. As examples to the contrary I point to the Local Supercluster which can hardly be considered as something that has accidentally struck ones eye without having a physical reality, and to the Coma supercluster which, in my opinion, is well isolated in the CPA position/velocity plots. In a somewhat different category there is the enormous supercluster found by N. Bahcall and Soneira. Then there is the more subtle but very important phenomenon of the preferential orientation of cluster axes in the direction of their nearest neighbour cluster, which points rather forcibly to a relation between objects separated by tens of Mpc.

PEEBLES: I certainly would not want to deny the reality of the Local Supercluster. What I have in mind is the question of how the clustering pattern formed. Suppose clumps of galaxies, each with diameter 20 Mpc (for H=100), were placed down at random. We would find here and there empty spaces appreciably broader than 20 Mpc and concentrations of galaxies more than 20 Mpc across. These would be real voids and, after gravity takes hold, real superclusters that formed out of noise. Is that the way our universe was produced? I think it must be counted as one possibility.

GIOVANELLI: I think the comparison of graphs of model distribution of galaxies with those obtained from observed samples with very deep limiting magnitudes, like the Lick counts, forces somewhat the star-chain interpretation. Observations suggest that filaments have volume overdensities of maybe 10 or 20: if you project them on a very deep background, the projected overdensity will be very small, and filament will be pratically washed out. I agree with you, filaments in the Lick survey map observe little credibility. I would like instead to see a comparison of model distributions with observed distributions drawn from shallower samples, like the Zwickly catalog so that filaments in the nearby universe, if they exist, would appear more prominently displayed.

PEEBLES: A systematic study of the clustering pattern at relatively shallow depths would indeed be interesting. In the model I made to compare to the CfA redshift sample no very striking filaments appear but then none is evident in the CfA maps either.

PERCOLATION STUDIES OF GALAXY CLUSTERING

S.P. Bhavsar and J.D. Barrow
Astronomy Centre
University of Sussex
Falmer Brighton BN1 9QH

SUMMARY: A percolation cluster analysis is performed on equivalent
regions of the CFA redshift survey of galaxies and the 4000 body
simulations of gravitational clustering made by Aarseth, Gott & Turner
(1979). The observed and simulated percolation properties are compared
and, unlike correlation and multiplicity function analyses, favour high
density ($\Omega = 1$) models with n = -1 initial data. Our results show that
the three-dimensional data are consistent with the degree of filamentary
structure present in isothermal models of galaxy formation at the level
of percolation analysis. We also find that the percolation structure of
the CFA data is a function of depth. Percolation structure does not
appear to be a sensitive probe of intrinsic filamentary structure.

1. INTRODUCTION

    We present the results of applying percolation analysis to the CFA
data set (Davis et al 1983). We compare these results with a percolation
analysis of 4000 body simulations of pure gravitational clustering. We
also try to assess the effectiveness of percolation analysis as a measure
of the degree of intrinsic filamentary structure in the data sample.
Although the 2,3 and 4 point correlation function (Peebles 1980)
analyses of the Shane-Wirtanen catalogue are consistent with
hierarchical clustering of galaxies, the eye picks out filamentary
structure. However, the eye is (presumably for evolutionary reasons)
over-sensitive to linear features and simulations of points
distributed in hierarchies of between eight and twelve levels give
similar optical impressions when projected on to a plane (Groth et al
1977). This raises the question of whether the filaments we discern
in the distribution of galaxies are statistically significant and, if
so, what they are telling us about the origin of galaxies?

    We would like to have an objective, quantitative means of
analysing distributions for the presence or absence of intrinsic
filamentary patterns. Unfortunately, the statistics of shape and
pattern recognition are not well-developed subjects in the open literature
(Cliff & Ord 1973; Ripley 1977, 1981; Broadbent 1980; Kendall &

F. Mardirossian et al. (eds.), Clusters and Groups of Galaxies, 415–421.
© *1984 by D. Reidel Publishing Company.*

Kendall 1980) and there appear few general techniques that will identify
the presence of a non-specific pattern in three dimensions. Kuhn
and Uson (1982) examined the deviation of nearest neighbour paths
through the Shane-Wirtanen maps from those generated by two-
dimensional random walks. This technique is sensitive to linear
structure but was only used on the two-dimensional data. Also it
appears to be sensitive to the initial direction vector of the random
walk. Einasto et al (1982) and Zeldovich, Einasto & Shandarin (1982)
have also applied percolation techniques to galaxy distributions
but since they do not define their procedure it is not clear whether
our results are directly comparable to theirs. However, there are
significant differences in methodology; whereas they take a cubic
sample including the local supercluster and compare this to samples
centred on rich clusters in a simulation, we shall examine data within
regions of complete sky coverage. This includes various cones out
to a 200 Mpc radius encompassing a volume $4.88.10^{6}$Mpc^{3}. Unlike other
investigators we have scaled our analysis to allow for selection
probabilities at differing depths and we have compared the
observations with equivalent, unbiased samples from N-body simulations
of gravitational clustering.

2. METHOD

To perform the percolation analysis on simulations and data samples
a computer code was developed to enclose individual galaxies by
spheres of radius x and compute the end to end length and
multiplicity of overlapping spheres as x increases from a value where
all spheres are disjoint until the entire sample of galaxies is linked
by overlapping spheres. Typically, a distribution of points and their
enclosing spheres is characterized by some critical value of x at
which the length, along one or more dimensions, of the longest single
connected chain of linked spheres may increase exponentially. If this
happens the system is said to percolate (Hammersley & Welsh, 1980) at
this radius. Note that a limitation of the analysis is that it
characterizes the length of percolated systems by $|z - y|$ where z and y
are coordinates of the end galaxies. If the filamentary patterns of
galaxies involve significantly curved filaments this could underestimate the length of the largest linkages.

The CFA Northern sky survey consists of a 50° cone centred on the
North Galactic Pole out of which a region is cut out by the plane
$\delta = 0^{\circ}$ passing through the apex. Thus all galaxies in the remaining
region have coordinates $\delta > 0^{\circ}, b^{II} > 40^{\circ}$. This defines a region of
1.83 steradians of sky. We have analysed the data out to depths of
80,120,160 and 200 Mpc, we take $H_{o} = 50 kms^{-1} Mpc^{-1}$ and label these distances
as R_{c}.

The 80 Mpc sample is complete to -18.5 Zwicky magnitudes but the
three deeper samples are weighted by the selection functions given by
Davis & Huchra (1983) (in effect, the galaxy counts are adjusted to
include all the galaxies expected to contribute to the total volume

sampled, if all galaxies brighter than −18.5 absolute magnitude were
detected out to the edge of that volume). This means that the radii
of spheres employed to evaluate a particular state of the percolation
process are weighted to allow for the incompleteness of the deeper
counts.* With regard to the N−body simulations we take a sub−region
from the simulation sphere identical in shape to the data. The
simulations contain 4000 bodies[†] evolving under a softened gravitational
potential for two different values of the cosmological density
parameters, $\Omega = 1$ and $\Omega = 0.1$, each for two choices of initial spectral
index $n = 0$ and $n = -1$, where[¤] the initial density inhomogeneity
spectrum with mass, M, is defined as

$$\frac{\delta \rho}{\rho} \propto M^{-1/2-n/6} \tag{1}$$

In order to compare the percolation properties of the simulations
and the data samples it is important to scale the radius of the
spheres x, surrounding each galaxy appropriately. We define the
dimensionless radius.

$$r = \frac{x}{(R)} \tag{2}$$

where (R) is the mean inter−particle distance, $(V/N)^{1/3}$ for a sample
of N points in a volume V. One expects values of x ~ (R) to produce
large linkages; for example, a Poisson distribution enclosed by spheres
with x ~ 0.5(R) should form connected regions linking almost all the
sample points. This scaling procedure is necessary if the simulations
are to be meaningfully compared with the data. The length, λ between
the most widely separated points of a linkage is also scaled to
facilitate comparisons between data and simulations as

$$L = \frac{\lambda}{R_c} \tag{3}$$

where R_c is the distance from the apex to the edge of the sample cones
described above. A system that percolates will have a linkage with
L ~ 1 just beyond its percolation length.

The four CFA samples correspond to (R) = 8.61, 9.86, 9.82 and
10.58 Mpc where, for the incomplete samples (R_c = 120,160 and 200 Mpc)
we have calculated (R) by taking the total number of particles as

$$N = \Sigma \text{ weights} \tag{4}$$

3. RESULTS

In Fig. 1 we present the results of a percolation analysis of the
CFA data out to depths of 80,120,160 and 200 Mpc respectively, within
the cone described above. The 80 Mpc sample is complete whilst the
others are weighted by the appropriate selection functions. For

comparison we have also included the results of percolating two
realizations of an equivalent Poisson distribution. The percolation
properties of three-dimensional Poisson distributions can be
calculated analytically (Hammersley & Welsh 1980) and should percolate
when r = 0.43, a value in good agreement with that found
'experimentally' in Fig. 1. The shaded region can be regarded as a
rough measure of the expected error in the statistical experiment.
According to the definition of percolation, the 160 and 200 Mpc samples
percolate whereas the 120 and 80 Mpc ones do not. The flattening of
the steady exponential rise in the percolating examples beyond L $\cong$ 1 is
not an indication of major behavioural changes. At this scale, the
shape of the L(r) curve is dependent on the orientation of the largest
clump within the cone of data and the presence of 'field' galaxies that
may, or may not, connect up two very long chains in preferred
directions.

     The failure of the 120 and 80 Mpc samples to percolate could be
either because the latter samples are too small to give a fair sample
of the universe or perhaps the clustering really is hierarchical out
to a scale between 120 and 160 Mpc, and only on very large scales
$\sim$ 160-200 Mpc is the linear and filamentary structure overt. Note
that the 200 and 160 Mpc data behave in similar fashion and the 120 and
80 Mpc samples are also similar to each other. In subsequent figures
we reproduce the 80 and 200 Mpc CFA results as representative of the
data. In Figs 2 and 3 we show the results of percolating the final
state of the 4000 body simulations for the possible choices of Ω and n.
Specifically, Fig. 2 displays the results of Poisson (n=0) initial
data with two realizations of the final state analysed for each Ω
value. The span of the two different realizations give some measure
of the significance level of these statistics. Fig. 3 displays the
same results for the (n=-1) initial data favoured by other theoretical
models (Zeldovich et al. 1982).

     The results reveal a number of features that differ from past
analysis of the Aarseth simulations by means of correlation (Gott,
Turner & Aarseth 1979; Aarseth et al. 1979) and multiplicity functions
(Bhavsar, Gott & Aarseth 1981). We note the following three features.
Firstly, percolation analysis does differentiate between the final
state of N-body simulations with different initial conditions. The
n=-1 models clearly percolate whilst the n=0 models do not. Secondly,
the Ω = 1 models provide a reasonable fit to the CFA data but the
Ω = 0.1 models do not and the Ω = 1 models possess much more
connected structure on intermediate scales and thirdly, the 200 Mpc
and 160 Mpc data favour the Ω = 1, n = -1 hierarchical model.

     These features are particularly interesting when compared with the
correlation and multiplicity function analyses of the same simulations
(Gott et al. 1979; Aarseth et al. 1979; Bhavsar et al. 1981). The
observed two-point correlation function is best-fitted by the (Ω=0.1,
n=-1) and (Ω=1, n=0) models and the multiplicity function is more sensitive
to n than Ω and favours the n =-1 case. The percolation properties

appear more sensitive to Ω than n and favour the high density models.
We see that, at the level of percolation analysis the gravitational
clustering simulations are able to generate the same degree of linked
structures as the real sky. This may be showing that percolation
measures are not very sensitive indicators of linear structure. We
also found that the percolation structure of initial data aligned along
random rods gave an L(r) trend insignificantly different to the Poisson
distribution of Fig. 1.

We also performed analyses in which the N-body simulations were
analysed in a manner more closely related to the observations by
calibrating distances by the velocities of galaxies in the simulation.
An 'observer' at the apex of the conical sample in the simulations has
two angular coordinates giving 'sky' position and a distance from
the radial velocity. We analyse the corresponding positions in the
same way as before and the results are shown in figures 4,5 and 6.

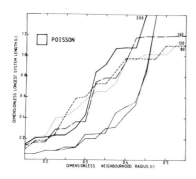

Figure 1

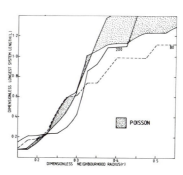

Figure 2

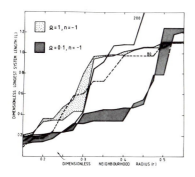

Figure 3

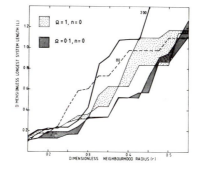

Figure 4

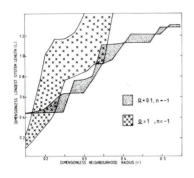
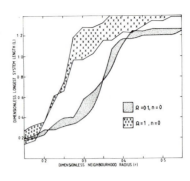

Figure 5   Figure 6

Figure 1: Dimensionless lengths L, of the longest linked system of
spheres with dimensionless radius r each centred on a galaxy versus
dimensionless sphere radius for a region of the CFA survey out to
depths of 80, 120, 160 and 200 Mpc. The shaded region is bounded by
results of Poisson distributions.
Figure 2: As Fig.1 but for a conical region of the 4000 body
simulation of Gott et al (1979) with n = 0 initial conditions evolved
to the present day. Two percolation analyses were performed for
each (Ω,n) model and the shaded region is spanned by the two results.
The CFA results for 80 and 200 Mpc depths are included; Ω is the
cosmological density parameter; n is defined by equation (1).
Figure 3: As Fig.2 but with n = -1 initial conditions.
Figure 4: As figure 1 but with distances calculated from radial
velocities and 80 and 200 Mpc data shown only.
Figure 5: As figure 3 but with distances from radial velocities.
Figure 6: As figure 2 but with distances from radial velocities.

     We have shown that although percolation analysis is able to
extract information from galaxy maps which differs from that
contained in correlation and multiplicity functions it does not
appear to give a sensitive measure of intrinsic filamentary structure.

     The segments of the CFA Northern Sky survey we have analysed
possesses a very similar percolation structure to some gravitational
clustering simulations. The difference in appearance between the CFA
survey and the simulations that is evident to the naked eye appears to
be only weakly reflected in the percolation properties.

Acknowledgments

     S. Bhavsar was supported by the SERC of the UK whilst this work
was carried out. The authors would like to thank J. Bather, D. Cooke,
C. Frenk, C. Goldie and B. Ripley for helpful discussions, M. Davis for
providing a tape of the CFA survey, S. Aarseth for a copy of his
clustering simulations and R. Ellis for suggesting improvements to
the manuscript.

References

Aarseth, S.J., Gott, J.R. & Turner, E.L., 1979. Astrophys. J., 228,664
Bhavsar, S.P., Gott, J.R. & Aarseth, S.J., 1981. Astrophys. J., 246,656
Broadbent, S., 1980. Jl R. statist. Soc. A.,143, 109.
Cliff, A. & Ord, J.K., 1973. Spatial Autocorrelation, Pion, London.
Davis, M., Huchra, J., Latham, D.W. & Tonry, J., 1983. Astrophys. J.
    Suppl. Ser., 52, 89.
Davis, M. & Huchra, J., 1983. Astrophys. J., 254, 437.
Einasto, J., Klypin, A.A., Saar, E. & Shandarin, S.F., 1982. Tartu
    Observatory preprint.
Gott, J.R., Turner, E.L. & Aarseth, S.J., 1979. Astrophys. J., 234, 13.
Groth, E.J., Peebles, P.J.E., Seldner, M. & Soneira, R.M., 1977. Sci.
    Amer., 237 (5), 76.
Hammersley, J.M. & Welsh, D.J.A., 1980. Contemp. Phys., 21, 593
Kendall, D.G. & Kendall, W.S., 1980. Adv. Appl. Prob., 12, 380.
Kuhn, J.R. & Uson, J.M., 1982. Astrophys. J. Lett., 263, L47.
Peebles, P.J.E., 1980. The Large Scale Structure of the Universe,
    Princeton University Press, New Jersey.
Ripley, B.D., 1977. J.R. Statist. Soc. B., 39, 172.
Ripley, B.D., 1981. Spatial Statistics, Wiley, New York.
Zeldovich, Y.B., Einasto, J. & Shandarin, S.F., 1982. Nature, 300, 407.

*   The weighting employed is based on the frequently made assumption that galaxies possess a universal luminosity function. Consequently, any subsample limited by a brightness cut-off should be representative of the galaxy distribution. The adjustment of the sphere radii keeps our procedure self-consistent throughout the sample volume.

‡   These are the models, I, II, III and IV described in Gott et al. 1979.

¤   The number distribution of masses is chosen to mimic the Turner-Gott luminosity function $N(m) \propto m^{-1} \exp(-m/m_*)$ and the simulation contains mass points in a range from 0.037 to 25.2 mass units; m_* is the average mass.

THE SPATIAL DISTRIBUTION OF DARK MATTER IN THE COMA CLUSTER

Yoel Rephaeli
Department of Physics and Astronomy, Tel Aviv University
Tel Aviv 69978, Israel.

ABSTRACT: Dark matter in clusters can be either more or less centrally concentrated than galaxies, depending on whether it was once bound to galaxies in halos. If the distributions of galaxies and dark matter are represented by similar profiles, but with different core radii, then the radial variation of the velocity dispersion in the Coma cluster and the degree of efficiency of gravitational drag on galaxies are somewhat more consistent with the dark matter being more centrally concentrated than galaxies.

I. INTRODUCTION

The large values of mass-to-luminosity ratio in rich clusters, $200 - 600$ $M_\odot/L_\odot$, are commonly thought to strongly suggest the existence of dark matter. The nature of this dark matter is not yet known, the possibilities being elementary particles, low mass stars and masssive black holes. Various considerations set some constraints on the allowable mass range of these dark matter candidates, such as phase-space arguments in the case of fermions, and tidal effects on cluster galaxies generated by massive black holes.

The spatial distribution of the dark matter is also not yet known. In principle, the dark matter can be either more or less centrally concentrated than the galaxies, depending on its nature, the cluster formation process, and the dynamical stage at which halos were torn off galaxies. For example, if neutrinos of mass $0(10$ eV$)$ dominate the cluster mass, then clusters collapse before galaxies form, and since these form dissipationally one expects the neutrinos to be less centrally distributed than galaxies. On the other hand, if the dark matter is in the form of Population III stars, then clustering is more likely to be hierarchical, in which case galaxies lose their massive halos during the compressional phase of cluster collapse, and since tidal tearing of halos is more efficient in the cluster core, the dark matter concentrates there. In this case, it is not unreasonable to expect dark matter concentration to be higher than that of galaxies. This would also apply if the dark

F. Mardirossian et al. (eds.), Clusters and Groups of Galaxies, 423–428.

matter consists of massive black holes which formed before cluster
collapse.

Obviously, the various possibilities for the dark vs. visible
matter distribution imply different dynamical conditions which might
have observable consequences, mainly in the cluster central region. We
briefly discuss here the nature of the dark matter, mainly to motivate
our consideration of the possibility of inferring its spatial distribution from examination of the galaxy velocity dispersion and the
efficiency of gravitational drag of galaxies by the dark matter.

II. THE NATURE OF DARK MATTER

The dark matter whose existence in galactic halos is deduced from
flat rotation curves, as expressed by over-stellar mass-to-luminosity
ratios, roughly in the range 10 - 30 $M_\odot/L_\odot$ (see the review by Faber and
Gallagher, 1979), is identical in nature to that in clusters.

This is the simplest and most reasonable assumption to make.
Elementary particles (e.g., massive neutrinos, photinos, gravitinos,
etc.), Population III low mass stars (m $\lesssim$ .01 $M_\odot$), and massive black
holes are possible forms of dark matter. In most respects, for our
discussion here, we need think of only two classes of dark matter candidates: elementary particles, an architype being a massive neutrino,
and low luminosity objects examplified by "Jupiters". The basis for
this broad categorization is that weakly interacting particles of low
enough mass much smaller than $\sim$1 keV [and, therefore, excluding gravitino if its mass is 0(1 keV)] would have over-galactic Jeans mass scales
and perturbations in their density will be most directly described by
the adiabatic theory of growth of structure in the universe, while low
luminosity sub-galactic objects represent pregalactic formation most
directly described in the context of isothermal clustering. Note that
in either case dark matter clustering in galaxies is free of dissipation.
It is interesting that the two classes of dark matter candidates are
linked to the two distinct theories of the origin and evolution of
density perturbations.

Phase space considerations (Gunn, 1977; Tremaine and Gunn, 1979)
set a lower limit on the mass of weakly interacting fermions, which is
about 30 eV, if these are bound to galaxies. This lower limit can be
much higher, perhaps as high as 0(1 keV) if the process of galaxy formation is (essentially) collisionless and is well described by Lynden-
Bell's (1967) theory of violent relaxation (Rephaeli, 1982). As neutrino
masses are not yet known, with only preliminary experimental indication
that m $\lesssim$ 30 eV (Lyubimov et al., 1980), it is not clear that galaxies
can have neutrino halos. Numerical studies of neutrino clustering in
the context of the adiabatic pancake theory (Melott, 1982; Bond et al.,
1983) suggest that neutrinos can be captured by galaxies forming in the
planes pancakes, but it is not yet clear if large enough fraction of the
neutrino mass can be captured to galaxy halos (before these were torn off

by collisional processes). This also leaves the possibility open that theoretically it will be difficult to explain how field galaxies and galaxies not formed in the planes of pancakes could have massive halos if these are to be composed of low mass fermions.

There are no (obvious) phase space considerations to restrict the mass of dark objects in galaxies. Other considerations apply, however: for example, if these are black holes then their mass is upper bound by 10^6 $M_\odot$, since centrally located larger mass holes would have been dragged to galactic centers in less than $\sim 10^{10}$ yr. In the intracluster space, massive black holes cannot be heavier than $\sim 10^8$ $M_\odot$ if they are not to tidally disrupt galaxies (van den Bergh, 1969) but if these once dominated galactic halos, then the lower upper limit of 10^6 $M_\odot$ applies. Jupiters are optically dark, which is canonically taken to mean that $m \lesssim .01$ $M_\odot$. It has recently been claimed (Hegyi and Olive, 1983) that if the initial mass function extends down to .04 $M_\odot$ with a unique power law index, then one can exclude Jupiters as likely candidates for the dark matter based on the observational limits on radiation, in the J and K Johnson spectral bands, from NGC 4565. This conclusion, though, seems to be premature in view of it being based on an ad-hoc assumption and observations of only one galaxy.

III. THE SPATIAL DISTRIBUTION OF DARK MATTER

What can we expect the dark matter distribution to be? If it is in the form of particles which do not bind in galaxies, then, since during cluster collapse the baryonic component which forms galaxies undergoes some dissipation, the cluster dark mass will be less centrally distributed than galaxies. If these particles were initially bound to galaxies, then after a relatively short time, perhaps already in the compressional stages of cluster collapse, galactic halos were torn off galaxies (Merritt, 1983). Indirectly, there is evidence for early tearing of halos in the work of Farouki, Hoffman and Salpeter (1983), who have shown that if all cluster mass remains in galaxies for $\sim 10^9$ yr, then segregation of light and heavy galaxies could have developed, an effect for which there is no observational evidence. It is unlikely that the tidal tearing of halos was so early that the dark matter participated in the initial violent relaxation phase, so that its distribution is expected to be more centrally concentrated than that of galaxies, since tidal tearing is more efficient in the denser inner regions of clusters. We see that if the dark matter consists of particles its distribution can be either more or less central than that of galaxies, depending essentially on whether most of it was once bound to galaxies.

If the dark matter is in the form of low luminosity objects, then its distribution is probably more centrally concentrated than that of galaxies. And if we require, as mentioned earlier, that dark galactic matter be of the same composition as dark intracluster matter, in spite of the possible theoretical difficulty with binding of some particles to galaxies, we are led to the result that the dark matter is more centrally concentrated.

The distribution of dominant dark mass can be deduced from the
dynamics of galaxies. It is usually <u>assumed</u> that galaxies are adequate
tracers of the cluster potential well. Some justification for this was
given by Rood et al. (1972) who argued, convincingly, that the two
distributions cannot be much different without drastically affecting
cluster dynamics. The variation across the cluster of the velocity
dispersion can yield $M(r)/r$ in virialized systems. Now, cluster veloci-
ty-dispersion profiles have not been determined yet with sufficient
accuracy. As an example, we here use what seems to be an established
feature of one such profile: For the most studied of the rich nearby
clusters, the Coma cluster, the velocity dispersion is found to decrease
with radius by about a factor of 2 over a distance of $\sim 3°$ (Rood et al.,
1972; Kent and Gunn, 1982). This is a slow decrease, over a significant
fraction of the total cluster radius, which indicates that the dark
matter is nearly isothermally distributed.

While the galaxy and dark matter density profiles may be similar,
the length scales characterizing the two distributions may not be equal.
For suppose the galaxy density profile is represented by the very good
analytic fit $\rho_g \propto (1 + r^2/r_g^2)^{-3/2}$, where r_g is the core radius (Rood
et al., 1972). If the dark matter profile is similar, it can be represented by $\rho_d \propto (1 + r^2/r_d^2)^{-3/2}$, r_d being the dark matter core radius.
Now, (the mean square line-of-sight velocity dispersion) $\langle\sigma^2\rangle \propto M(r)/r$,
whose profile is the function

$$\mu(y) = \{\ln [y+(1+y^2)^{\frac{1}{2}}] - y/(1+y^2)^{\frac{1}{2}}\} / y \qquad (1)$$

where $y = r/r_d = x/\alpha$, and $x = r/r_g$. We can easily see that this
profile can also be deduced from the hydrostatic equation if we use the
above forms for ρ_g and ρ_d. To obtain a <u>rough</u> estimate on the possible
range of values of α, consider the shape of $\langle\sigma^2\rangle^{\frac{1}{2}} \propto \mu^{\frac{1}{2}}(y)$. This function
varies linearly with y for $y \ll 1$, peaks at $y \cong 2.9$, and thereafter
decreases monotonically. Kent and Gunn (1982) have recently reanalyzed
the dynamics of galaxies in the Coma cluster. From their work we adopt
$r_g \cong 9' = 360$ kpc ($H_0 = 50$ km s^{-1} Mpc^{-1}), and $\langle\sigma^2\rangle^{\frac{1}{2}} \cong 600$ km s^{-1} at
$r = 160'$. The velocity dispersion remains approximately constant
throughout the central region, decreasing by a mean factor $f\sim 2$, but
certainly less than ~ 2.5, over 160'. If we require $\mu^{\frac{1}{2}}(y)$ to decrease by
a factor $f = 2$, then $\alpha \sim 1/3$, and with $f < 2.5$, $\alpha > 1/5$. For comparison,
the maximum variation of f over the region 9' to 160' is 1.4, if $\alpha = 1$.
Although this seems too small a variation, $\alpha \sim 1$ cannot be ruled out when
it is remembered that the procedure leading to the deduced velocity
dispersion profile probably leads to an error beyond the nominal ~ 100
km s^{-1} in just the velocity dispersion. In addition, obviously, there
is some uncertainty in our simple-minded approach, mainly in the choice
of the density profiles. A value of $\alpha \cong 1/3$ corresponds to $r_d \sim 120$ kpc,
which would mean that the central density is 3×10^{-24} g cm^{-3}, if the
total mass within 3° is 3×10^{15} M$_\odot$. Smaller values of α are probably
unreasonable. We note, in passing, that the qualitative result, that

faster decrease of the velocity profile is more consistent with the
dark matter being more concentrated than galaxies, holds also if the
galaxy density profile is less steep than that taken here, such as the
one suggested by Peach (1984).

The distribution of dark matter might also affect galactic orbits
by dynamical friction (Chandrasekhar, 1942; Ruderman and Spiegel, 1971;
Rephaeli and Salpeter, 1980). The time rate of change of the kinetic
energy of a galaxy, mass m, moving in a sea of lighter objects is

$$\frac{d\,(V^2/2)}{dt} = -\frac{4\pi G^2 m}{V}\rho_d \ln\Lambda,\tag{2}$$

where V is the galaxy velocity, $\Lambda = b_{max}/b_{min}$ is the ratio of maximum to
minimum impact parameters. The effect of this gravitational drag on the
motion of massive galaxies in clusters has already been estimated (Lecar,
1975) and has (increasingly) been explicitly taken into account in
numerical studies of cluster evolution (see, e.g., White, 1976; Barnes,
1983; Merritt, 1983). In most previous studies the fate of massive
galaxies, $m \gtrsim 10^{12}\,M_\odot$, was of interest. Now the timescale associated
with this drag is $\tau \propto V^3/\rho_d m$, i.e. more central dark matter concentration
might also have observational consequences on the motion of galaxies
which have already lost their halos, but the increased central mass
density also increases V.

One can readily estimate the extent of the region within which the
drag is effective in appreciably changing galactic orbits by calculating
$\tau_{1/2}$, the time it takes to drag a galaxy to 1/2 its initial distance,
r_o, from the cluster center. Lecar (1975) calculated the function
$\tau_{1/2}(r_o)$ assuming circular galaxy orbits. With $m \simeq 2.5\times10^{12}\,M_\odot$, Lecar
found that $\tau_{1/2} \lesssim 10^{10}$ yr for galaxies which were initially at $r_o \sim 600$
kpc from the center. We have repeated the calculation with our adopted
distribution for $\rho_d(r)$, taking m to be only 1 - 3 x $10^{11}\,M_\odot$. We find
that $\tau_{1/2} \lesssim 10^{10}$ yr yields values of $r_o(\alpha)$ which are actually lower for
$\alpha < 1$ than those for $\alpha > 1$. For example, with $m = 10^{11}\,M_\odot$, $r_o \lesssim 1/4$ Mpc,
and for $m = 3\times10^{11}\,M_\odot$, $r_o \lesssim 1/3$ Mpc, if $\alpha \lesssim 1$. If $\alpha = 3$, a galaxy whose
mass is 3 x $10^{11}\,M_\odot$ has a drag time shorter or comparable to 10^{10} yr if
it is within $r_o \sim 1/2$ Mpc. This result, that the drag is more effective
to larger distances for less concentrated dark matter, is not surprising:
the decrease in the drag time due to higher central dark matter density
is more than compensated by the increase in V^3. Obviously, this is so
as long as $V^2 \propto M/r$, i.e., not only when orbits are circular.

What are we to learn from these estimates? Deductions based on the
effectiveness or ineffectiveness of the drag on galaxies are not very
definite. Technically, the drag on a finite size object is smaller by a
factor of ~ 2 than estimated from Chandrasekhar's (1942) formula (Hausman,
1981). More importantly, it is not completely clear what is to be avoided
by requiring that $\tau_{1/2}$ is not to be smaller than 10^{10} yr within some
non-negligible fraction of the core radius of the galaxy distribution.

If a galaxy initially moves at a velocity $V > <v^2>^{1/2}$ higher than the velocity dispersion of the dragging objects, it can lose kinetic energy until it is slowed down to $<v^2>^{1/2}$, or to somewhat lower velocity. Thereafter, its drag time will be longer since the fractional number of dragging objects is reduced. This can happen if falling in the gravitational field of the cluster does not necessarily mean speeding up; the constancy of $<\sigma^2>^{1/2}$ in the inner region is an indication for this. It is also not clear whether the frequency of the cD phenomenon (in 20% of rich clusters (Bahcall, 1977)) in general, or the presence of such a galaxy in a given cluster, can be taken to imply efficient drag, because of the possibility that cD galaxies form through merging of galaxies.

REFERENCES

Bahcall, N.A. 1977, Ann. Rev. Astron. Ap. 15, 505.

Barnes, J. 1983, Mon. Not. Roy. Astron. Soc. 203, 223.

Bond, J.R., Szalay, A.S. and White, S.D.M. 1983, Nature 301, 584.

Chandrasekhar, S., 1942, Principles of Stellar Dynamics (Chicago: University of Chicago Press).

Faber, S.M. and Gallagher, J.S. 1979, Ann. Rev. Astron. Ap., 17, 135.

Farouki, R.T., Hoffman, G.L. and Slataper, E.E. 1983, preprint.

Gunn, J.E. 1977, Ap. J. 218, 592.

Hausman, M. 1981, preprint.

Hegyi, D.J. and Olive, K.A. 1983, Phys. Lett. 126B, 28.

Kent, S.M. and Gunn, J.E. 1982, Astron. J. 87, 945.

Lecar, M. 1975, in IAU Symposium 69, Dynamics of Stellar Systems, ed. A. Hayli (Dordrecht: Reidel).

Lynden-Bell, D. 1967, Mon. Not. Roy. Astron. Soc. 136, 101.

Lyubimov, V.A., Novikov, E.G., Nozik, V.Z., Tretyakov, E.F. and Kosik, V.S. 1980, Phys. Lett. 94B, 266.

Melott, A.L. 1982, Phys. Rev. Lett. 48, 894.

Meritt, D. 1983, preprint.

Peach, J.V. 1984, this volume, p. 89.

Rephaeli, Y. 1982, Phys. Rev. D26, 770.

Rephaeli, Y. and Salpeter, E.E. 1980, Ap. J. 240, 20.

Rood, H.J., Page, T.L., Kintner, E.C. and King, I.R. 1972, Ap. J. 175, 627.

Ruderman, M.A. and Spiegel, E.A. 1971, Ap. J. 165, 1.

Tremaine, S. and Gunn, J.E. 1979, Phys. Rev. Lett. 42, 407.

Van den Bergh, S. 1969, Nature 224, 891.

White, S.D.M. 1976, Mon. Not. Roy. Astron. Soc. 174, 19.

NONLINEAR EVOLUTION OF LARGE-SCALE DENSITY STRUCTURE

A. R. Rivolo: Space Telescope Science Institute, Baltimore MD

A. Yahil: State University of New York, Stony Brook NY

1. Introduction

The current observational facts on the very nonlinear regions of the universe on scales > 1 Mpc may be summarized as follows:

(1) The peak density contrast is always of order $\delta\rho/\rho \sim 10^{4\pm1}$.

(2) The velocity dispersion in these regions is normally distributed, with a remarkably constant $\sigma(v) \sim 700 \times (2)^{\pm1}$ km s^{-1} (1-dim).

(3) The systematic velocity deviation from pure Hubble flow (infall) near these regions (at least in one case--the Local Supercluster) is of the order of 300 km s^{-1} at radii of $H_o r \sim 1500$ km s^{-1}.

(4) The radial density profiles appear to be good power-laws "softened" at the origin. Typically $\delta\rho/\rho \propto$ (const. + r)$^{-2.5\pm.5}$.

(5) The velocity dispersion very far from the highly nonlinear regions is of order $\sigma(v) \sim 100 \times (2)^{\pm1}$ km s^{-1} (1-dim).

These observational data are begining to reveal a texture to the universe at large. This texture may be described as an infinite "mountain range" in density-space, each mountain being very similar to its neighbors in overall size. Some of the mountains may not be very symmetric, and some may have multiple peaks, but the basic object is the mountain.

We have addressed the question of whether it is possible to make such a nonlinear object as a "mountain" using the simplest of scenarios--collisionless, gravitational instability of a dispersive (hot) medium under the constraint of spherical symmetry. The answer to this question is an unqualified yes, and details of the model may be found in Rivolo and Yahil (1983). Here we restrict our attention to the generic qualities of the evolving distribution function (DF) in the phase-space of the collapse.

2. The Phase-Space of Collapse

From a cursory inspection of the physical problem one is tempted to approach the situation with a classical "stellar" dynamics in terms of a DF, whose time-independent form may be expressed in terms of constants of the motion. This bias, carried over from work on globular

429

F. Mardirossian et al. (eds.), Clusters and Groups of Galaxies, 429–434.

clusters, is what has been largely responsible for the conceptual-
ization that rich clusters of galaxies are isolated Emden spheres with
well defined time-independent DFs. This point-of-view ignores the
intimate coupling that a supercluster-scale perturbation ($M \sim 10^{15}$ $M_{\odot}$)
must have with the cosmological expansion throughout its growth. In
contrast, the scale of globular clusters ($M \sim 10^6$ $M_{\odot}$) is completely
isolated from the cosmological expansion. The DF of a supercluster-
scale perturbation must transition, in a continuous fashion, from a
hot virialized core to the cool homologous expansion of the Hubble
flow at its periphery. Clearly such a requirement precludes any
description in terms of an equilibrium configuration DF.

   Since an analytical treatment of the collisionless Boltzmann
equation under such non-equilibrium conditions is not tractable, we
have chosen to evolve the DF numerically using a large number of test-
masses to populate the phase-space. These test masses are then moved
under the influence of their own collective potential using Newtonian
dynamics. With this scheme, the local phase-space density (DF) is
everywhere defined, within $N^{1/2}$ statistics, by sampling a number of
test masses in the vicinity of the desired location.

   Figure 1 is a time sequence of frames showing two of the three
phase-space planes for a typical collapsing supercluster. The v_t-r
plane has been excluded for brevity. The time is parametrized by the
redshift, z, and is shown in each frame.

   The first frame shows the system as the perturbation enters the
nonlinear regime ($\delta\rho/\rho \sim 1$). At this epoch the velocity field is in
near Hubble expansion (smooth line) with a superposed isotropic, Gaussian dispersion field. One technical note must be made here. The
test masses are merely tracers of the density and velocity fields,
whose values are independent of the individual masses. Thus we have
chosen a mass distribution such that the number of points is constant
per unit radial increment. This allows us to attain spatial resolution in the core of the system where most of the action occurs. In
Fig. 1, the range of individual masses spans six decades, and it
should be kept in mind while looking at the diagrams that the points
do not represent galaxies.

   The next five frames show the DF of the system at various intermediate stops including the final configuration. The following
generic features should be noticed:

(1) The density contrast grows very slowly over a large redshift
interval, then transitions rapidly to a nearly constant profile.

(2) The transition occurs at a relativily recent epoch (here z $\sim$
1.0). This point may easily be delayed to smaller redshifts by
decreasing the initial perturbation amplitude, but earlier collapses
are difficult to achieve, since very little time is available at large
redshift (i.e. at large redshift $dt/dz \propto (1+z)^{-5/2}$).

(3) The final density profile can be fit by a reasonable King model.

(4) The velocity field is little affected until the collapse transition save for adiabatic cooling due to the cosmological expansion,
and the adiabatic heating due to central compression.

(5) The radial velocities do not fill the phase-space ergodically.
Streaming motion (with well defined orbits in phase-space) dominate

the central mass distribution. This point can visciate any virial analyses, especially using projected components.

(6) Most of the mass in the perturbation at the present epoch is in the initial infalling stream. The dominance of this infall should be used to define a "supercluster" in dynamical terms.

(7) A well defined characteristic radius exists (tangent point to the outermost orbits in the velocity diagram), and may be used to separate, dynamically, the central cluster from the surrounding super-cluster. This radius may be identified with the dynamical transition radius of $6°$ in the Virgo cluster.

(8) A secondary maximum arises in the density profile at late epochs due to the "bunching" in velocity-space.

(9) Athough the density profile is smooth, and looks like a reason-able Emden isothermal profile, the velocity field is fundamentally different from an isothermal sphere, or any equilibrium polytrope.

3. Variations of Model Parameters

In addition to the obvious normalization of the mass scale (and with it the length and velocity scales), the problem has two free dimensionless parameters: z_1, the redshift at which the perturbation enters the nonlinear regime, and Ω_o, the cosmological density parameter. In principle these dynamical models may discriminate between possible choices of these parameters. In practice this appears to be difficult at present, since the models exhibit little sensitivity to these parameters, and the observational data are poor.

Figure 2 shows the phase-space of three models constructed to fit the Local Supercluster. Models (A) and (C) are $\Omega_o = 0.1$ cosmologies, and vary only in age (z_1). Model (B) represents a "best fit" subject to the constraint of $\Omega_o = 0.5$. The arrows mark the two special radii: the inner central cluster radius, and the outer radius enclosing a mean density contrast $\langle \delta\rho/\rho \rangle = 3$. In comparing (A) to (C) one can see that (C) is a bit more centrally condensed, and considerably more ergodic in the core. Such differences, due to age only, may account for observed morphological variations from supercluster to super-cluster (c f. Virgo and Coma).

The effect of varying Ω_o is immediatly obvious in (B). The central velocity dispersion is much larger and may well be in conflict with the observational data. Overall the reader should notice that the density profiles are poor discriminators of dynamics, and more emphasis should be placed on obtaining better velocity field maps.

4. Comparison With Observation

The velocity diagrams of Fig. 2 are very informative, but such diagrams cannot be constructed by an Earth-bound observer. To remedy this situation we have transformed the spatial velocity maps of Fig. 2 into line-of-sight projected velocity maps. Figure 3 shows the three velocity fields of Fig. 2 as seen by an internal observer located at the outer radius (i.e. r for $\langle \delta\rho/\rho \rangle = 3$, corresponding to the location of the Sun within the Local Supercluster). Notice the "pinched waist"

appearance due to the systematic infall. These diagrams easily separate the different dynamical regimes of the supercluster. It should
be possible in the near future to construct such diagrams with sufficiently large counts to allow verification of the dynamics.

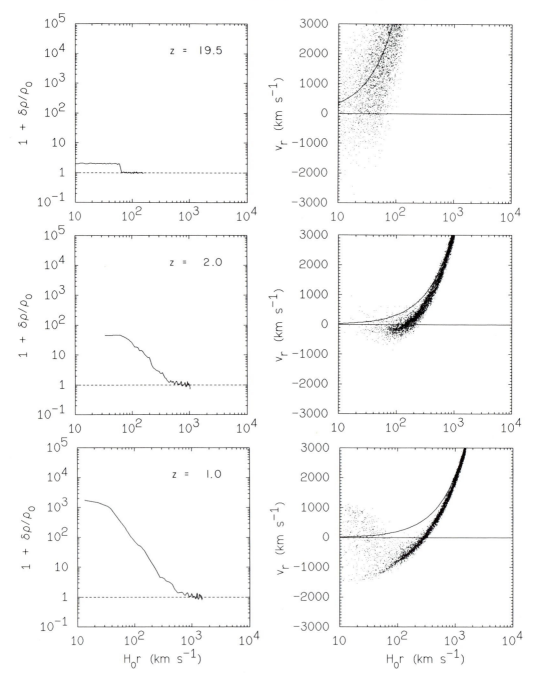

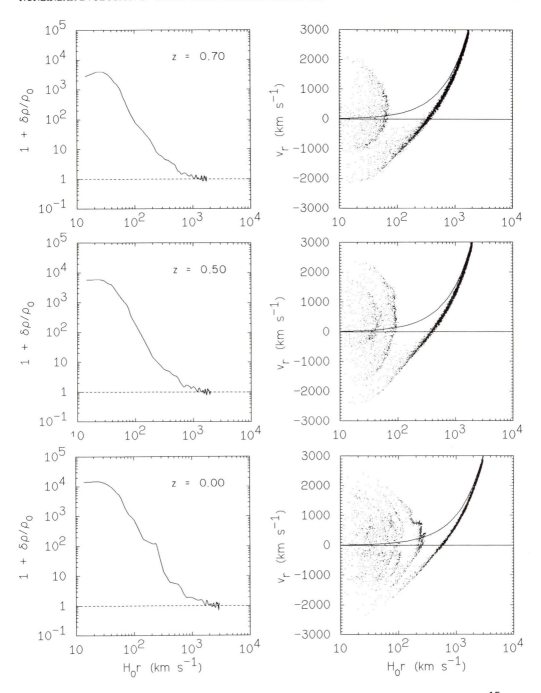

Figure 1. Time sequence of phase-space densities for a $M \sim 10^{15}\ M_\odot$ perturbation at six epochs during the evolution. The total mass has been distributed among 5000 test masses ranging in mass over six decades such as to keep the number of particles per unit radius constant.

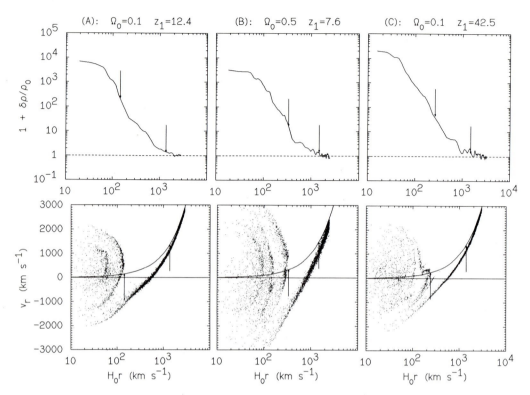

Figure 2.   Present epoch configurations for three models showing effects of age (initial epoch), and density parameter.

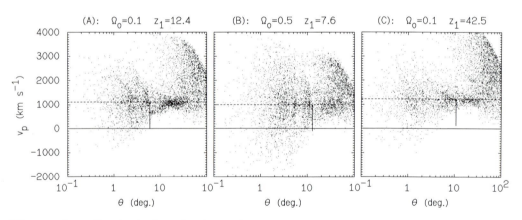

Figure 3.   Projected velocity phase-space as seen by an observer at the periphery of the collapsing systems.

Reference

Rivolo, A. R. and Yahil, A., 1983, Ap. J., <u>274</u>, (in press: 15 Nov).

GROWTH OF PERTURBATIONS BETWEEN HORIZON CROSSING AND MATTER DOMINANCE:
IMPLICATIONS FOR GALAXY FORMATION

Joel R. Primack
Board of Studies in Physics and Santa Cruz Institute for
Particle Physics, University of California,
Santa Cruz, CA 95064

George R. Blumenthal
Board of Studies in Astronomy and Astrophysics,
Lick Observatory, University of California,
Santa Cruz, CA 95064

The dark matter (DM) that appears to be gravitationally dominant
on all astronomical scales larger than the cores of galaxies can be
classified, on the basis of its characteristic free-streaming damping
mass M_D, as hot ($M_D \sim 10^{15} M_\Theta$), warm ($M_D \sim 10^{11} M_\Theta$), or cold ($M_D <
10^8 M_\Theta$) (Bond and Szalay, 1983; Primack and Blumenthal, 1983). For the
case of cold DM, the shape of the DM fluctuation spectrum is determined
by (a) the primordial spectrum (on scales larger than the horizon),
which is usually assumed to have a power spectrum of the form $|\delta_k|^2 \propto k^n$
(inflationary models predict the "Zeldovich spectrum" $n = 1$); and (b)
"stagspansion", the stagnation of growth of DM fluctuations which enter
the horizon while the universe is still radiation-dominated. Stagspansion flattens the fluctuation spectrum for $M \lesssim 10^{15} M_\Theta$. We report here
the results of a numerical evaluation of the fluctuation spectrum in
which all relevant physical effects have been included (Blumenthal and
Primack, in preparation), together with implications for galaxy formation (Blumenthal, Faber, Primack, and Rees, in preparation; hereafter
BFPR).

1.   WHY COLD DM IS INTERESTING

If a species of neutrino is the gravitationally dominant component
of the universe, its mass $m_\nu = 100 \, \Omega h^2 \, eV$ (where $h = H_0/100 \, km \, s^{-1} Mpc^{-1}$
lies in the range $1/2 \leq h \leq 1$) implies a free-streaming damping mass
$M_D \sim 10^{15} M_\Theta$ corresponding to hot DM. Probably because this type of DM
has been the most intensively studied, a number of potential problems
have been identified − for example the late formation of supercluster
"pancakes", at $z_p \lesssim 2$ (see, e.g., Frenk, White, and Davis, 1983), which
later fragment into galaxies, in contradiction to the observed abundance
of QSOs with $z > 2$, which implies earlier galaxy formation.

F. Mardirossian et al. (eds.), Clusters and Groups of Galaxies, 435–440.
© *1984 by D. Reidel Publishing Company.*

We would like to call attention to another discrepancy between the hot DM hypothesis and observations, including data presented at this conference: namely the absence of evidence for growth of the DM/baryon ratio from the scale of ordinary galaxies to rich clusters. If the DM is composed of neutrinos, which first collapse gravitationally on super-cluster scales with corresponding velocities $\sim 10^3$km/s, only a small fraction of the neutrinos can be captured to form the DM halos of galaxies having escape velocities typically $\sim 10^2$km/s, while a much larger fraction will be bound in rich clusters. One-dimensional simulations (Bond, Szalay, and White, 1983) confirm this expectation. The most relevant observational parameter is not the oft-discussed M_{tot}/L but rather $M_{tot}/M_{\ell um}$, where M_{tot} is the total mass determined dynamically and $M_{\ell um}$ is the mass in stars plus that in x-ray emitting gas. The available data, including that on our galaxy (which imply that its massive halo extends at least to ~ 50kpc), on small spiral groups, on small cD groups, and on rich clusters, is all consistent with $M_{tot}/M_{\ell um} \approx$ 10 (BFPR).

Yet another problem for the hypothesis that the DM is neutrinos is the recent prediction (Faber and Lin, 1983) that the dwarf spheroidal galaxies near the Milky Way have heavy halos; this is supported by recent observational determinations of the velocity dispersion of Draco (Aronson, 1983; Lin and Faber, 1983) and Carina (Aronson, private communication to Faber). The phase space constraint (Tremaine and Gunn, 1979) implies that $m_\nu \gtrsim 500$ eV. This is of course inconsistent with the cosmological density bound on m_ν, implying that the DM in these dwarf spheroidal halos is not hot. It is probably not warm DM either. Warm DM first collapses on a scale $\sim 10^{12} M_\Theta$ with $\sigma \sim 10^2$ km/s (Blumenthal, Pagels, and Primack, 1982; Bond, Szalay, and Turner, 1982), and too little could be captured by dwarf spheroidals, having $\sigma \sim 10$ km/s, to form the heavy halos indicated by the observations.

Besides the evidence just summarized against hot and warm DM, a further reason to consider cold DM is the existence of several plausible physical candidates, including <u>axions</u> of mass $\sim 10^{-5}$ eV (Preskill, Wise, and Wilczek, 1983; Abbott and Sikivie, 1983; Dine and Fischler, 1983; Ipser and Sikivie, 1983); heavy stable particles, such as the <u>photino</u>, with a mass ~ 10 GeV and very weak interactions; and <u>primordial black holes</u> with 10^{17} g $\lesssim m_{PBH} \lesssim M_\Theta$.

2.   CALCULATION AND IMPLICATIONS OF THE FLUCTUATION SPECTRUM

We will follow the current conventional wisdom and assume that the primordial fluctuations were adiabatic. In the standard formulation (e.g. Peebles, 1980), fluctuations $\delta \equiv \delta\rho/\rho$ grow as $\delta \sim a^2$ on scales larger than the horizon, where a = scale factor = $(1 + z)^{-1}$. When a fluctuation enters the horizon in the radiation-dominated era, the photons (together with the charged particles) oscillate as an acoustic wave, and the neutrinos (assumed to have negligible mass) free stream away. As a result, the main driving terms for the growth of δ_{DM}

disappear, and the growth of δ_{DM} stagnates ("stagspansion") until matter dominates (Guyot and Zeldovich, 1970; Meszaros, 1974; Groth and Peebles, 1975); see Fig. 1. Matter domination first occurs at $z = z_{eq}$, where

$$z_{eq} = 4.2 \times 10^4 \, h^2 \, \Omega \, (1 + 0.68 \, n_\nu)^{-1}$$

$$= 2.5 \times 10^4 \, h^2 \, \Omega \text{ for } n_\nu = 3$$

(1)

The first study of the growth of cold DM fluctuations was the numerical calculation of Peebles (1982), who for simplicity ignored neutrinos: $n_\nu = 0$ in (1). We have done numerical calculations including the effects of the known neutrino species ($n_\nu = 3$, $m_\nu \approx 0$) both outside and inside the horizon. Numerically, the largest effect of including neutrinos is the change in z_{eq}.

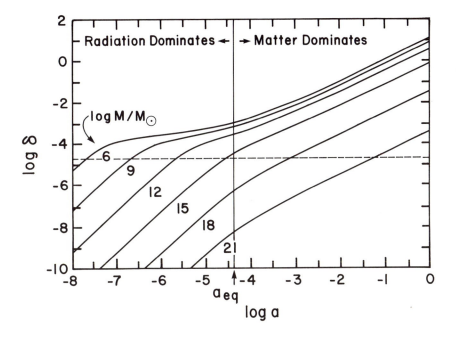

Figure 1. Numerical results for the growth of $\delta = k^{3/2} \delta_k$ versus scale factor a for fluctuations of various masses $M = 4 \pi^4 k^{-3} \rho_c / 3$. The curves are drawn for n = 1, $\Omega = h = 1$, and a baryonic to total mass ratio of 0.1. The vertical line represents the value of a when the universe becomes matter dominated, and the dashed line shows the (constant, for n = 1) value of δ when each mass scale crosses the horizon. These curves illustrate the stagnation of perturbation growth after small mass scales cross the horizon and show why at late times $\delta(k)$ is nearly flat for large k (small M).

It is instructive to make the further approximation of setting $\delta_{\gamma+b} = \delta_\nu = 0$ once a fluctuation is inside the horizon. Then one can analytically match the solution for $a > a_{horizon}$ (Peebles, 1980, p.58)

$$\delta_{DM}(a) = A_1 D_1(a) + A_2 D_2(a) = \delta_1 + \delta_2, \tag{2}$$

$$D_1 = 1 + 1.5y, \quad \text{where } y = a/a_{eq}, \tag{3a}$$

$$D_2 = D_1 \ln \left[\frac{(1+y)^{\frac{1}{2}} + 1}{(1+y)^{\frac{1}{2}} - 1} \right] - 3(1+y)^{\frac{1}{2}}. \tag{3b}$$

to the growing mode $\delta_{DM} \sim a^2$ for $a < a_{horizon}$. Matching the derivatives requires δ_2 comparable to δ_1 but opposite in sign. For $a \gg a_{horizon}$ only the growing mode D_1 survives, which explains the moderate growth in δ_{DM} between horizon crossing and matter dominance. In the limit of large k, one finds $\delta_k \propto k^{n/2-2} \ln k$. Correspondingly, for $M \ll M_{eq} \approx 10^{16} M_\odot$, the rms fluctuation in the mass within a random sphere containing average mass M is $\delta M/M \propto |\ln M|^{3/2}$. Turner, Wilczek, and Zee (1983) considered only the Meszaros solution (3a) and erroneously inferred that the fluctuation spectrum would be essentially flat for $M < M_{eq}$ for a Zeldovich primordial spectrum, which would then be inconsistent with observations.

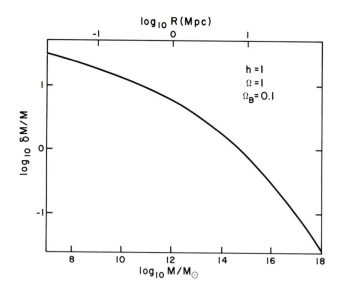

Figure 2.   The r.m.s mass fluctuations within a randomly placed sphere of radius R in a cold DM universe. The curve is normalized at 8 Mpc and assumes an initial Zeldovich (n = 1) fluctuation spectrum, and h = Ω = 1.

Our numerical results for $\delta M/M$ are shown in Fig. 2 for $\Omega = h = 1$, assuming a Zeldovich ($n = 1$) spectrum (reflected in $\delta M/M \propto M^{-2/3}$ for $M > M_{eq}$). We have followed Peebles (1982) in normalizing $\delta M/M = 1$ at $8h^{-1}$ Mpc. For either h or Ω less than unity, $\delta M/M$ is somewhat flatter.

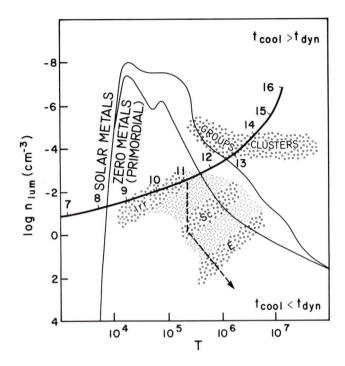

Figure 3. The baryonic density versus temperature as average perturbations having total mass M become nonlinear and virialize. The numbers on the tick marks are the logarithm of M in solar units. This curve assumes $n = 1$, $\Omega = h = 1$, and a baryonic to total mass ratio of 0.07. The region where baryons can cool within a dynamical time lies below the cooling curves. Also shown are the positions of observed galaxies, groups, and clusters of galaxies. The dashed line represents a possible evolutionary path for dissipating baryons.

Fig. 3 (Primack and Blumenthal, 1983) shows the density of ordinary (baryonic) matter vs. internal kinetic energy (temperature) of typical fluctuations of various sizes, just after virialization, calculated from $\delta M/M$ of Fig. 2. This is superimposed upon the Rees–Ostriker (1977) cooling curves (for which cooling time equals gravitational free fall time) and data on galaxies (with kinetic energy determined from rotation velocity for spirals and velocity dispersion for ellipticals). Fluctuations that start with greater amplitude than average will turn

around earlier, at higher density, and thus lie below the virilization
curve on Fig. 3. As the baryons in a virialized fluctuation dissipate,
their density will initially increase at constant T within the surrounding isothermal halo of dissipationless material (DM), and then T
will increase as well when the baryon density exceeds the DM density,
as suggested by the dashed line in the figure. Note that the galaxy
data lies below the virialization curve (the curve is somewhat higher
for $h = \frac{1}{2}$, $\Omega = 1$). In this respect, the Zeldovich primordial spectrum is
more consistent with the data than an $n = 2$ (or $n = 0$) primordial spectrum,
which lies too low (too high) on the figure compared to the galaxies
(see BFPR for more details). With the Zeldovich spectrum, the important
conclusion is that one should observe dissipated systems with large
halos having total mass $10^8 M_\odot \lesssim M \lesssim 10^{12} M_\odot$. This is essentially the
range of observed galaxy masses.

REFERENCES

Abbott, L., and Sikivie, P.: 1983, Phys. Lett. 120B,pp.133-136.
Aronson, M.: 1983, Astrophys. J. (Lett.) 266,pp.L11-L15.
Blumenthal, G.R., Pagels, H., and Primack, J.R.: 1982, Nature 299,pp.37-38.
Bond, J.R., and Szalay, A.S.: 1983, Astrophys. J. (in press).
Bond, J.R., Szalay, A.S., and Turner, M.S.: 1982, Phys. Rev. Lett. 48, pp.1636-1639.
Bond, J.R., Szalay, A.S., and White, S.D.M.: 1983, Nature 301,pp.584-585.
Dine, M., and Fischler, W.: 1983, Phys. Lett. 120B,pp.137-141.
Faber, S.M., and Lin, D.N.C.: 1983, Astrophys. J. (Lett.) 266,pp.L17-L20.
Frenk, C., White, S.D.M., and Davis, M.: 1983, Astrophys. J. (in press).
Groth, E.J., and Peebles, P.J.E.: 1975, Astron. Astrophys. 41,pp.143-145.
Guyot, M., and Zeldovich, Ya. B.: 1970, Astron. Astrophys. 9,pp.227-231.
Ipser, J., and Sikivie, P.: 1983, Phys. Rev. Lett. 50,pp.925-927.
Lin, D.N.C., and Faber, S.M.: 1983, Astrophys. J. (Lett.) 266,pp.L21-L25.
Meszaros, P.: 1974, Astron. Astrophys. 37,pp.225-228.
Peebles, P.J.E.: 1980, *The Large Scale Structure of the Universe* (Princeton University Press, Princeton, NJ).
Peebles, P.J.E.: 1982, Astrophys. J. (Lett.) 263,pp.L1-L5.
Preskill, J., Wise, M., and Wilczek, F.: 1983, Phys. Lett. 120B,pp.127-132.
Primack, J.R., and Blumenthal, G.R.: 1983, *Formation and Evolution of Galaxies and Large Structures in the Universe*, ed. J. Audouze and J. Tran Thanh Van (Reidel, Dordrecht, Holland); *Proceedings of the Fourth Workshop on Grand Unification* (Birkhäuser, Boston).
Rees, M.J., and Ostriker, J.: 1977, M.N.R.A.S. 179,pp.541-559.
Tremaine, S.D., and Gunn, J.E.: 1979, Phys. Rev. Lett. 42,pp.407-410.
Turner, M.S., Wilczek, F., and Zee, A.: 1983, Phys. Lett. 125B,pp.35-40 and (E)p.519.

RADIATION PERTURBATIONS IN A THREE COMPONENT UNIVERSE

N. Vittorio[1], F. Occhionero[1], F. Lucchin[2], S. Bonometto[2,3]
1. Istituto Astronomico, Università di Roma
2. Dipartimento di Fisica, Università di Padova
3. International School for Advanced Studies, Trieste

ABSTRACT: We study numerically the evolution of initial adiabatic perturbations in a Universe with critical density containing three components: massive neutrinos or the like, baryonic matter and radiation. The evaluation at decoupling of the residual perturbation in the radiation field shows that the common expectations on the neutrino influence on the monopole and the dipole components are likely to be modified. Thus, the lack of detection of distortions in the cosmic microwave background may remain an unresolved problem.

1. INTRODUCTION

It has been suggested that a weakly interacting hot massive particle (WHIP) component of the cosmic medium (such as massive neutrinos) may solve a number of problems in cosmology (Doroshkevich et al., 1980; Bond et al., 1980; Klinkhamer and Norman, 1981; Wasserman, 1981). In particular one such problem is the lack of detection of small scale ($\sim 10'$) distortions in the cosmic microwave background (CMWB).
In fact, while in the canonical Universe (composed of radiation and baryonic matter only) the solution may consist in a special choice of the spectral index n of the primordial fluctuations (Bonometto et al., 1983a), in the presence of the WHIP component it is argued that gravitational instability on the large scales originates WHIP condensations well before matter-radiation decoupling. For the paradigmatic case of three massive neutrinos of 30 eV each – to which we will adapt our entire numerology – and for the scales of interest, this occurs at a redshift z of the order of 40,000.
Thus, while matter and radiation, coupled by Thomson scattering, oscillate at almost constant amplitude, the amplitude of the WHIP density fluctuations grows by a factor 40 (assuming that decoupling occurs instantaneously at z = 1000). The same large factor is therefore available to reduce the amplitude of the baryonic matter fluctuations needed to produce bound structures at a reasonable redshift. Under the assumption of adiabaticity, this is believed to reduce automatically by the same factor the amplitude of the radiation density fluctuations and to avoid the conflict with observations.

441

F. Mardirossian et al. (eds.), Clusters and Groups of Galaxies, 441–446.
© *1984 by D. Reidel Publishing Company.*

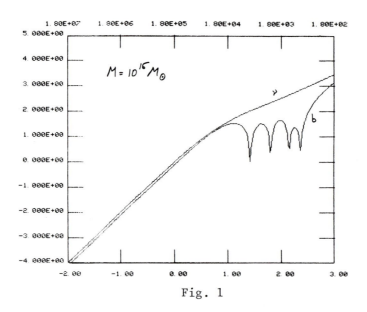

Fig. 1

We have undertaken to verify quantitatively how the interaction among the three components affects this scenario.

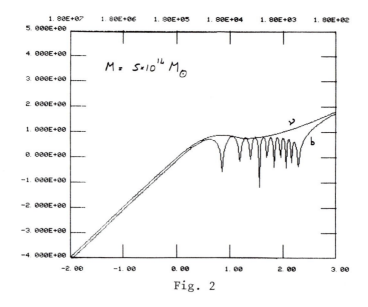

Fig. 2

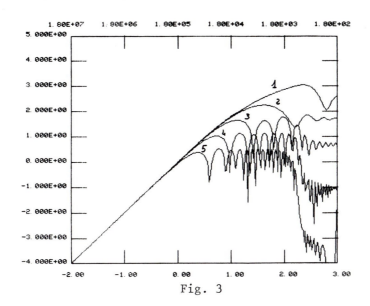

Fig. 3

2. METHOD AND RESULTS

Computations in this area have recently appeared in the literature
(Bond and Szalay, 1981; Peebles, 1982; Szymanski and Jaroszynski, 1983):
here we improve upon published work by assuming the simultaneous existence of the three components in the Universe (WHIP's, photons and
baryons) and treating each of them in the most accurate way. Specifically,
we describe the three components as follows:

i) WHIP's are treated via the numerical integration of the perturbed
collisionless Boltzmann equation written by Peebles (1982). We use however a slightly different method from his, in treating the angular
dependence of the distribution function with a Legendre polynomial expansion;

ii) photons are described via the kinetic approach (Silk and Wilson,
1980; Wilson and Silk, 1981; Bonometto et al., 1983a);

iii) baryons are treated as a fluid.

On decoupling we use previous work (Bonometto et al., 1983b) showing
that the ionization fraction, evaluated after taking into account the
WHIP dynamical role on the expansion of the Universe, is half an order
of magnitude bigger (for the same baryonic content) than in the case of
massless WHIP's.
We follow the evolution of an initial adiabatic perturbation for several
masses between 10^{14} and 10^{19} $M_\odot$, beginning at a redshift of the order of
10^7, where all these scales are well outside their horizons, and halting
the integration at a redshift of the order of 10^2. An equivalent time
parameter is the expansion factor a normalized to unity when the WHIP's
become non-relativistic.

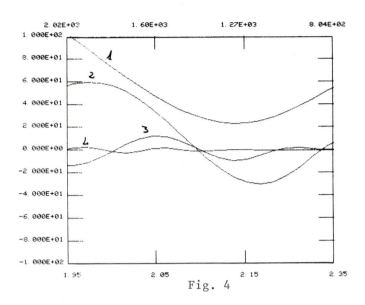

Fig. 4

Finally we assume an Einstein-de Sitter Universe in which the present baryonic density is 3% of the total.

Fig. 1 shows on a log-log plot the a- and z- dependence (bottom and top horizontal scales) of the neutrino (ν) and baryon (b) density contrasts for a mass of 10^{16} $M_\odot$. In the neutrino curve the analytical trends $(1+z)^{-2}$ and $(1+z)^{-1}$ for large and small redshifts are evident. The baryon curve shows several oscillations right after entering the horizon and the subsequent infall into the neutrino gravitational wells after decoupling.

In Fig. 2 the same quantities are plotted for $M = 5 \times 10^{14}$ $M_\odot$. The neutrino curve shows the expected Landau damping and the later recovery of growth; the baryon curve has the same qualitative behaviour of Fig. 1. However, for smaller still total masses (not shown here) it occurs on the contrary that it is the baryon gravity to drive the neutrino perturbation, which by itself would have damped more strongly. Further results, concerning transmission factors, mass variances, dependence on the initial spectral index, etc., are given in detail elsewhere (Bonometto et al., 1983c) where radiation is treated with the fluid approximation.

Fig. 3 shows, again on a log-log plot, the evolution of the monopole component of radiation over the entire range of integration for masses of: 1) 10^{18}; 2) 10^{17}; 3) 10^{16}; 4) 10^{15}; 5) 10^{14} $M_\odot$.

Fig. 4 shows on a semilog plot the evolution of the same monopole component but only during decoupling, i.e. for 2000 > z > 800 and for masses of: 1) 10^{17}; 2) 10^{16}; 3) 10^{15}; 4) 10^{14} $M_\odot$. The presence of an external gravitational engine shows up in the fact that the "barycenter" of the oscillations is pushed toward positive values, which increase with M. This is the analog of the Sachs and Wolfe (1967) effect on the small scales.

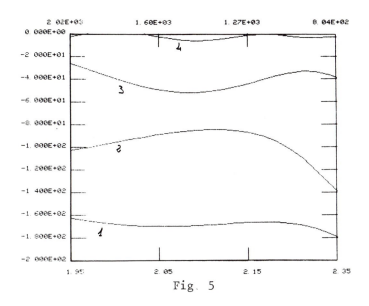

Fig. 5

Fig. 5 shows, again on a semilog plot and only during decoupling, the
evolution of the dipole component for masses of: 1) 10^{17}; 2) 5×10^{16};
3) 10^{16}; 4) 10^{15} $M_\odot$. The coupling of radiation to the WHIP gravitational
field is shown by large velocities which never invert their sign for
large masses ($\sim 10^{15}$ $M_\odot$).
  From the results of Figs. 4 and 5 it seems that the canonical scenario
outlined in the beginning must be treated with caution: the problem of
the CMWB fluctuations may not be solved by WHIP's if the velocity fields
are as high as it appears here. Work is in progress to evaluate quantitatively the temperature fluctuations.

REFERENCES

Bond, J.R., Efstathiou, G., and Silk, J.: 1980, Phys.Rev.Lett., 45, 1980

Bond, J.R. and Szalay, A.S.: 1981, in Proceedings of Neutrino '81, 1, 59

Bonometto, S., Caldara, A., and Lucchin, F.: 1983a, A.A., in press

Bonometto, S., Lucchin, F., Occhionero, F., and Vittorio, N.: 1983b,
A.A., 123, 118

————id.————: 1983c, submitted

Doroshkevich, A.G., Khlopov, M. Yu., Sunyaev, R.A., Szalay, A.S., and
Zel'dovich, Ya.B.: 1980, in Tenth Texas Symposium on Relativistic Astrophysics, eds. R. Ramaty and F.C. Jones, New York Academy of Sciences,N.Y.

Klinkhamer, F.R. and Norman, C.A.: 1981, Ap.J.Lett., 243, L1

Peebles, P.J.E.: 1982, Ap.J., 258, 415

  Sachs, R.K. and Wolfe, A.M.: 1967, Ap.J. 147, 73

Silk, J. and Wilson, M.L.: 1980, Phys. Scripta, 21, 708

Szymanski, M. and Jaroszynski, M.: 1983, MNRAS, in press

Wasserman, I.: 1981, Ap.J., 248, 1

Wilson, M.L. and Silk, J.: 1981, Ap.J., 243, 14

PANCAKES AND GALAXY FORMATION

Paul R. Shapiro
Department of Astronomy
University of Texas
Austin, Texas

ABSTRACT

The pancake scenario for the formation of large-scale structure
in the Universe by the growth of a spectrum of primordial, adiabatic
density perturbations is reviewed. The rôle of possible, finite mass,
stable, weakly interacting elementary particles -- the "darkinos" (e.g.
neutrinos, photinos, gravitinos, axions) -- in dominating the total mass
density of the universe and resolving the difficulties of the original
Zel'dovich pancake scenario is discussed, in particular. Results of
detailed, numerical hydrodynamical calculations of the combined growth
of (baryon) + (darkino) pancakes in a post-recombination Friedmann
universe dominated by such collisionless particles are summarized.
Implications of these results and of other, non-hydrodynamical, purely
collisionless simulations (i.e. which ignore the baryon component) for
the viability of the pancake model are assessed.

I. INTRODUCTION

The Universe is homogeneous and isotropic only in an average sense
and only on the very largest scales. Galaxies represent inhomogeneities
which are themselves clustered and superclustered in inhomogeneous
groupings of up to 10^{15} $M_\odot$ or more, and on scales as large as from 20
to 100 Mpc (e.g. Davis et al. 1982; Kirshner et al. 1981; Tarenghi
et al. 1980; Einasto, Joever, and Saar 1980; Gregory and Thompson 1978).
Much work has been directed towards explaining the presence of this
structure in the universe as the effect of the growth of either isothermal or adiabatic density perturbations present in the early universe.

Zel'dovich and his collaborators have advocated the view that
galaxies and clusters formed when purely adiabatic perturbations, whose
spectrum is truncated at short wavelengths by radiative damping prior
to recombination and matter-radiation decoupling, grew by gravitational
instability in the post-recombination epoch (e.g. Zel'dovich 1970, 1978;
Sunyaev and Zel'dovich 1972; Doroshkevich et al. 1980). In the original
picture, known as the "pancake", or adiabatic, scenario, large-scale

447

F. Mardirossian et al. (eds.), Clusters and Groups of Galaxies, 447–478.

perturbations on a scale greater than of the order of the critical
damping length (which encompasses a mass well in excess of that of a
single galaxy) must reach the nonlinear stage of growth before galaxies
can form. During this growth, initially small anisotropies in the
perturbation spectrum are strongly amplified into highly flattened
structures, or "pancakes". If the matter infalling during pancake
formation is pressure-free, then as it accumulates in the midplane of
the pancake a caustic surface eventually forms at this midplane where
streamlines intersect, making the density momentarily infinite there.
For a collisional fluid (like the baryon-electron component of the
universe), shock waves must develop in the midplane, rather than a
caustic (actually, one shock on either side of the midplane), which
dissipate the kinetic energy of infall by heating the gas, enabling it
to cool radiatively in the post-shock flow to a temperature $\leq 10^4$ K.
The high density and low temperature of this thin, central sheet
ultimately results in its fragmentation and, eventually, in galaxy
formation.

In its original form, the Zel'dovich pancake scenario is difficult
to reconcile with observations of the microwave background, the present
spatial distribution of galaxies, and the light element abundances. It
is now known that primeval adiabatic perturbations in a baryon-dominated
universe which are large enough in amplitude to reach the nonlinear
stage of growth by the present violate the observed upper limits to the
angular fluctuations in the temperature of the microwave background and
the observational constraints on the galaxy covariance function, unless
$\Omega \approx 1$ and the perturbation spectrum is highly contrived (Silk and Wilson
1981; Wilson and Silk 1981; Peebles 1981a). An $\Omega \approx 1$ baryon-dominated
universe, however, is in conflict with the upper limits to the baryon
density ($\Omega_b \leq 0.1$) inferred from Big Bang nucleosynthesis arguments
concerning the abundance of He, D, and ^{7}Li (e.g. Olive et al. 1981;
Steigman 1983; Gott et al. 1974). On the other hand, evidence based
upon dynamical arguments suggests that $\Omega > 0.1$, perhaps as high as
$\Omega \sim 1$, implying the existence of dark, or unseen, matter with average
density in excess of that allowed by these nucleosynthesis arguments
for the baryonic component of the universe (cf. Peebles 1981b; Aaronson
et al. 1982; Davis and Huchra 1982; Davis and Peebles 1983).

Recent developments in particle physics have suggested a way to
avoid the difficulties of the original pancake scenario while at the
same time resolving the apparent discrepancy between the dynamical
estimates of the mass-density of the universe and the primordial
nucleosynthesis limits on the baryon density. Central to this suggestion
is the existence of some stable, weakly interacting, finite mass,
elementary particle whose number density is sufficient to allow that
particle to dominate the total mass-density of the universe. In its
earliest incarnation, this suggestion involved a finite mass neutrino
of mass in the range of tens of eV (for a review and references, see
Doroshkevich et al. 1981). With such a mass per particle, the number
density of neutrinos which freezes out when the temperature of the
universe drops below a few MeV would be sufficient to close the Universe.

In that case, the neutrinos could provide the "dark" matter seen
dynamically on a variety of scales without requiring the baryon density
to exceed the nucleosynthesis limit (e.g. Schramm and Steigman 1981).

From the point of view of the Zel'dovich pancake scenario, neutrinos
with a mass in this range can provide a natural way, as follows, for
adiabatic density perturbations to explain the large-scale structure in
the universe without violating the constraints on the microwave background anisotropy. For any reasonable primordial perturbation spectrum,
the maximum neutrino effective "Jeans mass" ($M_{J,max} \sim 10^{17} [M_\Theta/m_\nu(eV)]^2$,
where $m_\nu(eV)$ is the sum of the mass of the different neutrino types in
eV; Wasserman 1981) defines the mass scale of the first systems to condense out of the smooth Friedmann background. This is of the same order
as galaxy supercluster masses for $m_\nu \gtrsim 10$ eV. Just as radiative damping
was responsible in the original scenario for eliminating power on small
wavelengths in the perturbation spectrum of the baryons, so collisionless
processes damp density perturbations in the neutrinos on wavelengths
smaller than that which corresponds to this $M_{J,max}$. Only after the
baryon-electron component recombines and decouples from the radiation,
however, do the neutrino density perturbations on the scale of $M_{J,max}$
induce growing density perturbations in the baryons on a baryonic
mass scale $M_b = (\Omega_b/\Omega_\nu)M_{J,max}$. The combined neutrino-baryon perturbation
then undergoes the same nonlinear, aspherical collapse as envisaged in
the original Zel'dovich pancake scenario. The fact that the neutrino
fluctuations grow between the earlier epoch at which the neutrinos go
nonrelativistic and the later epoch of photon-matter decoupling, while
the baryon and radiation fluctuations do not grow, then alleviates the
difficulty mentioned earlier of making adiabatic perturbation amplitudes
large enough in the baryon-dominated case without violating the microwave background anisotropy limits.

Interest in this massive-neutrino solution to the problems of the
pancake scenario has been stimulated by recent developments in particle
physics. Application of grand unified theories (GUTs) of the strong,
weak, and electromagnetic interactions to the early universe has, for
the first time, provided a fundamental explanation of the entropy per
baryon of the universe measured today (e.g. Weinberg 1979). Independent
of the details of the models used to calculate this entropy to baryon
ratio, however, the idea that it is determined only by such basic
microphysics implies that, within the standard, isotropic Friedmann
cosmology, the entropy per baryon must have been the same everywhere
once the present baryon asymmetry was established. Hence, isothermal,
or entropy, perturbations cannot survive the baryosynthesis epoch, and
there are only adiabatic perturbations which remain to form galaxies,
just as the pancake scenario requires (cf. Press 1980).

GUTs have offered another stimulus to the neutrino-variant of the
pancake scenario by predicting that neutrinos do have a nonzero rest
mass (e.g. Magg and Wetterich 1980; Zee 1980; Witten 1980; Barbieri
et al. 1980a,b; Mohapatra and Senjanovic 1980; Mohapatra and Marshak
1980). Theory suggests only a lower bound to the mass which is too low

to be of cosmological significance, however. The experimental evidence, on the other hand, although inconclusive, is still consistent with a mass in the cosmologically interesting range. A review of this experimental evidence is beyond the scope of this paper. We note only that, while preliminary evidence for neutrino oscillation, which implied a mass greater than ~ 1 eV (Reines, Sobel, and Pasierb 1980) has subsequently been called into question (e.g. Kwon et al. 1981), claims are still made that the measurement of the tritium beta decay spectrum yields a mass for the electron neutrino as large as 14-46 eV (Lyubimov et al. 1980 and private communications concerning more recent, improved versions of the same experiment). Most of the limits that have been placed, in fact, concern the mass of the electron neutrino. Experimental upper limits to the mass of the μ and τ neutrino are much weaker (e.g. $m_{\nu_\mu} \leq 570$ keV, Daum et al. 1978; $m_{\nu_\tau} \leq 250$ MeV, Bacino et al. 1979). Cosmology, in principle, limits the sum of the masses of stable neutrino types to roughly 100 eV (e.g. Schramm and Steigman). That is, Ω_ν cannot much exceed unity, and the neutrino mass can be written as

$$m_\nu = 70.5 \; \Omega_\nu h^2 / (\alpha^2 \beta \theta^3) \; \text{eV} \tag{1}$$

(Peebles 1982), where α is the ratio of neutrino temperature to the background radiation temperature, conventionally taken to be $(4/11)^{1/3}$, θ is the present value of the radiation temperature in units of 2.7 K, β is the number of massive neutrino spin states and species, and h = $H_0 / (100 \; \text{km s}^{-1} \; \text{Mpc}^{-1})$. In short, within this limit, it is still possible that neutrinos have a mass of cosmological significance.

The development of supersymmetric GUTs has, in the meantime, resulted in the suggestion that, in addition to the ordinary neutrinos, there may be other weakly interacting, stable particles with a nonzero rest mass, such as the photino (Cabibbo, Farrar, and Maiani 1981; Weinberg 1983) or gravitino (Pagels and Primack 1982). Depending upon the uncertain details of these theories, it is possible that one of these so-called "inos", the finite mass, fermionic superpartners of more familiar bosons, such as photons or gravitons, has just the right mass and emerges from the early universe with just the right number density to dominate the total mass density of the universe. Such particles would behave as a collisionless gas coupled to other matter only by gravity, thus able to play a rôle analogous to that described above for the massive neutrinos in resolving some of the difficulties of the original pancake scenario. The suggestion has also been made that axions, if they exist, may have a mass per particle and a number density which qualify them as the collisionless dark matter responsible for making $\Omega \approx 1$ (e.g. Preskill, Wise, and Wilczek 1983; Abbott and Sikivie 1983; Dine and Fischler 1983; Ipser and Sikivie 1983). Hence, axions, which are bosons rather than fermions, might also make the pancake scenario viable. As we shall see, the crucial cosmological distinction to be made amongst these different elementary particles is that of the different rôle which collisionless damping plays in determining the structure which emerges in a universe dominated by one or the other of these particles (to which we shall henceforth refer collectively as "darkinos").

   In section II, which follows, I will sketch this range of differing
cosmological outcomes of the different darkino types. In § III, I will
describe the results of detailed numerical calculations of the nonlinear
growth of pancakes in a neutrino-dominated universe -- the prototype
case -- designed to explore the hypothesis that galaxies can form in
such a universe. In § IV, analytical results which approximate the
detailed numerical hydrodynamics will be summarized in order both to
explain the evolution of combined (baryon) + (darkino) pancakes and to
anticipate the dependence of this evolution on pancake size, collapse
epoch, and the like. The results of the detailed hydrodynamics calculations for a variety of pancakes scales and collapse epochs will
also be summarized in this section. Lastly, in § V, I will address the
potential difficulties which can presently be identified for the
neutrino-variant of the pancake scenario (including, for example, those
pointed out by Peebles elsewhere in this volume) and demonstrate, I
hope, that, contrary to the claims which have been widely made, the
neutrino-dominated universe is still a viable one, although tightly
constrained. In the process, I hope to identify some unanswered
questions which may ultimately decide the fate of this model.

II.   THREE KINDS OF PERTURBATION GROWTH IN A UNIVERSE DOMINATED BY
      DARKINOS

   The nature of the nonlinear structure which emerges in the pancake
scenario at late times in a universe dominated by darkinos is, in
principle, fully determined by the amplitude and shape of the adiabatic
perturbation power spectrum as measured in the linear stage of growth,
in the post-recombination epoch. The range of behavior expected for
this amplitude and shape of the linear perturbation spectrum after
recombination can be summarized as follows. [Different parts of this
description can be found in whole or part in a number of references,
including Doroshkevich et al. (1981) and references therein; Bond,
Efstathiou, and Silk (1980); Wasserman (1981); Bond, Szalay, and Turner
(1982); Blumenthal, Pagels, and Primack (1982); Peebles (1982, 1983);
Turner, Wilczek, and Zee (1983), Primack and Blumenthal (1983)].

   The darkinos are collisionless. One can define an effective Jeans
length and mass for the darkinos by replacing the square of the sound
speed c_s^2 in the expression for the Jeans mass in a collisional fluid
by $<v^2>/3$, the mean square random thermal velocity. As usual, perturbations on mass scales M greater than M_J, this effective Jeans mass,
can grow, while those on smaller scales cannot since particles stream
out of density fluctuations on times which are shorter than the local
gravitational time scale [i.e. $\sim (G\rho)^{-1/2}$]. When the random particle
velocities are relativistic (in the comoving frame), M_J is roughly equal
to M_H, the mass which fills the horizon at that moment, since the
horizon size and the free streaming length are comparable. Hence,
during this epoch, M_J increases with time [e.g. if the universe is
radiation-dominated at that time, then M_H increases as $M_H \propto t \propto T^{-2}$,

where t is time and T is the radiation temperature]. When the particle velocities are nonrelativistic, however, M_J decreases with time [e.g. if the universe is matter-dominated at that time, then $M_J \propto T^{3/2} \propto (1+z)^{3/2}$ for fermionic darkinos]. Hence, the Jeans mass reaches a maximum M_J^{max} roughly when $kT_X \sim m_X c^2$, where T_X is the kinetic temperature corresponding to the random thermal velocities of the darkinos (i.e. "X" particles).

The ability of the darkinos to stream freely also tends to damp out density fluctuations on length scales λ comparable to the mean displacement per particle during the age of the universe up to that moment. When the particles are relativistic, the <u>directional</u> dispersion in velocities accomplishes this smearing. When they are nonrelativistic, the faster particles move through larger distances than the slower ones, causing the smearing. In the end, this collisionless "phase-mixing" tends to eliminate power in the perturbation spectrum on all wavelengths shorter than λ_J^{max}, the Jeans length corresponding to M_J^{max}.

The different categories of darkinos may be grouped according to the rôle which this collisionless damping plays in determining the non-linear structure which emerges from the growth of the perturbation spectrum. There are three basic types of darkinos as follows.

(a) Large-Scale Damping (LSD)

Particles of this type decouple while they are still relativistic. This includes the case of massive neutrinos with $m_X \sim 10$'s of eV. Furthermore, the transition from relativistic to nonrelativistic particle velocities for these darkinos occurs at roughly the same time as the transition from radiation to matter domination of the total energy density of the Universe. As a result, M_J^{max} is roughly equal to M_{eq}, the mass which fills the horizon at the transition to matter domination. We can see this as follows.

For darkinos in this category, we require that decoupling occur at a temperature $T_{X,d}$ which is small enough that the number of other particle species present which can annihilate at temperatures $T < T_{X,d}$ is relatively small (as is the case for neutrino decoupling, after which only e^+-e^- annihilation occurs). This ensures that $\alpha \equiv (T_X/T_\gamma) \sim 1$, where T_X is the darkino temperature after darkino decoupling, T_γ is the radiation temperature measured at the same time, and α is a factor which takes account of the "heating" of the thermal background radiation relative to the darkinos which results from the annihilations of other species after the darkinos decouple. For neutrinos, for example, α is thought to be $(4/11)^{1/3}$, as mentioned earlier. In general, we can use the fact that, at decoupling, any of the darkinos which decouple when they are relativistic are described by a relativistic Fermi-Dirac distribution to show that, for a given present value of Ω (assumed to be dominated by the darkinos), we can write

$$g_X m_X \alpha^3 \simeq \Omega h^2 \theta^{-3}, \qquad (2)$$

where g_X is the darkino statistical weight, m_X is the darkino mass per particle, and θ and h were defined previously. Our assumption that $\alpha \sim 1$ for darkinos in this category, then, fixes the value of the quantity $g_X m_X$ which will make $\Omega \sim 1$ at 10's of eV. Equation (1), it turns out, was merely a special case of equation (2) for the particular case of the neutrinos. In the meantime, the same reasoning which led to equation (2) can be used to write the mass which fills the horizon when $kT_X \sim m_X c^2$, the transition from relativistic to nonrelativistic particle velocities, with remarkable simplicity crudely as

$$M_H(kT_X \sim m_X c^2) \approx g_X M_{P\ell}^3 m_X^{-2} \alpha^6, \qquad (3)$$

where $M_{P\ell}$ is the Planck mass $[M_{P\ell} \equiv (\hbar c/G)^{1/2} = 2 \times 10^{-5}$ g]. The transition from radiation to matter domination, on the other hand, occurs roughly at a redshift z_{eq} given by

$$1 + z_{eq} \sim (\Omega \rho_{crit} c^2)/(aT_{\gamma,0}^4) \approx 4 \times 10^4 \, \Omega h^2 \theta^{-4}, \qquad (4)$$

where $a = 7.56 \times 10^{-15}$ erg cm^{-3} deg^{-4}. The mass M_{eq} which fills the horizon at this epoch is just

$$M_{eq} = (4\pi/3)(ct_{eq})^3 (1+z_{eq})^{-3} (\Omega \rho_{crit}) \qquad (5a)$$

$$\gtrsim 10^{15} (\Omega h^2)^{-2} M_\Theta. \qquad (5b)$$

For the $(g_X m_X)$ value of 10's of eV determined above for this category of darkino, therefore, we see that $M_H(kT_X \sim m_X c^2)$ in equation (3) roughly equals M_{eq} in equation (5), as stated above. Since $M_J^{max} \sim M_H(kT_X \sim m_X c^2)$, according to our previous discussion, we can further state that the maximum effective Jeans mass for these darkinos is just the mass which fills the horizon when matter first dominates,

$$M_J^{max} \sim M_{eq}. \qquad (6)$$

In order to see what nonlinear structures will emerge in this case, we must consider the shape of the perturbation spectrum. It is customary to assume that the primordial perturbation power spectrum was a scale-free power-law over many decades of wavenumber $k = 2\pi/\lambda$, where λ is the comoving wavelength, according to

$$|\delta_k|^2 \propto k^n, \quad -3 < n \leq 4 \qquad (7)$$

where δ_k is the Fourier transform of the density perturbation $(\delta\rho/\rho)$ and n is an integer (cf. Peebles 1980). In that case, the density enhancement measured at the moment that a particular wavelength fills the horizon can be written as

$$(\delta\rho/\rho)_H \propto M^{-(n-1)/6}, \qquad (8)$$

where M is the mass scale encompassed by that wavelength. The case of n = 1, sometimes referred to as the Zel'dovich spectrum, is special in that it implies that $(\delta\rho/\rho)_H$ is independent of scale and, moreover, that space curvature fluctuations accompanying $\delta\rho/\rho$ diverge only as $\log \lambda$ at large and small scales, eliminating the need to specify the cutoffs to the power-law in equation (7).

Damping eliminates power in the darkino perturbation spectrum in this case for all scales $M < M_{eq}$, while the spectrum on scales $M > M_{eq}$ retains its primordial shape, transformed to

$$\delta\rho/\rho \propto M^{-(n+3)/6} \quad (M > M_{eq}), \tag{9}$$

at any given time for all those scales M which have already entered the horizon. The end result of this kind of spectrum is the following: Nonlinear structure develops first on characteristic mass scales $M_{eq} \geq 10^{15} M_\odot$, the mass of a supercluster. These are the pancake masses referred to earlier, corresponding to comoving length scales of 10's of Mpc as measured at the present. Smaller structures like galaxies must result later from the fragmentation of the large-scale structure. This description applies to all values of n > -3. Since all scales smaller than M_{eq} are damped and M_{eq}, therefore, uniquely characterizes the scale of the nonlinear structure which emerges, we shall refer to this case as the "large-scale damping" case ("LSD").

(b) Intermediate-Scale Damping ("ISD")

A different situation arises if the darkinos decouple while relativistic but are particles which interact even more weakly than do neutrinos. In particular, if these particles decouple at a temperature $T_{X,d}$ greater than the temperature at which the quark-hadron phase transition is believed to occur ($T_{QH} \sim 200$ MeV), then there will be a relatively large number of relativistic species present which will annihilate some time after T drops below $T_{X,d}$. This will result in the heating of the radiation without increasing the number density of the weakly interacting darkinos. If we let $g_*(T_{X,d})$ be the total effective number of degrees of freedom of all relativistic species in thermal equilibrium at darkino decoupling, then the quantity α in equation (2) will be roughly given by $\alpha \sim [g_*(T_{X,d})]^{-1/3} \ll 1$ for $g_*(T_{X,d}) \gg 1$. The current standard model of particle physics, together with the simplest grand unified theories [e.g. minimal SU(5)], predict that $g_*(T_{X,d}) \sim 10^2$ over an enormous range of values of $T_{X,d}$ roughly between T_{QH} and $T_{GUT} \sim 10^{15}$ GeV. Current models of supersymmetry increase g_* only by a factor of order unity. According to equation (2), therefore, the mass per particle which allows these darkinos to make $\Omega \approx 1$ is correspondingly larger than the value found in the LSD case above by a factor ≥ 10. For these darkinos, it can be shown, $(g_X m_X) \sim (\Omega h^2 \theta^{-3})$ keV. The value of M_J^{max} thus determined for these particles by equation (3), on the other hand, is orders of magnitude smaller than that for the LSD particles. Apparently, unlike the LSD case, the transition from relativistic to nonrelativistic particle velocities for these darkinos occurs significantly earlier than the transition from radiation to

matter domination. A more careful treatment than that which led to
equation (3), in fact, gives $M_J^{max} \lesssim 10^{12}$ $M_\odot \ll M_{eq}$ if the darkinos
which make $\Omega \approx 1$ have $m_X \sim 1$ keV. Since this is substantially smaller
than the damping scale for LSD particles, we shall refer to this case
as that of "intermediate-scale damping" ("ISD").

One suggestion of a candidate for the ISD particles was that of
1 keV gravitinos, the spin 3/2 supersymmetric partner of the graviton
(Pagels and Primack 1982). The gravitino mass is related to the supersymmetry breaking mass scale M_{ss} according to $m_X \sim (M_{ss}^2/M_{P\ell})$, however.
Hence, this suggestion depended upon the assumption that $M_{ss} \sim 10^6$ GeV,
a value which is no longer favored by current theories of supersymmetry
(e.g. $M_{ss} \sim 10^{11}$ GeV and $m_X \sim 10^2$ GeV for the gravitino are now
fashionable). In another suggestion, the darkino was a photino, the
spin 1/2 superpartner of the photon, with mass $m_X \lesssim 500$ eV and $T_{X_d} \sim$
200 MeV (Sciama 1983).   Once again, the currently popular models of
supersymmetry suggest that photinos are stable as the lightest R-odd
particle and may very well contribute $\Omega \approx 1$, but with $m_X \gtrsim 1$ GeV,
instead (e.g. Weinberg 1983; Goldberg 1983). As such, these photinos
would not be ISD particles, as we will discuss below. Given the uncertainties that presently accompany the particle physics theories, however,
one should consider the question of candidates for the ISD darkinos as
still open.

In the meantime, the shape of the perturbation spectrum which
results from the effects of damping on the primordial spectrum is
quite different in the ISD case from that in the LSD case described
above. Once again we have power damped away for $M < M_J^{max}$, except that
M_J^{max} now corresponds to the scale of a large galaxy, rather than a
supercluster. Once again, for $M > M_{eq}$, we find the primordial shape

$$(\delta\rho/\rho) \propto M^{-(n+3)/6} \quad (M > M_{eq}) \tag{10}$$

as in equation (9). The scales which come across the horizon between
M_J^{max} and M_{eq}, however, find their growth suppressed by the fact that
radiation still dominates the total energy density. As a result, the
spectrum on these scales is flattened relative to the primordial shape
in equation (10) to

$$(\delta\rho/\rho) \propto M^{-(n'+3)/6} \quad (M_J^{max} \lesssim M \lesssim M_{eq}), \tag{11}$$

where $n' \cong n-3$. This means that power tends to be distributed much
more broadly between the scales 10^{12} $M_\odot$ and 10^{15} $M_\odot$, roughly. Of course,
for $n \sim 4$, the damping mass $M_J^{max} \lesssim 10^{12}$ $M_\odot$ still defines the characteristic scale of the first nonlinear structure to form, and the conven-
tional pancake scenario still adequately describes the formation of
structure in this case. For $n \sim 0$, however, equation (11) indicates
that nonlinear structure may develop almost simultaneously on all scales
between M_J^{max} and M_{eq}. In this case, the first structures to form may
look like pancakes on a variety of mass scales, or the situation may be
more complicated.

(c) No Damping ("ND")

    Some candidate collisionless particles have been suggested which
decouple when they are nonrelativistic. These particles are either
much more massive than 1 keV or else, while not describable by a thermal
distribution, are nevertheless quite "cold" in the sense of having very
small velocity dispersion at all times. An example of the former is
the stable photino of mass $m_X > 1$ GeV referred to above, which could
annihilate almost completely at high temperatures, leaving just enough
in number density to dominate the total mass density of the universe
(e.g. Weinberg 1983; Goldberg 1983). An example of a "cold" particle
in the latter category is the axion, a spin zero Pseudo-Goldstone boson
which has been proposed to explain the absence of CP violation in strong
interactions within the context of the standard model of particle
physics, quantum chromodynamics. It has been suggested that, if they
exist, then when the universe cools below roughly 1 GeV, axions may
comprise a gas of collisionless, nonrelativistic particles with a mass
$m_X \gtrsim 10^{-5}$ eV and a number density sufficient to close the universe,
interacting only by gravity (e.g. Preskill, Wise, and Wilczek 1983).
These axions would have essentially zero momentum.

    For darkinos of this type, the collisionless damping processes
referred to above for LSD and ISD particles are essentially irrelevant
on all scales of cosmological interest, since the free-streaming length
is so small. Hence, this case is referred to here as that of "no
damping" ("ND").

    The nonlinear structure which emerges in this case is not yet well
understood. Just as for the LSD and ISD cases, mass scales $M > M_{eq}$ in
the ND darkino perturbation spectrum still have $(\delta\rho/\rho) \propto M^{-(n+3)/6}$, once
they come within the horizon. However, even without damping, all scales
which enter the horizon before M_{eq} cannot grow very rapidly until
$M_H = M_{eq}$, since radiation dominates the total energy density until then.
The result is a <u>flattening</u> of the primordial spectrum for all scales
$M < M_{eq}$. For scales corresponding to $\lambda \lesssim 1$ Mpc, in fact, we can write
crudely that

$$(\delta\rho/\rho) \propto M^{-(n'+3)/6} \quad (M \lesssim 10^{11-12} M_\odot), \qquad (12)$$

where $n' = n-4$, according to the fit to detailed, numerical results
reported by Peebles (1982). According to equation (12), the Zel'dovich
($n = 1$) spectrum yields a flat dependence of $(\delta\rho/\rho)$ on mass scale for
these short wavelengths. This means that fluctuations on all such
scales would grow to nonlinearity at roughly the same time. Actually,
the smaller scale fluctuations may reach nonlinearity somewhat earlier
than the larger scale ones. This is discussed in more detail by Peebles
(1983, and elsewhere in these proceedings). In any case, the important
thing to note is that, unlike the LSD and ISD cases, power is ultimately
distributed quite broadly over all mass scales below the bend in the
spectrum at $M_{eq} \sim 10^{15} M_\odot$.

The nonlinear structure which emerges in the ND universe may, therefore, be quite unlike that in the standard pancake scenario. If small scale fluctuations reach nonlinearity first, then the minimum scale is determined by the fact that baryons will not be able to cluster in units much smaller than their post-recombination Jeans mass, ~ 10^6 $M_\odot$. Such small mass units might then function as the building blocks in a kind of hierarchical clustering model. It might be difficult for such a model to account for the observed largest-scale structure in the present spatial distribution of galaxies. However, a fully nonlinear simulation of this ND case, even ignoring the baryon component, has yet to be performed with a combination of resolution and dynamic range which is adequate to follow the growth of structure over the full range from the smallest to the largest scales (e.g. ~ 10^6 $M_\odot$ up to $\geq 10^{16}$ $M_\odot$) of relevance in the problem.

III. THE NEUTRINO-DOMINATED UNIVERSE

We focus in this section on the LSD case, embodied by the massive neutrino-dominated universe. This will serve as our prototype for the pancake scenario in a universe dominated by arbitrary darkinos. The central question before us is that of whether galaxies as we observe them today can form in such a universe, within the context of the pancake scenario. We will reduce this question to two, much simpler questions, as follows: (1) What is the structure which evolves in such a universe? (2) Can the shock-heated baryon-electron component radiatively cool in time to permit fragmentation and galaxy formation? To answer the first question, it is necessary to perform three dimensional (3-D) collisionless particle numerical simulations of the growth of the appropriate linear perturbation spectrum into the non-linear stage. To answer the second, we must perform detailed, numerical hydrodynamical calculations of the coupled growth of the collisional baryon plus collisionless neutrino perturbations into the nonlinear stage, including the effects of ionization, recombination, radiative and Compton cooling, and thermal conduction. Fortunately, as we shall see, the answer to question (1) above indicates that we can reduce the great complexity of this hydro-dynamical calculation by limiting our attention to one-dimensional (1-D), plane-symmetry.

(a) 1-D Pancakes from 3-D Perturbations

Several calculations have been performed to date of the 3-D, collisionless, gravitational clustering of the neutrinos in the post-recombination Friedmann universe, starting from a perturbation spectrum with the appropriate short wavelength damping cutoff (Shapiro, Struck-Marcell, and Melott 1983; Centrella and Melott 1983; Klypin and Shandarin 1983; Frenk, White, and Davis 1983). I will briefly summarize the results of the simulation presented in Shapiro, Struck-Marcell, and Melott (1983). This utilized the so-called cloud-in-cell method. The equations were solved on an Eulerian grid of cubic cells, with the total grid comprising 32 cells per side, or 32^3 cells in all, with one particle per cell. The initial condition at $z = 10^3$ was taken to be that of a spectrum of density perturbations given by

$$\vec{S}(\vec{q}) = 2 \times 10^{-3} \sum_{i=0}^{4} \sum_{j=0}^{4} \sum_{k=0}^{4} \left(\frac{i^2+j^2+k^2}{16}\right)^{n/4 - 1/2}$$

$$X \frac{\vec{k}_{ijk}}{|\vec{k}_{ijk}|} \sin (\vec{k}_{ijk} \cdot \vec{q} + \psi_{ijk}) \tag{13}$$

where $\vec{S}(\vec{q})$ is the vector displacement of every particle from its
unperturbed position $\vec{q}$, the initial particle velocities were assumed
to be unperturbed, the quantity $\vec{k}_{ijk} \equiv (2\pi/32d)(\sigma_1 i\hat{x} + \sigma_2 j\hat{y} + \sigma_3 k\hat{z})$, d
is the length of a side of a cubic cell, $\sigma_{1,2,3} = \pm 1$ (with the sign
chosen randomly), and the phases ψ_{ijk} are equal to 2π multiplied by a
random number between 0 and 1. A Zel'dovich primordial spectrum (n =
1) was assumed. The short-wavelength cutoff corresponds to eight cell
widths. Periodic boundary conditions were imposed.

Figure 1 shows the spatial distribution and velocity directions
of the collisionless particles for a sequence of contiguous spatial
slices at a single redshift, z = 4, after the appearance of nonlinear
structure. Each slice is one cell-width thick. Each arrow represents
one particle, the size of each arrow points in the direction of that
particle's velocity in the plane of the slice. Accordingly, any
density enhancements in Figure 1 which extend along a line in one of
the planar slices and which also appear simultaneously in several
adjacent slices can be interpreted as sheets, or pancakes, seen edge-on.
Figure 1 clearly demonstrates that such linear structures are prominent
and do repeat from one slice to the next. In fact, the most prominent
pancake in Figure 1 extends throughout roughly eight slices (i.e. one
damping length), including the four adjacent slices not shown. Space
slices taken at earlier and later redshifts (e.g. z = 7 and z = 2)
exhibit the same pancakes, with density enhancement increasing with
decreasing z. Furthermore, as Figure 1 shows, the velocity arrows of
the particles comprising the pancakes tend to point toward the plane
of the pancakes, along the direction normal to this plane.

These results indicate that 1-D pancakes are a prominent and
generic form for the nonlinear structure in the growth of a fully 3-D
initial spectrum of density fluctuations. There are other forms, too,
of course. Higher-order singularities are indicated, as well,
apparently at the intersections of pancakes. If we think of the overall structure as cellular, in fact, then the pancakes represent the
cell walls. Eventually, larger density enhancements than those at the
centers of the pancakes can be found in the form of filaments extending
along the lines of intersection of two pancakes. Larger density
enhancements still are the knots that eventually form at the vertices
where more than two pancakes intersect. Near the edges of pancakes,
therefore, particles eventually tend to move along the pancake planes,
towards the filaments and knots. At no time, in fact, do individual
pancakes collapse (relative to the uniform Hubble expansion) along
their planes. Hence, it would not, for instance, be correct to view

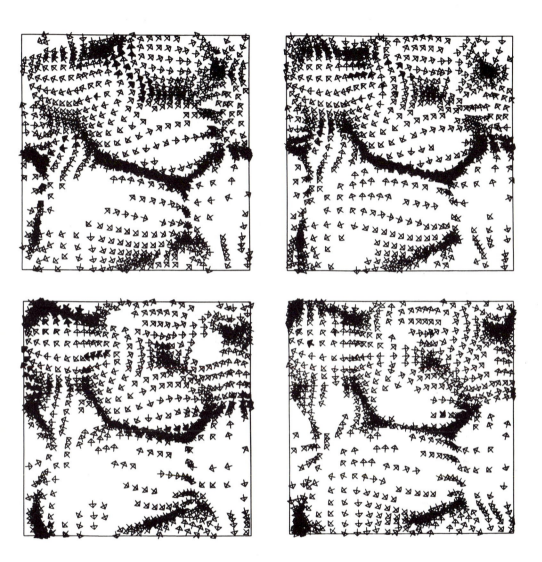

Fig. 1 - Four consecutive two-dimensional space-slices of the 3-D distribution of neutrinos at z = 4 in the evolution of the 3-D perturbation spectrum. Each arrow represents one particle in the simulation and points in the direction of the component of the neutrino velocity in the plane of the page. The order of the space-slices is upper left, upper right, lower left, lower right.

pancakes, as has been done in the past, as isolated, oblate, homogeneous
ellipsoids which are collapsing simultaneously in <u>all</u> directions.

These results are consistent with the expectations generated by
earlier work, ranging from the discussions of the pancake instability
by Lin, Mestel, and Shu (1965), to the first suggestion of the pancake
scenario by Zel'dovich (1970), to the discussion of the singular
geometries predicted for a pressure-free universe using catastrophe
theory by Arnold, Shandarin, and Zel'dovich (1982), to the two-
dimensional, detailed cloud-in-cell simulations (Doroshkevich <u>et al</u>.
1980; Melott 1983). Some of the 3-D simulations mentioned above,
those by Frenk, White, and Davis (1983) and by Klypin and Shandarin
(1983), were unable to see the pancakes described above, although they
did see the filaments and knots. The N-body approach of the former
simply did not have a dynamic range adequate to show the sheets, whose
density contrast is lower than that of the filaments and knots. The
cloud-in-cell approach of the latter, on the other hand, with somewhat
lower resolution than that shown in Figure 1 was simply not displayed
as in Figure 1 in order to reveal the prominence of the sheets. In
the meantime, however, Centrella and Melott (1983) have performed a
much larger cloud-in-cell calculation, with 27 particles per cell and
32 cells per side, or a total of $\sim 9 \times 10^5$ particles. The results of
their much improved dynamic range supports the conclusions described
above concerning the appearance of pancakes as a prominent and generic
form for the nonlinear structure.

(b) The Growth of Coupled (Baryon) + (Neutrino) Pancakes

We shall henceforth limit our attention to the 1-D, plane-
symmetric pancakes described above. Ultimately, the collisionless
simulations cannot tell us how the baryonic component forms galaxies.
To understand this we must couple the collisionless neutrino component
to the dissipative baryon-electron gas.

In the discussion which follows, I will summarize the detailed,
numerical calculations presented in Shapiro, Struck-Marcell, and Melott
(1983). This calculation follows the growth of the coupled baryon-
neutrino pancake by solving the hydrodynamical conservation equations
for the baryons simultaneously with the Poisson equation for the combined baryon-neutrino gravitational potential and the Vlasov equation
for the collisionless neutrinos, within a homogeneous, isotropically
expanding universe described by the Robertson-Walker metric. The
hydrodynamical equations are finite-differenced by a Lagrangian scheme,
while the neutrinos are followed by an N-body approach, both of which
are described more fully in Shapiro and Struck-Marcell (1983a,b). [A
similar calculation, using an Eulerian scheme to difference the hydrodynamical equations, instead, has been described by Bond <u>et al</u>. (1983).]
The effects of radiative cooling due to thermal bremsstrahlung,
radiative recombination, dielectronic recombination, and electron
collisional line excitation in an optically thin gas of primordial
composition comprised of H and He, with the abundance of He atoms equal

to 10% of that of H by number have been explicitly included. This
necessitates the solution of the ionization balance equations for H
and He simultaneously with the other flow equations. We have assumed
that the ionization balance is given by a collisional ionization
equilibrium (i.e. the coronal approximation). We have also included
classical, electron thermal conduction, as corrected for the effects
of saturation. Compton cooling by the cosmic background radiation
photons is included in the cases to be described in the next section,
but is unimportant for the neutrino case described here.

As described in § II, the neutrino perturbation spectrum is well
approximated, for all reasonable initial spectra, by one in which
almost all of the power is concentrated at wavelengths near the critical
damping wavelength λ_{min}. Peebles (1982a) gives λ_{min} as
12.8 α^4 β k^{-1} $\theta^2/(\Omega_\nu h^2)$ Mpc, where k is a dimensionless wavenumber
whose value lies somewhere roughly between 0.25 and 0.5 for the
parameters assumed here. If we take $\theta = 1$, $h = 0.75$, $\alpha = (4/11)^{1/3}$,
$\Omega_\nu = 0.9$, and $\beta = 2$ (e.g. one massive and two massless neutrino types,
each with two spin states), then equation (1) gives $m_\nu \approx 49$ eV for the
massive neutrino, while λ_{min} is between 20 and 50 Mpc. We have assumed,
therefore, that the initial condition for the neutrino density at
$z = 10^3$ is given by a 1-D, plane-symmetric, sinusoidal perturbation at
a single dimensionless wavelength $\tilde{\lambda} = \lambda_{min}/r_0 = 2\pi/\tilde{k}$, according to the
following:

$$\rho_\nu = \bar{\rho}_\nu [1 - \delta \cos (\tilde{k}\vec{\tilde{q}}_\nu \cdot \hat{x})]^{-1} \qquad (14)$$

where $\bar{\rho}_\nu$ is the average neutrino density, $\vec{\tilde{q}}_\nu$ is the unperturbed
Lagrangian coordinate of the neutrinos (in units of r_0), and r_0 is a
constant with dimensions of length. This density corresponds to a
dimensionless displacement function

$$\vec{S}(\vec{\tilde{q}}_\nu) = -[\delta/\tilde{k}][\sin(\tilde{k}\vec{\tilde{q}}_\nu \cdot \hat{x})]\hat{x}, \qquad (15)$$

where δ is the amplitude of the perturbation at $z_i = 10^3$. The peculiar
velocity field (i.e. excluding the average, Hubble expansion) of the
neutrinos at this time is given by

$$\vec{v}_\nu(\vec{\tilde{q}}_\nu) = 2(1+z_i)^{1/2} \vec{S}(\vec{\tilde{q}}_\nu) \qquad (16)$$

[in units of $(r_0 H_0)/2$], as it would be in the exact solution of the
neutrino equation of motion for a pure neutrino, matter-dominated
Einstein-de Sitter universe with a density perturbation at z_i given by
equation (14). We can safely ignore the small effects of finite
temperature for the neutrinos.

The baryons at z_i are assumed to be unperturbed [i.e. $\rho_b = \bar{\rho}_b$,
$v_b = 0$, $T_b = 3000$ K], which is consistent with the effects of radiative
damping prior to and during the recombination era and with the limits
imposed by the microwave background anisotropy measurements. We have
chosen a "canonical" set of parameters for the neutrino-dominated

universe, with $\Omega = 1$, $\Omega_\nu = 0.9$, $\Omega_b = 0.1$, and $\lambda_{min} = 40$ Mpc (i.e. $m_\nu \approx$ 49 eV, as mentioned above). This maximizes Ω_b within the limits set by the primordial nucleosynthesis arguments and, as we shall see, thereby maximizes the likelihood of success in forming galaxies. We have set the amplitude δ at 7.8×10^{-3}, which makes the neutrino caustic and baryon shocks form at $z_c = 6$ [i.e. $(1+z_c) \cong (1+z_i) \Omega_\nu \delta$].

The results of this calculation are briefly summarized in Figures 2 and 3. The quantities plotted are dimensionless, comoving baryon and neutrino densities $\tilde{\rho}_{b,\nu} = (\rho_{b,\nu}/\rho_{crit})(1+z)^{-3}$, baryon temperature T, dimensionless Eulerian position $\tilde{x} = (x/r_0)(1+z)$, and dimensionless, peculiar, baryon and neutrino velocities [i.e. actual velocity $\vec{u} = H(t) \cdot \vec{x} + \vec{v}$, where $H(t) = \dot{a}/a$, $a = (1+z)^{-1}$, $\vec{v} = $ peculiar velocity] $\tilde{v}_{b,\nu} = (2v_{b,\nu}/r_0 H_0) \cdot (1+z)^{-1}$, where $r_0 = 20$ Mpc. These plots are of the case in which thermal conduction is ignored, as it should be if it is suppressed by even a weak magnetic field. Later, I will describe how our runs which include conductivity differ.

At a redshift $10^3 > z \gg 10$, earlier than is shown in the figures, during the linear stage of neutrino perturbation growth, the baryons "catch up" to the neutrino perturbation. From this time until $z_c \gtrsim 6$, the neutrinos and baryons have virtually identical velocity and density profiles, with the neutrino density everywhere (Ω_ν/Ω_b) times the baryon density. During this period, the collapse of the baryons toward the central plane (at $\tilde{x} = 0$) is virtually adiabatic and is superposed on the general adiabatic expansion. Hence, while the temperature and density increase adiabatically near the center, they actually decrease across much of the perturbation away from the central plane.

Eventually, at $z = 6$, a single density spike forms in the neutrino distribution at $\tilde{x} = 0$, as expected from our previous discussion of caustic formation. The highly nonlinear, virtually adiabatic compression of the baryons near the center, however, succeeds in raising pressure forces near $\tilde{x} = 0$ to a level capable of offsetting gravity. The result is the formation of a strong shock just outside of the central plane (i.e. one shock on other side of the plane, symmetrically placed, but we henceforth limit our discussion to the half-space $\tilde{x} \geq 0$). This shock, it can be shown, forms very close to the central plane, and only a very tiny mass fraction ($\tilde{q} \lesssim 2.5 \times 10^{-3}$) is interior to the shock at that moment. Outside the shock, the infall is highly supersonic.

Figures 2a and 2b show the density, velocity, and temperature profiles of the baryons and the density and velocity profiles of the neutrinos at $z = 5.9$, shortly after the shock and caustic form. The neutrino velocity plot shows the first neutrino counterstream emerging from the other side of the central plane as it moves back out through the rest of the infalling neutrinos. A strong, adiabatic shock is in evidence, in the meantime, in all three of the baryon plots in Figures 2a and 2b, as it overtakes an ever-increasing mass fraction (~ 10% by this point), staying strikingly close to the position of the first neutrino counterstream. The innermost mass fraction ($\tilde{q} \sim 2.5 \times 10^{-2}$)

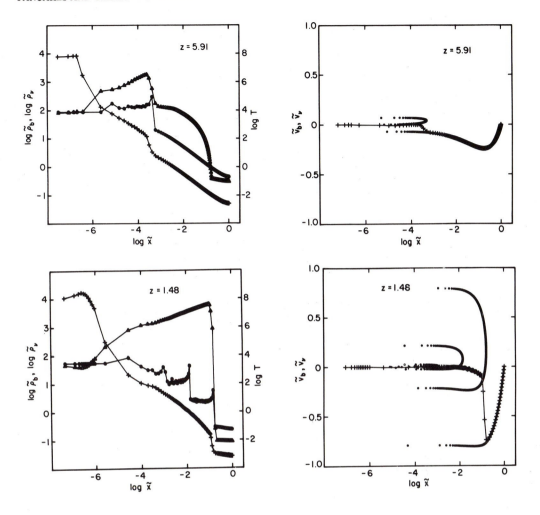

Fig. 2 – (a) The dimensionless, comoving baryong density $\tilde{\rho}_b$ (plus signs)
and neutrino density $\tilde{\rho}_\nu$ (circles), and the baryon temperature (triangles)
versus the logarithm of the Eulerian position $\tilde{x}$ at z = 5.9. Each symbol
marks the position of a zone center. There are 200 baryon zones at this
time step. Note that the "bulge" in the temperature profile outside the
shock is the artifact of artificial viscosity. This has a negligible
effect on the dynamics of the shock and of the postshock flow. (b) The
dimensionless, comoving baryon velocity $\tilde{v}_b$ (plus signs) and neutrino
velocity $\tilde{v}_\nu$ (dots) versus Eulerian position $\tilde{x}$ at z = 5.9. Note that
each dot marks the position of a single neutrino sheet, except for
positions $\tilde{x}$ to the right of the first neutrino counterstream (i.e. the
counterstream with the greatest extent along the $\tilde{x}$-axis), where only a
sample of the sheets is plotted. (c) Same as Fig. 2a, except that z =
1.5 and there are now 100 baryon zones. (d) Same as Fig. 2b, except
that z = 1.5 and there are now 100 baryon zones.

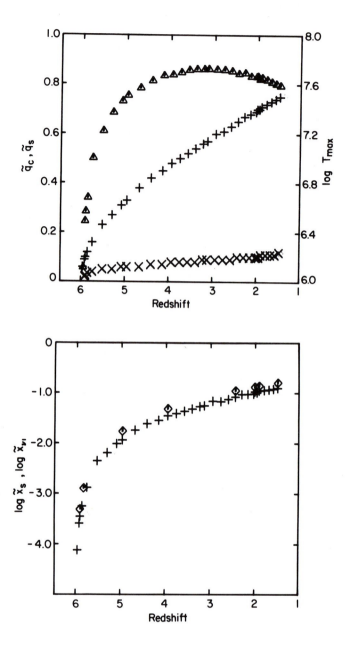

Fig. 3 - (a) The baryon mass fraction interior to the shock, $\tilde{q}_s$, or equivalently, the Lagrangian position of the shock (plus signs), the radiatively cooled mass fraction, $\tilde{q}_c$ (crosses), and the postshock temperature (triangles) versus redshift z. (b) The Eulerian positions $\tilde{x}_s$ (plus signs) and $\tilde{x}_{\nu,1}$ (diamonds) of the shock and the leading edge of the first neutrino counterstream (i.e. the counterstream with the greatest extent along the $\tilde{x}$-axis), respectively, versus redshift.

was shock-heated to a low enough temperature ($T \lesssim 3 \times 10^5$ K) that radiative cooling was extremely rapid. Even in the short time between $z = 6$ and $z = 5.9$, therefore, this mass fraction has succeeded in radiatively cooling and recombining down to $T < 10^4$ K. The post-shock temperature is an increasing function of time, however, and the cooling rate declines sharply with temperature once $T \gtrsim 3 \times 10^5$ K and the plasma is fully ionized, leaving only thermal bremsstrahlung to provide the cooling. Hence, the mass-fraction which is shock-heated after the innermost few percent takes considerably longer to cool, with the post-shock temperature already as high as ~ 3×10^6 K by $z = 5.9$. In the meantime, the post-shock pressure is kept quite uniform throughout the region behind the shock. Hence, the gas which cools does so nearly isobarically, resulting in a very large compression as it cools, as indicated by the large, central density enhancement of the innermost cool zones relative to the hotter post-shock zones in Figure 2a.

As Fig. 3 indicates, the post-shock temperature eventually climbs to $T \sim 5 \times 10^7$ K. By this point, a large fraction of the baryons have been shocked to a temperature which is too high to allow cooling within the age of the universe. As Figures 2c and 2d indicate for a redshift as late as $z = 1.5$, only the inner ~ 12% of the gas has succeeded in radiatively cooling, thereby forming a cold, thin, dense layer at the origin. By this time, more than 75% of the baryonic mass is interior to the shock, and the shock is still advancing (eventually, it will overtake ~ 90% of the baryons).

In the meantime, the neutrino density and velocity plots (Figures 2c and 2d) clearly show that three counterstreams have developed in the neutrinos, with a fourth about to emerge. This behavior of the collisionless neutrinos as they "wind up" the spiral in phase space in the process of violent relaxation is similar to that found earlier in 1-D calculations of a purely collisionless gas (with no baryons included) in the expanding universe, starting from similar initial conditions (e.g. Doroshkevich et al. 1980; Melott 1983a). The baryons, it turns out, tend to increase somewhat the rate at which new windings appear in the neutrino phase space plot compared with a purely collisionless calculation, by enhancing the potential very near the central plane where the baryons dominate the total mass density. The effect, which is relatively minor here, is quite pronounced in other runs of ours in which the ratio of Ω_ν/Ω_b is substantially higher.

As depicted in Figure 3, the principal conclusion to be drawn from this calculation may be stated as follows: For a canonical case with $\Omega_\nu = 0.9$, $\Omega_b = 0.1$, $h = 0.75$, a characteristic neutrino perturbation scale length of 40 Mpc, and a perturbation amplitude large enough to bring about shock and caustic formation by $z = 6$, only $\approx 12\%$ of the baryons form, by $z \sim 1$, a cold, thin, dense layer capable of fragmenting and forming galaxies. This implies that, at most, $\Omega_{galaxies} \sim 0.01$ in luminous matter, which is marginally consistent with observation. Most of the baryon gas (~ 80% by the present epoch) is left behind as a hot ($10^6 \lesssim T \lesssim 5 \times 10^7$ K), diffuse, intergalactic gas with $\Omega_{IGM} \sim 0.1$, which occupies ~ 10% of the total volume and does not violate any of the familiar constraints on such an intergalactic medium. The remaining baryons fill most of the volume at very low density and are never overtaken by the pancake shocks.

The effect of thermal conductivity is depicted in Figure 4 shown in the next section. Its primary effect is to halt the increase in mass-fraction which cools and joins the thin, dense central layer at some intermediate redshift. The net result is that only ~ 6% of the baryon gas cools by z ~ 1 in this case, rather than ~ 12% as described above. The important point to be made is that the gas which is able to cool and compress into the central layer cannot subsequently be evaporated away. Once cold, this gas stays cold, since, with its high density and T < 10^4 K, it is very efficient at radiating away the heat conducted into it by the adjacent, hot, shocked gas. Small increases in T in the vicinity of 10^4 K, that is, result in orders of magnitude increase in the radiative cooling rate, as a result of H Lyman line collisional excitation and the increase in free electron density. This is a barrier which the saturated conduction cannot surmount. The conduction saturates in this case because the temperature scale height $T/|\vec{\nabla}T|$ at the interface between the cold, central layer and the hot, post-shock gas is smaller than the electron collision mean free path.

IV. PANCAKES IN GENERAL: ANALYTICAL APPROXIMATIONS AND NUMERICAL RESULTS

In order to anticipate and understand the results for the growth of pancakes of different sizes, collapse epochs, baryon densities, and the like, as may emerge in the cases other than the neutrino-dominated universe as well as the neutrino-dominated case, itself, we briefly summarize an approximate analytical description of this pancake growth. The following results are quoted in part in Shapiro, Struck-Marcell, and Melott (1983) and described in full detail in Shapiro and Struck-Marcell (1983b). Our detailed numerical results justify the following assumptions: (1) the pre-shock gas is approximately pressure-free and adiabatic; (2) the shock is very strong and adiabatic, so the adiabatic, strong shock jump conditions apply; (3) except at very late times, the pressure interior to the shock is quite uniform at any given time; (4) the shock position is always very close to the origin. With these assumptions, it is possible to derive the position of the shock as a function of time, the immediate post-shock values of the dependent gas variables, and the run of these variables throughout the post-shock flow. I will simply quote a few of the results.

The Eulerian position of the accretion shock is approximately given by

$$\tilde{x}_s \cong (\pi^2/36)\tilde{q}_s^{\,3} \qquad\qquad (\tilde{q}_s < 1), \qquad\qquad (17)$$

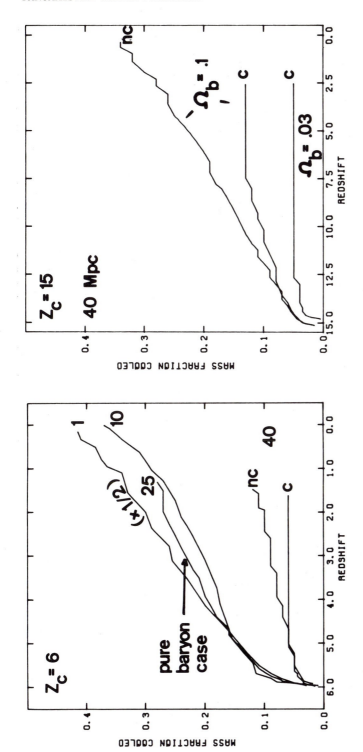

Fig. 4 – (a) The mass fraction of the baryon-electron gas which radiatively cools to $T < 10^4$ K after passing thru the pancake shocks is plotted against redshift for a variety of cases. The numbers labeling the curves refer to the present comoving value of the pancake scalelength in Mpc. The labels "c" and "nc" mean "with" and "without" thermal conduction, respectively. The absence of either label means that conduction was unimportant for that case. The curve labeled "(x 1/2)" has been divided by 2. All curves assume $\Omega_{total} = 1$ and $z_c = z(caustic) = 6$. The curve labeled "pure baryon case" has no collisionless particles included. The other curves assume that $\Omega_b = 0.1$ and, hence, $\Omega(collisionless) = 0.9$. (b) Same as Fig. (a), except that $z_c = z(caustic) = 15$ and $\lambda = 40$ Mpc for all cases shown. Ω_b is as labeled.

in units of $\lambda/2$, where $\tilde{q}_s$ is the mass fraction interior to the shock. Hence, $\tilde{x}_s \approx 0$ is a very good approximation thru much of the pancake growth after the shock forms.

The post-shock temperature is given by

$$T_2(\tilde{q}_s) \cong 1.8 \times 10^7 (h/0.75)^2 (\lambda/40 \text{ Mpc})^2 [(1+z_c)/7]\tilde{q}_s \sin \pi\tilde{q}_s \text{ °K}, \quad (18)$$

which, in the limit of $\tilde{q}_s \ll 1$, gives

$$\tilde{q}_s(\leq T_2) = 0.05 \ (T_2/10^6 \text{K})^{1/2} (h/0.75)^{-1} (\lambda/40 \text{ Mpc})^{-1} [(1+z_c)/7]^{-1/2}$$

$$(\tilde{q}_s \ll 1), \quad (19)$$

where $\tilde{q}_s(\leq T_2)$ is the mass fraction shocked to a temperature $\leq T_2$. Hence for example, if the pancake size is changed from λ to $\lambda' = g\lambda$, then $\tilde{q}_s(\leq T_2)$ varies roughly _inversely_ with g.

The post-shock pressure is given approximately by

$$P_2(\tilde{q}_s) \cong 4.8 \times 10^{-16} (\Omega_b/0.1) (h/0.75)^4 [(1+z_c)/7]^{-1} (\lambda/40 \text{ Mpc})^2 f(\tilde{q}_s) \ \cdot$$

$$(1+z)^5 \text{ erg cm}^{-3}, \quad (20)$$

where

$$f(\tilde{q}_s) = (\pi\tilde{q}_s)^2/\{[\sin(\pi\tilde{q}_s)/(\pi\tilde{q}_s)] - \cos(\pi\tilde{q}_s)\}, \quad (21)$$

a slowly varying function of $\tilde{q}_s$. In the limit of $\tilde{q}_s \ll 1$, this becomes

$$(nT)_2 \approx 10(\Omega_b/0.1) [(1+z_c)/7]^{-1} (\lambda/40 \text{ Mpc})^2 (1+z)^5 \text{°K cm}^{-3}$$

$$(\tilde{q}_s \ll 1), \quad (22)$$

where n is the particle number density, and where the redshift at which a mass fraction $\tilde{q}_s$ is shock-heated is given roughly by

$$(1+z_s) \cong (1+z_c) [\sin(\pi\tilde{q}_s)/(\pi\tilde{q}_s)] \quad (23a)$$

$$\sim (1+z_c) \quad (\tilde{q}_s \ll 1). \quad (23b)$$

The post-shock baryon mass density is given by

$$\rho_2(\tilde{q}_s) \cong 1.5 \times 10^{-27} (\Omega_b/0.1) (h/0.75)^2 [(1+z_c)/7]^3 F(\tilde{q}_s) \text{ g cm}^{-3} \quad (24)$$

where

$$F(\tilde{q}_s) = [\sin(\pi\tilde{q}_s)/(\pi\tilde{q}_s)]^4/\{[\sin(\pi\tilde{q}_s)/(\pi\tilde{q}_s)] - \cos(\pi\tilde{q}_s)\}. \quad (25)$$

In the limit of $\tilde{q}_s \ll 1$, this becomes

$$\rho_2(\tilde{q}_s) \approx 4.6 \times 10^{-28}(\Omega_b/0.1)(h/0.75)^2[(1+z_c)/7]^3\tilde{q}_s^{-2} \ g \ cm^{-3} \qquad (26)$$

We can use these expressions to get a crude idea of how the cooled baryon mass fraction $\tilde{q}_c$ scales with a change of pancake characteristics. Suppose the key to whether some shocked fluid element radiatively cools in the age of the universe is whether the thermal bremsstrahlung cooling time t_c is shorter than the Hubble time t_H. It is easy to show that the ratio $t_c/t_H \propto T_2^{1/2} (1+z_c)^{3/2} h \rho_2^{-1}$ if we further suppose that only the immediate post-shock value of this ratio is important and limit ourselves to small mass fractions interior to the shock. The fraction of the baryons that cool in the post-shock flow should correspond roughly to some critical value of t_c/t_H. Using the small $\tilde{q}_s$ limits of T_2 and ρ_2 in equations (18) and (26) and setting t_c/t_H equal to a constant, we can see that the cooled mass fraction must scale crudely according to

$$\tilde{q}_c \simeq \lambda^{-1/3} \ \Omega_b^{1/3} \ (1+z_c)^{1/6}, \qquad (27)$$

as long as $\tilde{q}_c$ is not too close to unity. Hence, decreasing the pancake size from the 40 Mpc value described in the previous section to 10 Mpc should increase $\tilde{q}_c$, measured at the same redshift, by roughly less than two, while decreasing it from 40 Mpc to 1 Mpc should increase it by more than three. On the other hand, increasing λ would have just the opposite effect. Increasing Ω_b, in the meantime, has as strong an effect as decreasing λ, but changing the collapse epoch will only weakly alter the cooled fraction.

Actually, if $z \gtrsim 10$, then inverse Compton scattering of cosmic background photons by plasma electrons dominates over thermal bremsstrahlung as the dominant cooling process. In that case, $t_c/t_H \propto (1+z)^{-4}$ for a fully ionized gas, in which case, for $\tilde{q}_s \ll 1$ again, we find that this ratio $t_c/t_H \propto h (1+z_c)^{-5/2}$, independent of λ and Ω_b! Hence, once $z_c \gtrsim 10$, the earlier the collapse occurs, the more rapid the cooling will be and the larger the cooled fraction. Moreover, the dependence of the relative importance of Compton and bremsstrahlung cooling on the assumed pancake characteristics can be estimated by holding the ratio of Compton to bremsstrahlung cooling fixed and asking how the mass fraction which cools more rapidly by one process than the other scales. We find that the mass fraction which cools more rapidly by bremsstrahlung than by Compton cooling decreases roughly as $\lambda^{-1/3} \ \Omega_b^{1/3} \ h^{1/3} \ (1+z_c)^{-2/3}$, for small values of $\tilde{q}_s$. Hence, the earlier the collapse occurs, the larger the mass fraction for which Compton cooling is more important, if it is important at all.

Lastly, a scaling law for the degree of compression which accompanies the isobaric radiative cooling can be crudely stated. In the bremsstrahlung cooling case, a critical post-shock temperature T_c can be identified, according to equations (19) and (27), corresponding to the temperature of that fluid element which just manages to cool in a Hubble time. Since this fluid element, if it cools, will cool to $T \lesssim 10^4$ K isobarically, the ratio ξ of density after cooling to density before cooling is ~ $T_c/10^4$ K, or

$$\xi \simeq h^2 \, \Omega_b^{2/3} \, (1+z_c)^{7/3} \, \lambda^{4/3}, \tag{28}$$

for $\tilde{q}_c \ll 1$. Hence, if λ is decreased from 40 Mpc to 10 Mpc and no other parameters are changed, we should expect to see a central density in the cold, central layer which is roughly an order of magnitude <u>smaller</u> relative to the mean density than in the 40 Mpc case, while if λ is decreased from 40 Mpc to 1 Mpc, the reduction in compression should be by more than two orders of magnitude. This degree of compression is significant, since the fragmentation of the cold layer will depend upon it.

We have performed detailed, numerical hydrodynamical calculations of the combined (baryon) + (darkino) pancake growth for a variety of assumed pancake sizes, collapse redshifts, and baryon densities in order to explore the effects of departures from the canonical parameter set in § III. This was intended, as well, to indicate what differences might arise in a universe dominated by darkinos in categories other than the LSD case. In the ISD case, for example, pancake scales may range from 1 Mpc to 10's of Mpc, depending upon the initial spectrum. Even in the ND case, there may be some pancaking which occurs.

Our results, which are described in detail in Shapiro and Struck-Marcell (1983a,b), are summarized in Figure 4. Figure 4 indicates the mass fraction of the baryons which cooled to $T < 10^4$ K in each case versus redshift, in rough agreement with the scaling laws sketched above. In fact, the detailed results in each case agree fairly well with the approximate solutions described above, although differences of detail occur in the cases which included thermal conductivity, since we made no attempt to include this in our analytical results. As Fig. 4a indicates, the smaller the pancake, the more efficient is radiative cooling, holding other parameters fixed. For example, the 10 Mpc pancake yields about twice the cooled fraction as the 40 Mpc pancake (without conductivity), comparing the two cases at the same redshift. The case of $\lambda = 1$ Mpc, which would correspond to a 1 keV ISD darkino and an $n = 4$ initial spectrum, for example, results in virtually <u>all</u> of the post-shock gas cooling to $T < 10^4$ K in the post-shock flow, as anticipated by our scaling laws. As the scaling laws further anticipated, the cases shown in Figure 4b of collapse at $z_c = 15$ show substantially more cooling than did their $z_c = 6$ counterparts, for the 40 Mpc case. The earlier collapse epoch, in fact, is enough to compensate for the lower baryon density in the two cases with $\Omega_b = 0.03$. We have also displayed a case with only baryons and $\lambda = 25$ Mpc, since this corresponds roughly to one of the characteristic pancake masses identified by Press and Vishniac (1980) for a pure baryon universe, although the value of $\Omega_b = 1$ which is necessary to reconcile this case with the microwave background constraints is not consistent with the nucleosynthesis limits, as discussed earlier.

V. POTENTIAL DIFFICULTIES FOR THE NEUTRINO-DOMINATED UNIVERSE AND THEIR
    RESOLUTION

I shall conclude by briefly addressing several arguments which have
been made recently challenging the viability of the LSD, or neutrino,
pancake scenario, including some of those made by Jim Peebles elsewhere
in this volume. I will then add some of my own to this list of potential
flaws in the model.

(1) The 3-D collisionless particle simulations of Frenk, White, and
Davis (1983) and Centrella and Melott (1983) indicate that the mass
distribution in the pancake model can match the observed slope and
coherence length of the galaxy two-point correlation function, but only
for a narrow range of epochs. At earlier times, the slope is too flat;
at later times, the slope is too steep. If the epoch of this match-up
is assumed to be the present, then nonlinear structure must have appeared more recently than $z \sim 2$. Hence, galaxy formation is forced to
begin at uncomfortably recent times.

I suggest the following possible resolution of this apparent
dilemma. The dilemma is based upon the assumption that the mass distributions of the dark and luminous matter are the same. However, the
dark and luminous matter may very well be segregated. Not all kinds
of segregations will help in resolving the dilemma, of course. However,
if galaxies form preferentially in the sheet-like structures that appear
in this kind of universe, rather than in the knots and filaments which
also appear, then the effect would be to flatten the galaxy distribution
correlation function relative to that of the dark matter. The sheets
are of substantially lower density contrast than the knots and filaments,
that is. The two-point correlation function scales as the square of the
local density, however. As such, it gives preferential weight to small,
dense knots over larger, less dense sheets. Hence, the mass autocorrelation function of the neutrinos might very well be steeper than the
correlation function of galaxies at present, if the galaxies which
dominate the observed correlation function formed less efficiently in
knots and filaments than in sheets.

(2) Dekel (1983) concluded on the basis of an N-body simulation
of a single, isolated pancake that, even without dissipation, pancakes
would show too great a degree of flattening in comparison with the
Local Supercluster, unless the pancake is just now collapsing. As such,
this argues against the neutrino scenario, which forms galaxies only
after this collapse occurs. The resolution of this apparent difficulty
is straightforward if we recall the results of the fully 3-D, collisionless particle simulations described in § III. These simulations, such
as those of Centrella and Melott (1983), clearly show that the interaction of the pancake structures prevents the continued flattening of
sheets. The sheets eventually expand towards the intersections with
other sheets. In any case, the description in terms of a single,
isolated pancake with no external interactions does not apply to the
structures which emerge in the neutrino-dominated universe.

(3) Peebles (1983) argues that (i) galaxies formed at $z \gtrsim 3.5$, (ii) therefore, galaxies formed when the mean density was more than 100 times its present value, but (iii) gravity tends to <u>increase</u> binding energy with time, and (iv) therefore, the fact that most galaxies are not <u>now</u> in regions with $\delta\rho/\rho > 100$ argues against the pancake scenario. Once again, I appeal to the 3-D simulations for my counterargument. These show that, while matter does <u>eventually</u> flow into higher density regions at the intersections of pancakes, the lower-density-contrast sheets (which may be the earliest generic structure to form) never <u>contract</u> relative to Hubble expansion. In fact, they <u>expand</u> along their planes relative to the general Hubble expansion. Therefore, I am led again to postulate that galaxies formed preferentially in the sheets, rather than the filaments and knots.

Once pancake fragmentation produces the first bound objects which undergo star formation, these objects may very well inject a significant amount of explosion energy (e.g. supernovae) capable of opposing the tendency of matter to pile up in regions of (previously) negative energy. In fact, this explosive injection may be at the heart of the postulated turning-off of the galaxy formation process in the denser knots and filaments. For example, spherically symmetric collapse, such as may characterize the dense knots, tends to be completed in a time interval which is much smaller than the continuous infall which characterizes the plane-symmetric collapse of the sheets. As a result, the first objects to form and explode (e.g. quasars) may very well be located in the knots and may subsequently inject enough energy into a small enough volume that reversal of the infall of matter into these regions occurs.

(4) Pancakes have sometimes been criticized on the grounds that they should show collapse <u>in the plane</u>, relative to Hubble expansion, which is not observed. The response to this argument is already contained in my discussion of points (2) and (3) above.

(5) It has sometimes been said that the collapse of large-scale structures in the pancake scenario must imply larger infall velocities (~ 1000 km/sec) than are observed (i.e. ~ hundreds of km/sec in the Local Supercluster). However, such large velocities do not emerge in the pancake scenario since the pancake shocks <u>stop</u> matter which is infalling onto the central plane of symmetry <u>before</u> galaxies form from this matter. Furthermore, it is only <u>that</u> fraction of the gas which never fell very far (i.e. never achieved large v_{infall}) which radiatively cools and forms galaxies.

(6) Lastly, Aaronson (1983), Faber and Lin (1983), and Lin and Faber (1983) have argued that dwarf spheroidal galaxies have large velocity dispersions, implying high mass-to-light ratios and the presence of substantial amounts of dark matter. The reported M/L would require that the dark matter violate the Tremaine-Gunn phase-space constraints on the neutrinos unless the particle mass is several hundred eV, which would, in turn, violate the cosmological upper bounds on Ω_ν. This argument depended upon a very limited data set for Draco. Based upon

the velocity dispersion reported for Draco by Aaronson (1983), Lin and
Faber (1983) predicted a large velocity dispersion for the more luminous
Fornax dwarf galaxy assuming a similar, large M/L. However, Cohen (1983)
has subsequently reported a much lower velocity dispersion than this in
observations of the Fornax globular clusters, implying a much lower M/L
than was predicted. Moreover, it may yet turn out that, even if the
original Draco measurements ultimately are found to give the correct
velocity dispersion, the explanation lies in the disruption of the Draco
galaxy as suggested by Hodge and Michie (1969), rather than the presence
of a large amount of dark matter. While this is a very worthwhile line
of investigation, much more data will be necessary before a serious
dilemma can be assumed to exist for the neutrino-dominated pancake
scenario.

There are at least two more potential difficulties for the neutrino-
dominated picture, however, which I do not think have been emphasized
before. I will conclude by stating them as cautionary notes for the
future. (i) The surface mass density of cold, baryonic matter in the
pancakes is orders of magnitude less than that of a single, typical
galaxy. For example, for the "canonical" case plotted in Figures 2 and
3, the cooled pancake layer has baryonic surface density $\Sigma \sim 2 \times 10^{11}$
$(1+z)^2$ $M_\odot/(Mpc)^2$, as compared to $\Sigma_{galaxy} \sim 10^{15}$ $M_\odot/(Mpc)^2$. A reasonable
suggestion for improving the apparent mismatch might be the requirement
that the pancakes collapse at higher z, since this would increase the
cooled mass fraction and provide a factor $(1+z)^2$ by which to substantially increase Σ, as well. In that case, whatever further Σ increase
is required could then be accomplished by the dissipative interaction of
the pancake fragments. (ii) However, even though the neutrinos do
apparently serve to resolve the old problem of making galaxies in the
original pancake scenario without violating the anisotropy limits to
the microwave background, nevertheless we are not free to postulate the
formation of neutrino pancakes at arbitrarily high, post-recombination
redshift. For the "canonical" case discussed here, in fact, with $\Omega = 1$,
h = 0.75, and z_c = 6, a value of $(\delta\rho/\rho) \sim 2 \times 10^{-4}$ is implied for a
Zel'dovich spectrum of primordial perturbations, for all scales larger
than the damping length, as measured when each scale enters the horizon.
This is necessary in order for caustic formation to occur at z_c = 6. The
Sachs-Wolfe effect, however, must then introduce a temperature fluctuation $(\delta T/T) \sim 10^{-4}$ on the angular scale of the horizon at recombination
$(\Delta\theta \sim$ degrees) and on greater scales. The observed upper limits to
$(\delta T/T)$ on such intermediate and large angular scales already constrain
$(\delta T/T)$ to be no larger than this, however. Hence, for a Zel'dovich
spectrum, at least, it is not permissible to increase the perturbation
amplitude on the pancake scale at $z = 10^3$ by a factor large enough to
raise z_c by more than a factor of order unity. Of course, since the
predicted $(\delta T/T)$ decreases with increasing angular scale θ in proportion
to $\theta^{(1-n)/2}$ (Silk 1981), we can greatly weaken this constraint by considering values of n > 1. (iii) Ultimately, of course, there is always
the possibility that the neutrino does not have a mass large enough to
be of cosmological significance. In that case, we shall either have to
appeal to another candidate for the LSD darkino, or else restrict our

attention to particles of the ISD and ND types. Aside from these
cautionary remarks, however, I think it would be premature for us to
abandon the neutrino pancake scenario altogether just yet.

REFERENCES

Aaronson, M. 1983, Ap. J. Letters, 266, L11.
Aaronson, M., Huchra, J., Mould, J., Schechter, P.L., and Tully, R.B.
    1982, Ap. J., 258, 64.
Abbott, L., and Sikivie, P. 1983, Phys. Lett., 120B, 133.
Arnold, V.I., Shandarin, S.F., and Zel'dovich, Ya.B. 1982, Geophys. Ap.
    Fluid Dyn., 20, 111.
Barbieri, R. et al. 1980a, Phys. Letters, 90B, 91.
_____ 1980b, Phys. Letters, 90B, 249.
Blumenthal, G.R., Pagels, H., and Primack, J.R. 1982, Nature, 299, 37.
Bond, J.R., Centrella, J., Szalay, A.S., and Wilson, J.R. 1983, preprint.
Bond, J.R., Efstathiou, G., and Silk, J. 1980, Phys. Rev. Letters, 45,
    1980.
Bond, J.R., Szalay, A.S., and Turner, M.S. 1982, Phys. Rev. Lett., 48,
    1636.
Cabibbo, N., Farrar, G.R., and Maiani, L. 1981, Phys. Letters, 105B, 155.
Centrella, J., and Melott, A.L. 1983, Nature, 305, 196.
Cohen, J.G. 1983, Ap. J. Letters, 270, L41.
Davis, M., and Huchra, J. 1982, Ap. J., 254, 437.
Davis, M., Huchra, J., Latham, D.W., and Tonry, J. 1982, Ap. J., 253,
    423.
Davis, M., and Peebles, P.J.E. 1983, Ap. J., 267, 465.
Dekel, A. 1983, Ap. J., 264, 373.
Dine, M., and Fischler, W. 1983, Phys. Lett., 120B, 137.
Doroshkevich, A.G., Khlopov, Yu., Sunyaev, R.A., Szalay, A.S., and
    Zel'dovich, Ya.B. 1981, in Tenth Texas Symposium on Relativistic
    Astrophysics, eds. R. Ramaty and F.C. Jones (Ann. N.Y. Acad. Sci.,
    375, 32).
Doroshkevich, A.G., Kotok, E.V., Novikov, I.D., Polyudov, A.N.,
    Shandarin, S.F., and Sigov, Yu.S. 1980, M.N.R.A.S., 192, 321.
Einasto, J., Joever, M., and Saar, E. 1980, Nature, 283, 47.
Faber, S.M., and Lin, D.N.C. 1983, Ap. J. Letters, 266, L17.
Frenk, C.S., White, S.D.M., and Davis, M. 1983, Ap. J., 271, 417.
Goldberg, H. 1983, Phys. Rev. Lett., 50, 1419.
Gott, J.R., Gunn, J.E., Schramm, D.N., and Tinsley, B.M. 1974, Ap. J.,
    194, 543.
Gregory, S.A., and Thompson, L.A. 1978, Ap. J., 222, 784.
Hodge, P.W., and Michie, R.W. 1969, A. J., 76, 587.
Ipser, J., and Sikivie, P. 1983, Phys. Rev. Lett., 50, 925.
Klypin, A.A., and Shandarin, S.F. 1983, M.N.R.A.S., in press.
Kirshner, R.D., Oemler, A., Schecter, P.L., and Schectman, S.A. 1981,
    Ap. J. (Letters), 248, L57.
Lin, C.C., Mestel, L., and Shu, F.U. 1965, Ap. J., 142, 1431.
Lin, D.N.C., and Faber, S.M. 1983, Ap. J. Letters, 266, L21.
Magg, M., and Wetterich, Ch. 1980, Phys. Letters, 94B, 61.
Melott, A.L. 1982, Phys. Rev. Letters, 48, 894.

Melott, A.L. 1983a, Ap. J., 264, 59.
_____ 1983b, M.N.R.A.S., 202, 595.
Mohapatra, R.N., and Marshak, R.E. 1980, Phys. Rev. Letters, 44, 1316.
Mohapatra, R.N., and Senjanovic, G. 1980, Phys. Rev. Letters, 44, 912.
Olive, K.A., Schramm, D.N., Steigman, G., Turner, M.S., and Yang, J.
      1981, Ap. J., 246, 557.
Pagels, H., and Primack, J.R. 1982, Phys. Rev. Letters, 48, 223.
Peebles, P.J.E. 1980, The Large-Scale Structure of the Universe
      (Princeton University Press).
_____ 1981a, Ap. J., 248, 885.
_____ 1981b, in Tenth Texas Symposium on Relativistic Astro-
      physics, eds. R. Ramaty and F.C. Jones (Ann. N.Y. Acad. Sci., 375,
      157).
_____ 1982a, Ap. J., 258, 415.
_____ 1982b, Ap. J. (Letters), 263, L1.
_____ 1983, preprint.
Preskill, J., Wise, M., and Wilczek, F. 1983, Phys. Lett., 120B, 127.
Press, W. 1980, Phys. Scripta, 21, 702.
Press, W.H., and Vishniac, E.T. 1980, Ap. J., 236, 323.
Primack, J.R., and Blumenthal, G.R. 1983, in Proc. of Fourth Workshop
      on Grand Unification, in press.
Schramm, D.N., and Steigman, G. 1981, Ap. J., 243, 1.
Sciama, D. 1983, preprint.
Shapiro, P.R., and Struck-Marcell, C. 1983a, in preparation.
_____ 1983b, in preparation.
Shapiro, P.R., Struck-Marcell, C., and Melott, A.L. 1983, Ap. J., 275,
      in press.
Silk, J. 1981, in Tenth Texas Symposium on Relativistic Astrophysics,
      eds. R. Ramaty and F.C. Jones, Ann. N.Y. Acad. Sci, 375, 188.
Steigman, G. 1983, Proceedings of the ESO Workshop on Primordial Helium
      (Munich, 1983 February), in press.
Sunyaev, R.A., and Zel'dovich, Ya.B. 1972, Astr. Ap., 20, 189.
Tarenghi, M., Chincarini, G., Rood, H.J., and Thompson, L.A. 1980,
      Ap. J., 235, 724.
Tremaine, S.D., and Gunn, J.E. 1979, Phys. Rev. Letters, 42, 467.
Turner, M.S., Wilczek, F., and Zee, A. 1983, Phys. Lett., 125, 35.
Weinberg, S. 1979, Phys. Rev. Letters, 42, 850.
_____ 1983, Phys. Rev. Letters, 50, 387.
Wilson, M.L., and Silk, J. 1981, Ap. J., 243, 14.
Zel'dovich, Ya.B. 1970, Astr. Ap., 5, 84.
_____ 1978, in The Large-Scale Structure of the Universe,
      IAU Symposium No. 79, eds. M.S. Longair and J. Einasto (D. Reidel
      Dordrecht, Holland), p. 409.

DISCUSSION

PEEBLES: I agree that if you make galaxies in the lower density sheets rather than in denser knots and filaments you vitiate my objection, but what do you do with the denser knots and filaments? In a dense $10^{15}M_\odot$ potential well surely some gas would be left behind or accreted and be observable as an extra-cluster X-ray source?

SHAPIRO: This assumes that the baryon-electron gas passes thru an accretion shock and continues its infall into the denser knots and filaments, following the same spatial distribution as the dark, collisionless matter which gathers in these regions. It is quite possible, however, that the gaseous infall is slowed with respect to the dark matter, halted, or even reversed, as follows. If the first bound objects to form in the knots and filaments inject a reasonable amount of kinetic energy into the surrounding gas, as a result of energy-releasing phenomena like supernovae and quasars, then the total energy of the surrounding gas may become positive. This can halt or reverse the infall, thus shutting-off the galaxy formation process in these regions and preventing the further build-up of density which would made the X-ray emission from this gas copious enough to be observable in comparison with either the X-ray emission from galaxy clusters or the cosmic X-ray background.

PRIMACK: There is additional unpublished data on velocity dispersions of stars in Draco, Carina, and Ursa Minor which suggests that all of them have heavy halos (private communication from Marc Aronson), as expected from the Faber-Lin tidal argument. (Many of the stellar velocities have been remeasured and known binaries excluded.) Cohen's data is consistent with the Faber-Lin tidal estimate of the velocity dispersion of Fornax, and perhaps -- taking measurement accuracy into account -- not inconsistent with a fairly heavy halo.

REES: The data we have on the spatial distribution of galaxies consist of two spatial coordinates transverse to our line of sight and a radial coordinate proportional to redshift. It would be good if the results of N-body simulations could be granted in this format (rather than in terms of 3 spatial coordinates). The resultant pictures may look different if there are significant transverse velocities across the filaments.

SHAPIRO: You are quite right. However, I would caution that any direct comparison between the purely collisionless simulations and observed spatial distribution of galaxies may be misleading, since it is not the collisionless matter which forms galaxies but rather the dissipate, hydrodynamical, baryonic component that does. Unfortunately, we are presently very far from understanding the ultimate segregation of dark and luminous matter in these models.

CORBYN: What sort of diameter do your numerical simulations suggest for filamentary structures? And can you (by numerical simulation or theory) distinguish between thicker filaments which maybe should be termed "fingers" and thinner ones, which seem to have been observed, and should perhaps be termed "filaments" proper?

SHAPIRO: The 3-D, collisionless particle simulations which I have described show filament diameters which are of the order of a few grid spacing. Hence, for the 4C Mpc damping length chosen to characterize the neutrino-dominated (i.e. large scale damping) universe, the filament thickness is of the order of 10 Mpc. However, this may reflect the limited resolving power of the finite grid-spacing. In any case, these simulations do not take account of hydrodynamical dissipation. For the 1-D, plane-symmetric calculations described here, for example, of the combined growth of (baryon)+(collisionless matter) pancakes, the cold, dense baryonic layer from which galaxies would have to form occupies only $\sim 10^{-6}$ of a wavelength in thickness (for the neutrino-dominated case), or only 10's of parsecs. It is difficult , at present, therefore, to use the 3-D simulations of collisionless matter only to predict the observed thickness of filamentary structure in the spatial distribution of galaxies.

REES: It is interesting to consider all possible observable effects of pancakes at $z \simeq 2$. Bond and Szalay have shown that the contributions to ultraviolet and soft X-ray background are not important, but that the Sunyaev-Zel'dovich dips might be detectable. These could be distinguished from microwave background fluctuations originating at $z \simeq 1000$ by two frequency observations, since the Sunyaev-Zel'dovich effect reverses sign at frequency where the background spectrum peaks

SHAPIRO: If pancakes are the order of 40 Mpc in size at present, as in the neutrino-dominated picture, then an observation in any direction must sample many randomly oriented pancakes, over a wide range of redshift. In this case, it would be necessary to detect the angular fluctuation in the Sunyaev-Zel'dovich effect relative to the average cumulative Sunyaev-Zel'dovich effect of all the pancakes along the line sight. The amplitude of this angular fluctuation which will be on scales ranging from $\sim 10'$ to $\sim 1°$), however, will be much smaller than this average cumulative effect, smaller, in fact, by at least $\sim N^{1/2}$, where N is the average number of pancakes per line of sight, a large number. One can estimate from the results described in this talk for the neutrino-dominated case that the dimensionless parameter $y= \tau_{es} \cdot (4kT/m_e c^2)$, which indicates the average fractional frequency shift of a microwave photon traversing a single pancake, is of the order of 10^{-6}; with $N \sim (40Mpc)^{-1} (c/Ho) \sim 10^2$, this implies that the average net effect of the pancakes will correspond to $< \Delta \nu /\nu > \sim Ny \sim 10^{-4}$. However, the angular fluctuation in this effect, which can be interpreted as a temperature fluctuation $\Delta T/T$, will be $\Delta T/T < N^{1/2} y \lesssim 10^{-5}$, which is currently below the limits of detectability. However these estimates are crude, it is possible that the actual effect is larger by as much as an order of magnitude, which would make detection more likely. Care must be taken not confuse the Sunyaev-Zel'dovich effect due to pancakes with that due to the hot gas inside clusters of galaxies, an effect which can reach an amplitude at centimeter wavelengths which corresponds to $\Delta T/T \geq 10^{-5}$ on angular scales of 3'-20'.

CURVATURE FLUCTUATIONS AS PROGENITORS OF LARGE SCALE HOLES

N. Vittorio, P. Santangelo, F. Occhionero
Istituto Astronomico, Università di Roma

ABSTRACT: We extend previous work to study the formation and evolution of deep holes, under the assumption that they arise from curvature or energy perturbations in the Hubble flow. Our algorithm, which makes use of the spherically symmetric and pressureless Tolman-Bondi solution, can embed a perturbation in any cosmological background. After recalling previous results on the central depth of the hole and its radial dimension, we give here specific examples of density and peculiar velocity profiles, which may have a bearing on whether galaxy formation is a dissipative or dissipationless process.

1. INTRODUCTION

Observations of the large scale structure of the Universe have shown recently the existence of large holes (Davis et al., 1982; Kirshner et al. 1981; Tarenghi et al., 1980; Einasto et al., 1980; Gregory and Thompson, 1978). This cellular structure of the Universe seems direct evidence against the dissipationless theory of hierarchical clustering and in favor of the dissipative theory of galaxy formation (Zel'dovich, 1978; Arnold et al., 1982) or of collective explosion models (Ostriker and Cowie, 1981). It is nevertheless argued that the dissipationless dynamics of primordial negative density perturbations can account self-consistently for the formation and evolution of the voids in the same way as positive density perturbations are believed to be the progenitors of clusters (Hoffman and Shaham, 1982). In particular, Peebles (1982) shows that dissipationless dynamics can be reconciled with the lack of observations of high peculiar velocities perpendicular to the plane of superclusters. Aarseth and Saslaw (1982), on the other hand, argue that the issue is basically linked to the dimensions of the hole; small holes (20 or 30 Mpc) are compatible with gravitational clustering, large holes (100 Mpc) are not. Another interesting point concerns the trace that should have been left in the cosmic microwave background by the (negative) density perturbations at decoupling (clearly under the assumption of no further ionization of the cosmic medium after the canonical decoupling). Hoffman et al. (1983) find that the expected temperature fluctuations on the small scale ($\sim 10'$) tend to be in conflict with the corresponding observed upper limits

479

F. Mardirossian et al. (eds.), Clusters and Groups of Galaxies, 479–484.

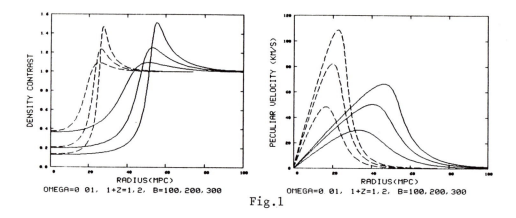

OMEGA=0 01, 1+Z=1,2, B=100,200,300   OMEGA=0 01, 1+Z=1,2, B=100,200,300

Fig.1

(e.g. Partridge, 1981) for the case of cavities, as it was known to be
true for condensations (Boynton, 1978).
In the present paper we address ourselves to these problems by generali-
zing a model (Occhionero et al., 1983a; hereafter Paper I), whereby we
studied the non-linear evolution of spherical cavities originated by
localized energy excesses (rather than density depletions) in an Einstein-
de Sitter cosmological model; further references to the non-linear evolution of energy or curvature perturbations (as opposed to density perturbations) are given in Paper I. The simplifying assumption of spherical
symmetry is clearly much less stringent for cavities than it is for condensations, where on the contrary any initial asphericity amplifies
enormously (Lin et al., 1965; White and Silk, 1979).
A general feature of our cavity models – as well as of the models of the
cited authors – is that each cavity is surrounded by an overdense shell
that expands in the umperturbed cosmological model with a speed somewhat
in excess of the Hubble one. Thus, given two nearby cavities, the corresponding shells are bound to collide with each other; it is tempting to
imagine that the formation of superclusters occurs at the site of these
collisions. Whether the process is dissipationless or not, will then
depend on whether or not galaxy formation has proceeded to any substantial
degree during the emptying of the cavity.

2. RESULTS AND DISCUSSION

In extending the work of Paper I we recall first some results established
in previous work (Occhionero et al., 1982; Paper II). In the first place,
we have shown that the sharpness of the density spike in the expanding
shell and the magnitude of the residual velocities around the holes depend strongly on the steepness of the gradient in the assumed energy
perturbation. Therefore, in order to keep the peculiar velocities as low
as possible and avoid conflict with the observations in the case of very
large holes, we limit ourselves here to the mildest of the simple analytical energetics given in Paper II.
Secondly, two features of our models are instead easily parametrizable:
these are the central depth and the radial dimension of the hole. Concer-

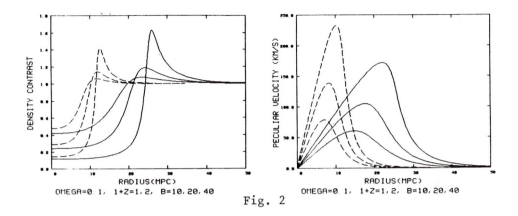

Fig. 2

ning the former, we find convenient to introduce a dimensionless parameter which defines the central strength of the perturbation in suitable
units: $B \cong (\delta\rho/\rho)_1 (1+z)_1$ where $(\delta\rho/\rho)_1$ and $(1+z)_1$ are the density contrast
and the redshift at a reference epoch when we make the observations (typ-
ically the epoch of decoupling). On account of the above argument on the
fluctuations in the microwave background, we are not allowed to accept
for $(\delta\rho/\rho)_1$, values larger than 10^{-3} and consequently we cannot accept
for B values in excess of unity; in fact it should be much less. When
this constraint is investigated on the (σ_0, q_0)-plane, σ_0 and q_0 being
defined as usual, it is found that it can be satisfied only for background models with positive spatial curvature and $\Lambda > 0$. On the contrary,
low density models show the greatest difficulty from this point of view.
Coming to the radial dimensions of the holes, we find that they are always
of the same order of magnitude of the characteristic length,
$L = 10$ Mpc $(m_{15}/\Omega_0 h^2)^{1/3}$, where m_{15} is the mass in units $10^{15}M_\odot$, Ω_0 is
the density parameter and h is the Hubble constant in units of 100 km/s/
Mpc. Therefore, if the observations push toward radii of the order of
50 Mpc, we must choose either $\Omega_0 h^2 \sim 0.01$ or $m_{15} \sim 100$. In the first case
we are in conflict with the observations on the microwave background;
the second choice might instead be acceptable since the unusually large
mass is in reality to be compared with the sum of the masses of the superclusters which make up the walls of the cavity.
Now our algorithm evaluates from the field equations the evolution in
time of the relevant physical quantities such as the density contrast and
the peculiar velocity field. Depending on the assumed curvature perturbation we produce either a condensation surrounded by a spherical cavity
or a cavity surrounded by a mass ridge; the first possibility was explored by Occhionero et al., 1981, a and b; the second possibility is of
interest here. A complete description of the method will be given elsewhere (Occhionero et al., 1983b). Our present results are given as snapshots at the redshifts $1+z = 1,2$ of the mentioned quantities. In Fig. 1
we consider $\Omega_0 = 0.01$ and $B = 100,200,300$; full lines refer to the present,
broken lines to $1+z = 2$; on the left the density profiles vs. radius in
Mpc show that the hole depth is an increasing function of B, that the
present hole radius is of the order of 50 Mpc and that the present

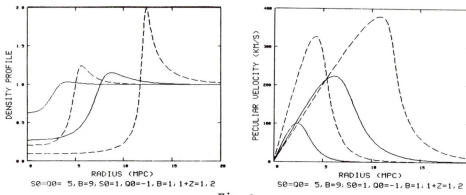

SØ=QØ= 5, B=9; SØ=1, QØ=-1, B=1; 1+Z=1, 2   SØ=QØ= 5, B=9; SØ=1, QØ=-1, B=1; 1+Z=1, 2

Fig. 3

density profile was of course frozen long ago. On the right, the residual velocities cool down with the Hubble expansion and, being of the
order of a few hundreds of km/s, are well within the observational limits.
In Fig. 2 we give the cases Ω_0 = 0.1 and B = 10, 20, 40; the cavity
radius has shrunk to 20 Mpc.
We come now to the cavities evolving in background models with $\Lambda > 0$
and positive curvature. There are at least three reasons for taking these
models into account: i) the incertitude on the Hubble time; ii) the possibility that the cosmic density may be close to critical due to the
presence of a dark component (such as massive neutrinos; Zel'dovich and
Sunyaev, 1980); iii) the indication that the growth of condensations
both in the linear and the non-linear regime is favored in such models
(Occhionero et al., 1980 and 1982). In Figs. 3 and 4 we give examples
of such computations again for 1+z = 1,2; now however broken lines refer
to the Einstein-de Sitter model which is taken as a reference. On the
latter it is remarked that it takes B = 9 in order to dig a cavity whose
present density is 10% of the asymptotic one. In Fig. 3 we have σ_0 = 1,
q_0 = -1; in Fig. 4 we have σ_0 = 1, q_0 = -2; in both cases B = 1. One
difficulty seems to arise: the peculiar velocities may – as in the case
of Fig. 4 – grow beyond plausibility. In order to circumvent this difficulty we might have to postulate that dissipational mechanisms are at
work. In particular, if galaxy formation has not proceeded to completion
in the ridge and the latter consists largely of gaseous material when it
collides with the nearby ridge, then the collision itself will be highly
dissipative.

REFERENCES

Aarseth, S.J. and Saslaw, W.C.: 1982, Ap.J.Lett., 258,L7

Arnold, V.I., Shandarin, S.F., and Zel'dovich, Ya.B.: 1982, Goephys.
Astrophys. Fluid Dynamics, 20, 111

Boynton, P.E.: 1978, in The Large Scale Structure of the Universe, ed.
M.S. Longair and J. Einasto, D. Reidel Publ. Co., Dordrecht, Holland

Davis, M., Huchra, J., Latham, D.W., and Tonry, J.: 1982, Ap.J. 253; 423

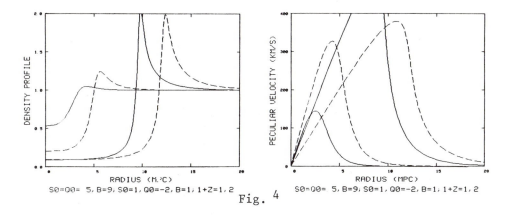

Fig. 4

Einasto, J., Jôeveer, M., and Saar, E.: 1980, MNRAS, 193, 353

Gregory, S.A. and Thompson, L.A.: 1978, Ap.J., 222, 784

Hoffman, G.L., Salpeter, E.E., and Wasserman, I.: 1983, Ap.J., 268, 527

Hoffman, Y. and Shaham, J.: 1982, Ap.J.Lett. 262, L23

Kirshner, R., Oemler, A., Schecter, P., and Schechtman, S.: 1981, Ap.J. Lett., 248, L57

Lin, C.C., Mestel, L., and Shu, F.: 1965, Ap.J., 142, 1431

Occhionero, F., Vittorio, N., Carnevali, P., and Santangelo, P.: 1980, A.A. 86, 212 and 1982, A.A. 107, 172

Occhionero, F., Veccia-Scavalli, L., and Vittorio, N.: 1981 a and b, A.A. 97, 169 and 99, L12

Occhionero, F., Santangelo, P., and Vittorio, N.: 1982, in Clusters of Galaxies, ed. D. Gerbal (Paper II)

——— id. ———: 1983a, A.A., 117, 353 (Paper I)

——— id. ———: 1983b, in preparation

Ostriker, J.P. and Cowie, L.: 1981, Ap.J.Lett., 243, L127

Partridge, R.B.: 1981, in The Origin and Evolution of Galaxies, ed. B.J.T. Jones, and J.E. Jones, D. Reidel Publ. Co., Dordrecht, Holland

Peebles, P.J.E.: 1982, Ap.J. 257, 438

Tarenghi, M., Chincarini, G., Rood, H.J., and Thompson, L.A.: 1980, Ap.J., 235, 724

White, S.D.M. and Silk, J.: 1979, Ap.J., 231, 1

Zel'dovich, Ya.B.: 1978, in The Large Scale Structure of the Universe, eds. M.S. Longair and J. Einasto, D. Reidel Publ. Co., Dordrecht, Holland

Zel'dovich, Ya.B. and Sunyaev, R.A.: 1980, Sov. Astron.Lett. 6, 249

DISCUSSION

PEEBLES: May I add, as a supplement to your very interesting paper, the comment that the "Bondi-Tolman" solution was discovered by Georges Lemaître in 1933.

OCCHIONERO: Thank you.

CLUSTERS AND THEIR PRECURSORS

Martin J. Rees
Institute of Astronomy, Madingley Road
Cambridge CB3 OHA, England

ABSTRACT

Three possible forms are discussed for the post-recombination spectrum of primordial fluctuations. Attention is then given to models in which small scale systems (of typically $\sim 10^6$ $M_\odot$) condense out before clusters — adiabatic fluctuations in a universe dominated by "cold" non-baryonic matter, or isothermal perturbations in the baryon distribution. Uncertainties in the fragmentation process are emphasised (§ 2) — the first bound systems could form very massive objects or low mass stars, neither of which can yet be ruled out by observations (§ 3). If these pregalactic objects emit UV radiation, the universe could be reionized before more than $\sim 10^{-4}$ has condensed. This leads to a negative feedback effect; in the "cold dark matter" model, where the fluctuation spectrum is very flat, this effect generates surprisingly long range correlations, so that proto-clusters and proto-voids on scales $\sim 10^{15}$ $M_\odot$ could start to evolve differently even at epochs as early as $z \simeq 50$ (§ 4). In § 5, some secondary processes for generating large-scale density or velocity perturbations are discussed. Finally, in § 6 it is argued that if the initial fluctuations were isothermal, population III objects could contribute the full cosmological density without violating the conventional cosmological constraints on light element synthesis.

1. THREE POSSIBLE FORMS FOR THE POST-RECOMBINATION SPECTRUM OF FLUCTUATIONS

Several speakers have discussed the observed distribution of matter, and the properties of clusters and superclusters. A prime aim of theorists is to understand how the Universe can have developed this structure from its postulated amorphous beginnings. Given an initial spectrum of linear perturbations, it is relatively straightforward to calculate how various wavelengths evolve - how they enter the horizon, and then undergo mass-dependent growth or damping, affected by viscosity, free-streaming, etc. On the relevant scales the

F. Mardirossian et al. (eds.), Clusters and Groups of Galaxies, 485–497.
© *1984 by D. Reidel Publishing Company.*

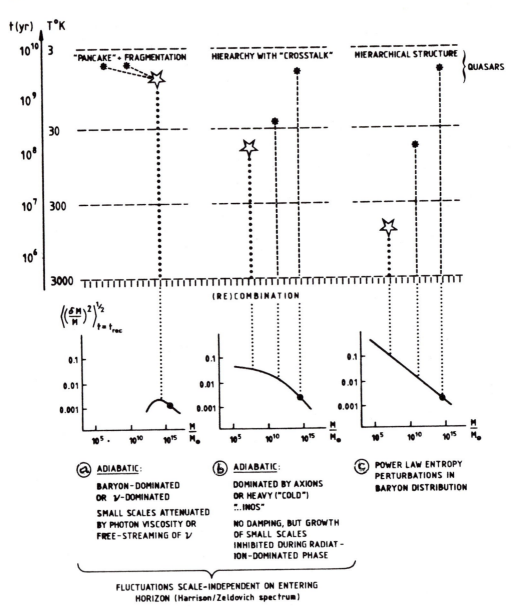

Fig. 1. Cosmogonic processes after (re)combination depend on the spectrum of fluctuations which have survived damping processes, etc. at earlier times. In case (a), (super)clusters are the first systems to condense, and they form quite recently. In case (b), baryonic systems of $\lesssim 10^6 M_\odot$ condense in potential wells produced by slow moving heavy "..inos" or axions. The third case shows the spectrum that might arise from primordial entropy perturbations. In cases (b) and (c), subgalactic systems would form *before* galaxies; their energy input could generate "secondary" perturbations on larger scales which swamp the genuinely primordial ones. This diagram is merely a schematic illustration of three possibilities.

fluctuations are still linear at recombination, but thereafter the
contrast grows until bound systems separate out. The cosmogonic process
depends crucially on the nature of the initial fluctuations (presumably
imprinted by exotic physics at some early cosmic epoch) and on the
nature of the dominant matter in the present day Universe - baryons,
neutrinos of mass about 10 ev, or so-called "cold dark matter" such as
primordial black holes or axions.

Figure 1 shows schematically how the order in which various scales
of structure form depends on which option we favour. Moreover, the
redshift at which things first go non-linear is highly uncertain. In
the neutrino picture (option (a)), discussed by P. Shapiro at this
meeting, the Universe remains quiescent, consisting of almost uniform
neutral gas, until it is more than a billion years old (i.e. $z \lesssim 3$).
But in other schemes, pregalactic activity may occur at much larger
redshifts. The right hand part of the diagram (option (c)) shows
schematically the kind of spectrum which some authors have envisaged
for isothermal fluctuations: fluctuations of this kind have fallen out
of fashion recently, because of the belief that the photon-to-baryon
ratio should have a universal value fixed by Grand Unified Theories of
baryon synthesis.

The intermediate form of spectrum (shown as option (b)) is one
which has only recently received attention - it is a consequence of
the hypothesis that the hidden mass is non-baryonic, but is "cold", in
the sense that free streaming does not erase fluctuations on small
scales (as it does for neutrinos). It thus permits small scale irregularities to survive, even if the initial fluctuations are purely adiabatic. If these initial fluctuations have a constant curvature spectrum
(Harrison 1970, Peebles and Yu 1970, Zeldovich 1972) then the fluctuations surviving after recombination have a characteristic calculable
form, as discussed by Peebles and by Primack at this meeting. This
post-recombination spectrum has the property that on scales up to a
galactic mass, the quantity $\Delta(M) = <(\delta M/M)^2>^{\frac{1}{2}}$ is flat, apart from a
logarithmic dependence; for larger masses it "rolls over" towards an
$M^{-2/3}$ law. Ordinary gas will fall into these fluctuations, except on
scales smaller than the baryonic Jeans mass; in option (b), therefore,
as in (c), the first baryonic systems to go non-linear will be objects
of subgalactic scale ($\sim 10^6$ $M_{\odot}$), and clusters of galaxies arise much
later.

2.   THE FIRST BOUND BARYONIC SYSTEMS: POPULATION III

The nature of the first bound systems is of crucial importance in
determining all subsequent stages of cosmic evolution in schemes (b)
and (c), and for the "hidden mass" problem. However, theory provides
little firm guidance. Kashlinsky and I (1983) addressed this question,
and concluded that $\sim 10^6$ $M_{\odot}$ baryonic clouds condensing after recombination could either fragment into very low mass stars ("Jupiters"),
or form single supermassive stars — it is quite possible that some

fraction ends up in each form. We are not optimistic about narrowing down this uncertainty until a good deal more is understood about star formation and fragmentation processes in a more general context. Although we did not include non-baryonic matter, the uncertainties are equally great for case (**b**), where clouds of baryons would settle into potential wells due primarily to "cold" non-baryonic material (in this connection see also Carr and Rees (1984)).

We know too little about collapse and fragmentation to assess whether these pregalactic systems will form low mass stars or super-massive objects. There are however some observational constraints, reviewed in detail by Carr *et al.* (1984). All the hidden mass could be in $\lesssim 0.1$ $M_\odot$ stars, or in remnants of very massive stars (VMOs) heavier than a few hundred solar masses. Masses $\gtrsim 10^6$ $M_\odot$ are ruled out (at least in galactic halos, if not in intergalactic space) because their dynamical friction effects would cause excessive velocity dispersion in disc galaxies. A severe constraint on the fraction of mass processed through ordinary massive stars, of the kind which explode as supernovae, comes from the requirement that heavy elements should not be overproduced. If we require that these objects produce no more than a Population II abundance, then $< 10^{-4}$ of the critical density could have been processed through them.

[If the evolution of the pregalactic $\gtrsim 10^6$ $M_\odot$ clouds were bimodal – with some fragmenting into low mass stars and some forming VMOs – then one might envisage that a Population III and (some) Population II stars may be coeval – in other words, that some of the presently-observed Population II stars may be "pregalactic", in the sense that they formed in $\sim 10^6$ $M_\odot$ clusters which only later became incorporated (via hierarchical clustering) in a galaxy. The intergalactic medium could then, in principle, have been contaminated with heavy elements, even at $z \cong 100$, to a level far exceeding the extreme Population II abundances. For such a "bimodal" scheme to be consistent with present-day galaxies, the relative proportions of clouds developing into supermassive and ordinary stars would need to be about 10:1.]

3. DOES BARYONIC MATTER CONTRIBUTE ALL THE "UNSEEN" MASS?

Despite the many exotic candidates for dark matter, ordinary baryons — which have the unique virtue of being known to exist — are not yet discredited as a possibility. Indeed, if the Hubble constant is ~ 50 km s^{-1} Mpc^{-1} (i.e. $h \cong \frac{1}{2}$) primordial nucleosynthesis arguments based on the standard hot big bang model require $\Omega_b \cong 0.1$ (Yang *et al.*, 1984); the observed luminous matter gives a factor ~ 10 less than this, implying that in this model there *must* be some baryonic dark matter. (In § 6 I will suggest that the nucleosynthesis constraints could even allow $\Omega_b = 1$). The extra baryons could all be in a very hot inter-cluster gas, provided that the thermal history of this medium is suitably constrained (Fabian and Kembhavi, 1982); but the most plausible option is Population III remnants.

Any observational handle on the masses of these objects would be very valuable - in particular, it would be interesting if one could discriminate between "Jupiters" and VMO remnants (black holes of 10^3 - 10^6 $M_\odot$). Searches for gravitational lensing may offer a genuine prospect of doing this. Galactic-mass lenses can yield images separated by a few arc seconds and several instances of this phenomenon are now known (Turner, these proceedings). The effects are not of any profound interest for gravitation theory - the deflection is about the same amount (an arc second or so) as the gravitational light-bending within the Solar System, and is therefore not a manifestation of strong-field gravity. The study of gravitational lens images will, nevertheless, when more instances of the phenomenon are found, be of great potential importance in probing the distribution and nature of the dark matter in galactic halos and clusters.

If the halo material is in stars or other discrete objects of characteristic mass m_* there will be fine structure in each image due to this granularity with angular scale $\theta_f \cong 10^{-6} (m_*/M_\odot)^{\frac{1}{2}}$. This will affect the magnification if the intrinsic size of the source is $\lesssim \theta_f$, and lead to non-intrinsic variations as the individual stellar-mass objects move across the quasar (Chang and Refsdal 1979, Gott 1981, Young 1981). If $m_* \cong 1$ $M_\odot$; micro-arc-second resolution would be required to resolve the images; but then the non-intrinsic variations are rapid enough to be detectable. The line-emitting region would probably be too large to be affected by this small-scale "graininess"; the central continuum may well be $\lesssim 10^{-7}{''}$ and would be affected even by Jupiter-sized masses (Gott 1981); only a halo of massive neutrinos (individual masses $\sim 10^{-32}$ gm) is smooth enough for this complication to be ignored. If the "hidden mass" was in the form of 10^6 $M_\odot$ black holes, the fine-stucture in the lens images, with a milli-arc-second scale, would affect the broad line region and VLBI radio structure as well as the optical continuum. Canizares (1982) argues that $\Omega \cong 1$ in low mass stars or black holes can perhaps already be excluded, because these objects would act as efficient lenses for the quasar continuum source but not for the (more diffuse) line emitting region, and would thereby create an unacceptably large scatter in the equivalent widths of lines in samples of quasars.

4. FEEDBACK EFFECTS ON UNCONDENSED GAS

If Population III objects were all low-mass "Jupiters", they would have no significant non-gravitational effects on the subsequent course of galaxy and cluster formation. On the other hand, pregalactic high-mass stars or VMOs could have two important effects:

(i) Even if only $\sim 10^{-4}$ of the primordial matter went into massive stars, enough processed material could be ejected to contaminate all the rest of the Universe with a Population II abundance of heavy elements. The extra cooling due to heavy elements would certainly affect the complex processes of agglomeration and fragmentation by which galaxies subsequently form.

(ii)   Ultraviolet radiation from massive stars or VMOs would photoionize uncondensed gas (Hartquist and Cameron 1974). The details of this process depend on the redshift at which it occurs, and on how long-lived the UV sources are (see Carr *et al.* (1984) for a fuller discussion), but in general only $10^{-4} - 10^{-5}$ of the mass, if converted into objects radiating at near the Eddington limit, would suffice to ionize the entire Universe. The intergalactic medium would have cooled to $T_g \lesssim 100°K$ by the time ($z \lesssim 50$) when the first bound systems form in "case (b)" [see Figure 1]. If the gas were heated to $\sim 10^4°K$, the Jeans mass ($\propto T_g^{3/2}$) would rise by $\gtrsim 10^3$. The actual factor may not be as drastic as this, because the photoionized gas may cool again due to Compton and molecular cooling. Nevertheless, it will certainly not end up on such a low adiabat as before the heating occurred. Consequently the photoionization would terminate (or at least exert a negative feedback on) the formation of systems of the kind which are expected to form first in cases (b) and (c): there would then be a pause in the cosmogonic process until larger systems (exceeding the new Jeans mass) went non-linear.

These considerations suggest that feedback processes may modify the primordial state of the gas (and change the scale or character of bound systems forming later) after a fraction $f \lesssim 10^{-4}$ has condensed out - if there were a (gaussian?) spread in ($\delta M/M$), these would be the rare ($> 3\sigma$) high amplitude peaks on a mass scale $M_1 \simeq 10^6 M_\odot$. Doroshkevich *et al.* (1967) were the first to discuss an effect of this kind; they argued that the feedback process would generate a natural "secondary" scale $\sim f^{-1} M_1$, which they related to galaxies. They implicitly assumed that the first systems to turn around would be randomly distributed. This is probably not a bad assumption for a spectrum where $\Delta(M) = <(\delta M/M)^2>^{\frac{1}{2}}$ decreases steeply with M, as in case (c) (though for isothermal perturbations there is, incidentally, no good reason for postulating a gaussian distribution of amplitudes on a given mass-scale). However, it is not difficult to see that for a flat spectrum, as in case (b), the first bound systems would tend to occur in clumps; the consequences of this may be relevant to clustering on larger scales.

When the initial spectrum is very flat, "cross-talk" between different scales has important effects: for the "cold dark matter" spectrum sketched in Figure 1(b) (and discussed in detail in Primack's contribution), the 3σ peaks at $10^{10} M_\odot$ are already non-linear when the average $\sim 10^6 M_\odot$ perturbation is turning around. This makes the build-up towards larger scales more rapid and chaotic than the self-similar hierarchical process expected for the power-law spectrum of Figure 1(c). A corollary (cf. Peebles (1983)) is that the first ($\sim 10^6 M_\odot$) bound systems will tend to occur in groups. This is because the long-wave fourier components are not negligible in amplitude, and there is a significantly higher chance of getting large over-densities (on mass-scales of, say, $10^6 M_\odot$) if these are riding on a peak (rather than being in a trough) of a wave mode with scale $\sim 10^{10} M_\odot$. This effect is still more marked if we are concerned with very rare fluctuations on the steeply falling high-amplitude tail of the gaussian distribution. ($f \leq 10^{-4}$ corresponds to $\geq 3.72\sigma$).

Let me give a — purely illustrative — numerical example of the resultant effects. Suppose that the first bound systems, of mass $M_1 \cong 10^6 \ M_\odot$, fragment into massive stars, which produce UV radiation and heavy elements, and that all the remaining gas can be photoionized when a fraction $f = 10^{-4}$ has condensed into such systems (i.e. $f = 10^{-4}$ is enough to turn the Universe into an HII region). Suppose further that the (more massive) systems which form from the reheated gas evolve differently, and in a manner which depends on whether they have been contaminated with heavy elements.

We can then ask: what is the largest scale M_2 on which the subsequent evolution will be inhomogeneous, owing to clumping of the first $M \cong M_1$ ($\sim 10^6 \ M_\odot$) bound systems, and consequent inhomogeneous deposition of the first heavy elements? Let us calculate what difference it makes if these systems lie within a 2σ peak (protocluster) rather than a 2σ trough (protovoid) on a much larger mass-scale M_2. The density contrast between these regions is equivalent to a difference of $4(\Delta(M_2)/\Delta(M_1))$ standard deviations in the amplitude on scale M_1. Thus a 3.72σ peak on scale M_1 (corresponding to $f = 10^{-4}$) in the protocluster has as great an overdensity as a $[3.72 + 4(\Delta(M_2)/(\Delta(M_1))]\sigma$ peak in the protovoid: the first bound systems and the production of heavy elements, occur preferentially in the protoclusters. A value $f = 5 \times 10^{-5}$ corresponds to 3.89σ . So, even on mass-scales M_2 such that $\Delta(M_2)/\Delta(M_1)$ is as small as $0.25 \times 0.17 \cong 0.042$, there would be a factor 2 difference between the heavy element abundance produced before photoionization choked off the process. Taking the "cold dark matter" spectrum calculated by Blumenthal and Primack (1983) for a model with $\Omega = 1$ and Hubble parameter $h = 0.5$, we find that the value of M_2 is $\sim 10^{15} \ M_\odot$.

This example shows how pregalactic processes at $z \cong 50$, involving gravitationally bound systems whose mass is only $M_1 \cong 10^6 \ M_\odot$ could be responsible for an important "environmental" parameter, the pregalactic heavy element abundance, and that this abundance could vary by a factor ~ 2 on scales $10^{15} \ M_\odot$ — scales which would still not have turned around even at the present epoch. If the process of galaxy formation (or the stellar IMF in galaxies) were sensitive to this abundance, and consequently proceeded differently in protoclusters and in "voids", then the contrasts in the density of bright galaxies could be greater than the overall density contrast in baryons. This possibility (or analogous ones based on different feedback processes) would be important if we were in a high density ($\Omega = 1$) universe, as it would help to reconcile the low non-Hubble velocities of galaxies with the large-scale nonuniformities in their distribution.

5. FORMATION OF "SECONDARY" LARGE-SCALE FLUCTUATIONS BY PREGALACTIC OBJECTS.

Pregalactic objects could in principle generate "secondary" fluctuations on larger scales which have a bigger amplitude than the primardial fluctuations on those scales and thus give rise to galaxies and clusters of glaaxies indirectly. Several scenarios have been suggested (see Carr and Rees 1984). One important possibility is that bound objects can produce energy in some way — energy which may in principle be used to generate large-scale peculiar velocities. A simple argument shows that the energy input required can be far less than the binding energy of the system which the secondary fluctuations eventually form.

After decoupling, the peculiar velocities in a growing perturbation vary with expansion factor as $R^{\frac{1}{2}}$ (because δ goes as R, whereas the differential Hubble velocity across a comoving sphere goes as $R^{-\frac{1}{2}}$), so the energy required to generate these peculiar velocities at red-shift z_i is just (z_B/z_i) times the *eventual* binding energy, where z_B is the redshift at which the perturbation stops expanding. Equivalently, the peculiar velocities generated as z_i can be smaller than the virial velocity of the final bound system by a factor $(z_B/z_i)^{\frac{1}{2}}$. One consequence of this is that, if the specific binding energy of a virialised system is εc^2, then it entered the horizon at a redshift $\varepsilon^{-1} z_B$. If we regard this as the largest redshift when the velocity perturbations could have been generated, then the specific energy input required is $\varepsilon^2 c^2$. This result assumes that the system entered the horizon after t_{eq}. Otherwise, it is optimal to feed the energy in just after t_{eq} (or just after decoupling) in a hot Universe) and the energy input required is $\varepsilon c^2 (z_B/z_{eq})$.

As a crude illustrative example, a "galaxy" of 10^{12} $M_\odot$ and velocity dispersion 300 km/sec (whose present binding energy is 10^6 $M_\odot c^2$) could have been produced "causally" by the injection of only 1 $M_\odot c^2$. Of course, this is unrealistic because we have seen that galactic mass perturbations do not necessarily grow uninterruptedly after entering the horizon. On the other hand dissipative effects may imply that the binding energy of *protogalaxies* was smaller than that of present-day galaxies. By the same argument, a "globular cluster" of 10^6 $M_\odot$ and velocity dispersion 10 km/sec could have been generated from an energy input of only 10^{-12} $M_\odot c^2$.

Carr and Rees (1984) have explored and reviewed various ways in which pregalactic seeds can generate fluctuations on galactic and cluster scales which exceed the amplitude of any fluctuations that might have been present initially on those same scales. The reader is referred to that paper for a full discussion.

One interesting possibility is that an energy input comes from galactic explosions. Any star with initial mass M above about 10 $M_\odot$ but below a critical value $M_c \cong 200$ $M_\odot$ (Bond *et al.* 1984) should end up exploding. The explosive energy generated can be expressed as ϵMc^2 where the efficiency parameter ϵ is in the range $10^{-4} - 10^{-5}$ over all mass ranges (Bookbinder *et al.* 1982). Stars larger than M_c probably collapse to black holes, although it is possible that some of the envelope may be ejected even in this case due to hydrogen shell burning (Bond *et al.* 1984). Supermassive stars in the range $M > 10^5 \, M_\odot$ could explode only if they had non-zero metallicity, but this would not be a characteristic of the first stars.

As Doroshkevich *et al.* (1967), Ostriker and Cowie (1980) and Ikeuchi (1981) have emphasised, the energy input from exploding stars — or clusters of them — could induce the formation of new bound objects. One might envisage a shock front sweeping up a spherical shell which expands adiabatically until its cooling time becomes comparable to the cosmological expansion time. (For $z > 5$ the dominant cooling process is inverse Compton cooling off the 3K background).

For a runaway process to be initiated, the shell must sweep up more than the seed's initial mass; also, the shell must be able to fragment into the second-generation seeds. One can thereby initiate a bootstrap process which progressively amplifies the fraction of the Universe in stars. Carr and Rees (1984) discuss under what conditions, in terms of seed mass and redshift range, this type of effect might occur.

One general point about the explosion scheme is that, though it may work in principle, it is not "efficient" in using energy to generate fluctuations in the sense discussed at the beginning of this section. A lot of energy is wasted in cooling. Moreover, most of the momentum goes into generating density contrasts > 1. To generate fluctuations early, one needs to produce velocities of only $V_{pec} = V_{virial} (1 + z)/(1 + z_B)^{-\frac{1}{2}}$, where z_B denotes the redshift when the system condenses out and binds. However, the differential Hubble velocity across the relevant scale is $V_{pec}(1 + z)/(1 + z_B) \gg V_{pec}$, so the mean propagation speed of the agent generating the peculiar velocity must be much larger than the velocity that is generated. This is not possible for blast waves, except at very late stages after their momentum (and, *a fortiori*, their energy) has been reduced by the cosmological expansion (cf. Schwartz, Ostriker and Yahil 1975: Ikeuchi, Tomisaka and Ostriker 1983).

This suggests that it may be easier to generate large-scale fluctuations from some sort of dissociation front: we need the "front" to change the sound speed of the gas by a small fraction of its propagation speed. The resulting pressure gradients then create peculiar velocities. The efficiency of converting radiation into kinetic energy can be high provided that cooling is not too serious. A specific mechanism highly efficient in this respect, invented by Hogan (1983), involved dissociation of H_2, which releases about 6 ev per molecule. If the primordial hydrogen were largely molecular, this could be a highly efficeint mechanism. Ionization of atomic hydrogen by UV can be equally effective if the UV is released in a pulse lasting less than the recombination time of $\sim 10^5 n_e^{-1}$ yrs (which is itself much less than the Hubble time at early epochs). When the pulse is over, the gas recombines, but remains on a higher adiabat than material that was never ionized. However, if the ionization is caused by UV from an object at the Eddington limit, rather than by an explosion with $L \gg L_{Edd}$, then the kinetic energy generated corresponds only to the luminous output over a recombination time.

A related proposal (Hogan 1983; Hogan and Kaiser 1983) is that inhomogeneities are generated by gradients in radiation pressure arising from randomly-distributed sources. If the sources were sufficiently intense to ionize the primordial gas, the resultant fluctuations could extend up to the Jeans mass of $\sim 10^{17} M_\odot$ (this being the largest scale over which a wave driven by radiation pressure can propagate), with a natural cut-off on still larger scales.

6.   IS PRIMORDIAL NUCLEOSYNTHESIS OF LIGHT ELEMENTS COMPATIBLE WITH $\Omega_b \cong 1$?

Primordial nucleosynthesis in the standard hot big bang yields gratifying agreement with the observed abundances of light elements (Yang *et al.*, 1984), provided that the baryon density is fairly low, with $\Omega_b h^2 \lesssim 0.035$. The dynamically inferred hidden mass in clusters and groups, discussed in Geller's contribution to this meeting, contributes $\Omega \gtrsim 0.1$ - marginally consistent with the orthodox nucleosynthetic inference if $h \cong 0.5$, but apparently requiring some non-baryonic matter if $h \cong 1$. Advocates of inflationary cosmology favour $\Omega \cong 1$, which is compatible with the dynamical evidence if the hidden mass is less clumped than the galaxies even on scales as large as $10 h^{-1}$ Mpc.

It might seem that baryons cannot provide the critical density without violating the nucleosynthesis constraints. This conclusion is, however, neither mandatory nor clear-cut, because if one indeed hypothesises a large amount of dark baryonic matter, one has (to be self-consistent) to explore other new effects that could influence the abundance of light elements. The baryonic hidden mass must now be in Population III objects or their remnants (contributing a high M/L). In a baryon-dominated hot big bang universe, large amplitude inhomogeneities in the baryon-photon ratio are prerequisites for the formation of Population III. If the scales of structure now observed have formed hierarchically via graviational clustering of smaller units since the recombination time t_{rec}, then the masses of these units would have to be in the range $10^6 - 10^8 M_\odot$ (the precise value depending on the slope of the fluctuation spectrum shown in Figure 1(c)). In a baryon-dominated universe, this scenario requires isothermal fluctuations frozen in before recombination. Moreover, the amplitude may be $\gtrsim 1$ on the overdense regions. In consequence, the gas that escapes incorporation into Population III objects and remains available for condensation into luminous matter would come selectively from underdense regions   The yields from primordial nucleosynthesis depend on

the *local* value of the photon-baryon ratio. Therefore, the material
available for present-epoch observations (i.e. the fraction of baryons
that escape incorporation into Population III) would not be a fair
sample of all baryonic matter, and could have D, He and Li abundances
characteristic of low n_b/n_γ, even if $\langle n_b/n_\gamma \rangle$ were high enough to contribute $\Omega_b = 1$.

Large amplitude isothermal fluctuations were discussed by Hogan
(1978). If we define $\delta = (n_b - \bar{n}_b)/\bar{n}_b$, then a region with $\delta > 1$ can
still be treated as a small-amplitude perturbation of the cosmological
model - in the sense of being a small perturbation of the metric -
provided that, when the background temperature is T,

$$\delta \times (n_b/n_\gamma) \left(\frac{kT}{m_p c^2} \right)^{-1} (\ell/\ell_H)^2 \ll 1.$$

ℓ being the lengthscale of the perturbation and ℓ_H the horizon scale.
Perturbations with δ as large as 10 - 100 satisfy this requirement
right back to the nucleosynthesis stage even on mass scales as large as
$10^6 M_\odot$. Isothermal perturbations of mass $> 10^6 \delta^{-\frac{3}{2}} M_\odot$ would condense
out immediately after recombination, having maintained constant ampli-
tude ("frozen in") until that time - they would form Population III
objects (either single very massive objects (VMOs), or clusters of low-
mass stars). Only the underdense material between such perturbations
would survive to form the luminous content of galaxies. It is plainly
possible to envisage an $\Omega_b \cong 1$ universe with isothermal perturbations
such that (for instance) half the volume of the universe has
$n_b/n_\gamma \cong 3 \times 10^{-8}$ and half has $n_b/n_\gamma \cong 3 \times 10^{-10}$. If the high-density
matter were in lumps of $10^5 - 10^6 M_\odot$ then only the chemical composition
of the remainder (characteristic of a low density universe) might now
be measurable.

No very satisfactory model for isothermal fluctuations yet exists
(irregular deposition of entropy in a phase transition? spatial variations
in the CP-violation parameter?); there is certainly no theoretical basis
for specifying the spectrum such fluctuations might have. The above
example is merely illustrative of the principle that a high initial
n_b/n_γ or Ω_b can be compatible with primordial production of D, ^{3}He and
^{7}Li.

ACKNOWLEDGEMENTS

I am grateful to many colleagues, especially Bernard Carr,
Craig Hogan and Sasha Kashlinsky, for discussions or collaboration on
topics discussed in this paper.

REFERENCES

Blumenthal, G. and Primack, J.: 1983, Ap.J., in press.
Bond, J.R., Arnett, W.D. and Carr, B.J.: 1984, Ap.J., in press.
Bookbinder, J., Krolik, J., Cowie, L.L., Ostriker, J.P. and Rees, M.J.:
        1980, Ap.J., 237, p. 647.
Canizares, C.R.: 1982, Ap.J., 263, p. 508.
Carr, B.J., Bond, J.R. and Arnett, W.D.: 1984, Ap.J., in press.
Chang, K. and Refsdal, S.: 1979, Nature, 282, p. 561.
Doroshkevich, A.G., Zeldovich, Y.B. and Novikov, I.D.: 1967, Sov.
        Astron., 11, p. 233.
Fabian, A.C. and Kembhavi, A.: 1982, in "Extragalactic Radio Sources"
        ed. Heeschen and Wade (Reidel).
Gott, J.R.: 1981, Ap.J., 243, p. 140.
Harrison, E.R.: 1970, Phys. Rev. D.1, p. 2726.
Hartquist, T.W. and Cameron, A.G.W.: 1977, Astrophys.Sp.Sci., 48, p. 145
Hogan, C.J.: 1978, M.N.R.A.S., 188, p. 781.
Hogan, C.J.: 1983, M.N.R.A.S., 202, p. 1101.
Hogan, C.J. and Kaiser, N.: 1983, Ap.J., 274, p. 7.
Ikeuchi, S.: 1981, Publ.Astron.Soc.Japan, 33, p. 211.
Ikeuchi, S., Tomisaka, K. and Ostriker, J.P.: 1983, Ap.J., 265, p. 583.
Kashlinsky, A. and Rees, M.J.: 1983, M.N.R.A.S., 205, 955.
Ostriker, J.P. and Cowie, L.L.: 1981, Ap.J. (Lett), 243, L127.
Peebles, P.J.E.: 1983, Proc. 3rd Moriond Astrophysics Conference, in
        press.
Peebles, P.J.E. and Yu, J.T.: 1970, Ap.J., 162, p. 815.
Schwartz, J., Ostriker, J.P. and Yahil, A.: 1975, Ap.J., 202, p. 1.
Yang, J. *et al.*: 1984, Ap.J. in press.
Young, P.J.: 1981, Ap.J., 244, p. 756.
Zeldovich, Y.B.: 1972, MNRAS, 160, p. 1.

DISCUSSION

PEEBLES: i) I would urge that we not use the phrases "primevally
adiabatic" and "primevally isothermal" as equivalent to the "fragmentation" and "hierarchy" scenarios. As you note, either initial
state can lead up to either scenario. ii) As the Vaucouleurs remarks,
we should not take too seriously the density parameter $\Omega \sim 0.2$ derived
from the virgocentric motion of the Local Group under the assumption
that mass and galaxies are equally distributed because the estimate
of Ω assumes spherical symmetry and the Local Supercluster is far
from spherically symmetric.

REES: i) I agree — the "cold" non-baryonic matter hypothesis leads
to a hierarchical build-up of cluster, even if the initial perturbations are purely adiabatic. ii) Yes. Hofmann and Salpeter have
also emphasized how uncertain this inference is if the galaxies

are not good tracers of the overall mass distribution.

DRESSLER: You suggested that Pop. II and Pop III might form coevally. Could you clarify the role of Pop. III in producing the heavy element abundance of Pop. II and/or the much higher abundance of the intra cluster gas?

REES: My own ideas are very unclear on these topics! The only point I made which is in any way novel or unconventional is that the heavy element abundance could have been built up to a higher level than that in Population II by pregalactic processes: this could happen if Population II stars formed in subgalactic units which subsequently agglomerated into galaxies, at the same time as a (massive) Population III. I don't think the high intracluster abundances actually require this kind of explanation, however.

THE DYNAMICAL EVOLUTION OF CLUSTERS AND GROUPS OF GALAXIES

A. Cavaliere (1), P. Santangelo (2), G. Tarquini (2) and
N. Vittorio (2)
(1) Istituto di Astronomia, Università di Padova, Italy
(2) Istituto Astronomico, Università di Roma, Italy

Abstract: Computer simulations of the evolution of clusters and of groups
(including degrees of freedom internal to galaxies) test how the dissipationless clustering scenario meets the challenge posed by the observed
variety of morphologies and structures.

Clusters of galaxies show in X-rays (Forman and Jones 1982) and in
the optical band (Geller and Beers 1982) a remarkable variety of morphologies including some smooth, regular configurations but many more mul-
ticomponent structures indicative of quite unrelaxed conditions; the
range is essentially coeval (all morphological types are represented in
the local environment $z < 0.1$), and the types are apparently uncorrela-
ted with mass. We address the question of whether dissipationless hierarchical clustering from reasonable initial conditions can meet the challenge by brute morphology over the scales $10^{12} - 10^{15}$ $M_\odot$.

For cluster simulations, we use a canonical N-body integrator to e-
volve dynamically the initial conditions, thus producing the global gravitational potential $V(\underline{r},t)$ with resolution ≥ 50 kpc. We give two different representations of the system's state (compared in the caption to
Fig.1): isopleths of the projected galaxy counts; isophotes of the projected X-ray brightness from thin Bremsstrahlung emitted by intracluster plasma in hydrostatic equilibrium, valid over times longer than the
relevant plasma time scales. Our ensemble of 20 initial conditions (an
extension of that analyzed by CDSV 1983) is comprised of different random samples from a homogeneous sphere of 500 generally equal point-mas-
ses with Hubble-like velocities but negative total energy (expansion factor 6); occasionally, we use a Schechter mass distribution, or add a small
random $\underline{v}$ field. The initial t corresponds to cosmic epochs $z \cong 10 - 20$.

The resulting isopleths and maps reproduce with gratifying realism
on scales 100 kpc - a few Mpc all the morphological types observed (clumpy, bimodal, relaxed, cf. Figs. 1,3,2) in an evolutionary sequence whose
effective time scale has itself a large variance (range $\cong 6$ at constant
total mass). In fact, the cluster evolution may be understood in terms of
two coupled sequences: (i) general expansion, halt and recollapse; (ii)

F. Mardirossian et al. (eds.), Clusters and Groups of Galaxies, 499–504.
© *1984 by D. Reidel Publishing Company.*

early generation of much structure in the form of groups and subclusters, to be gradually erased through coalescence of lumps and slow increase of their specific binding energy (White and Rees 1978). The coupling acts in the sense that subclustering much developed on certain scales inhibits or delays higher clustering levels, including overall collapse; so, well detached multiple components last long against general recollapse (cf. Fig. 3). Much of this non-linear evolution is predictable from the map of the initial residual binding energies: $b_i = v_i^2/2 - \Sigma_{i \neq j} V_{ij}$ – contribution at $\underline{r}_i$ of the homogeneous expanding sphere. Figs. 6 and 7 compared with Figs. 1, 2 and 3 visualize how slow these systems are in forgetting the initial imprints. Ellipticities tend to persist well after the canonical virialization time. Effects of a realistic external environment conceivably increase the variance in morphology and time scales found here.

Focussing then on the internal evolution of subgroups, we simulate aside compact, initially expanding groups of tens of galaxies each endowed with 10^2 internal degrees of freedom. In particular, we compare the galaxy merging processes in the absence and in the presence of a dark, collisionless and diffuse component (ses CCS 1981 and CSTV 1983 for details). Fast formation obtains of a large central merger with realistic phase-space structure (cf. Fig. 10, and Thuan and Romanishin 1981), but only when the diffuse component is not dominant, and the system's $\langle v^2 \rangle$ does not exceed considerably the galactic internal one. Both conditions are bound to be less and less fulfilled in a subcluster as the clustering proceeds: $\langle v^2 \rangle$ then increases and a substantial fraction of the galactic halos gets stripped off (cf. White and Rees 1978) thus slowing down and eventually stopping any merging process. In fact, simulations of the post-collapse evolution (Merritt 1983, Richstone and Malumuth 1983) stress the difficulties for growing a cD body in a formed cluster, and even for maintaining it except than at a minimum of the potential $V(\underline{r})$, where an extended envelope may accrue (cf. Dressler 1984). In the present picture, cD bodies form in small aggregates of galaxies; cDs in clusters signal limited interference of the forming cluster with the internal evolution of its subclusters.

The compounded body of the present results linking the evolutions of diverse structures: cDs, groups and clusters, support as agent pure gravitational instability working since early z on density fluctuations with a broad power spectrum white-noise-like as for securing early galaxy formation and little linear interference of scales $>10^{13} M_0$ with their substructures. Presence of much mass in a truly diffuse form at the subclustering stage would suppress the merging efficiency necessary to build up even a few cD bodies against the limiting processes effective in clusters; at earlier epochs, it would weaken the subcluster structure and the related time scale variance. Sub-cluster structure might be started at late z by collapse and fragmentation of supercluster masses in a hot adiabatic scenario (cf. Zeldovich et al. 1982), but this is in several respects orthogonal to the present one, and should be tested in turn for morphologies and timings on all scales. On the basis of dissipationless instability, a unified picture including supercluster texture requires a broad spectrum with more power than white noise on scales $>10^{15} M_0$ (cf. Frenk et al. 1983, Primack and Blumenthal 1983), yet such as to conserve progressive and well-spaced (on the average) collapse times for increasing masses.

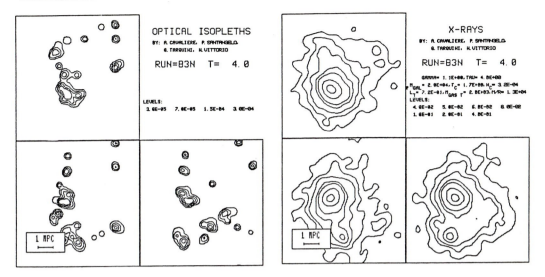

Fig. 1. A clumpy configuration at a relatively early stage of evolution in one of our cluster simulations. Optical isopleths and X-ray isophotes are shown. Note that the correspondence is good but not complete because isopleths stress local enhancements and include superposition effects, while the X-ray maps show true features of the large-scale gravitational potential with a stress on deeper minima. As total emission is mapped here, the contrasts tend to be enhanced realative to a soft X-ray telescope image; notice also the (large) scale of the isopleths as plotted here.

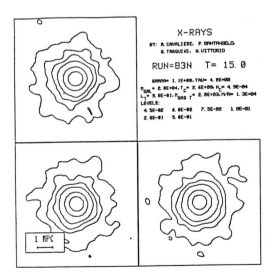

Fig. 2. The same run as in Fig. 1 takes on a smooth, relaxed configuration at a later stage, well after the canonical recollapse time t_c (our time unit is $t_c/2\pi$).

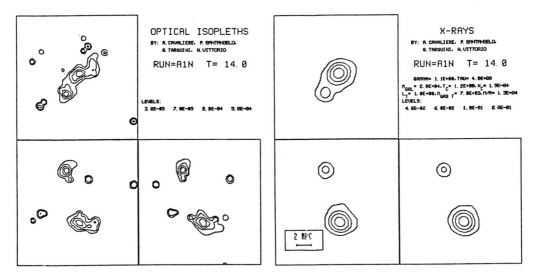

Fig. 3. A run from different initial conditions maintains for long times a sharp bimodal configuration, with widely separated lobes. Their centers of mass will then approach with large velocity, about 1000 km/s at a separation of 3 Mpc.

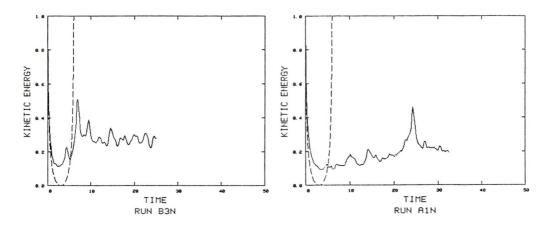

Figs. 4,5. Kinetic energy vs. time for the two runs in Figs. 1,2 and 3 (solid lines) compared with the corresponding canonical model of a homogeneous sphere (dashed lines). These runs develop from two different random choices of the initial positions for 500 equal point masses within a homogeneous sphere. Note the longer effective time scales for the run A1N that develops into a sharp bimodal configuration.

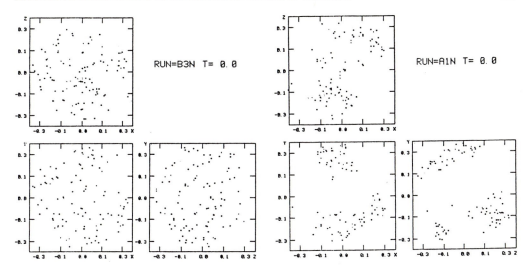

Fig. 6,7. Maps of the initial residual binding energy b_i for the two runs illustrated in Figs. 1,2 and 3; only the most bound points are shown. Note for the run A1N the close correspondence with the bimodal configuration well developed at times around $T \simeq 14$, cf. Fig. 3.

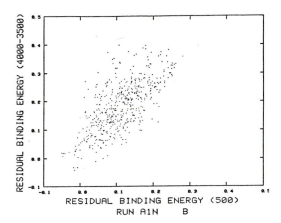

Fig. 8. To show how the residual binding energies b_i are altered when the original 500 points of run A1N are surrounded by 3500 points randomly placed in a shell with outer radius twice as large.

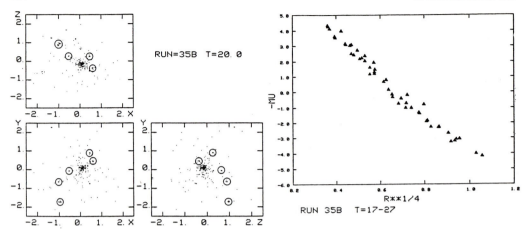

Fig. 9. Simulation of group dynamics: a late configuration taken by 20 galaxies originally with 60 internal degrees of freedom and mass spec- trum of Schechter's type (details in CSTV). The galaxies surviving the orgy of early merging are encircled. Diffuse mass is absent.

Fig. 10. The surface density of the central merger in Fig. 9 follows a de Vaucouleurs profile over a range of 9 equivalent magnitudes (M/L = const. throughout), as observed in group cDs. In this run $\langle v^2 \rangle$ increases out to 100 kpc.

References

Carnevali, P., Cavaliere, A. and Santangelo, P.: 1981, Ap.J. 249, p.449 (CCS
Cavaliere, A., De Biase, G., Santangelo, P. and Vittorio, N.: 1983, in
    "Clustering in the Universe", Gerbal and Mazure eds., Edition Fron-
    tières, p.15 (CDSV)
Cavaliere, A., Santangelo, P., Tarquini, G. and Vittorio, N.: 1983, loc.
    cit., p.25 (CSTV)
Dressler, A.: 1984, this volume, p. 117.
Forman, W. and Jones, C.: Ann. Rev. Astron. Astrophys. 20, p.547
Frenk, C.S., White, S.D.M. and Davis, M.: 1983, Ap.J. 271, p.417
Geller, M.J. and Beers, T.C.: 1982, PASP 94, p.421
Merritt, D.:1983, Ap.J. 264,p.24
Primack, J. and Blumenthal, G.: 1983, preprint of the invited talk at
    4th Workshop on GUT, April 1983
Richstone, D.O. and Malumuth, E.M.: 1983, Ap.J. 268, p.30
Thuan, T.X. and Romanishin, W.: 1981, Ap.J. 248, p.439
White, S.D.M. and Rees, M.J.: 1978, MNRAS 183, p.341
Zel'dovich, Ya.B., Einasto, J. and Shandarin, S.F.: Nature 300, p.407.

We acknowledge grants from CNR and MPI.

THE NON-LINEAR EVOLUTION OF THE CLUSTERING LENGTH IN THE "PANCAKE" SCENARIO

K. Górski   Warsaw University, Institute of Theoretical
                Physics, Hoża 69, 00681 Warsaw, Poland and
                Copernicus Astronomical Center, Warsaw
R. Juszkiewicz Copernicus Astronomical Center, Bartycka 18,
                00716 Warsaw, Poland

If the primordial perturbations were adiabatic and if neutrinos are massless, all fluctuations on scales smaller than the Silk length $\lambda_D=2(0.036+\Omega_o h^2)^{-1}$ Mpc are dissipated (here Ω_o is the density parameter and h_o the Hubble constant in units 100 $\text{kms}^{-1}\text{Mpc}^{-1}$), see ref. 1. The presence of massive neutrinos increases the damping length to $\lambda_{D\nu}=40\ M_{30}^{-1}$ Mpc, where M_{30} is the average neutrino mass in units 30 eV (2). In the first order perturbation theory, the gravitational instability manifests itself by the growth of the clustering length (i.e. the lag for which the correlation function has unit value). Therefore, it seems resonable to expect that the present mass clustering length is $r_o \gtrsim \lambda_D$, $\lambda_{D\nu}$. Since for the observed galaxy two-point correlation function $r_o=5/h_o$ Mpc (1), this condition is violated for most plausible values of the parameters $\Omega_o h^2$ and M_{30}. This is the so-called clustering length problem and it has been suggested as an argument against the "pancake" scenario of galaxy formation (3),(4).

In this paper we briefly present the results of an investigation of the developement of one-dimensional density perturbations   with a shortwave cutoff. The details of our calculations will be published elsewhere (10). Using the Zel'dovich approximation (5), we show that non-linear effects lead to a significant shortening of the predicted value of r_o.

The solution of Zel'dovich (5) relates the position of every fluid element in space $\underline{r}$ (i.e. its Eulerian coordinates) to its initial comoving (Lagrangian) coordinates : $\underline{r}=(ax-a^2\Psi(x),ay,az)\equiv(r,u,v)$. Here a is the scale factor, and we assume $\Omega_o=1$. This solution combines the Hubble flow (components $\propto a$) with the influence of one-dimensional density perturbations described by the vector $(-a^2\Psi,0,0)$. The density $\rho(\underline{r},t)$ of a fluid element in the neighbourhood of $\underline{r}$ at time t is given by the Jacobian $\partial(r,u,v)/\partial(x,y,z)$ of the transformation between the Eulerian and Lagrangian coordinates (5).

In our model we consider the density field in a periodic box $0\leq x\leq1$. The initial spectrum is flat, with a sharp long-wave cutoff at k=2π and short-wave cutoff at $k=k_D=200\pi$. More specifically, we calculate $\Psi(x)$

505

F. Mardirossian et al. (eds.), Clusters and Groups of Galaxies, 505–507.
© *1984 by D. Reidel Publishing Company.*

from the formula

$$\Psi(x) = \alpha \sum_{\ell=1}^{100} (A_\ell \cos k_\ell x - B_\ell \sin k_\ell x)/k_\ell ,$$

where $k_\ell = 2\pi\ell$; A_ℓ and B_ℓ are Gaussian random variables with zero mean and unit variance, determined by a random number generator. The parameter α measures the initial value of the correlation function at zero lag. For every realization (i.e. for every fixed set of A_ℓ and B_ℓ) we calculate the correlation function $\xi(r)=<\rho(r_1)\rho(r_2)>/<\rho>^2-1$, where $r=|r_1-r_2|$. The brackets $<...>$ denote spatial average over the box, followed by the averaging over five realizations.

It is very well known from comparisons with the N-body experiments (see references in (8)) that the Zel'dovich approximation gives a very accurate description of the evolution of the large-scale matter distribution, especially at early times, corresponding to $z>z_c$ (z_c is the redshift at which caustics start to develop in the model; the value of α in our model was adjusted so, that $z_c \cong 5$). After the pancakes have formed, this approximation fails to reproduce the dense knots which develop at the intersection of sheets and filaments. Therefore we have evolved ξ only up to $z=10$.

Our results are shown in figure 1. We have computed the evolution of $\xi(r)$ with redshift. At large redshifts ξ evolves self-similarly in agreement with linear theory, so that coherence length is time independent. Here the coherence length is the comoving separation, for which $\xi(x)/\xi(0)=.5$, (4). However, as time goes on, the slope of ξ increases while the coherence length shrinks. The clustering length grows less rapidly and its predicted present value r_o is shorter then expected in linear theory. Similar effects were recently observed in N-body experiments (6) and in calculations based on second-order perturbation theory (7). The mechanism responsible for the shortening of r_o was discussed earlier (7). It is the steepening of the density profiles $\rho(\underline{r},t)$, which transform from smooth sine waves into "N-waves" as $z \to z_c$.

Our results show that the inclusion of non-linear effects may resolve the clustering length problem, or at least make it less severe. However, one must bear in mind that the results reported here refer to a one-dimensional model and therefore can provide only a qualitative analysis of the non-linear clustering. In particular, one cannot exclude the possibility that the correct value of r_o will be shorter than in the linear theory but still too large to be consistent with observations (9). To investigate this problem quantitatively, realistic numerical simula - tions, including dissipative gas dynamical effects are necessary (the N-body models are dissipationless and therefore become inaccurate at $z<z_c$).

We conclude that before the development of the shape of ξ is investigated in realistic numerical simulations it would be premature to rule out the adiabatic scenario because of the clustering length problem apparent in linear theory.

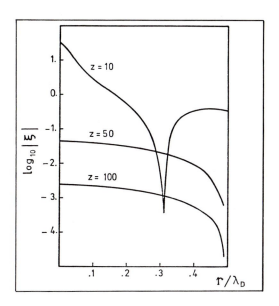

Figure 1. The computed correlation function $\xi(r/a)$ at redshifts z=100, 50 and 10. The comoving separation r/a is measured in units of the comoving cutoff wavelength λ_D/a. At large redshifts ξ evolves self-similarly in agreement with linear theory. However, as time goes on the slope of ξ steepens and its coherence length (or half-width) shrinks due to non-linear effects. The clustering length (i.e. the lag for which $\xi=1$) grows less rapidly than expected in linear theory and its final value r_o is shorter.

REFERENCES

1. Peebles P.J.E. (1980) ,*The Large Scale Structure of the Universe*
                         (Princeton U.P., NJ)
2. Bond J.R., Efstathiou G., Silk J. (1980), *Phys. Rev. Letters* <u>45</u>, 1980
3. Peebles P.J.E. (1982), in *Astrophysical Cosmology*, ed. Bruck H.A.,
                         Coyne G.V., Longair M.S. (Pontifica Acad. Scient.)
                         p. 165
4. Peebles P.J.E. (1981), *Ap.J.* <u>248</u>, 885
5. Zel'dovich Ya.B. (1970), *Astr.&Astrophys.* <u>5</u>, 86
6. Frenk C.S., White S.D.M., Davis M. (1983), *Ap.J.*,<u>271</u>, 417
7. Juszkiewicz R., Sonoda D.H., Barrow J.D. (1983), *Mon.Not.R.Astr.Soc.*
                         (in press)
8. Efstathiou G., Silk J. (1983), *Fundamentals of Cosmic Physics*, (in press
9. White S.D.M., Frenk C.S., Davis M. (1983), talk given at the third Mo-
                         riond astrophysics meeting
10. Górski K., Juszkiewicz R. (1983), *Mon.Not.R.Astr.Soc.* (submitted)

THE INHOMOGENEITY OF INVISIBLE MATTER IN AN EXPANDING UNIVERSE

L.Z. Fang
Dipartimento di Fisica, Università degli Studi di Roma
"La Sapienza", Rome, Italy
and

Astrophysics Research Division,
University of Science and Technology of China,
Hefei, Anhui, China[+]

1. Introduction

The density distributions of visible and invisible matter in the universe are quite different from each other. Visible objects are obviously clustered on the scales of galaxies, clusters and superclusters, but invisible matter is distributed fairly uniformly[1].

The flatness of the rotation curves of galaxies indicates that the length scale of the distribution of invisible matter around galaxies is larger than that of visible matter in galaxies. The growth of M/L with the size of the system also indicates the small magnitude of the inhomogeneity of invisible matter. The differences in the value of the deceleration parameter q_0 given by different methods are due mainly to the different hypotheses underlying the methods used. In the mean-mess density method, we assume that the distribution of all matter is proportional to galaxies, clusters or superclusters, and that invisible matter is distributed no more uniformly. Therefore, the systematic differences in q_0 probably imply the existence of a quasi-uniform distribution of invisible matter[2].

These results force cosmology to face a new question: how are such differences in the distribution of visible and invisible matter formed? That is, why, on all scales and especially on smaller scales, does visible matter cluster, while invisible matter has up to now remained quasi-uniform?

In this article we will show that, only if the densities and dispersion velocities of both visible and invisible matter satisfy appropriate conditions, can visible and invisible matter cluster differently, even in the stage of the Jeans instability.

[+] Permanent address

F. Mardirossian et al. (eds.), Clusters and Groups of Galaxies, 509–513.
© 1984 by D. Reidel Publishing Company.

2. Mechanism

In a universe with two components of matter, 1 and 2, there are four relevant time scales with respect to a density perturbation of length scale l, namely, two damping times defined by

$$t_d^1 \sim l/v_1 , \qquad t_d^2 \sim l/v_2 \tag{1}$$

and two growing times defined by

$$t_g^1 \sim (4 \pi G \rho_1^{(0)})^{-1/2} , \qquad t_g^2 \sim (4 \pi G \rho_2^{(0)})^{-1/2} \tag{2}$$

In (1) and (2), v_1, v_2 denote the dispersion velocities of 1 and 2 respectively, and $\rho_1^{(0)}$, $\rho_2^{(0)}$ the corresponding densities in the ground state. According to the relationship between the time scales, the responses to perturbations can be divided into the following cases:

a) growth in both components

$$t_g^1 < t_d^1 , \qquad t_g^2 < t_d^2 \tag{3}$$

b) damping in both components

$$t_g^1 > t_d^1 , \qquad t_g^2 > t_d^2 \tag{4}$$

c) growth in 1, but damping in 2

$$t_g^1 < t_d^1 , \qquad t_g^1 , t_g^2 > t_d^2 \tag{5}$$

d) growth in 2, but damping in 1

$$t_g^2 < t_d^2 , \qquad t_g^2 , t_g^1 > t_d^1 \tag{6}$$

e) growth in 1, slow growth in 2

$$t_g^1 < t_d^1 , \qquad t_d^2 > t_g^2 \gg t_d^1 \tag{7}$$

f) growth in 2, slow growth in 1

$$t_g^2 < t_d^2 , \qquad t_d^1 > t_g^1 \gg t_d^2 \tag{8}$$

Therefore, the responses to a perturbation l should be different. For instance, if the two components of matter satisfy the following conditions

$$\rho_1^{(0)} \ll \rho_2^{(0)} , \qquad \lambda_{1J} \ll \lambda_{2J} . \tag{9}$$

where the Jeans lengths are defined by $\lambda_{iJ} = (v_i^2/4 \pi G \rho_i^{(0)})^{1/2}$, then the response to a perturbation with $l \gtrsim \lambda_{1J}$ belongs to case (5) and $l \gtrsim \lambda_{2J}$ belongs to (8). This means that, under the conditions of (9), a significant inhomogeneity with a scale of about λ_{1J} can arise in 1, but not in 2. The component 2 can only have large-scale ($\sim \lambda_{2J}$) inhomogeneity.

3. Numerical results

In the Jeans stage of clustering, the growth or damping of a perturbation can be described by the development matrix $D_{ij}(k,t)$, defined by

$$
\begin{pmatrix} \delta_{1k}(t) \\ \delta_{2k}(t) \end{pmatrix} = \begin{pmatrix} D_{11}(k,t) & , & D_{12}(k,t) \\ D_{21}(k,t) & , & D_{22}(k,t) \end{pmatrix} \begin{pmatrix} \delta_{1k}(0) \\ \delta_{2k}(0) \end{pmatrix}
\tag{10}
$$

where $\delta_{ik}(0)$ and $\delta_{ik}(t)$ are the density inhomogeneities with length scale $\lambda = 2\pi/k$ in component i at initial time t_0 and t respectively.

As an example we have calculated the development matrix of a universe in Newtonian cosmology with two kinds of collisionless gas 1 and 2. The results are shown in Figs. 1 and 2, in which we take

$$\rho_1^{(0)} / \rho_2^{(0)} = 0.1 , \qquad \lambda_{1J} / \lambda_{2J} = 0.1$$

and t is in unit of $(4\pi G \rho_2^{(0)})^{-1/2}$, k in $2\pi/\lambda_{2J}$.

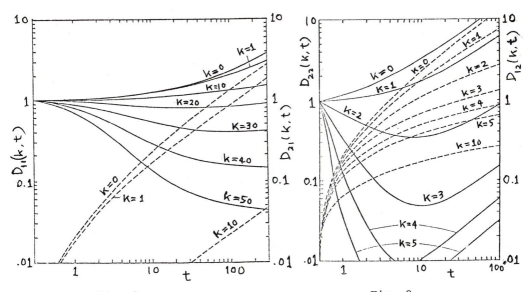

Fig. 1

The development matrix elements $D_{11}(k,t)$ (real lines) and $D_{21}(k,t)$ (dotted lines) as functions of t.

Fig. 2

$D_{22}(k,t)$ (real lines) and $D_{12}(k,t)$ (dotted lines).

It can be seen from Figs. 1 and 2 that the clusterings of components 1 and 2 are quite different. In the early period, the perturbations with $1 \lesssim k \lesssim 10$, namely $\lambda_{2T} \gtrsim l \gtrsim \lambda_{1T}$, are growing only in 1, but are damped in 2. The perturbations with $k \lesssim 1$, i.e. $l \gtrsim \lambda_{2j}$, are growing in both components, but much less in 2 than in 1; that is, we always have

$$\delta_{1k} > \delta_{2k} \quad \text{or} \quad \delta_{1k} \gg \delta_{2k} \qquad (11)$$

The evolutions of the spectrum of inhomogeneities are shown in Figs. 3 and 4. The initial perturbations in Fig. 3 are taken as a white spectrum in 1, but as no perturbation in 2, i.e.

$$\delta_{1k}(0) = 1 , \qquad \delta_{2k}(0) = 0 \qquad (12)$$

In Fig. 4, they are taken as

$$\delta_{1k}(0) = 0 , \qquad \delta_{2k}(0) = 1 \qquad (13)$$

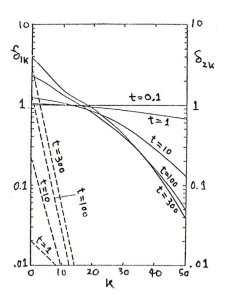 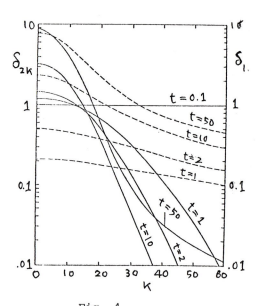

Fig. 3

The evolution of the spectrum of inhomogeneities $\delta_{1k}(t)$ and $\delta_{2K}(t)$. Initial spectrum $\delta_{1K}(1) = 0, \delta_{2k}(0) = 0$

Fig. 4

$\delta_{1k}(t)$ and $\delta_{2k}(t)$. Initial spectrum $\delta_{1k}(0) = 0, \delta_{2k}(0) = 1$

4. Conclusions

From Figs. 1-4, we find that, regardless of whether the initial perturbations exist in 1 or 2 or whether the length scales of perturbations are large or small, the clustering of 1 is always stronger than that of 2 (only excepting very long length perturbations). This result does not depend on the detail of the dispersion mechanism, but only on the conditions of (9). We find similar results for a universe with two components described by hydrodynamics or a universe consisting of a collisionless gas plus a fluid.

Conditions (9) can be satisfied in a neutrino dominated universe. If the rest mass of the neutrino is larger than several ev, we have $\rho_\nu \gg \rho_B$, ρ_ν and ρ_B being the mass densities of neutrinos and baryons respectively. Moreover, after the recombination epoch, the Jeans length of the neutrino component $\lambda_{\nu J}$ is much larger than that of the Baryonic component[3]. Therefore, if the invisible matter consists mainly of massive neutrinos, the observed differences in inhomogeneities can occur.

Finally, we should stress that conditions (9) are closely related to the properties of invisible matter. Thus, from the clustering feature of invisible matter, one can infer some restrictions on the composition of invisible matter.

References

(1) L.Z. Fang, S.P. Xiang, Acta Astrophys. Sinica, to be published
(2) L.Z. Fang , T. Kiang, F.H. Cheng, F.X. Hu, Quarterly J. Roy. Astr. Soc. 23, 363 (1982)
(3) L.Z. Fang, Y.Z. Lin, Lett. Nuovo Cimento, 32, 129 (1981)

FORCES BETWEEN OVERLAPPING PLUMMER-MODEL GALAXIES

Farooq Ahmad
Department of physics, University of Kashmir
Srinagar-I90 006, Kashmir,India
and
International centre for theoretical physics
34I00,Trieste,Italy

In groups and clusters of galaxies interpenetrating collisions are common.In the coma cluster,for example, a typical galaxy may undergo three interpenetrating collisions in 10^{10}yrs. It has been found convenient to represent density distribution in galaxies(Fujimoto et al I977) and in clusters of galaxies(White I976) by Plummer-model. Estimates for energy transfer in a collision between two identical Plummer-model galaxies was made by Toomre(I977), and Ahmad(I979) extended this work for non-identical cases.

In above calculations we need gravitational potential energy between such systems. The mutual potential energy due to gravitational interaction of two galaxies of masses M_1 and M_2 and scale lenghts α_1 and α_2 respectively is given by Ahmad(I979):

$$|W| = \frac{3GM_1M_2}{4\pi} \int_0^{2\pi}\int_0^{\pi}\int_0^{\infty} \frac{r_2^2 \sin\theta_2\, d\varphi_2\, d\theta_2\, dr_2}{(\alpha_1^2+r_1^2)^{1/2}(\alpha_2^2+r_2^2)^{5/2}}, \qquad \ldots(I)$$

r_1 and r_2 are related by:

$$r_1^2 = r^2 + r_2^2 - 2\,r_2 r \cos\theta_2, \qquad \ldots(2)$$

where r is the sepration between two galaxies. For the cases when r=0, analytical expressions are given by Ahmad(I979). For any separation the mutual potential energy can be obtained analytically by solving the triple integral (eq.I) after few transformations and is given by:

$$|W| = \frac{GM_1M_2}{(\alpha_1+\alpha_2)} A(\alpha_1,\alpha_2,r)\,; \quad A \equiv \frac{\alpha_2^2(\alpha_1+\alpha_2)}{2r}(I_1+I_2) \qquad \ldots(3)$$

where I_1 and I_2 are given by:

$$I_1 = \frac{(\mu_2-\mu_1)}{\sqrt{[(\mu_2-r)^2+\alpha_1^2](\mu_1^2+\alpha_2^2)^3}}\left[(r-\mu_2)\gamma(r)J_{11}+(\mu_1+\mu_2-2r)\sqrt{\gamma(r)}\,J_{12}+(r-\mu_1)J_{13}\right]\ldots(4)$$

and

$$I_2 = \frac{(\mu_2-\mu_1)}{\sqrt{[(\mu_2-r)^2+\alpha_1^2][\mu_1^2+\alpha_2^2]}(\mu_1^2+\alpha_2^2)}\left[(r-\mu_1)\gamma(0)J_{21}+(\mu_1+\mu_2-2r)\sqrt{\gamma(0)}\,J_{22}+(r-\mu_2)J_{23}\right]\ldots(5)$$

F. Mardirossian et al. (eds.), Clusters and Groups of Galaxies, 515–516.
© 1984 by D. Reidel Publishing Company.

where

$$\gamma(\ell) = \frac{(\mu_2 - \ell)^2 + \alpha_1^2}{(\mu_1 - \ell)^2 + \alpha_1^2} \qquad \text{and} \qquad \gamma(0) = \gamma(\ell) \quad \text{when} \quad r=0 \qquad \dots(6)$$

$J_{11}, J_{12} \dots, J_{23}$ are given by the relations:

$$J_{11} = \sum_{i=1}^{2}(-1)^i \left[F(\beta_i, K) - D(\beta_i, K) - \frac{\sin\beta_i \cos\beta_i}{\sqrt{1 - K^2 \sin^2\beta_i}} \right] / (1-K^2) \Bigg\}$$

$$J_{12} = \sum_{i=1}^{2}(-1)^i / K \sqrt{1 - K^2 \sin^2\beta_i} \quad , \quad J_{13} = \sum_{i=1}^{2}(-1)^i \sqrt{1 - K^2 \sin^2\beta_i} \qquad \dots(7)$$

and

$$J_{22} = \sum_{i=3}^{4}(-1)^i D(\beta_i, K) \quad , \quad \bar{J}_{22} = \sum_{i=3}^{4}(-1)^i \sqrt{1 - K^2 \sin^2\beta_i} \Bigg\}$$

$$J_{23} = \sum_{i=3}^{4}\left[F(\beta_i, K) - D(\beta_i, K) \right] \cdot (-1)^i \qquad \dots(8)$$

$D(\beta_i, K)$ is defined as:

$$D(\beta_i, K) = F(\beta_i, K) - E(\beta_i, K)/K^2 \qquad \dots(9)$$

$F(\beta, K)$ and $E(\beta, K)$ are elliptical integrals of first and second kind respectively. β_i's are given by:

$$\tan\beta_1 = \frac{\mu_1}{\mu_2}\sqrt{\gamma(\ell)} \; , \; \tan\beta_2 = \sqrt{\gamma(\ell)} \; , \; \tan\beta_3 = \frac{\mu_1}{\mu_2 \sqrt{\gamma(0)}} \; , \; \tan\beta_4 = \frac{1}{\sqrt{\gamma(0)}} \qquad \dots(10)$$

μ_1, μ_2 and K are by the expressions:

$$\mu_{1,2} = \left[(\alpha_1^2 - \alpha_2^2 + \ell^2) \pm \sqrt{(\alpha_1^2 - \alpha_2^2 + \ell^2)^2 + 4\alpha_2^2 \ell^2} \right] / 2\ell \; , \; \alpha_1 > \alpha_2 \quad \dots(11)$$

and

$$K^2 = 1 - \gamma(\ell)/\gamma(0) \qquad \dots(12)$$

For any case when K^2 becomes negative and $|K| > 1$, then replace $F(\beta, K)$ and $E(\beta, K)$ by new F and E's as indicated by Gradshyteyn and Ryzhik(1980). The numerical results of W can be easily computed for various α_1, α_2 and r from equation(3). Knowing W, the parabolic velocity of escape V_{esc} of galaxies in a cluster or a group is readily obtained from:

$$\frac{1}{2}\frac{M_1 M_2}{(M_1 + M_2)} V_{esc}^2(\ell) + W(\ell) = 0 \qquad \dots(13)$$

The change in the internal energy ΔU and orbital energy ΔE of galax- due to the tidal interaction can be estimated using eq(3).

REFERENCES

Ahmad,F.:1979, Ap.space Sci.,60,493
Fujimoto,M.,Sofue,Y.and Jogorku,T.:1977,Publ.Astr.Soc.Japan,29,1
Gradshteyn,I.S. and Ryzhik,I.M.:1980, Tables of integrals,series
        and products (Academic press) P 172-174.
Toomre,A.:1977, in the Evolution of Galaxies and Stellar Populations
        eds. B.M.Tinsley and Larson (Yale Uni.Observatory) P401
White, S.D.M.:1976, Mon.Not.R.Astr.Soc. 174,90

COSMIC STRING–LOOPS AND GALAXY FORMATION –
ANGULAR MOMENTUM AND THE HUBBLE SEQUENCE

Piers Corbyn
Queen Mary College, London

The cosmic string-loop model is developed whereby super-massive string-loops form nucleation centres for galaxies. Estimation of the interaction of string-loops with matter gives acceptable decay times and radiative power for (proto) galaxy cores. The modes of formation and angular momentum properties of string-loops suggest an explanation for the elliptical/spiral discontinuity and a modified Hubble tuning-fork sequence.

THE MAJOR PROBLEMS

The cosmic string model appears to overcome major problems of the fragmentation ("pancake") and aggregation theories.

(i) It provides natural "originating density perturbations" which (unlike the pancake theory) do not significantly upset the micro-wave background.

(ii) Unlike the pancake theory it enables galaxies to cluster in a scale-free way <u>after</u> the formation of proto-galaxies around string loops.

(iii) It provides a natural origin for the rotation of spiral and SO galaxies, and explains in terms of loop types why ellipticals have low angular momentum.

LOOP SIZE, LIFETIME AND POWER

Extending the work of Kibble (1976), Zel'dovich (1980) and Vilenkin (1981) the theoretical existence of long-lasting loops was demonstrated by Kibble and Turok (1982) and an equation for a typical loop found (Turok 1983a). Loops of radius ~ 10pc could be formed in the early universe in the right numbers and give the right density perturbations to nucleate galaxies.

517

F. Mardirossian et al. (eds.), Clusters and Groups of Galaxies, 517–519.
© *1984 by D. Reidel Publishing Company.*

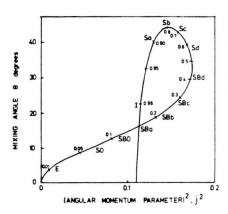

FIG. 1: Mixing angle for spinning
string loop as a function of (angular
momentum parameter)^2. Values of
α ($0 < \alpha < 1$) are marked along the curve
with estimates of what galaxy types
they may correspond to.

FIG. 2: A new physical sequence for galaxies. This modifies the standard Hubble
sequence.

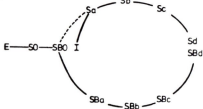

The slowing down time, T_L, and the corresponding power, P_L,
radiated from a pulsating and shrinking string loop can be estimated
by Chandrasekhar drag on the surrounding infalling gas of density
ρ (Corbyn 1984):

$$T_L^{-1} \simeq 3\pi^2 \mu G^2 \beta \hat{\rho} a / c^3; \quad P_L \sim \pi c^2 \mu \, a \, T_L^{-1} \tag{1}$$

where βc is the string speed and $\mu = 10^{20}$ kg m^{-1} is its' effective
mass per unit length (Turok 1983b). For $\beta = 1$, a=10pc and $\rho = 10^{-10}$kg m^{-3}
we get $T_L \sim 2\times10^9$yr and $P_L \sim 10^{38}$ W which compares well with typical
Seyfert core powers. It may be that various types of collapsing
strong-loop, some of which may become black holes, could account
for such sources as N radio galaxies, quasars and Seyfert cores.

LOOP ANGULAR MOMENTUM, BARRING AND HUBBLE TYPE

The three basic loops (I,II,III) of very low, intermediate
and high angular momentum could possibly nucleate E, SO and S galaxies
respectively. Detailed calculation (Corbyn 1984) gives loop angular
momentum:-

$$J_L = \pi \mu c a^2 j \quad ; \quad 0 < j < 0.4 \tag{2}$$

The approximate probability distribution of j for type III loops corresponds to the observed distribution of angular momentum of spiral galaxies and typical angular momenta size also compare well.

The time varying loop field gives a preferred plane containing two antiparallel preferred directions for the attraction of material. Providing J_L is large enough this favours the growth of two armed rotating planar structures which we suggest may form bars if the mixing angle θ between J_L and the normal to the preferred plane is small enough. We present θ (j^2) in Fig. 1 with suggested positions for corresponding galaxy types, and the resulting modified Hubble sequence, Fig. 2.

REFERENCES

Kibble, T.W.B. and Turok, N. 1982 Phys Lett <u>116B</u>, 141.
Turok, N. 1983a Phys Lett 126B, 437 (1983b: Private comunication).
Corbyn, 1984 In preparation.

THE INFLUENCE OF ENVIRONMENT ON GALAXY FORMATION

E. J. Shaya and R. Brent Tully
Institute for Astronomy, University of Hawaii

A schema of galaxy formation is developed in which the environmental influence of large-scale structure plays a dominant role. This schema was motivated by the observation that the fraction of E and S0 galaxies is much higher in clusters than in low-density regions and by an inference that those spirals that are found in clusters probably have fallen in relatively recently from the low-density regions. It is proposed that the tidal field of the Local Supercluster acts to determine the morphology of galaxies through two complementary mechanisms. In the first place, the supercluster can apply torques to protogalaxies. Galaxies which collapsed while expanding away from the central cluster decoupled from the external tidal field and conserved the angular momentum that they acquired before collapse. Galaxies which formed in the cluster while the cluster collapsed continued to feel the tidal field. In the latter case, the spin of outer collapsing layers can be halted and reversed, and tends to cancel the spin of inner layers. The result is a reduction of the total angular momentum content of the galaxy. In addition, the supercluster tidal field can regulate accretion of fresh material onto the galaxies since the field creates a Roche limit about galaxies and material beyond this limit is lost. Any material that has not collapsed onto a galaxy by the time the galaxy falls into a cluster will be tidally stripped.

The angular momentum content of that part of the protogalactic cloud which has not yet collapsed continues to grow linearly with time due to the continued torquing by the supercluster and neighbors. Galaxies at large distances from the cluster core can continue to accrete this high angular momentum material until the present, but galaxies that enter the cluster are cut off from replenishing material. Consequently, low angular momentum systems formed in the environment that produced the initial cluster because of incipient tidal locking and were subsequently deprived of fresh infall material because of a restrictive Roche limit. High angular momentum galaxies formed where the tidal spin-up mechanism was uninhibited and where accretion continued for a long time, perhaps until today.

F. Mardirossian et al. (eds.), Clusters and Groups of Galaxies, 521–522.
© 1984 by D. Reidel Publishing Company.

The distinctive morphological components, disks and spheroids, are determined by the specific angular momentum of a system. If the angular momentum content is high, then the collapse factors required to achieve centrifugal equilibrium are not large and disks can be formed. When the angular momentum content is low, collapse factors are larger, central densities are expected to be higher, and if star formation occurs before the equilibrium condition is achieved, a spheroid is formed.

MASS ESTIMATION OF CLUSTERS WITH UNSEEN MASS

Haywood Smith, Jr.
Department of Astronomy, University of Florida

Unseen mass can bias cluster mass estimates using the virial theorem, as implied by Limber's (1959) modification of the virial theorem and confirmed by the author (Smith 1980) using N-body experiments. The virial mass M_V becomes

$$M_V = M_G \left(1 + K_2 \frac{M_B}{M_G} \right)$$

where M_G is the mass of visible matter (e.g., galaxies) and M_B is the mass of background (unseen) matter. The factor K_2 depends on the distributions of galaxies and background; if they are the same K_2 has the value unity, whereas if the background is more centrally concentrated than the galaxies $K_2 > 1$.

Further analysis of the numerical experiments indicates (perhaps not surprisingly) that the value of K_2 and therefore that of M_V/M_T (where $M_T = M_B + M_G$) do not depend on the detailed form of the distributions but mainly on the ratio of the scale lengths characterizing them. In particular we have used the median radius in projection r_m. The results from the numerical experiments are shown in Fig. 1.

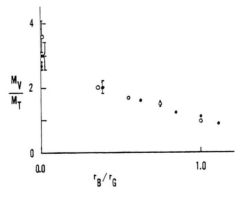

Figure 1. M_V/M_T vs. r_B/r_G, where r_B and r_G are the median radii of background and galaxies, respectively. <u>Open circles:</u> equal-mass systems; <u>filled circles:</u> systems with an inverse-square mass spectrum.

F. Mardirossian et al. (eds.), Clusters and Groups of Galaxies, 523–524.
© *1984 by D. Reidel Publishing Company.*

There appears to be a well–defined relation regardless of the background's shape (point, Plummer model, sphere).

The effect of unseen mass can be reduced by using the radial velocity profile to estimate the potential and thereby M_T, as with the potential-estimation (PE) method (Smith 1982). The method has the advantage of not requiring special assumptions about the nature of the unseen matter; however, it does not work if applied to the inner region (in projection) of a cluster. It is of interest to see whether it works using only the most luminous (massive) members. Results for samples limited by luminosity are given below.

BACKGROUND

Sample	None	Sphere	Point
Brightest $\frac{1}{4}$	0.96 ± 0.07 (m.e.)	1.28 ± 0.08	1.35 ± 0.50
Brightest $\frac{1}{2}$	0.91 ± 0.07	1.23 ± 0.09	1.28 ± 0.14
All	0.94 ± 0.09	1.23 ± 0.17	1.28 ± 0.28
			(est.)

The PE method is evidently insensitive to selection by magnitude.

REFERENCES

Limber, D.N. 1959, Astrophys. J., 130, 414.
Smith, H. 1980, ibid., 241, 63.
Smith, H. 1982, ibid., 259, 423.

ADIABATIC PERTURBATIONS IN A PHOTINO DOMINATED UNIVERSE

Riccardo VALDARNINI
International School for Advanced Studies, Strada Costiera 11
Trieste - Italy

In this paper the results are presented of a numerical integration for adiabatic perturbations in a photino (X) dominated Universe (Bond et al.,1982; Blumenthal et al.,1982), starting from an initial redshift $z_i=10^9$, down to $z_F=10^3$. The X perturbations are coupled to the matter radiation fluid through the gravitational field equations. For the X particles a fully Boltzmannian kinetic treatment is used, while matter and radiation perturbations are treated in the ideal fluid approximation. Our notation follows Peebles (1982) but the numerical scheme of integration is different. We start from the metric $ds^2=dt^2-a^2(t)(\delta_{\alpha\beta}-h_{\alpha\beta})$ (α , β =1,2,3) and from eqs.

$$\ddot{h}+2\dot{a}\dot{h}a^{-1}=6(\delta_{rad}+c_x e_x \delta_{x,\beta}) \quad \dot{h}_{33} -\dot{h}=6a^{-3}k^{-1}(4v/3+c_x e_x f_x) \qquad (1)$$

($h=h_{\alpha}^{\alpha}$) relating h with the matter velocity and δ (fluct. amplitudes); here the wave number k is related to λ =2πa(t)k^{-1} while $c_x=15\beta_x a_x^4/2\pi^4$ and $a_x=T_x/T_{rad}$; T_x is the X temperature; β_x is the sum of X spin states. In eqs.(1) $e_x = \int_0^{\infty} p^2 q dp(e^p +1)^{-1}$ with $q^2=p^2+(m_x/T_x)^2$, $e_x \delta_{x,\beta}=\int_0^{\infty}g_I^{(+)} q^{-1} (p^2+0.5a^2)p^3 e^p(e^p+1)^{-2} dp$, $e_x f_x =\int_0^{\infty}g_I^{(-)} p^4 e^p (e^p+1)^{-2} dp$. According to Peebles (1982) p is the proper momentum in units of T_x/m_x while

$$g_I^{(+)} =\frac{1}{2}\int_{-1}^1 g d\mu \quad , \qquad g_I^{(-)} =-\frac{i}{2}\int_{-1}^1 g \mu d\mu \quad , \qquad \mu =\cos\theta \qquad (2)$$

and g=g(t,μ ,p) is related to the perturbations of the X distribution, being

$$\partial g/\partial t +ik\mu pga^{-1}q^{-1} =((1-\mu^2)\dot{h}+(3\mu^2 -1)\dot{h}_{33})/4 \qquad (3).$$

Units are chosen such that a(t)=m_x/T_x(t). To integrate the system g is split into Legendre polynomials over the μ-space,the real part is composed by the even armonics

$$g=\sum_{\ell=0}^{\infty}(\sigma_{2\ell}\frac{4\ell+1}{2} P_{2\ell} +i\sigma_{2\ell+1}\frac{4\ell+3}{2}P_{2\ell+1}) \qquad (4).$$

Then eq.(2) becomes (1 $\gtrsim$ 0)

$$\dot{\sigma}_\ell =(-)^\ell kpa^{-1}q^{-1}(\sigma_{\ell+1} (1+1)+ \sigma_{\ell-1} 1)/(2l+1)+(\dot{h}/3)\delta_{\ell 0} +(\dot{h}_{33} -\dot{h}/3)\delta_{\ell 2} /5 \quad (5).$$

The series is truncated at some l_{max} such that $|\sigma_{\ell max}/\sigma_3| \lesssim 10^{-2}$ at z_F ($l_{max} \simeq 10^2$). the integration over p has been done with a set of points $p_1 ,p_2 ,\ldots,p_{N_P}$ using a Gauss-Laguerre quadrature (N_P =10). The routines which integrates the set of differential equations is a Merson one that is optimizing dt and guaranting high computational speed for small k, as well as high accuracy for large k. The values of k have been chosen such that the mass range $M_x =\frac{\pi}{6} \rho_{x_0} \lambda_0^3 =(10^9 -10^{20})M_{\odot}$ is

525

F. Mardirossian et al. (eds.), Clusters and Groups of Galaxies, 525–526.

swept with 5 values of k in each order of magnitude change in M_X .
We have taken Ω_o =1, H=100 Kmsec^{-1}Mpc^{-1} , T_{rad} (z=0)=2.7°K .
In Fig.s 1-3 the mass variance given by
$$(\delta M/M)^2 = \text{const} \times \int_o^{k(M)} k^2 \mid \delta_k (z=z_F)\mid^2 dk \qquad (6)$$
is plotted (a:n=-2; b:n=-1; c:n=0; d:n=1; e:n=2).
The following cases have been considered: $a^3_X = a^3_\nu$ =4/11, $\beta_X = \beta_\nu$ = 6
(Fig.1); a^3_X =1/25, β_X =2 (Fig.2); a^3_X =1/50, β_X=2 (Fig.3).
Acknowledgments. I am grateful to S.Bonometto for suggesting the field
of discussion and checking some results with those of a work he had in
progress.

References

Bond,J.R.,Szalay,A.S.,Turner,M.S.,1982,Phys.Rev.Lett.,48,1636
Blumenthal,G.R.,Pagels,H.,Primack,J.R.,1982,Nature,299,37
Peebles,P.J.E.,1982,Ap.J.,258,415

FIG.1   FIG.2

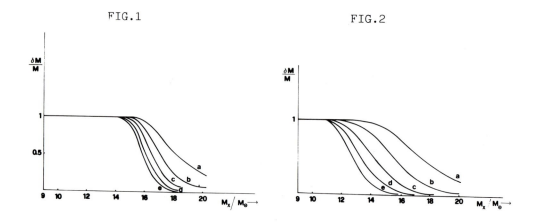

FIG.3

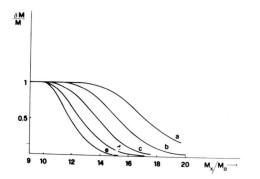

ESTIMATION OF THE VELOCITY LAW OF INFALLING GALAXIES IN A SPHERICALLY SYMMETRIC CLUSTER FROM OBSERVATIONAL DATA

Nikos Voglis
University of Athens, Department of Astronomy
Panepistimiopoli, Athens 621, Greece

Consider a cluster of galaxies with spherical symmetry centered at O (fig. 1). Define cylindrical coordinates (r, ϕ, ζ) with the origin at O, the axis ζ along the line of sight and (r, ϕ) on the plane of the sky. In fig. 1 the circle (O, R_{max}) represents the sphere of maximum expansion. Galaxies inside this sphere infall while outside still expand. Galaxies at points like M,P,M′ show no deviation in their redshifts from the redshift of the center. Therefore, there must exist points like N,N′ so that a galaxy there shows maximum deviation.

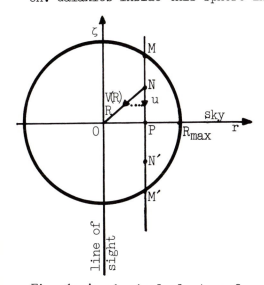

Fig. 1: A spherical cluster of galaxies and the sphere of maximum expansion. Galaxies at points N,N′ show maximum deviation in their redshifts from the redshift of the center O.

Let $\Delta Z = |Z - Z_c|$ be the absolute deviation of the redshift of a particular galaxy at (r, ζ) from the redshift of the center. If R is the distance of this galaxy from the center and V(R) its velocity relative to the center, the projection of V(R) on the line of sight is ($|u| = c\,\Delta Z$)

$$u = \frac{V(R)}{R}\,\zeta \qquad (1)$$

Inserting (1) into $R^2 = r^2 + \zeta^2 \Longrightarrow$

$$\frac{u^2}{V^2} + \frac{r^2}{R^2} = 1 \qquad (2)$$

Equation (2) maps all the galaxies at distance R on an ellipse (Palmer and Voglis, 1983). Those of the ellipses which correspond to $R < R_{max}$ have an envelope line. The condition for this line is obtained by differentiating (2) with respect to R (keeping u and r constants) and combining the result with (2). So we find

527

F. Mardirossian et al. (eds.), Clusters and Groups of Galaxies, 527–528.
© 1984 by D. Reidel Publishing Company.

$$\frac{d\ln V(R)}{d\ln R} = -\frac{r^2}{\zeta^2} \tag{3}$$

Calculating the partial derivative of u with respect to ζ from (1) and using (3) we get

$$\frac{\partial u}{\partial \zeta} = 0 \tag{4}$$

which is the condition for maximum $|u|$ or ΔZ as we explain in the first paragraph. Therefore, if we plot ΔZ vs r for many galaxies of the cluster, the diagram has to be as it is shown qualitatively in fig. 2. The line (a) in this figure is the envelope of the ellipses given by (2) (for infalling galaxies) and corresponds to condition (3). Similar diagrams have been produced by Struble, 1979, from observational data, and this theoretical prediction agrees with these observational diagrams very well.

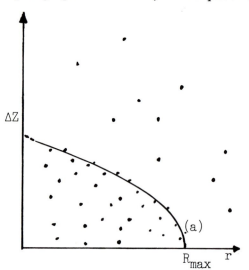

Fig. 2: The redshift deviations of galaxies in a spherical cluster. The envelope line (a) is the locus of maximum deviations of infalling galaxies. Observational data agree with this theoretical prediction very well.

Differentiating (1) with respect to r and using the condition (3) we find the following relation between the logarithmic derivatives holding along the line (a).

$$\frac{\partial \ln u}{\partial \ln r} \equiv \frac{d\ln \Delta Z}{d\ln r} = \frac{d\ln V(R)}{d\ln R} \tag{5}$$

So, if we plot ΔZ vs r or $\ln \Delta Z$ vs $\ln r$ using observational data of the cluster, and fit line (a) by a function we can find

$$\frac{d\ln V(R)}{d\ln R}$$

that is, the velocity law of infalling galaxies. Since the right-hand end of the line (a) points at R_{max} (fig. 2) we can also evaluate the radius of maximum expansion.

Obviously, if V(R) obeys a power law, R^n, $\Longrightarrow \frac{d\ln V(R)}{d\ln R} = n = $ constant and the line (a) is a straight line in logarithmic coordinates with slope equal to the exponent n.

The best candidates for applying this method are Virgo and Coma clusters.

REFERENCES

Palmer, P.L. and Voglis, N.: 1983, Mon. Not. R. Astron. Soc., in press.
Struble, M.: 1979, Astronomical Journal 84, p. 27.

THE HUBBLE DIAGRAM

G. A. Tammann*
Astronomisches Institut der Universität Basel
European Southern Observatory, Garching

I. INTRODUCTION

When Hubble (1929) made the first announcement of the linear expansion of the universe, he had estimated the distances to 24 galaxies (6 of which in the Local Group and 4 in the Virgo cluster) for which the "measured velocities" v were known. The velocity range of $-220 < v < 1100$ km s^{-1} gave in today's eyes only shaky evidence for an expanding "universe", although Hubble knew already Humason's (1929) velocity of $v = 3779$ km s^{-1} for NGC 7619 in the Pegasus cluster. Hubble's term "measured velocities" indicates that he did not doubt that the observed redshifts were due to a Doppler shift. He presented his data in a plot of velocities versus linear distance (Fig. 1). This diagram, which does not distinguish - appropriately enough for that time - between different kinds of distances, will not be referred to in the following as a Hubble diagram. The principal difficulty of a plot containing "linear distances" is illustrated by a paper by de Sitter (1930), in which he made surprisingly vague reference to Hubble (1929), and where he combined luminosity distances and diameter distances to demonstrate the expansion out to the Coma cluster.

A conceptionally clean way to present the expansion of the universe is to use a plot of log cz (z = redshift) versus apparent (K-corrected) magnitude or apparent (metric) diameter. The former relation is generally referred to as the "Hubble diagram", whereas the latter is due to Sandage (1972a).

The Hubble diagram and the log cz-apparent diameter diagram do not require linear distances, but only relative distances as indicated

---

* Visiting Associate Mount Wilson and Las Campanas Observatories

F. Mardirossian et al. (eds.), Clusters and Groups of Galaxies, 529–552.
© *1984 by D. Reidel Publishing Company.*

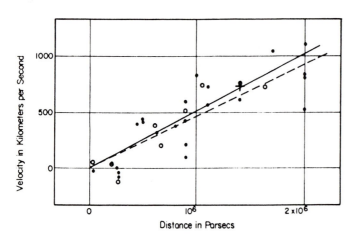

Fig. 1. Hubble's
1929 velocity-
distance diagram.
(Proc. Nat. Acad. Sci.).

by the apparent magnitudes (or diameters) of <u>standard candles</u> (standard
rods). The Hubble diagram of galaxies with widely different luminosities
would be a scatter diagram, because at any redshift galaxies would
occur over a wide range of absolute magnitude and hence of apparent
magnitude. A cosmologically useful Hubble diagram contains therefore
only extragalactic objects of constant absolute magnitude, i. e. standard
candles, whose apparent magnitudes are unbiased indicators of their
relative luminosity distance. It cannot be sufficiently stressed that the
application of the Hubble diagram implies the existence and knowledge
of standard candles.

II. THE IMPORTANCE OF THE HUBBLE DIAGRAM

For a number of reasons the Hubble diagram of standard candles
has played and still plays a dominant role in observational cosmology.
The major applications are related to the following topics:

- The Hubble diagram of first-ranked cluster galaxies, used as standar
candles, still provides the primary evidence of the large-scale expansion of the universe. Most importantly, it proves that the universe expands <u>linearly</u>, because the slope of 0. 2, expected for the ridge line
in the <u>linear</u> case from

$$\log cz = 0.2 \; m_{corr} + c, \tag{1}$$

(where m_{corr} is the apparent magnitude corrected for galactic absorption and the K-effect) is verified with high accuracy within the domain,
where other cosmological effects remain negligible (Sandage, Tammann,

and Hardy 1972; Sandage and Hardy 1973). Any non-linear overall expansion of the universe would give room only to a finite number of fundamental observers and hence contradict the Copernican principle.

- Hubble (1938) was apparently the first to realize that the Hubble diagram at large z-values could be used in principle to determine the
fundamentally important present deceleration parameter q_0. The necessary Friedmann redshift-magnitude relation for different values of
q_0 was known in closed form only since Mattig (1958), but already
earlier series expansions existed for the relation (Heckmann 1942):

$$m_{corr} = 5 \log z + 1.086 (1 - q_0) z + 0(z) + const,   \qquad (2)$$

where $0(z)$ are higher-order terms in z. The hope to determine q_0
from eq. (2) at sufficiently high values of z has stimulated to a large
extent the enormous observational effort which has been invested into
the Hubble diagram of first-ranked cluster galaxies.

- The possibility that first-ranked cluster galaxies could be degraded
as standard candles by their secular luminosity evolution was first
investigated, - probably inspired by the spurious Stebbins-Whitford
(1952) effect which suggested an excess reddening of galaxies, - by
Robertson (1955) and HMS. The possibility became a reality when it
was shown that much of the light of elliptical galaxies stems from giant
stars (Baldwin, Danziger, Frogel, and Persson 1973) and that they must
hence undergo rather rapid luminosity evolution. Although the effect of
luminosity evolution can be estimated from model calculations (e. g.
Tinsley and Gunn 1976), the hopes to determine q_0 from the Hubble diagram of brightest cluster galaxies has been considerably impaired. But
in partial compensation of the information loss the Hubble diagram can
provide valuable constraints on the effect of luminosity evolution.

- The hopes for ever determining q_0 from the Hubble diagram of first-
ranked cluster galaxies have been further dimmed by the expected
cannibalism between cluster galaxies: smaller galaxies are accreted
by larger ones (Ostriker and Tremain 1975; Gunn and Tinsley 1976;
Ostriker and Hausman 1977; Hausman and Ostriker 1978). Attempts to
model the ensuing magnitude change (e. g. Hoessel 1980) depend on the
assumption of (quasi-)homology of the galaxies involved, and because
this assumption is questionable (Schweizer 1981) this route to q_0 seems
to hold little promise. Concurrently with information on luminosity evolution, brightest cluster galaxies may, however, reveal their canniba-
listic habits, particularly if their Hubble diagram is compared with that
of n-th members in clusters.

- If the full scatter of brightest cluster galaxies about the ridge line of the Hubble diagram is read in $\Delta \log cz$ (instead of Δm caused by absolute magnitude scatter) one obtains upper limits for the peculiar or streaming velocities of cluster centers (Sandage 1972b). They can then be used to determine the density parameter Ω_0 ($\Omega_0 = 2 \, q_0$ in all Friedman models) from the cosmic virial theorem or from streaming models.

- The Hubble diagram is important to carry the distance scale out to large distances and hence to determine the global value of the present Hubble constant H_0 by means of one or a few standard candles, whose distances (and absolute magnitudes M_{SC}) can be determined nearby, no matter how insignificant their redshifts may be. This follows from eq. (1), where the constant term c is given by

$$c = -0.2 \, M_{SC} + \log H_0 - 5,$$

and hence

$$\log H_0 = \log cz - 0.2 \, (m_{corr} - M_{SC}) + 5. \tag{3}$$

Any pair of values of redshift z and apparent magnitude m_{corr}, - read off the ridge line of the Hubble diagram, - will provide H_0 from eq. (9), if M_{SC} is known from one or more nearby fellow standard candles. The method in this simple form should not be carried beyond $z \approx 0.1$ ($v_0 \approx 30\,000 \, km \, s^{-1}$), because cosmological effects become non-negligible. Its accuracy is limited by the intrinsic luminosity scatter σ (M_{SC}) and by the number of calibrating standard candles used. The method has been pioneered by Sandage (1968) using brightest cluster galaxies and the Virgo cluster as calibrator, and has again been applied for the 10 brightest cluster galaxies (Tammann 1977) and for supernovae of type I with two nearby supernovae as calibrators (Sandage and Tammann 1982).

The exceptional importance of the Hubble diagram of standard candles and the unique observational effort which has gone into the Hubble diagram of bright galaxies, may justify that some of the historical background is recalled in the following (cf. also Sandage 1975).

III. EARLY DEVELOPMENT

Within a span of only two years Humason had extended the observed redshift range by a factor of five (!), and Hubble and Humason (1931) published the first real Hubble diagram, which includes the Leo cluster at $cz = 19\,600 \, km \, s^{-1}$ (Fig. 2). The notion of standard candles was still somewhat vague, because the magnitudes were mean apparent

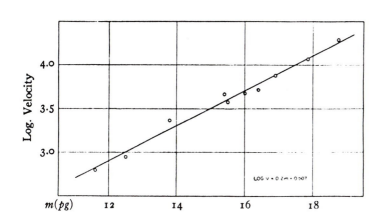

Fig. 2. The first
Hubble diagram
(1931) of Hubble
and Humason.
(Astrophys. J.).

magnitudes of the cluster members with known redshift. This point
was conceptionally improved by Humason (1936) who published a Hubble
diagram for the 5th brightest cluster members, after having measured
100 new redshifts, including the Ursa Major 2 cluster (discovered by
W. Baade in 1931) at cz = 42 000 km s^{-1} (Fig. 3). With this Humason had
rather exhausted the observational possibilities of the Mount Wilson
100 inch Telescope, and further progress stagnated until the 200 inch
Hale reflector went into operation in 1949. With the new instrument
Humason obtained many new redshifts, remeasured old ones, and set
with a 40 hour exposure a new record of cz = 61 200 km s^{-1} for the Hydra
cluster. These data were collectively published by Humason, Mayall,
and Sandage (1956; hereafter HMS).

With this Humason had reached a barrier. He attempted with his
unusual dedication to measure still higher redshift, but he remained
unsatisfied with the results. Only at his retirement he published (Humason 1957) the redshifts of four distant clusters, on which he had spent
hundreds of hours of exposure time and whose redshifts he could
estimate only from unmeasurably faint spectral features. As a tribut
to his perseverance the results of three of these clusters are compared
here with modern ones. The comparison for the first cluster is ambi-
guous because it has two components A and B, and interestingly Humason suggested two solutions. The other two clusters compare reasonably well.

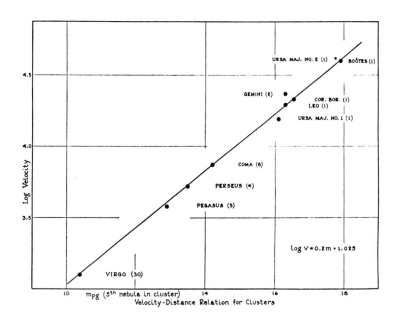

Fig. 3. Humason's (1936) Hubble diagram for the 5th brightest cluster
members. (Astrophys.J.). The same diagram was used also by Hubble
(1936a, 1936b), who gave in addition a diagram for field galaxies with,
as expected, much larger scatter.

Table 1
Humason's redshifts for very distant clusters

RA (1950)	δ (1950)	cz (Humason) km s^{-1}	cz (modern) km s^{-1}
$0^h24^m0^s$	+ 16°53'	103 000 (or 167 000)	A: 87 000[2] B: 117 600[3,4]
10 44 11	+ 9 20[1]	63 000	67 800[4]
14 47 33	+ 26 22	118 000	110 700[5]

[1] Abell 1093
[2] Baum (1962)
[3] Sandage (1972b)
[4] Sandage et al.(1976)
[5] Gunn and Oke (1975)

The situation of the Hubble diagram was reviewed by Hubble (1953)
in his George Darwin Lecture (Fig. 4), which was posthumously published

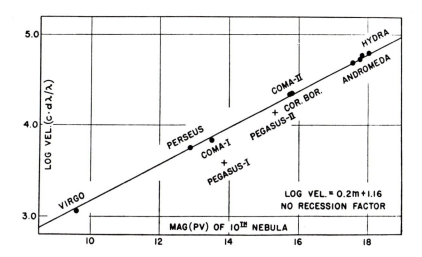

Fig. 4. Hubble's last Hubble diagram (Astrophys. J.). Note the small scatter, the perfectly linear relation, and the remark "no recession factor", i.e. no K-correction has been applied to the magnitudes.

from his notes by A. Sandage. Hubble used the 10th brightest cluster galaxies as standard candles to obtain a plot with surprisingly little scatter, and he labelled the ordinate of his plot as "log vel. $(c \cdot d\lambda/\lambda)$" and spelled out the relation "log vel. = 0.2 m + 1.16". But he added "no recession factor"! With this he meant, that no K-correction for the redshift effect was applied to the magnitudes. (To avoid confusion it is noted here that in the 1920's the word "K-term" was used for the expansion velocities, in analogy to Galactic astronomy; in the modern literature the word "K-correction" is reserved for the magnitude effect of redshifts). Why did Hubble not apply this correction? He over-estimated the accuracy of his magnitudes, and he realized that, if he applied the K-corrections, his diagram would curve upwards and re-quire an unreasonably large value of q_0. He had run into the same problem with his galaxy counts (Hubble 1936c), such that he begun to favor seriously the idea of non-velocity redshifts. It is curious that Hubble died with a doubt as to the nature of galaxy redshifts.

In the early 1950's the possibilities for further progress seemed very limited for a threefold reason: (1) very few distant clusters were known which were candidates for still higher redshifts. The then most distant Hydra cluster had been only discovered because it appeared on a plate taken by Hubble of the Shapley-Ames galaxy NGC 2713 with the

100 inch Telescope. (2) The photographic emulsions were too slow and too grainy to record the continuum spectrum of very faint galaxies against the night sky such that the absorption lines could be measured with any degree of reliability. (3) No reliable magnitudes were available for faint galaxies.

An attempt was made by W. A. Baum to subtract the night sky light on the photographic spectra of faint galaxies, but the success of this ingenious method had to wait for another 20 years until linear detectors came into use. Another innovation of his, however, became an immediate success: he determined the redshifts of clusters from photoelectric, pulse-counting, multicolor photometry. By this method he obtained z = 0. 4 for the cluster 1447 + 2619 (Baum 1958), which he improved later to z = 0. 36 (Baum 1962; modern value z = 0. 369 [Gunn and Oke 1975]), and he presented a Hubble diagram which extends out to 3C 295 at z = 0. 44 (Baum 1962).

The record-breaking redshift of 3C 295 had been measured already before by Minkowski (1960) on an o p t i c a l spectrum; this has been possible only because (1) the corresponding cluster had been noticed thanks to its radio source, and (2) the strong radio galaxy shows e m i s s i o n lines, of which Minkowski could measure the single [OII] line at 3727 Å.

In spite of the limited prospects in the early 1950's for further progress, the improvement of the Hubble diagram was systematically pushed at the Carnegie Office of the Mount Wilson and Palomar Observatories. A. Sandage scanned the Palomar Sky Survey for new distant clusters and measured positions of the brightest members for subsequent blind-offset spectroscopy at the 200 inch Telescope. The high accuracy required for the positions necessitated very laborious observations to obtain direct plates in the prime focus of the 200 inch, once with the Ross corrector lense out (to obtain the exact scale of the spectrograph) and once with the lense in position (to mark the east-west direction on the plates). This work resulted in an unpublished list of 50 clusters, for about 20 of which Humason could obtain redshifts. Later Sandage measured with the same technique also the offset position of Minkowski's 3C295. - Existing photoelectric magnitudes of galaxies (Stebbins and Whitford 1952) were greatly extended by Pettit (1954); for 11 fainter clusters the photoelectric magnitudes were secured by Pettit and Sandage. For the photometry of the six faintest clusters Sandage took plates with a jiggle camera (schraffierkassette), which he had mounted at the 200 inch Telescope.

The redshifts and magnitudes available by 1955 for field and cluster galaxies were collected by HMS (Humason, Mayall, and Sandage 1956). The paper marks thus the end of the development phase, and is at the same time the first modern paper on the subject. It not only contains a wealth of redshift data, but gives also for the first time magnitudes reduced to a uniform isophotal diameter; consistent K-corrections are applied to the magnitudes (it should be noted that Hubble's K-corrections have never been correct [cf. Sandage 1975]), and the effect of possible luminosity evolution is discussed. Various Hubble diagrams are given for field galaxies with the typically large scatter of non-standard candles, and for cluster galaxies using as standard candles a combined magnitude of the 1st, 3rd, 5th, and 10th brightest cluster members; the resulting scatter is only $\sigma(m) = 0.32$. An improved solution for H_O and a tentative one for q_O are offered. After 27 years of research of very exacting kind the linear expansion of the universe was firmly established out to more than one third of the velocity of light.

MODERN RESULTS

A new aera of observational cosmology was introduced by Sandage's (1961) paper, in which he investigated systematically the consequences of various theoretical world models on the observable parameters (redshift, magnitude, number, and metric and isophotal diameters) and subsequently also on the expansion age of the universe (Sandage 1961a). It turned out that the count-magnitude relation of galaxies was unexpectedly insensitive to q_O, and that the Hubble diagram offered still, in spite of all the difficulties, the best possibility to find q_O.

As a consequence Sandage set out to further improve the photometry of cluster galaxies in 1963 (cf. Sandage 1966, 1968a, 1970a, 1970b), and to extend the Hubble diagram with his collaborators to higher redshifts, - a program which so far came to an end in 1978. New growth curves of galaxies and the (weak) dependence of the aperture corrections on q_O were determined (Sandage 1972a). New B, V photometry of bright members in 41 clusters and of 3C radio sources were combined with data of Westerlund and Wall (1969) and Peterson (1970) and used to establish a Hubble diagram of first-ranked cluster galaxies with a scatter of only $\sigma(M_V) = 0^m.32$ (Sandage 1972d, 1972b; the preliminary data were also published by Peach [1970] and McVittie [1972]). The Hubble diagram of strong radio sources gave a scatter of $\sigma(M_V) = 0.49$ and indicated that these galaxies are fainter on average by $0^m.3$ than first-ranked cluster galaxies; from the Hubble diagram of quasars with their very high optical luminosities and broad luminosity scatter it was concluded that they are normal (radio) galaxies with an additional bright, blue, non-thermal, stellar-like component superposed (Sandage 1972c; for

a Hubble diagram of radio galaxies see also Smith 1977). The Hubble
diagram of the underlying galaxies in N systems proved that the redshifts of these mini-quasars are of cosmological nature (Sandage 1973a).
A new S20 photomultiplier was used, beginning in 1967, to measure
BVR magnitudes of several of the brightest members in clusters, groups,
and of radio E galaxies; from these data followed an improved Galactic
reddening law (Sandage 1973b, 1975a; cf. Sandage and Visvanathan 1978)
and a Hubble diagram for first-ranked galaxies in rich clusters and
sparse groups (Sandage 1973c). The dependence of the luminosity of
the three brightest cluster galaxies on the Bautz-Morgan (1970) class
and on the cluster richness was established (Sandage and Hardy 1973)
with the result that first-ranked galaxies are particularly bright, 2nd
and 3rd galaxies, however, faint for Bautz-Morgan class I; after
correction for the Bautz-Morgan effect and for a weak cluster richness
effect the scatter in the Hubble diagram could be reduced to $\sigma(M_V)$ =
$0\overset{m}{.}28$. The weakness of the richness correction was further studied with
a much increased sample, including the Southern Hemisphere, and it
was concluded that the stability in luminosity of brightest cluster galaxies required a nearly vertical cutoff at the bright end of the galaxian
luminosity function (Sandage 1975a, 1976). This conclusion was con-
tested by Geller and Peebles (1976), but Tremaine and Richstone (1977)
also found evidence, that the luminosity of the first-ranked galaxy is
determined by some physical effect rather than by a statistical luminosity function sampling.

With Westphal's linear SIT television detector sky subtraction
methods and the extension of the Hubble diagram toward higher redshifts became possible in 1974. Sandage and Kristian embarked on a
photographic survey, taken with the 48 inch Palomar Schmidt Telescope
on IIIaJ and red 127-04 plates in the northern and southern polar regions, to find new distant clusters; the ensuing sample was increased
by Abell clusters, 3C radio sources in clusters, and by Sandage's
earlier unpublished list of distant clusters. 87 redshifts were obtained
up to z = 0.75, and photoelectric BVR photometry (in some cases also
with the SIT camera) was secured for most of the first-ranking galaxies.
The resulting Hubble diagram (Fig. 5) has small scatter, $\sigma(M) \lesssim 0\overset{m}{.}3$,
and requires a formal value of q_0 = 1.6 ± 0.4 (Sandage, Kristian, and
Westphal 1976; Kristian, Sandage, and Westphal 1978).

In independent investigation of the Hubble diagram of brightest
cluster galaxies was launched by Oke and Gunn and collaborators; this
program is still in progress (Oke 1984). Oke's multichannel photometer
at the 200 inch Telescope made possible to measure sky-subtracted
energy distributions of distant galaxies and thus to measure high redshifts (Oke 1970). Spectrophotometry and redshifts were obtained for

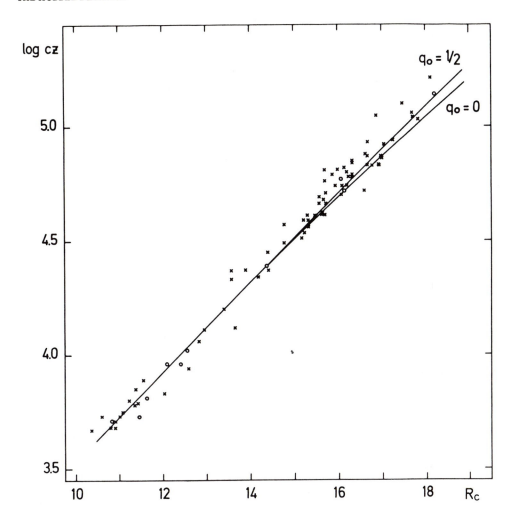

Fig. 5. The Hubble diagram from data by Kristian, Sandage, and West-phal 1976. The R magnitudes are aperture-corrected to a standard metric diameter with an assumed true value of $q_0 = 0$; Galactic absorption, K-, Bautz-Morgan, and richness corrections are applied. Open circles are for radio clusters.

distant first-ranked galaxies with $0.072 < z < 0.465$ (Gunn and Oke 1975). New g, r photometry was secured with a SIT camera and CCD detectors for a complete sample of nearby Abell clusters (Hoessel, Gunn, and Thuan 1980) and for 84 Abell clusters with $z < 0.3$ (Schneider, Gunn, and Hoessel 1983), as well as their redshifts. The discussion of these

data, although using a different formalism follows basicly that of
Sandage and collaborators. The correlation between absolute magnitude
of the first-ranked cluster galaxy with the Bautz-Morgan class was
closely confirmed, equally that on cluster richness. The resulting
scatter is again $\sigma(M) = 0\overset{m}{.}29$. The principle strategic difference is that
no K-corrections are applied, but that the apparent magnitude at the
rest wavelength of the galaxy is interpolated (for very high z-values
extrapolated) from the color data; in addition much emphasis is given
to the influence of selection effects of very distant clusters. For the
presently available data on first-ranked cluster galaxies it is possible
to recover the apparent magnitudes g_I (corrected for Galactic absorption,
redshift effects, Bautz-Morgan type, and richness); they are plotted
in Fig. 6 into a Hubble diagram. An eye fit to the data suggests a small
formal value of q_0. Indeed Hoessel et al. (1980) derived from the data
available in 1979 a preliminary formal value of $q_0 = -0.55 \pm 0.45$. It
will be very interesting tu see the final solution when the photometry
for the data discussed by Oke (1984) will become available.

It is disappointing to see that the two great programs for the Hubble
diagram of brightest cluster galaxies will apparently disagree in their
solutions for even the formal value of q_0; this is before any corrections
for luminosity evolution and dynamical evolution of cluster galaxies are
applied. One may therefore doubt whether it will be ever possible to determine q_0 from brightest cluster galaxies. To dim the hopes even
further, it should be stressed again that if the cosmological constant Λ
should be non-zero, a possibility which has gained attractiveness from
the inflationary universe theory, it is not possible to separate the effects
of q_0 and Λ unless the Hubble diagram is extended well beyond z = 1
(Sandage 1972b; Sandage and Tammann 1984a; for additional complications see e. g. Tammann 1974).

The question whether the Hubble diagram of brightest cluster galaxies is sensitive to selection effects is a very complex one. If these
galaxies have any luminosity scatter and if they are selected down to
an apparent magnitude limit it is in principle clear, that the most distant
ones should be overluminous. However, many of (Sandage's) first-ranked
cluster galaxies are radio selected, and because of the very wide radio
luminosity function a radio-strong galaxy is not necessarily optically
overluminous, - and indeed there is no clear difference in the Hubble
diagram of optical and radio selected cluster galaxies (or any such difference hinges strongly on the single case of 3C 295). More importantly,
first-ranked galaxies are not primarily selected on the basis of their
optical magnitude, but because of the detectability of their parent cluster. Once a cluster is detected, its brightest member should have perfectly random luminosity. An objection would be that some clusters

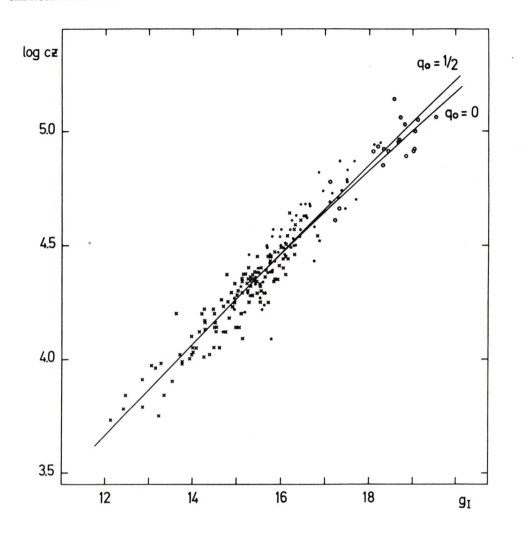

Fig. 6. The Hubble diagram from data by Gunn and Oke (1975; open symbols), Hoessel, Gunn, and Thuan (1980; x's), and Schneider, Gunn, and Hoessel (1983; crosses). g_I magnitudes are reduced to a galaxy radius of 16 kpc [H_O = 60; q_O (true) = 0.5], and are fully corrected.

may be only detected because they have an outstanding member; but such a cluster would be of Bautz-Morgan class I and would obtain the (hopefully) appropriate Bautz-Morgan correction. The greatest danger would seem to be that brightest cluster galaxies may be drawn from existing cluster lists because of their easy observability; clearly the clusters used should be a fair subsample of the clusters known.

It may be worthwile to make future experiments with cluster galaxies other than first-ranked ones. The luminosity of the latter may
still be correlated in some harmful way with the discovery chance of
the parent cluster. This would certainly not be the case for the 5th
brightest member for instance. Moreover fainter cluster members
commit much less cannibalism, which would reduce the enormous problem of dynamical evolution. They would also, probably, eliminate the
necessity of a Bautz-Morgan correction, and the remaining correction
for cluster richness exists anyhow for first-ranked cluster galaxies.

Several investigators are presently measuring infrared magnitudes
of cluster galaxies to establish a Hubble diagram at long wavelengths.
Smith (1983) has published such a diagram for corrected K-magnitudes
from data provided by M. J. Lebofsky, P. R. M. Eisenhardt, and G. H.
Rieke. The diagram contains optically and radio-selected galaxies and
there is no obvious difference between them. The objects extend to
well beyond $z = 1$ and suggest a formal value of q_0 in the order of unity.
The K-magnitude Hubble diagram of giant ellipticals in the 3C-catalog
suggests an even higher value of q_0 of 2.8 (Lilly and Longair 1982). As
for the optical Hubble diagrams these data may give above all information on the luminosity evolution of galaxies.

A sequence of record-breaking redshifts has been measured by H.
Spinrad with the Wampler scanner for radio-selected galaxies over the
past years. His latest published result is $z = 1.21$ for 3C 324, measured
with a CCD spectrograph (Spinrad and Djorgovski 1983). Still higher
values may become available soon. The information on the spectral
energy distributions emitted at an early cosmic epoch is most important, but in the absence of standard metric-diameter magnitudes the
data cannot be used for a Hubble diagram.

Hubble diagrams have been presented using other standard candles
than first-ranked cluster galaxies. Some of these are mentioned in the
following (cf. Section IV).

Weedman (1976) has used the mean nuclear magnitude of the five
and ten brightest galaxies in 9 clusters to derive a Hubble diagram with
very small scatter. Allowing for the fact that the definition of the nuclear
magnitudes is z-dependent, the magnitude scatter $\sigma(M)$ without Bautz-
Morgan and richness corrections is still about as small as for fully
corrected first-ranked cluster galaxies.

At maximum light supernovae of type I, which occurred in elliptical
galaxies (i.e. in dust-free galaxies), define a Hubble diagram with a
scatter of $\sigma[M_B(max)] = 0^m\!.4$ (Fig. 7), most of which is observational

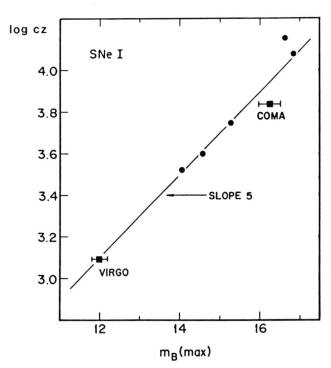

Fig. 7. The Hubble diagram for type I supernovae at maximum B light, which occurred in elliptical galaxies. The average magnitudes are shown for supernovae in the Virgo and Coma cluster. The straight line has the theoretical slope of 0.2 [cf. eq. (1)].

(Tammann 1982). Therefore these objects seem to be exceptionally good standard candles and to offer the best prospects for determining q_o, this even more so because they offer the advantage of point sources, they are unaffected by dynamical cluster evolution, and because one expects for them, if they are White Dwarfs driven over their Chandrasekhar limit, little secular luminosity evolution. To reach supernovae at sufficiently high redshifts requires, however, in view of their relative intrinsic faintness the Space Telescope (Tammann 1979) or a future Very Large Telescope (Tammann 1983).

Abell (1975 and references therein) has suggested that the logarithmic integrated luminosity function of galaxy clusters contained a well defined "knee" whose position, apparently governed by some universal physical process, could be used as a standard candle. A Hubble diagram with the apparent magnitudes of the knee was presented by Bautz and Abell (1973). Forcing the Virgo cluster on this relation, however, yields a peculiar velocity (Gudehus 1978) which is impossible in size and sign. It is therefore not clear whether the knee, which corresponds to Schechter's (1976) "characteristic luminosity" in his luminosity functions, can be determined with the necessary accuracy, and whether it may vary between different clusters. In any case it seems more practi-

cal to use the average magnitude of the n brightest cluster members as
a standard candle (Schechter and Press 1976).

Quasars with their wide optical luminosity function, already mentioned, give a degenerate Hubble diagram. It has been proposed, however, that on the basis of observable parameters quasar subsamples
could be chosen such that they had nearly constant luminosities. Particularly for flat-spectrum radio quasars Baldwin (1977) suggested that
a correlation existed between the strength of the Ly-α and CIV $\lambda 1550$
emission lines and the luminosity of the underlying continuum radiation.
Whether this relation can be used to reduce these objects to a common
standard luminosity and to derive then q_o, depends on the absence of
evolutionary effects (Wampler et al. 1983). Moreover, in case the still
not well determined intrinsic scatter of the relation is large and the
sample should be limited in any way by the optical apparent magnitude,
this kind of standard candle should be appropriately dealt with in the
next Section.

IV. DECEPTIVE STANDARD CANDLES

Some standard candles may have perfectly constant mean absolute
magnitudes, but large scatter about this mean. (For simplicity a
Gaussian scatter is assumed here, without loss of generality). If samples of such objects are limited by apparent magnitude, they yield degenerate Hubble diagrams, i. e. Hubble lines which curve upwards and
have a slope always larger than 0.2 as expected from eq.(1). The situation is explained in Fig. 8 and 9. If the objects have about constant space
density, their number increases with $\sim$ (distance)^3, and their large number at large distances will yield some objects which deviate by several
sigmas from the mean absolute magnitude, toward higher as well as
lower luminosities. If the objects are selected down to an apparent-
magnitude limit the intrinsically fainter objects are denied progressively with distance to the sample, and only exceptionally bright objects
can enter the catalog from large distances. The consequence is that the
mean luminosities of the sample increase with distance. The Monte
Carlo calculation in Fig. 8 is performed for a relatively nearby sample
of galaxies, but the scale of the diagram can be changed arbitrarily
(with some modification for large redshifts due to the K-correction).
All standard candles are in this sense "deceptive standard candles" if
their true intrinsic absolute-magnitude scatter, which is difficult to observe, is large [in practice $\sigma(M) \gtrsim 0^m_.5$]. If deceptive standard candles
(from an apparent-magnitude-limited catalog) are plotted into a Hubble
diagram (Fig. 9) they shall be more and more overluminous (shifted toward the left) as the redshift increases and this explains the degeneracy

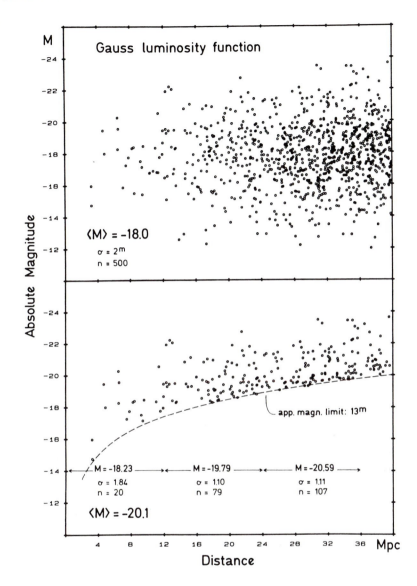

Fig. 8. Upper panel: Monte Carlo distribution in distance and absolute magnitude of 500 galaxies within 38 Mpc. Constant space density
and a mean absolute magnitude of < M > = -18ᵐ with a Gauss standard
deviation of σ_M = 2ᵐ are assumed. Lower panel: The same sample cut
by an apparent-magnitude limit of m = 13ᵐ.0. Note the increase of the
galaxian luminosities with increasing distance and the small effective
(observable) scatter σ_M within individual distance intervals. (By permission of A. Spaenhauer).

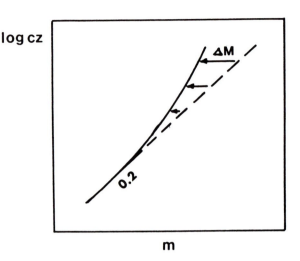

Fig. 9. Schematical
Hubble diagram for de-
ceptive standard candles,
i. e. for a magnitude-li-
mited sample of objects
with constant mean lumi-
nosity but large luminosity scatter. The resulting
luminosity increase with
distance yields always a
slope steeper than the
expected value of 0. 2.

of the diagram.

The effect described is well known in stellar statistics under the
term Malmquist effect, but it seemed worthwhile to discuss here in
some detail its consequence on the Hubble diagram.

The use of deceptive standard candles, or the underestimate of the
intrinsic luminosity scatter, has sometimes led to the conclusion that
the expansion velocities increased with the square of the distance
(Nicoll and Segal 1982, and references therein). This artefact could
therefore be coined as Segal's paradox.

Blunt examples of deceptive standard candles are for instance: (1)
unrestricted quasar samples with their wide range in optical luminosities, (2) strong radio galaxies if one chose to plot their apparent radio
fluxes instead of their apparent magnitudes, or (3) field galaxies
plotted with their apparent optical magnitudes or their 21 cm hydrogen
flux densities (Roberts 1972).

A less obvious case is presented by Sc galaxies of van den Bergh
luminosity class I. Although their apparent scatter in absolute magnitude is $\sim 0\overset{m}{.}6$, their true intrinsic scatter is much larger (Tammann,
Yahil, and Sandage 1979; Kraan-Korteweg, Sandage and Tammann 1984).
Why then could Sandage and Tammann (1975) claim that the Hubble diagram of ScI galaxies had the correct slope of 0. 2 (Fig. 10) ? The explanation is an accidental error compensation: the brighter galaxies are

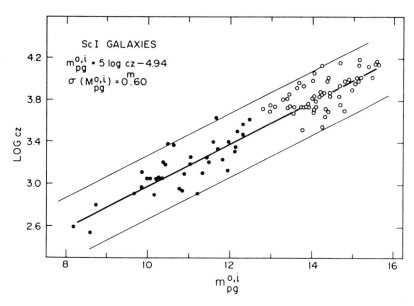

Fig. 10. The Hubble diagram of ScI galaxies. Dots are galaxies from the Shapley-Ames catalog, open symbols are from the Zwicky catalog. The galaxies seem to have reasonable scatter and to follow the theoretical line of slope 0.2. Cf. text. (From Sandage and Tammann 1975).

drawn from the magnitude-limited Shapley-Ames catalog, while the fainter galaxies come from the Zwicky catalog, which go to a fainter apparent-magnitude limit. The fainter, more distant galaxies should be intrinsically brighter, but this is compensated because they are sampled deeper into the luminosity function of ScI's.

The 21 cm line width-absolute magnitude relation (Tully and Fisher 1977) should make it possible to reduce sufficiently inclined spiral galaxies to a common line width and hence to a standard luminosity. The reliability of the resulting standard candles depends, however, on their true intrinsic scatter; results by Roberts (1978) suggest that it is non-negligible.

V. HUBBLE DIAGRAM, PECULIAR MOTIONS, AND HUBBLE CONSTANT

The peculiar velocities of galaxies within $\sim$ 10 Mpc are known to be small, i.e. $\sigma(v_p) < 100$ km s^{-1} for one dimension (Tammann, Sandage, Yahil 1980; Sandage and Tammann 1982a). However, our (three-dimensional) motion toward the 2.7 K microwave background, as seen from the centroid of the Local Group, is 630 ± 50 km s^{-1} in the direction $\alpha = 166 \pm 5^{\circ}$, $\delta = -26 \pm 5^{\circ}$ (Wilkinson 1983, and references therein), which clearly shows that larger peculiar motions occur over larger scales. Standard candles and their Hubble diagram are ideally suited to investigate peculiar motions over different scale lengths. (It should be noted that the determination of peculiar velocities does not require linear distances, but only <u>relative</u> distances).

If the scatter of the Hubble-diagram of fully corrected first-ranked cluster galaxies is read horizontally, one finds $\sigma(M) = 0\overset{m}{.}3$. If instead it is assumed, quite unrealistically, that all scatter is due to peculiar motions the corresponding value is $\sigma(\Delta z / z) = 0.115$ (Sandage 1972b). This gives a firm upper limit for the one-dimensional peculiar motions of cluster centers, i.e. $\sigma(v_p) < 115$, 230, and 345 km s^{-1} out to redshifts of $v_0 = 1000$, 2000, and 3000 km s^{-1}, respectively. It is unlikely that these peculiar velocities grow further with scale length, because the upper limit, at least for the Local Group, is given by the <u>three-dimensional</u> value of ~ 600 km s^{-1} mentioned above.

If instead one uses the relative distances, again from first-ranked galaxies, of the 10 nearest clusters and groups (all within $v_0 = 1800$ km s^{-1}) and assigns all scatter to peculiar velocities, one obtains $\sigma(v_p) < 110$ km s^{-1}. However, these objects exhibit additional direction-dependent velocity residuals, which can be removed if one assumes a local streaming velocity of $v_{VC} = 200 \pm 50$ km s^{-1} roughly in the direction of the Virgo cluster. This value is quite well determined and agrees closely with the best determinations of our Virgocentric infall velocity, which from other evidence give $v_{VC} = 220 \pm 50$ km s^{-1} (Sandage and Tammann 1984b, and references therein).

If one subtracts this Virgocentric velocity vector from our motion toward the microwave dipol, one finds a remaining vector of 460 ± 70 km s^{-1} in the direction $\alpha = 153 \pm 10^{\circ}$, $\delta = -42 \pm 10^{\circ}$. It is most interesting to note that this direction agrees within the errors with the position of the Hydra-Centaurus supercluster, which is with a mean redshift of about 3000 kms^{-1} probably the nearest supercluster from us. The conclusion from this is that the largest peculiar motion we partake of is not caused by the Virgo cluster, but by the nearby Hydra-Centaurus supercluster, and that these two aggregates together cause our motion toward the microwave background (Sandage and Tammann 1984a).

An attempt to prove our Hydra-Centaurocentric velocity directly by using first-ranked galaxies in clusters at $2500 < v_0 < 5000$ km s^{-1}

remains inconclusive, because the effect looked for makes only a
$\sim 10\%$ ($0.^{m}20$) signal and the sky distribution is very uneven. However,
the motion is visible at the 2σ-level in the velocities of spiral galaxies
with known 21 cm line width and within the relevant velocity range
(Sandage and Tammann 1984b).

The smallness of the peculiar motions requires a low value of the
density parameter $\Omega_0 = 0.1$ or $q_0 = 0.05$ (if $\Lambda = 0$). This result, which
rests partially on first-ranked cluster galaxies, is somewhat of a con-
solation for the failure to derive the true value of q_0 from the Hubble
diagram. It must be stressed, however, that the present value of Ω_0
(or q_0) rests on the assumption that all mass in the universe is clumped
at least as strongly as the visible galaxies. An underlying sheet of
mass would here remain undetected (Sandage and Tammann 1984a).

As discussed in Section II the global or cosmic value of the Hubble
constant can be determined from the Hubble diagram of standard
candles, if the absolute magnitude of at least one standard candle is
known. The error of this method is of course mainly controlled by the
intrinsic scatter $\sigma(M)$ of the specific Hubble diagram. Most authors
would agree today that the distance modulus of the Virgo cluster lies
in the range $31.^{m}0 < (m-M)^0 < 32.^{m}2$ (cf. Sandage and Tammann 1984c).
That restricts the absolute-magnitude range of the brightest Virgo
member (NGC 4472) to $-22.8 > M_V$ (fully corrected) > -24.0, and
inserting this into Sandage and Hardy's (1973) mean Hubble relation
yields, with allowance for the intrinsic scatter of $0.^{m}3$, $30 < H_0$ (cosmic)
< 75 (cf. Tammann 1977). A similar possible range for H_0 is obtained
when the Virgo cluster distance is used to calibrate the mean absolute
magnitude of the 10 brightest cluster members, or the mean absolute
magnitude of type I supernovae in ellipticals at maximum light.

In conclusion the Hubble diagram has proved to be a thorny vehicle
toward q_0, it has provided instead a wealth of other information, and
it still provides "the most fundamental aspect of the large-scale structure of the Universe" (Sandage 1975).

Support from the Swiss National Science Foundation is gratefully
acknowledged. Thanks are due to Mrs. M. Saladin for typing and editing
the manuscript.

REFERENCES

Abell, G. O. 1975, "Galaxies and the Universe", A. and M. Sandage
            and J. Kristian eds., Chicago: University of Chicago

Press, p. 601.

Baldwin, J. A. 1977, Astrophys. J. 214, 679.

Baldwin, J. R., Danziger, I. J., Frogel, J. A., and Persson, S. E. 1973, Astrophys. Letters 14, 1.

Baum, W. A. 1957, Astron. J. 62, 6.

Baum, W. A. 1962, "Problems of Extra-Galactic Research", G. C. McVittie ed., I. A. U. Symp. No. 15, p. 390.

Bautz, L. P., and Abell, G. O. 1973, Astrophys. J. 184, 709.

Bautz, L. P., and Morgan, W. W. 1970, Astrophys. J. Letters 162, L 149.

Geller, M. J., and Peebles, P. J. E. 1976, Astrophys. J. 206, 939.

Gudehus, D. H. 1978, Nature 275, 514.

Gunn, J. E., and Oke, J. B. 1975, Astrophys. J. 195, 255.

Gunn, J. E., and Tinsley, B. M. 1976, Astrophys. J. 210, 1.

Hausman, M. A., and Ostriker, J. P. 1978, Astrophys. J. 224, 320.

Heckmann, O. 1942, "Theorien der Kosmologie, Berlin: Springer, p. 64.

Hubble, E. P. 1929, Proc. National Acad. Sci. 15, 168.

Hubble, E. P. 1936a, Astrophys. J. 84, 270.

Hubble, E. P. 1936b, "The Realm of Nebulae, New Haven: Yale University Press.

Hubble, E. P. 1936c, Astrophys. J. 84, 517.

Hubble, E. P. 1938, "Observational Approach to Cosmology", Oxford: Oxford University Press.

Hubble, E. P. 1953, Monthly Not. Roy. Astron. Soc. 113, 658.

Hubble, E. P., and Humason, M. L. 1931, Astrophys. J. 74, 43.

Humason, M. L. 1929, Proc. Nat. Acad. Sci. 15, 167.

Humason, M. L. 1936, Astrophys. J. 83, 10.

Humason, M. L. 1957, "Carnegie Institution of Washington Year Book 56", Washington: Carnegie Institution, p. 62.

Humason, M. L., Mayall, N. U., and Sandage A. R. 1956, Astron. J. 61, 97 (HMS).

Kraan-Korteweg, R., Sandage, A., and Tammann, G. A. 1984, in press.

Kristian, J., Sandage, A., and Westphal, J. A. 1978, Astrophys. J. 221, 383.

Lilly, S. J., and Longair, M. S. 1982, "Astrophysical Cosmology", H. A. Brück, G. V. Coyne, and M. S. Longair eds., Vatican City: Pontificia Academia Scientiarum, p. 269.

McVittie, G. C. 1972, Monthly Not. Roy. Astron. Soc. 155, 425.

Mattig, W. 1958, Astron. Nachr. 284, 109.

Minkowski, R. 1960, Astrophys. J. 132, 908.

Nicoll, J. F., and Segal, I. E. 1982, Astrophys. J. 258, 457.

Oke, J. B. 1970, Astrophys. J. 170, 193.

Oke, J. B. 1984, this volume, p. 99.

Ostriker, J. P. , and Hausman, M. A. 1977, Astrophys. J. Letters 217,
        L 125.
Ostriker, J. P. , and Tremaine, S. D. 1975, Astrophys. J. Letters 202,
        L 113.
Peach, J. V. 1970, Astrophys. J. 159, 753.
Peterson, B. A. 1970, Astron. J. 75, 695.
Pettit, E.   1954, Astrophys. J. 120, 413.
Roberts, M. S. 1972, "External Galaxies and Quasi-Stellar Objects",
        I. A. U. Symp. No. 44, p. 12.
Roberts, M. S. 1978, Astron. J. 83, 1026.
Robertson, H. P. 1955, Publ. Astron. Soc. Pacific 67, 82.
Sandage, A. 1961, Astrophys. J. 133, 335.
Sandage, A. 1961a, Astrophys. J. 134, 916.
Sandage, A. 1966, "Atti del Conv. sulla Cosmologia", vol. II, tom. 3,
        Florence: G. Barbieri, p. 104.
Sandage, A. 1968, Astrophys. J. Letters 152, L 149.
Sandage, A. 1968a, Observatory 88, 91.
Sandage, A. 1970a, Physics Today 23, 34.
Sandage, A. 1970b, Pont. Acad. Sci. Scripta Varia 35, 610.
Sandage, A. 1972a, Astrophys. J. 173, 485.
Sandage, A. 1972b, Astrophys. J. 178, 1.
Sandage, A. 1972c, Astrophys. J. 178, 25.
Sandage, A. 1972d, Quat. J. Roy. Astron. Soc. 13, 282.
Sandage, A. 1973a, Astrophys. J. 180, 687.
Sandage, A. 1973b, Astrophys. J. 183, 711.
Sandage, A. 1973c, Astrophys. J. 183, 731.
Sandage, A. 1975, "Galaxies and the Universe", A. and M. Sandage
        and J. Kristian eds. , Chicago: University of Chicago
        Press, p. 761.
Sandage, A. 1975a, Astrophys. J. 202, 563.
Sandage, A. 1976, Astrophys. J. 205, 6.
Sandage, A. , and Hardy, E. 1973, Astrophys. J. 183, 743.
Sandage, A. , Kristian, J. , and Westphal, J. A. 1976, Astrophys. J.
        205, 688.
Sandage, A. , and Tammann, G. A. 1975, Astrophys. J. 197, 265.
Sandage, A. , and Tammann, G. A. 1982, Astrophys. J. 256, 339.
Sandage, A. , and Tammann, G. A. 1982a, "Astrophysical Cosmology",
        H. A. Brück, G. V. Coyne, and M. S. Longair eds. , Vatican
        City: Pontificia Academia Scientiarum, p. 23.
Sandage, A. , and Tammann, G. A. 1984a, "Large Scale Structure of
        the Universe, Cosmology and Fundamental Physics",
        First ESO-CERN Symposium, in press.
Sandage, A. , and Tammann, G. A. 1984b, in press.
Sandage, A. , and Tammann, G. A. 1984c, Nature, in press.
Sandage, A. , Tammann, G. A. , and Hardy, E. 1972, Astrophys. J.

172, 253.

Sandage, A., and Visvanathan, N. 1978, Astrophys.J.223, 707.

Schechter, P. 1976, Astrophys.J.203, 297.

Schechter, P., and Press, W.H. 1976, Astrophys.J.203, 557.

Schweizer, F. 1981, Astrophys.J.246, 722.

Sitter, W.de 1930, Bull.Astron.Netherlands 5, 157.

Smith, H.E. 1977, "Radio Astronomy and Cosmology", I.A.U.Symp. No. 74, p.279.

Smith, M.G. 1983, "Galactic and Extragalactic Infrared Spectroscopy", Proc.16th ESLAB Symposium.

Spinrad, H., and Djorgovski, S. 1983, Bull.Am.Astron.Soc.14, 959.

Stebbins, J., and Whitford, A.E. 1952, Astrophys.J.115, 284.

Tammann, G.A. 1974, "Confrontation of Cosmological Theories with Observational Data", I.A.U.Symp. 63, p.54.

Tammann, G.A. 1977, "Décalages vers le rouge et expansion de l'univers", I.A.U.Colloquium No.37, p.43.

Tammann, G.A. 1979, "ESA/ESO Workshop on Astronomical Uses of the Space Telescope", F.Macchetto, F.Pacini, and M. Tarenghi eds., Geneva: ESO, p.329.

Tammann, G.A. 1982, "Supernovae: A Survey of Current Research", M.J.Rees and R.J.Stoneham eds., Dordrecht: Reidel, p.371.

Tammann, G.A. 1983, "Wortshop on ESO's Very Large Telescope, Y.-P.Swings and K.Kjär eds., Garching: ESO, p.61.

Tammann, G.A., Sandage, A., and Yahil, A. 1980, Physica Scripta 21, 630.

Tammann, G.A., Yahil, A., and Sandage, A. 1979, Astrophys.J. 234, 775.

Tinsley, B.M., and Gunn, J.E. 1976, Astrophys.J. 203, 52.

Tremain, S.D., and Richstone, D.O. 1977, Astrophys.J.212, 311.

Tully, R.B., and Fisher, J.R. 1977, Astron.Astrophys.54, 661.

Wampler, E.J., Gaskell, C.M., Burke, W.L., and Baldwin, J.A. 1983, ESO Sci. Preprint No.233.

Weedman, D.W. 1976, Astrophys.J. 203, 6.

Westerlund, B.E., and Wall, J.V. 1969, Astron.J.74, 335.

Wilkinson, D.T. 1983, "Early Evolution of the Universe and its Present Structure", I.A.U.Symp.No.104, p.143.

INFORMATION ON GALAXY CLUSTERING FROM GRAVITATIONAL LENSES

Edwin L. Turner*
Princeton University Observatory

ABSTRACT

Observations of gravitational lens systems can provide information on
the distribution and perhaps the nature of dark matter in clusters of
galaxies. Lenses may prove to be one of the most powerful tools for
the study of a variety of cosmological problems.

1. Introduction

The study of gravitational lenses can provide information on three
general topics: the mass distribution in the lensing objects and along
the line-of-sight, the properties of the lensed sources, and the
cosmological background geometry in which the lensing occurs. Each
will affect the observable properties of lenses in independent though
not always easily separable ways. Before proceeding to consider two
potential gravitational lens applications in some detail, I simply list
several of the problems in each of these three categories to which
gravitational lens studies may be able to make significant contribution
contributions.

Lenses (= galaxies and clusters of galaxies usually)
1: Galaxy and cluster masses and surface densities +
2: Distribution of dark matter in clusters +
3: Nature of dark matter in galaxies and clusters
4: Existence of dark, massive objects (i.e., dark galaxies) −
5: Amplitude of very large scale inhomogeneities in the Universe −

Sources (= quasars usually)
1: Nature of the redshifts (cosmological or new physics) +
2: Explanation of quasar/nearby galaxy associations
3: Quasar evolution +
4: Intrinsic quasar angular sizes −
5: Detection and statistics of very faint, distant quasars

*Alfred P. Sloan Research Fellow

F. Mardirossian et al. (eds.), Clusters and Groups of Galaxies, 553–558.
© 1984 by D. Reidel Publishing Company.

Cosmology
1: Determination of H_0
2: Determination of q_0 -
3: Determination of Λ +
4: Test validity of Friedmann models -

  Many of these possible applications will be quite difficult to
realize and some may prove entirely impractical. Those which I judge
to be particularly promising are indicated by a "+" above while those
which seem likely to be rather difficult to realize are marked with a
"-". Although none of the three known gravitational lens systems
(Walsh, Carswell, and Weymann 1979; Weymann et. al. 1980; Weedman et.
al. 1982) have yet led to significant progress on any of these
problems, this fact probably should not be too discouraging. It might
be recalled that the first three pulsars discovered did not teach us
anything about neutron stars, stellar evolution, or General Relativity
but larger samples, the Crab pulsar, and the binary pulsar eventually
did. It may well be the same for gravitational lenses; major
discoveries will have to await statistical samples and particularly
simple or special objects.

2. Galaxy Cluster Surface Densities and Lens Image Separations

  One of the most easily determined properties of lenses is the
angular separation of the most widely separated pair of source images
produced by the intervening object. It turns out that this is also one
of the more easily calculated parameters for lensing due to ordinary
galaxies or to galaxies in the centers of rich galaxy clusters. At
Princeton we have recently carried out such a calculation (Turner,
Ostriker, and Gott 1984) at a fairly realistic level of approximation.
It is instructive to compare the expected image separations for galaxy
only lenses to those expected for composite galaxy plus rich cluster
lenses. It turns out that the expected image separations are extremely
sensitive to the central surface densities of rich clusters and may
give us our most accurate determination of this quantity.

  A particular model which has been calculated considers lensing in
a $q_0=0$ Friedmann universe in which there is no low redshift ($z<1$)
evolution of galaxies or clusters. The distribution of quasar
redshifts is taken from Burbidge and Hewitt (1980). Galaxies are
modeled as singular (i.e., zero core radius) isothermal spheres, and
clusters of galaxies are treated as uniform surface density sheets
which do not produce multiple images by themselves but which can
enhance the lensing due to individual galaxies very significantly.
These approximations are probably better than they first appear since
the maximum image separation due to an isothermal lens mass
distribution is very insensitive to the core radius and since the
uniform surface density approximation really amounts to the assumption
that inhomogeneities in the clusters mass distribution have scales
large compared to the image separation (typically a few arc seconds).
No attempt has been made to evaluate the absolute sampling bias which

favors the inclusion of lens systems in magnitude or flux limited samples due to the apparent brightness amplification produced by the lensing object. Nevertheless, the differential bias favoring high amplification lens systems over lower ones has been included.

The details of this calculation including the derivation of the relevant equations is described in Turner, Ostriker, and Gott (1984). Here I wish to emphasize the sensitivity of the expected image angular separation distribution to the central surface densities of rich clusters. In the above cited paper, it is shown that gravitational lensing involves a critical surface density

$$\sigma_{GL} = \frac{cH_0}{4\pi G} F(z_L, z_q, q_0) = 0.12h\ F(z_L, z_q, q_0)\ gm/cm^2$$

where $F(z_L, z_0, q_0)$ is a function which is quite insensitive to q_0 and to z_0 for $z_0 > 1$. The dependence of this function on z_L shows a broad minimum near ~ 7 for redshifts in the range ~ 0.2 to 0.8 but a steep increase for both smaller and larger values. The effect of viewing a normal single galaxy lens through a uniform density sheet (cluster) is to magnify the image of the source by a factor

$$\alpha = (1 - \frac{\sigma_c}{\sigma_{GL}})^{-1}$$

where σ_c is the sheet's (cluster's) surface density. This increases the apparent brightness amplification by a factor of α squared and produces an additional flux selection bias of α to the fourth power. It is thus obvious that clusters could greatly affect the image separation distribution if their surface densities approach σ_{GL}.

In fact the actual central surface densities of rich clusters are in the vicinity of σ_{GL} but are generally somewhat smaller. If clusters are modeled as isothermal spheres, their central surface densities are given by

$$\sigma_c = \frac{9V^2}{2\pi G r_c} = 0.69\ \frac{(V/1000\ km/s)}{(r_c/100\ kpc)}\ gm/cm^2$$

where V is the one dimensional velocity dispersion and r_c is the core radius. Another way to evaluate the near coincidence between σ_c and σ_{GL} is to consider the central region of the Coma cluster, the galaxies NGC4874 and NGC4889 and the cloud of companions which surround them. At a near optimum redshift of ~ 0.5 and with an assumed M/L of 400 in solar units, this system would have a mean surface density approximately one half of σ_{GL}. The detailed evaluation of this effect is complicated by the fact that very few rich clusters have their velocity dispersions and core radii well enough determined to reliably calculate α much less its fourth power! It should also be noted that σ_{GL} is so large that only a tiny fraction of galaxies will be affected by cluster enhancement, essentially only those in the cores of the very densest clusters.

Figure 1 shows the expected distribution of maximum image separations for several different models of the rich cluster enhancement effect. The dashed curve labeled "No Bias" indicates the expected distribution produced by galaxies alone with no contribution from clustering. The solid curve labeled "Amplification Bias" includes the effect of rich clusters by assuming that 3% of all elliptical and SO galaxies ($\sim$1% of all galaxies) reside in the cores of clusters with V=1200 km/s and r_c =0.2/h Mpc. The dotted curves show the effect of increasing or decreasing the derived cluster central surface density by 25% (i.e., a change in V of $\sim$12% for instance). None of these models is likely to be accurate in detail since there is clearly a wide range of cluster properties; moreover, few if any clusters other than Coma even have velocity dispersion measurements accurate to 12%. Nevertheless, these models do indicate the extreme sensitivity of the separation distribution to cluster properties, a result of the near coincidence of σ_{GL} with typical σ_c values. The discovery of lens systems with 2.3", 6.1", and 7.3" splittings and the absence of any small splitting cases (even in high resolution VLA surveys) already suggests that clusters are playing a statistically important role in lensing.

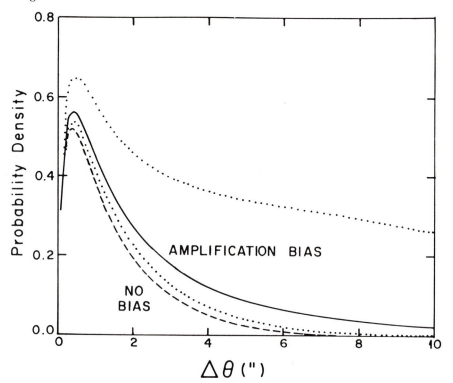

Figure 1. Expected image splitting distributions

3. Lens Redshift Distributions and 2345+007

Another parameter whose expected distribution has been calculated by Turner, Ostriker, and Gott (1984) is the lens redshift. It is much more difficult to determine observationally than image splittings since it requires detection and spectroscopy of a typically very faint galaxy seen near a bright quasar image. There are however other advantages to this parameter. Its distribution is relatively insensitive to assumed cosmology and to detailed properties of the lensing objects. In all cases the most likely lens redshift will be near 0.5 and the expected distribution will fall rapidly for redshifts <0.2 or >1.0 assuming only that the quasar redshift is significantly larger than one. This concentration of expected lens redshifts is particularly strong in the case of large splitting lenses since one then needs to obtain maximum effect from both the lensing galaxy and if possible from an enhancing cluster. Thus, almost all large splitting lens would be expected to be rich clusters with sufficiently small redshifts that they should be visible on deep images of the field. Absence of such a lensing cluster galaxy in a significant number or even one well studied, low quasar redshift case(s) would indicate the existence of dark condensed objects acting as lenses; this might be the only way in which such objects could be directly detected.

The gravitational lens 2345+007 (Weedman et. al. 1982) has an image splitting of 7" and a quasar redshift of 2.15. Our models indicate that such a system is the result of lensing by a galaxy in a rich cluster with a redshift <0.85 with 95% confidence. Deep PFUEI/5-meter CCD images of the field reveal some evidence for a rich cluster but no sign of any lensing galaxy close to the quasar images (Turner, Goldberg, and Gunn 1982). It is possible though by no means yet certain that 2345+007 is a lens system produce by a dark object. ST observations should determine the issue decisively if it proves impossible from the ground.

I wish to thank my collaborators J. P. Ostriker and J. R. Gott for allowing me to discuss some of our results in advance of their publication elsewhere. This work was supported in part by NSF grant AST82-16717.

References

Burbidge, G., and Hewitt, A. 1980, Ap.J.Supp., 43, 57.
Turner, E.L., Goldberg, R.E., and Gunn, J.E. 1982, Bull. AAS, 14, 974.
Turner, E.L., Ostriker, J.P., and Gott, J.R. 1984, Ap.J., submitted.
Walsh, D., Carswell, R.F., and Weymann, R.J. 1979, Nature, 279, 381.
Weedman, D.W., Weymann, R.J., Green, R.F., and Heckman, T.M. 1982, Ap.J. Lett., 255, L5.
Weymann, R.J., Latham, D., Angel, J.R.P., Green, R.F., Liebert, J.W., Turnshek, D.A., Turnshek, D.E., and Tyson, J.A. 1980, Nature, 285, 641.

DISCUSSION

PORTER: Do you think that the fact that only 3 lenses have been found in thousands of observations means that only the very highest density regions of the universe will lens?

TURNER: Only the highest density clusters but most galaxies can produce lensing effects. Our calculations suggest that the small total number found so far is consistent with what one would expect although their angular sizes are somewhat large.

DE VAUCOULEURS: What is the effect of individual stars or other small masses on the predicted lensing by galaxies?

TURNER: They will not produce observable effects on the multiple image structure of a lens but will produce changes of image brighnesses which can be distinguished from intrisic quasar variations since they will occur independently in each image. This effect could be used to distinguish between elementary particle and discrete dark object models, for the nature of the dark material in galactic halos.

THE CLUSTERING OF GALAXIES ABOUT POWERFUL RADIO SOURCES

R.M. Prestage
Department of Astronomy, University of Edinburgh
Blackford Hill, Edinburgh EH9 3HJ, United Kingdom

ABSTRACT

We have investigated the clustering environment of a sample of $\sim$300 powerful radio sources. We find that on average flat spectrum sources have clustering properties similar to those of galaxies in general. Steep spectrum sources are on average in regions of rather enhanced galaxy density; more so for FR I than FR II sources. The mean galaxy density around classical double sources is consistent with that of FR II sources as a whole.

INTRODUCTION

It is of particular interest in the study of radio sources to investigate the way in which their properties relate to the various factors which affect the basic radio source phenomenon. One aspect of this is to investigate the way in which properties such as radio morphology and luminosity are related to the source's cluster environment (i.e. the degree of clustering of normal galaxies in the vicinity of the radio source). Various authors have studied samples of radio sources selected due to their proximity to e.g. Abell clusters (Burns; Hanisch, these proceedings). While these sources have the advantage that their cluster environment is already known, such studies obviously cannot address the question of the clustering properties of the radio source population as a whole.

To undertake such a study in a systematic way, we must start with a radio-complete sample; we then have to parameterise the "degree of clustering" about each source. In an earlier study, Longair and Seldner (1979; LS) considered the relationship, for a sample of 3C galaxies, between the radio properties and the source cluster environment, as measured using the Lick counts of galaxies. They developed a method whereby the angular cross-correlation function of galaxies about a source was used to provide a measure of the local galaxy density. We are applying this technique to a statistical sample of powerful radio

559

F. Mardirossian et al. (eds.), Clusters and Groups of Galaxies, 559–564.
© 1984 by D. Reidel Publishing Company.

galaxies in the southern hemisphere, using U.K. Schmidt plates scanned by the COSMOS measuring machine: the resultant faint galaxy samples allow us to probe to higher redshifts than LS. The Lick-based analyses however are complementary to this work, in that low-precision results can be obtained quickly for a large number of sources. As a preliminary therefore, we have reworked the LS technique for a much larger sample of radio sources.

METHOD

Our method is similar to that of LS, who showed how knowledge of the galaxy luminosity function could be used to convert the observed angular correlation amplitude into a distance-independent spatial amplitude. In particular, they showed that if we assume the radio source to have a galaxy correlation function (defined in the standard way) of the form:

$$\xi(r) = B(r)^{-\gamma} \tag{1}$$

then at a redshift z, this will produce an observed angular correlation function:

$$\omega_z(\theta) = A_z(\theta)^{-(\gamma-1)} \tag{2}$$

where $A(z) = H(z)B$ and $H(z)$ is a function which can be calculated for a given galaxy luminosity function. In other words, if we measure a value of A, we can calculate the value of $H(z)$ appropriate to that redshift, and then obtain a value for the distance-independent parameter B, which is directly related to the number of excess galaxies near the source.

LS performed their counts out to fixed angular radius of 1 degree; if the assumed form for the galaxy excess held to an indefinitely large distance, then the observed value of B would be independent of the radius over which it was evaluated. In practise however, this cannot be the case; if we assume the existence of a "break" from the power law, then integrating past this distance will dilute the signal, and the greater the redshift of the source, the worse this effect will become. We have attempted to maximise the signal-to-noise by integrating the angular excess only out to a distance corresponding to 1 Mpc at the redshift of the source to obtain A. This is converted to B as above, and as our final measure of the degree of clustering about a given source, we use:

$$R = \frac{B}{B_{gg}} = X(z)\frac{A}{A_{gg}} \tag{3}$$

where B_{gg} is the spatial amplitude for galaxies in general, obtained from the value $A_{gg} = 0.068$ for the Lick counts derived by Groth & Peebles (1977). Besides normalising the statistic to the value expected for an "average" galaxy, using the ratio has the advantage that many uncertainties in the conversion from A to B, which are the same for both the radio source and galaxies in general, are cancelled out.

THE SAMPLE

     The radio sources used here all satisfy the conditions $-20^{\circ} < \delta < +90^{\circ}$, $|b| > 10^{\circ}$ and $0.01 < z < 0.15$. The limits on z are imposed because the correction factor $H(z)$ becomes large and uncertain outside these ranges; in fact A is measurable to larger redshifts. We have considered three samples:

A)   all identified sources with $S_{2.7} > 1$ Jy from Kuhr et al. (1981)
B)   additional sources from the Parkes 2.7-Ghz survey (see Wall 1977) with known redshifts
C)   additional sources from the Parkes 2.7-Ghz survey without redshifts

for the latter sample redshifts were estimated from the m-log(z) relationship for Parkes sources of known z.

RESULTS

     Fig. 1 shows the value of R obtained for all objects as a function of redshift. The mean value of R is about 3, while extreme values range

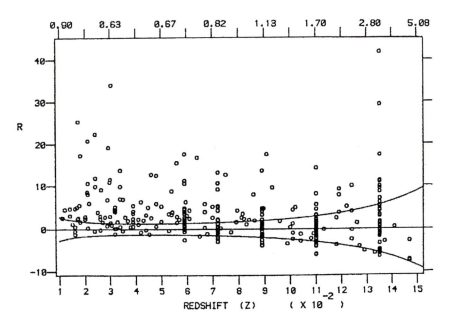

Figure 1.   R versus redshift for all sources. The solid line shows the value which would be obtained for a 1-sigma variation in the background. The numbers along the top z-axis are the corresponding values of X.

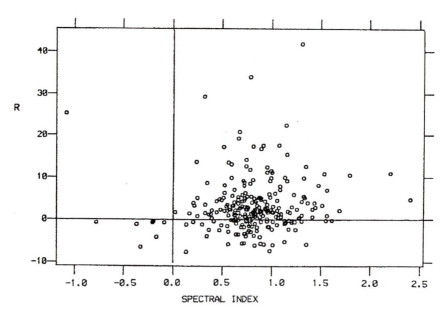

Figure 2.   R versus high-frequency spectral index $\alpha_{2.7}^{5}$

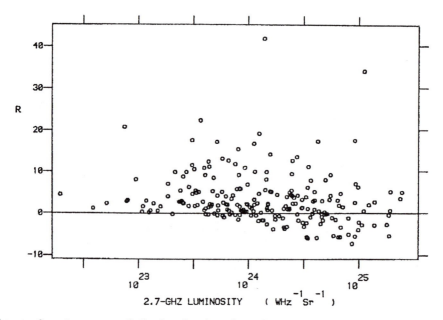

Figure 3.   R versus 2.7 Ghz luminosity for steep-spectrum sources $(\alpha_{2.7}^{5} > 0.5)$

from $\sim$25 to values typical of galaxies in the field. The solid line shows the value of R which would be obtained due to a one sigma fluctuation in the background within the counting circle. This is the dominant source of error in R for individual objects. Two of the sources with the largest values of R have estimated redshifts, and so could equally well be in the centre of the diagram. For comparison, Abell clusters of richness 0,1 and 2 produce values of R of $\sim$5, 10 and 15 respectively.

Ideally, we would like to compare values of R for various morphological types, but as structures are not available for the majority of the sources we use two properties which are indicators of radio morphology.

1) Spectral Index. Basically, all extended source have spectral index $\alpha_{2.7}^{5} > 0.5$. While the converse is not true, in that unresolved sources can have $\alpha_{2.7}^{5} > 0.5$, this is a useful means of dividing the samples.

2) Luminosity. Extended radio sources can be divided into two classes, known as Fanaroff-Riley classes I and II. FR I sources have their peaks of radio brightness separated by less than half the overall extent of the radio emission, while FR II sources have the reverse. It is found that this division corresponds to a division in radio luminosity, in the sense that sources weaker than $P_{2.7} = 10^{24} \text{WHz}^{-1} \text{Sr}^{-1}$ are basically FR I's, while those stronger are FR II's.

Fig. 2 shows the value of R versus spectral index; despite the scatter there is a definite trend in the sense that flatter spectrum sources have generally smaller values of R. Similarly, Fig. 3 shows the value of R versus 2.7-Ghz luminosity for all steep-spectrum sources. Again there is a clear trend to lower values of R at higher luminosity.

These correlations are shown more clearly in Table 1, which gives the weighted mean value of R for various subsets of the data. (The number of sources contributing to each value is shown in brackets.)

Table 1. Mean values of R for various subsets of the data

SAMPLE	WEIGHTED MEAN VALUE $<R>$	
	$\alpha_{2.7}^{5} < 0.5$	$\alpha_{2.7}^{5} > 0.5$
A	1.33 ± 0.62 (8)	3.72 ± 0.26 (51)
B	1.81 ± 0.81 (5)	3.02 ± 0.46 (18)
C	1.54 ± 0.46 (24)	2.78 ± 0.17 (152)
A+B+C	1.53 ± 0.33 (37)	3.05 ± 0.13 (221)
	STEEP-SPECTRUM SOURCES	
	$P_{2.7} < 10^{24} \text{WHz}^{-1} \text{Sr}^{-1}$	$P_{2.7} > 10^{24} \text{WHz}^{-1} \text{Sr}^{-1}$
A	5.26 ± 0.45 (15)	2.99 ± 0.31 (36)
B	2.75 ± 0.52 (10)	4.00 ± 0.99 (8)
C	3.21 ± 0.19 (70)	1.56 ± 0.32 (82)
A+B+C	3.44 ± 0.17 (95)	2.39 ± 0.22 (126)

These results may be summarised as follows:

a)   Flat-spectrum sources do not lie in regions of galaxy density significantly different from that around a normal galaxy selected at random; on the other hand, steep-spectrum sources are on average in regions of rather enhanced density.

b)   For steep-spectrum sources, the mean value of R for sources of lower luminosity (consistent with being FR I) is larger than those of higher luminosity (FR II). This result was suggested by LS, at a lower level of significance.

Finally, we have calculated the mean value of R for the eight sources described by LS as "classical doubles". For these objects $\langle R \rangle$ = 2.35 $\pm$ 0.64; this is not significantly different from the value for $\langle R \rangle$ for FR II sources as a whole. The LS result was probably biassed by a few high-z objects for which R is very uncertain.

CONCLUSIONS

Our most interesting result concerns the environment of the flat-spectrum sources; these are objects for which the local galaxy density is similar to that around an average galaxy. The suggestion of LS that this may also be true of some FR II objects now seems less likely; our results indicate that these sources reside in at least poor clusters. This is reassuring, since the extensive X-ray emission from these (supposedly) isolated galaxies predicted by LS has not been observed (Fabbiano et al. 1983). We may speculate that, for the flat-spectrum sources, it is the lack of a confining medium that influences the structure. Certainly our results are hard to account for in a scenario which produces flat-spectrum sources via orientation effects (Orr and Browne 1982).

REFERENCES

Fabbiano, G., Miller, L., Trinchieri, G., Longair, M.S. & Elvis, M., 1983. Astrophys. J. in press.
Kuhr, H., Witzel, A., Pauliny-Toth, I.I.K. & Nauber, U., 1981. Astron. Astrophys. Suppl. Ser., 45, 357.
Longair, M.S. & Seldner, M., 1979. Mon. Not. R. astr. Soc., 189, 433.
Groth, E.J. & Peebles, P.J.E., 1977. Astrophys. J., 217, 385.
Orr, M.J.L. & Browne, I.W.A., 1982. Mon. Not. R. astr. Soc., 200, 1067.
Wall, J.V., 1977. In Radio Astronomy and Cosmology, IAU Symp. 74, p. 269, D. Reidel, Dordrecht.

CORRELATION ANALYSES OF GALAXY CLUSTERING TO B $\sim$ 24

P.R.F. Stevenson, T. Shanks and R. Fong
Dept. of Physics, University of Durham, England.

ABSTRACT

Estimates of the angular two-point galaxy correlation function, $w(\theta)$, are presented as obtained from COSMOS machine measurements of 1.2m UK Schmidt telescope (UKST) and 4m Anglo-Australian telescope (AAT) plates. The UKST plates cover an area of sky of $\sim$ 170 square degrees, some four times larger than any previous study to these depths. The new estimate of $w(\theta)$ shows a break from its -0.8 power law behaviour at a scale corresponding to a spatial separation of $3h^{-1}$Mpc, (where h = Hubble's constant in units of 100 Kms^{-1}Mpc^{-1}) in agreement with the earlier results of Shanks et al (1980).

The AAT plates allow correlation analyses to be carried out to 24^m in the blue (b_J) passband, and 22^m in the red (r_F) passband, and at these faint magnitudes it may be possible to observe the effects of clustering evolution. It is found that the observed clustering amplitudes at these depths are lower than those predicted using well determined models that assume no clustering evolution. However the sampling errors are large and more 4m data is required in order to test the reality of this result.

1.   Introduction

At small angular scales corresponding to spatial separations less than $3h^{-1}$ Mpc the form of the galaxy two-point angular correlation function, $w(\theta)$, is well known to be a power law of slope -0.8 (see Peebles 1980). However at larger scales where the form of the correlation function is more sensitive to theories of galaxy formation and cosmology (Davis, Groth and Peebles, 1977 , Frenk, White and Davis, 1983) its form is not as well determined. From an analysis of the Lick catalogue Groth and Peebles (1977) found that $w(\theta)$ departed from its power law behaviour at a scale corresponding to $\sim 9h^{-1}$ Mpc. However from an analysis of deeper UKST plates Shanks et al. (1980; hereafter SFEM) observed the 'break' in $w(\theta)$ at a scale length of $\sim 3h^{-1}$ Mpc.

F. Mardirossian et al. (eds.), Clusters and Groups of Galaxies, 565–571.

In the present work we shall firstly report on an extension to the
SFEM study that now covers $\sim$ 170 square degrees to the depth of the
UKST surveys. This is an area of sky some four times larger than in
the previous study and with such a large area now analysed the reality
of the break can be tested with much greater confidence.

Secondly we shall report on new observations of $w(\theta)$ at fainter
4 metre AAT plate limits (to $b_J \sim 24$, and $r_F \sim 22$). At these faint
magnitudes the amplitude of clustering depends on both the redshift
distribution of galaxies and their possible clustering evolution. In
the SFEM study it was shown that the amplitude of clustering scaled
as expected from the brighter surveys of the Zwicky and the Lick catalogues to depths of $\sim$ 700 h^{-1} Mpc implying homogeneity and isotropy
of the Universe to those depths. With the present AAT data we can now
extend this analysis to depths of $\sim$ 3000 h^{-1} Mpc. Recently Shanks et
al (1983) have used the form of the galaxy number-magnitude, $n(m)$,
relation obtained from this AAT data in order to place constraints on
galaxy luminosity evolution and q_0. These constraints allow the redshift distribution of galaxies of faint magnitudes to be calculated,
and hence we are able to predict the expected clustering in these deep
samples. By comparing the observed and expected amplitude scaling
relation it may thus be possible to detect clustering evolution. We
finally compare our results with those of Ellis (1980),   Shanks
(1982), and Koo and Szalay (1983) who have also estimated $w(\theta)$ from
4m plates, but only in the b_J passband, in order to assess the uncertainty in the observed scaling relation.

The results described in this paper are preliminary and full
details of the analysis will be given in Stevenson et al (1983; hereafter SSF)

2.   UKST RESULTS

The ten UKST plates used in the present work (taken in two passbands denoted by b_J, for the blue band, and r_F, for the red band) were
measured by the COSMOS machine (Stobie et al. 1979) at the Royal
Observatory Edinburgh, and full details of the data reduction and
photometric procedures will be given in SSF. $w(\theta)$ for all of the galaxy
samples used here were estimated using the techniques discussed by
SFEM, but with no filtering of possible large scale gradients. In
Figure 1 the resulting $w(\theta)$ is shown for the ensembled UKST data at a
magnitude limit of $b_J = 21^m$. The error bars shown were calculated from
field-field variations for the eight separate (b_J) field making up the
ensemble. The individual $w(\theta)$'s obtained for each field are discussed
in SSF. It can be seen from Fig. 1 that a -0.8 power law slope gives
a reasonable fit to $w(\theta)$ at small angular scales. In general the UKST
$w(\theta)$ amplitudes measured here are in good agreement with those of SFEM,
but less flattening of the slope is seen at the faintest UKST limits
than was seen by SFEM at $b_J = 21^m.5$. However it should be noted that
the present data includes a complete re-reduction of the SFEM plate
material and if our analysis is restricted to just this area

(comprising $\sim$ 25% of the total area) we obtain the same result as SFEM. This suggests that the flatter slopes seen by SFEM were most probably caused by statistical fluctuations.

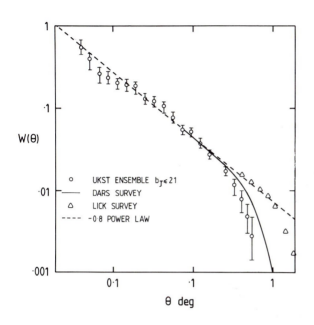

Figure 1: The angular correlation function, $w(\theta)$, at $b_J = 21^m.0$, estimated from eight UKST fields, together with the Lick catalogue $w(\theta)$ and that found from the projected Durham/AAT redshift survey $\xi(r)$, scaled to the same depth. A departure from the -0.8 power law (also shown) occurs at an angular scale corresponding to a spatial scale length of $\sim 3h^{-1}$ Mpc for the UKST and DARS survey $w(\theta)$. This is considerably smaller than the break scale of $9h^{-1}$ Mpc seen in the Lick catalogue $w(\theta)$.

At larger angular scales the $w(\theta)$ of Fig. 1 is seen to break from the power law at an angle of ~ 0.3 degrees, corresponding to a spatial separation of $\sim 3h^{-1}$ Mpc. Interestingly the $w(\theta)$ estimate from the large area studied here appears to break at the same scale as that found by SFEM, hence confirming their result. We note that the integral constraint, applying to our estimate of $w(\theta)$, has a negligible effect at all angular scales considered here, and so does not affect the shape or amplitude of $w(\theta)$. This result is therefore still in disagreement with the position of the break found in the Lick catalogue $w(\theta)$. (see Fig. 1).

Further evidence for the reality of the break found in the UKST $w(\theta)$ comes from the estimate of the 3-dimensional correlation function, $\xi(r)$, obtained from the Durham/AAT redshift survey. (Bean 1983, Shanks et al 1983). Although somewhat uncertainly because of the small sample size here, $\xi(r)$ shows a break from power-law behaviour at $r \sim 5h^{-1}$ Mpc. To allow a direct comparison with the projected $w(\theta)$ this estimate of $\xi(r)$ was converted into $w(\theta)$ using Limber (1953) projection formula. This result scaled to the depth of the UKST sample is shown as the solid line in Fig. 1 where it is seen to be in better agreement with our UKST $w(\theta)$ estimate than that found in the Lick catalogue. This consistency gives us good reason to believe that the break found in our estimate of $w(\theta)$ is real, suggesting that a similar feature occurs in

$\xi(r)$ at a scale of 3-5 h^{-1} Mpc.

3.   AAT RESULTS AND THE SCALING RELATION

In Fig. 2 the $w(\theta)$ estimate for the deepest, $b_J= 24^m$, AAT galaxy
sample is shown together with the ensemble UKST $w(\theta)$ from Fig. 1.   At
small angular scales the slope of the AAT $w(\theta)$ is still consistent
with a -0.8 power law (also shown in Fig. 2), but exhibits more noise
than the UKST sample.   The error bars in Fig. 2 were calculated by
dividing the AAT plate into quarters and estimating $w(\theta)$ for each
quarter separately.   The increase in the size of the error bars beyond
$\theta \sim 0.05$ degrees, indicates that there is little reliable information
about the position of the break in the AAT $w(\theta)$.

The observed amplitude scaling relation for both the UKST and AAT
data in the b_J and r_F bands are shown in Figs. 3a and 3b respectively. The amplitudes were calculated using a log-log least squares fit to $w(\theta)$, over a fixed spatial separation (so that the break was not included) and assuming a -0.8 power law slope. The resulting amplitudes were then corrected for contamination by stars using the procedure of SFEM and SSF. This is most critical at the UKST depths since at the faintest AAT limits the star/galaxy ratio is extremely small.

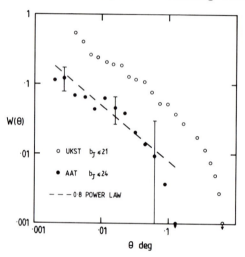

Fig. 2: The angular correlation function for the AAT data at $b_J= 24^m.0$, together with the UKST $w(\theta)$ from Fig. 1 for comparison.

The correlation function amplitude scaling relation has been predicted using models for the galaxy redshift distribution which gave
good agreement with the observed number-magnitude, n(m), counts. (See
Shanks et al 1984 for a detailed discussion).   Unfortunately the models
in the b_J band are very uncertain.   This is due to the large number of
galaxy types present in the blue samples whose k-corrections are uncertain, and, the large amounts of luminosity evolution required to fit
the b_J counts.   The two models shown in Fig. 3a assume equally acceptable choices of model parameters and give equally good fits to the b_J
counts, but show very different scaling relations.   Neither model
assumes any clustering evolution.   The difference is caused by the
different galaxy redshift distributions produced by each model.   Thus

until the b_J models are better determined it will be difficult to use the blue scaling relation to obtain constraints·on clustering evolution.

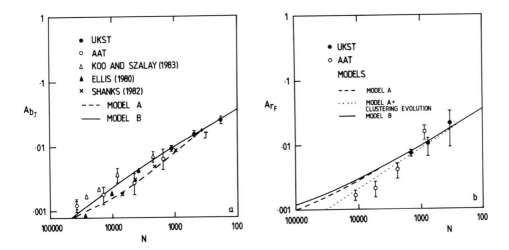

Figure 3: a) The observed and modelled scaling relations in the b_J band. Ab_J is the amplitude of clustering found at a surface density of galaxies, N, per square degree. The large dispersion in the observations of the various authors is most probably due to statistical fluctuations caused by the small samples observed.
b) As for a) but in the r_F band. The AAT data can be seen to be more consistent with the model which includes clustering evolution, although this result must be tentative due to the small area of sky surveyed.

In the red band however the models are much better determined due to the dominance of early type galaxies whose k-corrections are well known, and the small amount of luminosity evolution required to fit the r_F counts (see Shanks et al 1984). This is demonstrated in Fig. 3b where models A and B give very similar predicted scaling relations due to the same redshift distribution being predicted in both models. These two predictions represent reasonable variations in the parameters for models with q_0 in the range 0-0.5, with the constraint that the r_F counts are also fitted by the model. Interestingly both predictions lie systematically above the observed scaling relation. The dotted line in Fig. 3b shows the effect of adding clustering evolution (as in Phillipps et al 1978, where galaxies are less clustered in the past) and it can be seen to give a more reasonable fit to the observations. However the sampling errors on our observations may be larger than the error bars depicted in Fig. 3. This can be seen by considering the dispersion in the observed scaling relations of various authors in the b_J band shown in Fig. 3a. Unfortunately no other observations are available in the r_F band, but if we assume that the same dispersion is seen there as in the b_J band it may not be possible to rule out the non-evolving models. It must be concluded that more 4m data is required in the r_F band in order to obtain unambiguous estimates of the clustering amplitudes at these depths.

4.   CONCLUSIONS

      The conclusions of the present work may be summarised as follows:
a)   previous results of correlation analyses of galaxy clustering on UKST plates (SFEM) have been shown to be reproducible over much larger areas of sky. In this large sample the form of $w(\theta)$ shows the characteristic -0.8 power law at small angular scales and a break from the power law at large scales corresponding to a spatial separation of $\sim 3h^{-1}$ Mpc. This result is consistent with the form for $\xi(r)$ found in the Durham/AAT redshift survey, but the discrepancy with the $9h^{-1}$ Mpc break scale found in the Lick catalogues $w(\theta)$ still remains.

b)   The deeper AAT $w(\theta)$'s are also consistent with a -0.8 power law slope at small scales and at all magnitude limits. They enable us to observe the clustering amplitude scaling relation to $b_J = 24^m$ and $r_F = 22^m$, where in a comparison with model predictions tentative evidence for clustering evolution is obtained. However due to the large sampling errors the non-evolving model cannot yet be ruled out.

ACKNOWLEDGEMENTS

      We thank the COSMOS group at the Royal Observatory Edinburgh, especially Dr. H.T. MacGillivray for his help in ontaining the data used here. We also thank the U.K. Schmidt telescope and the Anglo-Australian Telescope for the use of their photographic plates. P.R.F. Stevenson was supported by the SERC during this work.

REFERENCES

Bean, J., 1983. Ph.D Thesis, University of Durham, England.
David, M., Groth, E.J. and Peebles, P.J.E., 1977. Astrophys.J.212, L107
Ellis, R.S., 1980. Phil. Trans. R. Soc. Lond. A296, 355.
Frenk, C.S., White, S.D.M., and Davis, M., 1983. Astrophys.J. 271, 417
Groth, E.J., and Peebles, P.J.E., 1977. Astrophys.J. 217, 385
Koo, D.C., and Szalay, A.S., 1983. Astrophys J. Submitted
Limber, D.N., 1953. Astrophys.J. 117, 134
Peebles, P.J.E., 1980. 'Large Scale Structure of the Universe',
  Princeton University Press.
Phillipps, S., Fong., R., Ellis, R.S., Fall, S.M. and MacGillivray, H.T.
  1978. Mon. Not. R. astr. Soc. 182, 673.
Shanks, T., 1982. 'Progress in Cosmology', Ed. A.W. Wolfendale, Reidel,
  Dordrecht, Holland.

Shanks, T., Fong. R., Ellis, R.S., and MacGillivray, H.T., 1980. Mon.
   Not. R. astr. Soc. 192, 209.
Shanks T., Bean, A.J., Efstathiou, G., Ellis, R.S., Fong. R., and
   Peterson, B.A., 1983. Astrophys. J. in press
Shanks, T., Stevenson, P.R.F., Fong. R., and MacGillivray, H.T. 1984,
   Mon. Not. R. astr. Soc. 206 in press.
Stevenson, P.R.F., Shanks, T., and Fong, R., 1983. In preparation.
Stobie, R.S., Smith, G.M., Lutz. R.K. and Martin, R., 1979. Image
   Processing in Astronomy, Ed., G. Sedmak, M. Capaccioli, R.J. Allen,
   Osservatorio Astronomico di Trieste.

CLUSTERING OF EARLY-TYPE GALAXIES IN THE SOUTHERN HEMISPHERE

Elaine M. Sadler
European Southern Observatory, Garching bei München
Federal Republic of Germany

ABSTRACT

   The results of a large observational study of early-type galaxies
are used to investigate the large-scale clustering of galaxies in a
region of sky only poorly covered by previous surveys. Redshifts and
magnitudes have been measured for all elliptical and S0 galaxies south of
declination -32° to a limiting magnitude of 14.0, and the data can be
used to study both angular (2D) and spatial (3D) clustering. Since early-
type galaxies tend to cluster more tightly than late-types, they are good
tracers for clustering studies. The two-point correlation function for
southern galaxies agrees closely with that already derived for the
north (δ > 0°).

1. INTRODUCTION

   The southern hemisphere, especially that part not accessible to
northern telescopes (δ < -20°), has been relatively neglected in studies
of galaxy clustering. There are several reasons for this, in particular
the absence of a galaxy catalogue corresponding to the Zwicky catalogue
in the north, so that most southern galaxies have no measured (or even
estimated) magnitude, and the poorer redshift coverage (there is as yet
no counterpart to the northern CfA survey (Huchra et al., 1983). This
accounts for the fact that many "maps" of clustering (e.g., Tully and
Fisher, 1977) have a prominent gap corresponding to the far south. A few
brave forays have now been made into this barely-explored region, aided
by the recent publication of the ESO/Uppsala Catalogue (Lauberts, 1982),
and hopefully this will continue, so that we can eventually have a
velocity map of the whole sky.

   Here, I shall describe some observations of a large sample of early-
type galaxies, originally collected for a different purpose, and use them
to say something about clustering in the south and how it compares with
previous studies in the north.

F. Mardirossian et al. (eds.), Clusters and Groups of Galaxies, 573–577.
© *1984 by D. Reidel Publishing Company.*

2.   OBSERVATIONS

A complete sample of 248 galaxies was taken from the ESO/Uppsala Catalogue (Lauberts, 1982). A magnitude was estimated for all galaxies south of $\delta = -32°$ classified as E, E: E-S0, S0 or S0: by inspection of the ESO (B) survey films, and a value of $m_{est} = 14.0$ used as the survey cut-off. Redshifts were measured by a cross-correlation method using the 1.9m telescope at Mt. Stromlo, and UBV aperture photometry obtained with the 1.0m and 0.6m telescopes at Siding Spring was used to derive accurate magnitudes (B_T) and colours for members of the sample using the method detailed in the Second Reference Catalogue (de Vaucouleurs et al., 1976). A more complete description of the observations (and a discussion of the sample completeness) is given elsewhere (Sadler, 1984a, b).

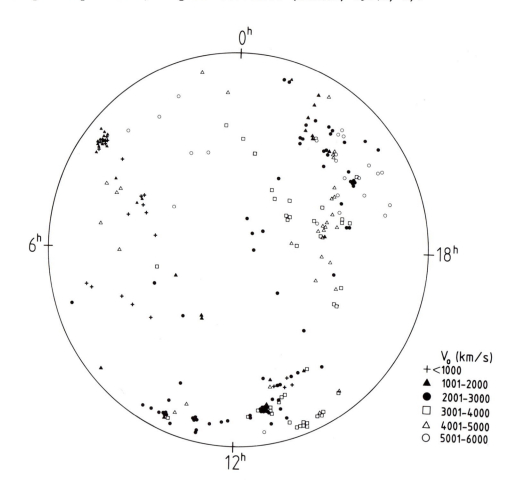

Figure 1.   The distribution of early-type galaxies with $B_T \leqslant 14.0$ and $V_o \leqslant 6000$ km/s in the southern hemisphere ($\delta < -32°$).

3. RESULTS

    The distribution of galaxies on the sky is shown in figure 1, where
bins in redshift are indicated by different symbols. The south celestial
pole (δ = -90°) is at the centre, with the boundary at δ = -30°. The
galactic plane runs almost horizontally across the lower part of the
plot, between about 9^h and 16^h R.A., and in the top left-hand quadrant
there is patchy obscuration due to the two Magellanic clouds. Several galaxies with radial velocity greater than 6000 km/s were observed, most of
them in the region between 0^h and 3^h R.A.; but they have been omitted
from figure 1 for clarity since the number of galaxies observed at this
distance is too small to give a realistic picture of their distribution.

    Several features are apparent in this picture. The "Local Super-
cluster" (de Vaucouleurs 1953, Tully 1982) is not prominent here, perhaps
because most of its members are late-type galaxies. The only remnant is
the group of nearby galaxies (+) around 12-13^h, -40°. Further out, a
chain of galaxies at 1000-2000 km/s ($\blacktriangle$) extends from the Fornax cluster
(prominent at 3^h30, -35°) towards the group at 12-13^h. This corresponds
to the "Southern Supergalaxy" identified by de Vaucouleurs (1953).

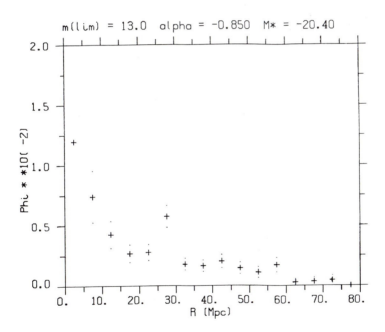

Figure 2. The space density, ϕ^*, of early-type galaxies in the southern
hemisphere (in galaxies/Mpc^3) as a function of distance R (taking H_0 =
100 km/s/Mpc), showing the excess of galaxies at R = 25-30 Mpc (V_0 =
2500-3000 km/s).

The region around the Centaurus cluster (12-13^h, -40°) is dynamically very complex and interesting. John Lucey (this meeting) has described
the unusual velocity structure of the cluster itself - in addition, it
seems to be a point of intersection for a number of large-scale features
and filaments.

Another feature of figure 1 is the excess of galaxies in the 2000-
3000 km/s bin. Figure 2, which is a plot of the space density of galaxies
as a function of distance, shows that this is a real density enhancement.
It may be indicative of superclustering, since the sample contains three
clusters in this velocity range, at 10^{h}30 -35°, 12-13^h -40° (Centaurus)
and 21^h -48°.

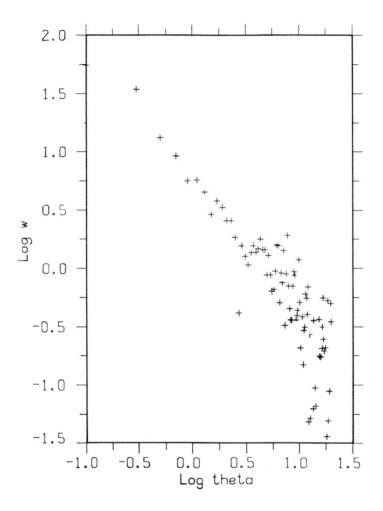

Figure 3. The two-point correlation function w(θ) derived for the sample
of galaxies discussed here.

4. THE TWO-POINT CORRELATION FUNCTION

In an attempt to compare the scale of galaxy clustering in the south with that in the north, Nigel Sharp and I derived the two-point (angular) correlation function (Peebles 1973) for this sample. The function is normally written in the power-law form $w(\theta) = a\theta^\beta$, where w is a function which describes the degree to which galaxies are clustered on an angular scale θ.

A least-squares fit to the data in figure 3 gave a slope $\beta = -1.11 \pm 0.07$, compared with a value $\beta = -1.10 \pm 0.09$ obtained by Davis and Geller (1976) for northern ellipticals from the UGC. Thus the spatial clustering of galaxies in the south appears reassuringly similar to that in the north. We are presently working on a study of the three-dimensional (spatial) correlation function in this region (Sadler and Sharp, in preparation).

5. CONCLUSION

The distribution of galaxies in the southern hemisphere shows a number of large-scale features which are prominent in both two and three dimensions. The two-point correlation function indicates that clustering scales in the south are closely similar to those in the north.

ACKNOWLEDGEMENTS

I should like to thank Nigel Sharp for introducing me to the study of galaxy clustering and for much practical help, including the calculation of the two-point correlation function in figure 3 and the luminosity function used to derive figure 2.

REFERENCES

Davis, M. and Geller, M.J. (1976) Astrophys. J. 208, pp. 13-19.
de Vaucouleurs, G. (1953) Astron. J. 58, pp. 30-32.
de Vaucouleurs, G., de Vaucouleurs, A. and Corwin, H.G. (1976) "Second Reference Catalogue of Bright Galaxies" (Austin, University of Texas Press).
Huchra, J., Davis, M., Latham, D. and Tonry, J. (1983) Astrophys. J. Suppl. Ser. 52, pp. 89-119.
Lauberts, A. (1982) "The ESO/Uppsala Survey of the ESO (B) Atlas" (Munich, ESO).
Peebles, P.J.E. (1973) Astrophys. J. 185, pp. 413-440.
Sadler, E.M. (1984a,b) Astron. J., in press.
Tully, R.B. and Fisher, J.R. (1977) in "The Large Scale Structure of the Universe", eds. M.S. Longair and J. Einasto (Dordrecht, Reidel), pp. 31-47.

PROGRESS REPORT ON A NEW SURVEY TO STUDY THE CLUSTERING PROPERTIES
OF QUASARS AT z ~ 2

Patrick S. Osmer
Cerro Tololo Inter-American Observatory
Casilla 603
La Serena, Chile

Quasars continue to be the only objects for which redshifts can be measured at z ~ 2. If they represent the large scale distribution of matter in the universe, such quasars will provide important information about the state of clustering when the universe was one third its present age. As we have been hearing at this meeting, the pancake theory of cluster formation on the one hand and the hierarchical clustering theory on the other make opposite predictions about the development of large scale clustering; the former has initial large scale structures while in the latter picture large scale clusters are recent developments. Even if the quasars are not representative of the large scale matter distribution, their space distribution will contain clues about how they formed, so that it is important to determine observationally what clustering properties they have.

Surveys done to date have yielded interesting cases of quasars with similar redshifts that are close together on the sky (e.g. Burbidge et al. 1980, Oort, Arp, and de Ruiter 1981, Osmer 1981). Such groups can occur on the scale of superclusters of galaxies or even be as small as galaxy clusters. However, it is not yet established if the groups are only chance fluctuations in a randomly distributed population or if they indicate that the quasar population as a whole clusters. The recent results on the distribution of clusters of galaxies and on large scale regions nearly devoid of galaxies do, however, raise our expectations of finding large scale structure in the quasar distribution.

To improve the sensitivity to clustering and to increase the statistical base, a new quasar survey is being carried out at CTIO with the following properties: 1) the survey is in a previously unstudied field at $12^h 10^m, -11°$ in the northern galactic hemisphere, 2) it covers a contiguous strip of sky $0°6$ wide by $13°$ long, 3) it has an area of 7.9 deg^2, and 4) it is made up of 23 CTIO 4m grism fields which are overlapped and recorded on IIIa-J emulsion. The use of an unstudied field will provide results statistically independent of previous work, where both problems of a posteriori statistics and interrelated fields (in several cases surveys were set up on quasars already found in

F. Mardirossian et al. (eds.), Clusters and Groups of Galaxies, 579–581.
© *1984 by D. Reidel Publishing Company.*

previous surveys) made it difficult to evaluate the significance of
some groups that were found. A related advantage is that a priori
predictions can be made for the expected distribution in the new field.
The field layout of a long thin strip also gives improved sensitivity
to clustering on scales of superclusters compared to the original CTIO
4m survey. The area covered is larger than the 5.2 deg^2 of the original
survey; this will contribute further to improved sensitivity to clustering.
Finally, the use of IIIa-J emulsion favors the selection of quasars with
z ~ 2, where the apparent space density is high. These points are
discussed in more detail by Osmer (1981), where numerical simulations
show that a gain of a factor of 1.4 in sensitivity to clustering is
expected in the new survey simply by using a connected thin strip for
the field layout compared to the original, disconnected survey fields.
Considering the larger area and the use of IIIa-J emulsion, a final
gain of nearly a factor of two is anticipated.

Note that the fields are overlapped so that they can be tied
together independently of the quasar found on the plates. By using
stars and the absorption features in their spectra in combination with
spectrophotometric measures, it will be possible to establish the
limiting sensitivity and spectral resolution of the plates, factors
that have been elusive in some of the original grism or objective-prism
surveys.

To date visual searches of the plates have yielded 125 quasar
candidates with emission lines in their spectra. Followup slit
spectroscopy with the SIT Vidicon system has been obtained for 40 of
the candidates, and it should be possible to cover the remainder in
1984. The followup spectroscopy is essential, for the three dimensional
distribution of the quasars is best determined from the accurate
redshifts it yields. Once these data are in hand, it will be possible
to make a complete analysis of the quasar space distribution in the new
survey. As an indication of what can be expected, the numerical
simulations mentioned above show that a value of the two point
correlation function of 0.6 will be detectable at the 2 sigma confidence
level at a scale of 25 h^{-1} Mpc (present epoch coordinates). This will
be adequate to detect the superclustering mentioned by Bahcall and
Soneira (1983) for galaxy clusters. Regardless of expectations,
however, it is important to establish more accurately what the space
distribution of quasars is at z ~ 2.

References:

Bahcall, N., and Soneira, R.: 1983, *Astrophys. J.* 270, 20
Burbidge, E. M., Junkkarinen, V. T., Koski, A. T., Smith, H. E., and
       Hoag, A. A.: 1980, *Astrophys. J. Lett.* 242, L55.
Oort, J. H., Arp, H., and de Ruiter, H.: 1981, *Astron. Astrophys.* 95, 7.
Osmer, P. S.: 1981, *Astrophys. J.* 247, 762.

DISCUSSION

MACGILLIVRAY: In view of the fact that different observers disagree
in quasar samples using the same plate material and that in a comparison of a visual sample with automatically determined sample, the
visual sample missed about 50% of the quasars, would you like to
comment on the meaningfulness of visual samples for statistical
investigations of the clustering of quasars?

OSMER: You raise important questions, and I am glad to have the
opportunity to comment on them. What is important for the clustering
problem is that the search process be uniform. In the present case
the plates are being searched by the same people to have internal
consistency. In addition the criterion for identification is the
presence of a conspicuous emission line. Because this criterion
and because of the use of prism material, which has higher dispersion
and higher signal to noise than UK Schmidt plates, for example,
I think that adequate completeness will be achieved for a first
look at the clustering properties. However, I completely agree that
automatic search procedures based on machine scans of the plates
should yield improved and larges samples of quasars, expecially
at fainter magnitudes and with weaker line strengths. A logical
next step in the program will be to scan the plates. In this case
the visually identified quasars and subsequent spectrophotometry
will be invaluable in calibrating the automatic search process.

PROPERTIES OF GROUPS AND CLUSTERS OF GALAXIES ASSOCIATED WITH QUASARS

H. K. C. Yee
Dominion Astrophysical Observatory, Herzberg
Institute of Astrophysics

Richard F. Green
Kitt Peak National Observatory

ABSTRACT. Images of fields around 108 quasars are used to study the association of galaxies with quasars. We find that the excess objects in the quasar fields are consistent with being galaxies at the cosmological distances implied by the redshifts of the quasars, and that quasars tend to prefer regions having higher than normal galaxy density.

To study statistically the association of galaxy groups and clusters with quasars, we have obtained images of fields around 108 quasars. The sample consists of radio-quiet quasars from the Palomar-Bright-Quasar Survey (Schmidt and Green 1983) and bright (m < 17) radio-loud quasars identified from 3C, 4C and Parkes surveys listed in the Burbidge, Crowne and Smith (1977) catalogue. Since we do not want to make any a priori assumptions about quasar redshift, the sample is selected without regard to redshift, which ranges from 0.05 to over 2.0. Images in the Gunn r band (6500 ± 450Å) were acquired using the SIT Vidicon direct camera at the Palomar 1.5 meter. For each quasar, two 1500-second exposures were obtained. One, the quasar field, is a 180" x 180" image centered on the quasar. The other, the "control field", was exposed in an identical manner, and centered exactly 1° North of the quasar. The control fields were acquired for the purpose of deriving an internally consistent calibration of background galaxy counts.

For each detected object, a position, magnitude and classification were derived. The classification into the categories of star or galaxy is performed by a simple "peak-value-versus-total-magnitude" scheme. The total galaxy counts versus magnitude obtained in the control fields compare well with those derived by Kron (1980) and Sebok (1982). Our average detection limit for galaxies is ~21.0 in r.

Simple galaxy counts show that there are excess galaxies associated with low redshift quasars. If we divide our quasars into subsamples with z < 0.45 and z ≥ 0.45, the ratios of galaxies counted in the quasar fields to these in the central fields are 1.61 ± 0.09 and 1.02 ± 0.09,

583

F. Mardirossian et al. (eds.), Clusters and Groups of Galaxies, 583–587.

respectively. Moreover, for the low redshift sample, the projected
surface density of galaxies shows a clustering toward the quasars. We
note that at z = 0.45, a 21st mag. galaxy is about one magnitude fainter
than a first ranked cluster elliptical.

An effective method to test the clustering of galaxies is the two-
point covariance function, W(θ), which, for galaxy-galaxy angular
correlation, has been found to have the form of A $^{(1-\gamma)}$, with $\gamma \simeq 1.77$
(see e. g., Groth and Peebles, 1977). Assuming the same form for the
quasar-galaxy covariance function, we derive A_{gq}, the angular
covariance amplitude, with a limiting r of 20.8 mag, for each quasar.
These are plotted in Figure 1 and illustrate a powerful argument for the
association of excess galaxies with quasars at the cosmological
distances of the quasars. The distribution of amplitudes is clearly a

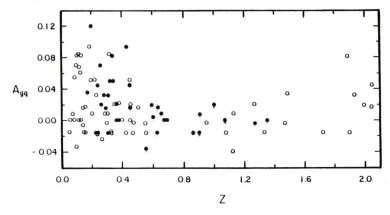

Figure 1. Angular covariance function amplitude A_{gq} versus
redshift. Closed dots (·): radio-loud quasars; open dots
(o): optical quasars.

function of redshift with a net positive average for low (z < 0.5)
redshift objects. Again, the cut-off in positive amplitude is
consistent with the detection limit of bright galaxies at the redshifts
of the quasars.

With the assumption that the excess galaxies are associated with
the quasars at the redshifts of the quasars, we can assign the quasar
redshifts to the galaxies and examine the physical properties of the
groups and clusters. In the following discussion, we also assume q_0 = 0
and H_0 = 50 kms^{-1} Mpc^{-1}.

Relative differential luminosity functions for the excess galaxies
in the optical and radio-loud subsamples in the low redshift range are
derived in the following manner. Absolute luminosities of the galaxies
in each quasar field are calculated by applying the redshift of the
quasar to the galaxies. K-corrections are approximated by assuming a
power law of the form r$^{-2.7}$ for the spectral energy distribution between

4500Å and 7000Å for galaxies. The galaxies are then binned by absolute magnitude and the counts of background galaxies are subtracted according to the counts in the corresponding apparent magnitude bins. Figure 2 shows the differential luminosity function obtained for the associated galaxies of optically-selected quasars with z < 0.30. The plotted lines are Schechter functions for local galaxies in r obtained by Sebok (1982).

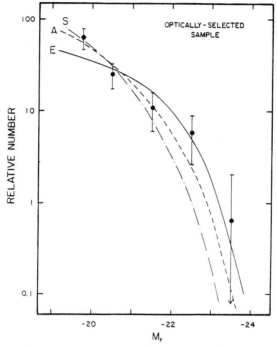

Figure 2. Differential luminosity function for galaxies associated with optical quasars with z < 0.30. Lines represent Schechter analytical forms for: E-elliptical galaxies, S-spiral galaxies, and A-combination of the two.

The radio-loud quasar sample gives a similar plot with perhaps a slightly brighter turn-over point. From the figure, we see that the brightness distribution of the excess galaxies is consistent with what one would expect from galaxies located at the cosmological distances of the quasars.

With the assumption of agreement in redshift, we can also examine the quasar-galaxy spatial covariance function. The angular covariance amplitudes derived earlier measure only the apparent surface density enhancement of galaxies around quasars, which is dependent on the redshift. To compensate for the dimming of the detected galaxies at greater distance, the luminosity function of galaxies must be taken into consideration. Following Longair and Seldner (1978), we convert the A_{gq}'s to B_{gq}'s, the spatial covariance function amplitudes. For the luminosity function of galaxies (LF) we assume a Schechter function with parameters from Krishner, Oemler and Schechter (1979) transformed to the r band. We derive a self-consistent normalization constant for the LF by integrating the function in z to produce a surface density of galaxies matched to our own background galaxy counts.

For comparison purposes, we form the ratio $R_{gg} = \langle B_{gq}/B_{gg} \rangle$ where $B_{gg} = 67.5 \pm 6.8$ (Mpc$^\gamma$) is the galaxy-galaxy coefficient derived by Davis and Peebles (1983). The results, separated into bins of redshift for both the radio-loud and optical samples, are given in Table 1.

TABLE 1

Spatial Covariance Amplitude Ratios

Redshift	Radio-loud Sample		Optically-selected Sample	
	N	R	N	R
0.00 - 0.15	0	-	16	1.89 ± 0.72
0.15 - 0.30	9	3.44 ± 1.26	15	1.96 ± 0.93
0.30 - 0.40	8	5.42 ± 2.23	8	2.42 ± 2.02
0.00 - 0.40	17	4.38 ± 1.23	39	2.03 ± 0.60

The combined sample of 56 quasars produces an R of 2.76 ± 0.57. There are several interesting points in these results. First, the environment of quasars is significantly richer than that of randomly chosen galaxies. However, on the average, the richness is not nearly as great as that of Abell richness class 1 clusters, which have R ~ 12.6. In fact, none of the clusters found is comparable to Abell 1 clusters. Second, the average cluster amplitude for radio-loud quasars is marginally larger than that of optical quasars. And third, we have not detected any evolutionary trend for quasar environments up to z = 0.4.

French and Gunn (1983), in a similar study, obtained R = $1.25 \pm .50$ for mostly radio-loud quasars. Among other differences in our data gathering and analysis, we note that they counted galaxies in an area having a radius 2 to 4 times larger than ours. The Stockton (1978) sample, also of radio-loud quasars, consists of areas slightly smaller than ours, and produces R ~ 4, similar to our results. Thus, the data indicate that there may be a steepening of the quasar-galaxy covariance function at small distances from the quasar. Further support of this idea comes from examining directly the surface density distribution of galaxies around quasars. We fit King models to the surface density of the richer associations. The results show that while the average cluster core radius of ~ 120 Kpc is much smaller than that of Abell richness class 1 clusters, the derived central space density of galaxies is comparable to that of Abell 1 clusters. This indicates that compact, high galaxy density regions are preferred sites for quasar activity.

In summary, we have examined fields around 108 quasars chosen from complete optical and radio-loud samples. We find that the excess galaxies in the quasar fields are consistent with being at the cosmological distances of the quasar redshifts. Moreover, compared to the average galaxy, quasars tend to have more galaxy neighbors, and seem often to prefer compact high galaxy density regions. A more detailed description of this work will appear in Green and Yee (1984) and Yee and Green (1984).

REFERENCES

Burbidge, G. R., Crowne, A. H., Smith, H. E.: 1977, Ap. J. Suppl., 33, p. 113.

Davis, M. and Peebles, P. J. E.: 1983, Ap. J., 267, p. 465.

French, H. and Gunn, J. E.: 1983, Ap. J., 269, p. 29.

Green, R. F. and Yee, H. K. C.: 1984, Ap. J. Suppl., in press.

Kirshner, R., Oemler, A., and Schechter, P. L.: 1979, A. J., 84, p. 451.

Kron, R. G.: 1980, Ap. J. Suppl., 43, p. 305.

Longair, M. S. and Seldner, M.: 1979, M.N.R.A.S., 189, p. 433.

Sebok, W. L.: 1982, PH.D. Thesis, California Institute of Technology, Pasadena.

Stockton, A.: 1978, Ap. J., 223, p. 747.

Yee, H. K. C., and Green, R. F.: 1984, Ap. J., in press.

THE DISTRIBUTION OF FAINT GALAXIES DETECTED ON DEEP UKST PLATES

H.T. MacGillivray[1] and R.J. Dodd[2]
1. Royal Observatory, Blackford Hill, Edinburgh, Scotland
2. Carter Observatory, P.O. Box 2909, Wellington, New Zealand

ABSTRACT

A programme is underway to investigate the large-scale projected distribution of faint galaxies from objective, machine-determined samples. Data are obtained from measurements made with COSMOS on deep photographs taken with the UK 1.2m Schmidt Telescope. The ultimate intention is to obtain a complete Lick-type galaxy catalogue from machine data for a large area of the southern sky. In this paper, we describe the methods and data reduction techniques used, and present some preliminary results. We find that the two-point correlation function goes to zero or negative values on scales of $6-7h^{-1}$ Mpc and that there is no evidence for the presence of filamentary structure in these objective samples.

1. INTRODUCTION

Our knowledge about the large-scale distribution of galaxies has been based in the past upon galaxy samples obtained by subjective means (i.e. visual scans of photographic plates), e.g. the Lick galaxy catalogue (Shane and Wirtanen 1967). Such samples are not ideal material with which to form conclusions about the nature of the distribution of galaxies because of biases inherent in them. With the COSMOS automatic, plate-scanning machine we are attempting to put this type of investigation on a more objective basis. Ultimately, it is our aim to construct a purely objective galaxy catalogue similar to that of the Lick survey. For this purpose, photographic plates of several fields of the SERC/ESO southern sky survey are being scanned and a mosaic of the sky pieced together. In the present paper we describe the methods used and present some preliminary results.

F. Mardirossian et al. (eds.), Clusters and Groups of Galaxies, 589–594.

2.   PLATE MEASUREMENT AND DATA REDUCTION

The advantages of using the COSMOS machine for this analysis are manyfold. The photographic plates can be scanned objectively, accurately and at high speed, and exceedingly large numbers of objects can be detected and processed efficiently. The area of plate scanned with COSMOS is a region of size 287 x 287 mm², corresponding to 5.35° x 5.35° on the sky. Since the field of view of the UK Schmidt Telescope (UKST) is 6.5° and the southern sky field centres are separated by 5°, then in the data there is a comfortable overlap of ∿ 1/3° on plates of adjacent fields. This overlap means that there are no missing gaps or strips in our sky coverage and enables a check to be made on the consistency of the magnitude system from plate to plate.

Image detection is carried out at the 7% night sky intensity level on these UKST plates (corresponding approximately to the B = 25.6 magnitudes/sq arcsec isophote) and the time taken to complete the scans is 6½ hours (with pixel size of 16μm). Star/galaxy separation is performed down to a limit of typically B = 21.5 by means of surface brightness criteria for objects fainter than B ∿ 16.5 (see MacGillivray & Dodd 1982b) and by means of geometric criteria for objects brighter than this limit.

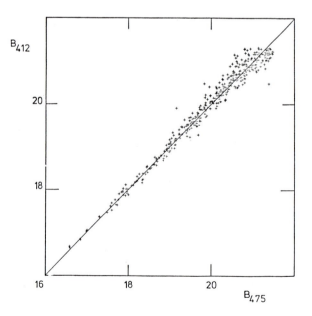

Figure 1. A comparison of Galaxy magnitudes in the overlapping region of fields 412 and 475. The line has slope of 45°.

3.   GALAXY PHOTOMETRY

It is important for homogeneity in the construction of a large-scale galaxy catalogue that magnitudes can be reliably determined and the magnitude system maintained over a large area of sky. In previous papers (MacGillivray & Dodd 1982a,b), we have carried out a comparison between galaxy photometry using COSMOS and the PDS. These comparisons indicate that magnitudes for galaxies can be reliably obtained from COSMOS measures at least in the range $13 \leq B \leq 23$, although at the faint end this necessarily requires image detection at lower isophotes (e.g. 1-2% of the night sky intensity level). Figure 1 shows a comparison of magnitudes for galaxies in common in the overlapping region of plates for fields 412 and 475. The magnitudes in each field were calibrated independently. The figure shows that a 45° line is a reasonable representation of the relationship between the points with no obvious effect of zero-point differences present. Thus, the evidence indicates that for our catalogue the magnitude system is indeed consistently maintained from field to field.

4.   RESULTS

Figure 2 shows an isoplethal map representing the distribution of galaxies detected down to $B = 21.8$ in the southern sky survey field 349. Parker et al (1984, these proceedings) have described work in this field in order to obtain objective-prism redshifts for galaxies and clusters for the purpose of examining the 3-D distribution of galaxies. In this paper we present results also obtained in this field. We have, however, analysed several other fields and obtained similar results in those. Two types of analysis have been used:- the two-point correlation function (Peebles & Hauser 1974) and methods for detecting filamentary structure (e.g. Kuhn & Uson 1982).

Figure 3 shows the two-point correlation function for the galaxies in the field. Note that $w(\theta)$ fluctuates about values of zero on scales greater than 1.5°. Parker et al have found that the majority of structures contributing to the isopolethal map in this field have peak redshifts in the range $z \sim 0.10 - 0.11$. If this is taken as the typical redshift for the objects in the data, then the linear scale at which $w(\theta)$ goes to zero is $\sim 6-7h^{-1}$ Mpc (with q_o = +1). This is smaller than the break point found for the Lick data ($9h^{-1}$ Mpc) by Groth & Peebles (1977), but yet this result is found consistently in other fields also. Indeed, in other fields $w(\theta)$ is mainly negative on scales larger than this value. Unfortunately, we have not yet been able to examine the two-point correlation function on scales of several tens of Mpc to discover whether the function remains zero or negative on larger scales or becomes positive again.

Kuhn & Uson (1982) have described a method for searching for filamentary structure in galaxy samples. Basically, their method searches for ridges in the galaxy distribution, a quantity $\langle \mu \rangle$ being

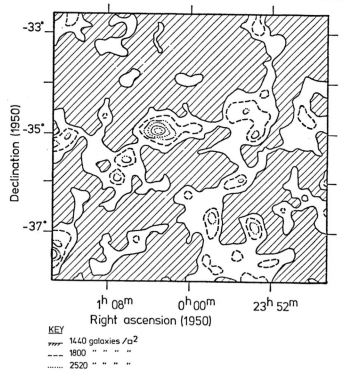

Figure 2.   Isoplethal map showing the distribution of galaxies down to
B = 21.8 in field 349.

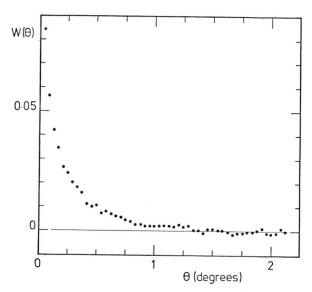

Figure 3.   The two-point correlation function for the distribution of
galaxies in field 349.

calculated which represents the mean cosine of the direction to cells along the ridge. The ridge is traversed by detecting high density neighbouring cells in the horizontal and vertical directions. A disadvantage of this technique is that it is mainly sensitive to filaments which are aligned either horizontally or vertically. We have developed a method which is similar to the Kuhn & Uson technique but which also allows high density cells in diagonal directions to be included. This method ought to be more sensitive to the presence of filaments aligned in any orientation. Also, since we retain the coordinates of all the galaxies, we can directly search for filaments using the galaxies themselves as centres rather than cells. This latter method is, however, rather expensive in terms of CPU time and so is only applied in practice to the brightest few hundred objects.

We have used all three of the above techniques with several different combinations of limiting magnitude and cell size. In no case have we detected any significant effect of the presence of filamentary structure. This is illustrated with the results for field 349:- using the Kuhn & Uson method and with the galaxies binned into cells of side 5' x 5' we obtain $\langle\bar{\mu}\rangle$ = 0.382, 0.382 and 0.386 with standard error of ±0.009 for galaxies limited at B = 21.8, 21.0 and 20.0 respectively. Compare this with the result of $\langle\bar{\mu}\rangle \cong$ 0.424 for the Lick data and $\langle\bar{\mu}\rangle \cong$ 0.396 for the random simulations.

The above should not be regarded as conclusive as yet since there are a number of factors which may have played a part. For example the methods used may not be particularly sensitive to the presence of filaments, any real filaments may have been completely swamped by the superposition of objects over large distances, or indeed filaments may only be found when we examine larger areas of sky. Nonetheless, the results are not encouraging for the objective verification of the presence of filamentary structure. It will be of interest to carry out the analysis on a much larger area of sky when possible.

REFERENCES

Groth, E.J., and Peebles, P.J.E.: 1977, Astrophys.J. 217, 385.
Kuhn, J.R., and Uson, J.M.: 1982, Astrophys.J. 263, L47.
MacGillivray, H.T., and Dodd, R.J.: 1982a, Observatory 102, 141.
MacGillivray, H.T., and Dodd, R.J.: 1982b, in Proceedings of the "Workshop on Astronomical Measuring Machines 1982", eds. R.S. Stobie and B. McInnes, Edinburgh, Scotland, p195.
Parker Q.A., MacGillivray, H.T., and Dodd, R.J.: 1984, in "Clusters and Groups of Galaxies" this volume, p. 595.
Peebles, P.J.E., and Hauser, M.G.: 1974, Astrophys.J.Suppl. 28, 19.
Shane, C.D., and Wirtanen, C.A.: 1967, Pub.Lick Obs. Vol. 22, Part 1.

DISCUSSION

VAN WOERDEN: In order to find filamentary structures in your material, I think you would need a distance indication for your galaxies.

MACGILLIVRAY: I agree that the filamentary structure, if present, would be detected easier in three-dimensional space rather than in two, and indeed we are attempting to obtain radial velocities for galaxies in our data for this purpose. However, there are problems with radial velocity data (c.g. the "finger of God" effect). Claims have been published concerning the detection of filamentary structure in projected galaxy distributions and so the basis for these claims is worthy of study also.

AN OBJECTIVE-PRISM GALAXY REDSHIFT SURVEY AT THE SOUTH GALACTIC POLE

Q. A. Parker[1], H. T. MacGillivray[2] and R. J. Dodd[3]
1.   Department of Astronomy, St. Andrews University, Fife, Scotland
2.   Royal Observatory, Blackford Hill, Edinburgh, Scotland
3.   Carter Observatory, PO Box 2909, Wellington, New Zealand.

ABSTRACT

A project has been initiated to obtain objective-prism redshifts for large numbers of faint galaxies in several fields around the South Galactic Pole. Measurements are made with the COSMOS machine on deep photographs taken with the UK 1.2m Schmidt Telescope. Ultimately, we intend to carry out a complete survey of an extensive area for the purpose of investigating the large-scale distribution of galaxies. Preliminary results from one particular field of the southern sky are presented in the present paper together with an assessment of the usefulness of the redshifts obtained for this type of analysis.

1.   INTRODUCTION

The technique used for obtaining galaxy redshifts from objective-prism plates has been described at length in other papers (Cooke et al 1981, 1982, 1984). Briefly, it relies upon the change with redshift of the location of the "4000Å feature" observed in galaxy spectra (at $\lambda = 3990$Å) from its rest wavelength position. Measurements are made with respect to the position of the emulsion cutoff (at $\lambda = 5380$Å for the unfiltered IIIaJ emulsion) which is taken as wavelength standard. Despite the low accuracy attainable $\Delta z \sim 0.01$), this technique provides a practical means for obtaining redshifts for several thousands of galaxies in a single Schmidt telescope field (6.5° x 6.5° on the sky).

2.   DATA REDUCTION

Scans were made with COSMOS in its Mapping Mode (see MacGillivray 1981) on an objective-prism plate, taken with the UK 1.2m Schmidt Telescope (UKST), of the SERC/ESO southern sky survey field number 349

595

F. Mardirossian et al. (eds.), Clusters and Groups of Galaxies, 595–600.

(1950 coordinates 00h 00m, -35°00'). The measurements were made with a 16μm pixel size (corresponding approximately to 1 arcsec on the plate). The semi-automatic technique described by Parker et al (1983) was used for the galaxy redshift determination. This is soon to be replaced by a completely automatic method (Cooke et al 1984). Note that for the purpose of our work, the results are based solely on the spectra of objects which are identified as galaxies on the direct plate of the field.

Isoplethal maps of the galaxy distribution on the direct plate down to a limit of B = 21.8 (see MacGillivray & Dodd, 1984) were used to identify rich clusters and other suitable areas for study. Two rich foreground clusters, identified from visual inspection of the direct plate, were also considered. Identification of these from the isoplethal plots was difficult due to their less concentrated nature. This method for cluster detection is far from ideal. However, attempts are being made to put the cluster identification on a more objective basis for the future.

3.   CONSISTENCY OF THE REDSHIFTS

Before any significance can be attached to the redshifts obtained, an appraisal of the accuracy and consistency of the results is necessary. Calibration of the results against slit spectra is still sparse but is so far consistent to within the errors (Cooke et al 1981). Although the low dispersion of the prism (2480Å mm^{-1} at Hγ) coupled with the small plate scale and blurring by seeing sets an intrinsic limit on the accuracy, the practical limitation for our purpose depends on the accuracy of the measurement of emulsion to cutoff distance. In the procedure used for this work, this accuracy is at worse ±1 pixel. The effect of this on the redshift accuracy is itself redshift-dependent since the dispersion of the prism varies with wavelength. The broken lines in figure 1 indicate the uncertainty in redshift estimation because of this source of possible error ($\Delta z = \pm 0.008$ at $z = 0.02$ increasing to $\Delta z = \pm 0.013$ at $z = 0.25$).

A large amount of data now exists enabling a check to be made of the repeatability of the technique. Figure 1 shows the comparison of redshifts obtained from two independent sets of measures of the rich cluster at 00h 03.5m, -35°00' (1950 coordinates). The measurements were made with separation of 1 year. As can be seen from this figure the points scatter about the 45° line and the majority lie within the expected uncertainty. This shows that the internal consistency of the measurements is high. The discrepant point is the result of ambiguity in the 4000Å feature identification, but note that in this case the redshift had been assigned low weight (a subjective weighting system, based on a number of factors such as the confidence attached to the 4000Å feature identification, is made during the procedure - see Parker et al 1982).

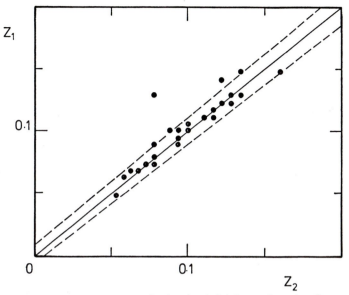

Figure 1. Repeat measures of the redshifts of galaxies showing the consistency of the results. The broken line represent the expected uncertainties.

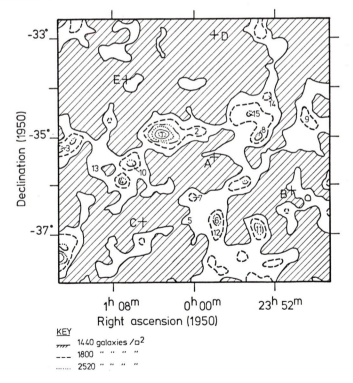

Figure 2. Isoplethal map of the distribution of galaxies in field 349. The clusters (1 - 15) and control areas (A-E) are indicated.

4. RESULTS

Objective-prism redshifts have been obtained for galaxies in the direction to 15 clusters and 5 control areas in field 349.   In figure 2 the isoplethal plot for the direct plate data is shown and on it are marked the clusters and control areas selected. These control areas, each of side 0.6° and chosen in regions where there were no obvious clusters present, were studied in an attempt to understand the likely field contamination in the direction to the clusters. In this way it was hoped to assess the significance of any redshift peaks.

The histograms representing the redshift distributions in the direction to each of the rich clusters are shown in figure 3. Examination of the isoplethal plot of figure 2 and the redshift histograms of figure 3 reveals an interesting observation. Apart from clusters 7, 8 and 9 (which are foreground clusters as confirmed in figure 3), most of the remaining clusters have peaks in the redshift interval $z \sim 0.10 - 0.11$. The peak is particularly strong in the case of clusters 1, 6, 13 and 15. Control areas A ad B also yield peaks in the same range. The conclusion we derive from these observations is of the presence of a structure at a redshift of $z \sim 0.10 - 0.11$ consisting of several rich clusters and a surrounding dispersed component. Corroboration for these objective-prism results is provided from a slit spectrum redshift of $z = 0.114$ obtained for the central cD galaxy of cluster 1 (Carter 1980), which is consistent to within the errors with the peak redshift determined herein. Assuming this large-scale structure extends for $\sim 4°$ on the sky, then this corresponds to a true physical extent of $\sim 15-20$ h^{-1} Mpc (with q_o = +1).

The histogram for the combined redshifts in the 5 control areas is shown in figure 4. Note in this diagram a peak is also evident at a redshift of $z \sim 0.10 - 0.11$ being mainly contributed by control areas A and B. To investigate whether this preference for a peak at $z = 0.10 - 0.11$ is an artefact of the objective-prism selection function, we used a Monte-Carlo simulation technique. The simulations have been described at length in other papers (MacGillivray et al 1982; MacGillivray & Dodd 1982). Basically, static snapshots of the Universe are generated in 3-D space and the galaxies projected onto an artificial photograph, all factors affecting the light from the distant galaxies being modelled and applied. The final data set was restricted to that which would appear in the observed sample under the same conditions. Results from 10 such simulations were averaged to produce the smooth curve in figure 4 (scaled to the area of the 5 control fields). Note that the modelled selection function does not match that from the observations leading us to conclude that the observed peak at $z = 0.10 - 0.11$ is a real feature and that there is a deficiency of galaxies compared to that which would be expected in the redshift range $z = 0.03 - 0.10$.

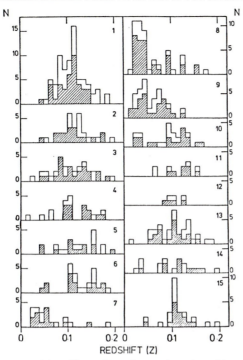

Figure 3. The redshift distributions in the direction to each of the 15 clusters. Shaded histograms represents a high confidence in the redshift obtained.

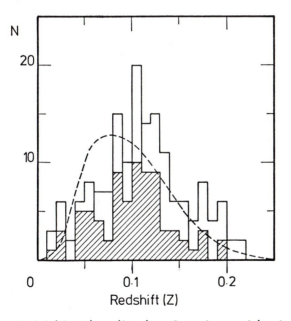

Figure 4. Redshift distribution for the combined control areas. The broken-line curve is the expected selection function (see text).

5. CONCLUSIONS

The field studied in this paper is interesting from the point of view that it contains a complexity of structures as evidenced by coincidences in the location of objects in both projected and spatial coordinates. We identify 3 main features in the redshift distribution:- a foreground supercluster at a redshift $z \sim 0.02 - 0.04$ consisting of clusters 7, 8 and 9 at least; an overall deficiency of galaxies in the redshift range $z \sim 0.03 - 0.10$; and a large-scale structure consisting of several rich clusters and dispersed component at a redshift $z \sim 0.10 - 0.11$.

Thus, although objective-prism redshifts are not sufficiently accurate to allow determination of internal cluster velocity dispersions, the results obtained herein show that they are capable of identifying structures separated in redshift space and of revealing the presence of regions deficient in galaxies.

ACKNOWLEDGEMENTS

QAP was supported by an SERC studentship during this work. We are grateful to Drs J A Cooke and S M Beard for making their software available.

REFERENCES

Carter, D.:   1980, Mon.Not.R.astr.Soc. 190, 307.
Cooke, J.A., Emerson, D., Kelly, B.D., MacGillivray, H.T., and Dodd, R.J.:   Mon.Not.R.astr.Soc. 196, 397.
Cooke, J.A., Emerson, D., Beard, S.M., and Kelly, B.D.:   1982, Proceedings of the "Astronomical Measuring Machines Workshop 1982", eds. R.S. Stobie and B. McInnes, Edinburgh, Scotland, 209.
Cooke, J.A., Kelly, B.D., Beard, S.M., Emerson, D.: 1984, in "Astronomy with Schmidt-type Telescopes" ed. M. Capaccioli, Asiago, Italy, D. Reidel Pub. Co., p. 401.
MacGillivray, H.T.:   1981, in "Astronomical photography 1981", eds. J.-L. Heudier and M.E. Sim, Nice, France, 277.
MacGillivray, H.T., Dodd, R.J., McNally, B.V., Lightfoot, J.F., Corwin, H.G.Jr, and Heathcote, S.R.:   1982, Astrophys.Sp.Sci. 81, 231.
MacGillivray, H.T., and Dodd, R.J.:   1982, Astrophys.Sp.Sci. 86, 437.
MacGillivray, H.T., and Dodd, R.J.: 1984, p. 589.
Parker, Q.A., MacGillivray, H.T., Dodd, R.J., Cooke, J.A., Beard, S.M., Emerson, D. and Kelly, B.D.: 1982, Proceedings of the "Astronomical Measuring Machines Workshop 1982", eds. R.S. Stobie and B.McInnes, Edinburgh, Scotland, 233.

COMPANIONS TO S0 GALAXIES

A. Busch, B.F. Madore and E.R. Anderson,
Department of Astronomy, David Dunlap Observatory,
University of Toronto, Toronto, Ontario, Canada, M5S 1A1.

ABSTRACT

We present statistical results on the distribution of dwarf companions to S0 galaxies. We find that the slope of the galaxy covariance function $\gamma = 1.60 \pm 0.10$, consistent with the value $\gamma = 1.8$ obtained at larger separations.

DISCUSSION

Galaxy counts were made for the environments around the 127 S0 and SB0 galaxies of the Revised Shapley-Ames Catalogue of Bright Galaxies out to a projected radius of 140 kpc. POSS plates and ESO films were examined. All galaxies with linear sizes greater than 1 kpc were noted. Based on size and morphology, a probability code (PC) of being a physical companion was assigned to each apparent companion. The PC's range from 0 (probable background/foreground) to 4 (probable physical companion). Assuming that the density of physical companions would decrease with radial distance from the parent galaxy, radial density distribution plots were used as a means of determining the number of physical companions above the background density. These plots also provided us with an independent check on the PC assignments.

The companions of PC 0 and 1 combined exhibit a flat radial density distribution indicating that they represent most of the sky background contamination. The radial density distribution of PC 3 and 4 galaxies combined (see Fig. 1) shows a smooth radial decrease and thus the majority of these galaxies are probably physical companions. We estimate the average number of dwarf companions per S0 galaxy within 100 kpc to be in the range $0.42 \leq \langle N \rangle \leq 1.56$. The upper limit is found by assuming that all PC 3 and 4 galaxies are physical companions The lower limit is found by assuming that the last bin of Figure 1 represents the background contamination.

F. Mardirossian et al. (eds.), Clusters and Groups of Galaxies, 601–602.
© *1984 by D. Reidel Publishing Company.*

Contrary to the results of Holmberg (1969), we find that the
PC 3 and 4 companions have a uniform angular distribution about the
axes of the edge-on galaxies of our sample.

Figure 2 shows a plot of log(surface density) vs log(radial
distance) for the physical dwarf companions. We assume that the
galaxy covariance function is a power law of the form $\Sigma \sim R^{1-\gamma}$. A
least-squares fit to the slope yields $\gamma=1.60\pm0.10$. Our value
of γ agrees well with the value of 1.52 ± 0.19 of Lake and Tremaine
(1980). Note that both of these values are consistent with the value
of 1.8 which has been determined at large scales by many investigators.

REFERENCES

Holmberg, E.: 1969, Arkiv. Astr. 5, pp. 305.
Lake, G. and Tremaine, S.: 1980, Ap.J. (Letters), 238, pp. L16.
Sandage, A. and Tammann, G.A.: 1981,
     'A Revised Shapley-Ames Catalogue of Bright Galaxies',
     Publ. 635, Carnegie Inst. of Washington, Washington D.C.

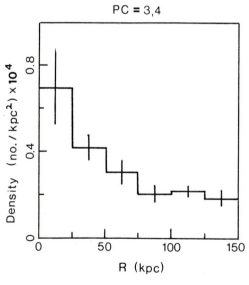

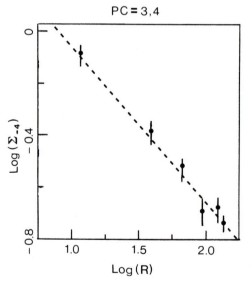

Figure 1. The combined radial
density distribution for galaxies
of PC 3 and 4.

Figure 2. The galaxy covariance
function.

STRÖMGREN PHOTOMETRY OF GALAXIES

Karl D. Rakos and Norbert Fiala

Institut für Astronomie der Universität Wien,
Türkenschanzstrasse 17, A-1180 Wien, Austria

In preparation for a program of observations of different types
of galaxies in distant clusters, we have calculated synthetic colors
from the spectrophotometry of radio galaxies published by Yee and Oke
(1978). The monochromatic magnitudes were calculated for the Strömgren
v, b and y bandpasses shifted into the rest frame of each galaxy. The
colors calculated were $(b-y)$ and $(v-b)-(b-y)=m_1$.

The Strömgren photometric system is well suited for this study
because these bandpasses are not usually affected by the hydrogen and
forbidden emission lines commonly found in the spectra of "active
galaxies." The H-delta line, which affects the v-bandpass, as a rule is
small in these spectra. Only the He II emission at 4686 Å may be very
strong, shifting m_1 to larger and $(b-y)$ to smaller values. The predicted
slope is indicated by an arrow in Fig. 1. A second reason to select the
Strömgren photometric system is its well established use in photometry
(Hauck and Mermilliod 1980, Gunn and Stryker 1983, Relyea and Kurucz
1978). The spectrophotometric scans of Yee and Oke (1978) were convolved
with the v, b, y response functions given by Crawford and Barnes (1970).
We also convolved in the same way similar spectrophotometric scans of a
selected number of standard stars (Breger 1976 a, b) covering the whole
spectral sequence. From a comparison of these results with v, b, y
photometry of the same stars, we get a set of transformations of the
form: $(b-y)_o = A+B(b-y)_{scan}$, $(m_1)_o = C+D(m_1)_{scan}+E(b-y)_o$. Synthesized color
indices derived in this way agree with measured photometric indices
within an accuracy of 0.02 magnitudes. The lower accuracy of extragalactic
spectrophotometry in any case generally does not allow higher accuracy.

From the optical spectra of the Yee and Oke sample it is possible
to discriminate the strong emission line objects with equivalent widths
of the N II - H-alpha blend greater than 100 Å from the objects with
moderately strong, weak or undetected emission lines. Yee and Oke also
published a spectrum of the so-called standard elliptical galaxy, a mean
of multichannel scans (40-80 Å resolution) of five first ranked elliptical
galaxies with average z values of the radio galaxy sample. The data were
corrected for galactic reddening. Only the reddening of Cyg A is contro-

F. Mardirossian et al. (eds.), Clusters and Groups of Galaxies, 603–604.
© *1984 by D. Reidel Publishing Company.*

versial, so we have excluded this galaxy from further discussion. We have also excluded 3C 305 classified as an Sa galaxy.

We calculated the indices (b-y) and m_1 in the rest frame of each galaxy. The standard deviations were derived from the published standard deviations of the monochromatic magnitudes in a single passband. One sigma error bars are drawn only for N-galaxies in Fig. 1. Sigma in (b-y) and m_1 for all other sources is in the order of 0.03 magnitudes. The diagram (Fig. 1) indicates that it is possible to discriminate N-galaxies (filled dots) from E- and D-types (open circles) by means of z-shifted v, b, y photometry. The square represents the standard galaxy. The photometric measurements, which we plan to obtain, are far less time-consuming than spectrophotometric measurements and therefore can be applied to a larger number of galaxies. The extremely low value of m_1 for four N-galaxies can be explained partly by the presence of nonthermal radiation. An additional influence of unknown broad emission lines, which are not conspicuous from direct inspection of the spectra, may also be in part responsible for the location of N-galaxies in the (b-y), m_1 diagram. No correlation can be found between N II - H-alpha blends and either m_1 or (b-y). In general E- and D-galaxies discussed here are "active ellipticals." They have mean values of (b-y)=0.57 and m_1=0.37 noticeably different from the standard (mean) galaxy. Any systematic difference between active ellipticals and normal first ranked ellipticals may be very important for the determination of the absolute magnitudes of "standard candles."

Onoura and Okoye (1983) showed that there is a strong correlation between radio luminosity and redshift for radio galaxies of the form: $P \sim (1+z)^4$. This prompted us to search for a similar correlation in our results. We found the correlation $m_1 = A(b-y)(1+z)^4 + B$ with A=-0.25 and B=0.40 with our sample of N-galaxies with a coefficient of correlation of 0.76 (Fig. 2). This relationship suggests that N-galaxies are a short-lived phenomenon occuring during the last 5×10^9 years. Unfortunately the sample of objects is too small for further statistical investigations.

Acknowledgements: This work has been supported by "Fonds zur Förderung der wissenschaftlichen Forschung, Projekt Nr. P5073."

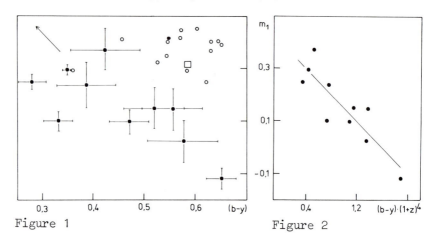

Figure 1   Figure 2

DYNAMICAL ENVIRONMENT AND RELIABILITY OF REDSHIFT MEASUREMENTS
IN STATISTICAL STUDIES OF BINARY GALAXIES

G. Tanzella-Nitti
Osservatorio Astronomico di Torino, I-10025 Pino Torinese TO

Many statistical works on the dynamics of binary galaxies usually employ observational material obtained by a single author. When large redshift Catalogues (Palumbo et al., 1983) are used to compare measures of velocities by different authors or to associate a reliable average velocity to the double galaxy orbital motion, a number of interesting results can be outlined. A weighted mean procedure can be performed employing the external errors of each optical reference that are obtained analyzing the distribution of the (V(HI)-Vopt) residuals. The error associated to the optical redshifts range from 60 km/s up to 150 km/s, while the radio redshift obtained at NRAO, Nançay and Arecibo are fully consistent within 20 km/s, in agreement with the finding of Rood (1982).

The evaluation of a projected orbital velocity $\Delta V = \langle Va \rangle - \langle Vb \rangle$ considering both radio and optical velocities on each component galaxy must take into account either radio beam confusion effects for doubles with apparent separation $\theta < 10'$ (see Fig. 1), which produce the under-estimation of ΔV differences obtained with radio measurements only, and the overestimation of optical velocities if they are obtained in

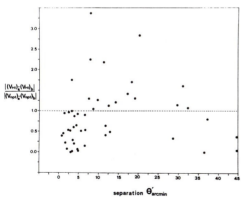

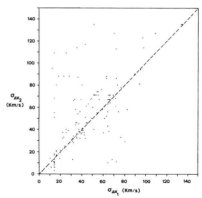

Fig. 1: Ratio between radio and optical orbital velocity ΔV versus separation of double galaxies.

Fig. 2: Errors associated to ΔV when average on each component ($\sigma_{\Delta V_1}$) or average of various ΔV_i ($\sigma_{\Delta V_2}$) are used.

F. Mardirossian et al. (eds.), Clusters and Groups of Galaxies, 605–606.

the blue region of the spectra in the velocity range 1500<V(km/s)<2100, due to the Roberts effect.

Concerning the internal dispersion of velocity measurements, Fig. 2 show that the formal errors associated to $\Delta V = \langle Va \rangle - \langle Vb \rangle$ are statistically smaller than the errors associated to the weighted average of various velocity differences ΔV obtained by different observers on the same double. In addition, the use of average procedures on single component galaxies does not imply the under-estimation of the final orbital velocity ΔV.

The complete environment classification of 586 doubles belonging to the Karachentsev, Peterson, Turner and Page samples has been performed using a survey of the literature as well as P.S.S. red prints visual inspections (about 250 fields). After the rejection of optical pairs (11%, 21% and 19% in Karachentsev, Peterson and Turner samples, respectively) 236 doubles were recognized as truly isolated pairs, 209 as located in groups or clusters and 42 with a nearby companion galaxy. Fig. 3 shows the distribution $N(\Delta V)$ obtained for truly isolat-ed binaries and all the other non-isolated physical pairs and the two distributions are different at the 99.3% significance level: the medians V are 82 and 121 km/s, respectively. This same result, found by White et al. (1983) for Turner's sample only is here confirmed for a more extended sample, though a non-negligible fraction of their environment classification has been revised due to the present analysis. Also spiral-spiral pairs have associated ΔV values significant lower than those of systems where at least an elliptical component exists: 113 and 164 km/s are the ΔV average values for $\sigma_{\Delta}<50$ km/s. This difference remains significant when we consider pure isolated binaries to avoid the redshift excess of group pairs, where ellipticals are mainly found. This result also holds for Turner sample, separately (White et al., 1983). Finally, the population of spiral-spiral pairs seems to have projected separations statistically larger than those of mixed or elliptical pairs. The average separations are 103 kpc for pure spiral doubles and about a half, 57 kpc, for the remaining doubles (Ho=50). This finding could state in the sense that total masses and total angular momenta have comparable values in both the two different classes.

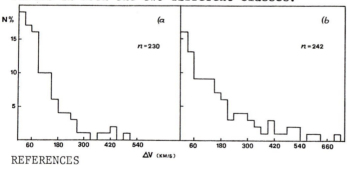

Fig. 3: Orbital velo-city distribution for isolated (a) and non-isolated (b) double galaxies.

REFERENCES

Palumbo, G., Tanzella-Nitti, G., Vettolani, G., 1983, A Catalogue of Radial Velocities of Galaxies, Gordon and Breach, New-York.
Rood, H.L., 1982, Astrophys. J. Suppl., 49, 111.
White, S.D.M., Huchra,J., Latham,D., Davies, M., 1983, MNRAS, 203,701.

SUMMARY

P. J. E. Peebles
Joseph Henry Laboratories
Princeton University

The study of the clustering of galaxies is hardly new. If we had
had this conference 25 years ago de Vaucouleurs could have lectured to
us on the structure and kinematics of the Local Supercluster, Abell on
the tendency of the great clusters of galaxies to appear within second-
order clusters, Zwicky on his concept of cluster cells with diameters
~ 50 Mpc (for $H = 75$ km s^{-1} Mpc^{-1}), and Shane on the "cloudy" nature
of the galaxy distribution seen on still larger scales in the Lick
survey (1,2,3,4). And Rubin, Limber, Neyman and Scott had already laid
out the program of statistical analysis of the distribution (5). What
has happened since then, as we have witnessed at this conference, is a
growth in the quality and quantity of the observations so great as to
represent a revolution. We are presented with a phenomenology so rich
and varied that we might have despaired of finding any simple patterns
in it, but yet such patterns are emerging and we are starting to see a
crystallization of opinion around some of the major points of
interpretation. Here is my impression of the main themes we have seen
develop at this conference.

Study of the large-scale galaxy distribution has been greatly
stimulated by the improvement of distance measures over that afforded
by apparent magnitude alone. We learned of impressive progress in the
technology of wholesale distance measurements, from the use of great
clusters as signposts to colors as luminosity indicators, from slitless
spectra to precision redshift measures. All this can only increase the
rate of discovery. What has been revealed so far on the very largest
scales, ~ 50 to 150 Mpc, has been variously described as cloudy, noisy,
frothy, cellular and filamentary. We heard some disagreement on which
term is most appropriate, but we also saw lots of good ideas for
sharpening our measures of what we are seeing, and I think we are well
on our way to a resolution.

Abell emphsized that the great clusters in his catalog that stand
out so prominently on scales ~ 0.1 to 2 Mpc are only the extreme peaks
in a rugged terrain. We have seen that these peaks often are complex
arrangements of sub- and super-clustering, but yet that they really

607

are special sites for the accumulation of plasma, radio sources, the cD phenomenon and even globular star clusters, all of which can be better indicators of the position of the cluster center than is the count of galaxies. The degree of irregularity of a great cluster has been attributed to a degree of immaturity. Lucy and Tully provided two beautiful cases where clustering patterns seem to be evolving on time-scales well below the Hubble time H^{-1}; de Vaucouleurs showed us that the Local Supercluster is evolving in its large-scale feature on a time-scale on the order of H^{-1}, and Huchra showed us how complicated is the redshift structure even within the Virgo cluster. Theoretical studies of rates of relaxation to a presumed simple equilibrium state have made important progress but are vexed by the uncertainty in the physical properties of the dominant mass component in clusters and also in the rate of disturbance of clusters.

It is an encouraging check that the mass distribution within a cluster derived on the assumption that the overall galaxy distribution is close to dynamical equilibrium is roughly in accord with the mass needed to hold in the intracluster plasma. This plasma might have been shed by the galaxies in the cluster, and indeed it was shown that in some groups X-ray plumes may reveal galactic winds. Of course, plasma might also have been contributed by galaxies that passed through the cluster in the past, and plasma might have dissipatively fallen into the cluster on its own. We do have a fascinating constraint on such speculation, that the abundance of the seen heavy elements is remarkably uniform from cluster to cluster at roughly half solar to solar. This demands tight control on processing through stars and on intermingling with unprocessed material such as might be present in an intercluster medium. We have also seen that an intercluster plasma may be important in the heat budget for the intracluster gas. On smaller scales the X-ray luminosities that are observed are so high compared to the heat capacities we think we know that the gas must be settling if it is not provided with a heat source. Particularly dramatic are the plasma pools around some giant galaxies. The pool around M87 has managed to dump its angular momentum and settle down to an exceedingly smooth shape that, as the temperature run seems to be slow, provides us with a magnificent tracer of a symmetric dark massive envelope. This brings us back to the great question, what is the material (or what are the materials!) that dominates the great clusters and the outskirts of some galaxies? On a slightly simpler level, taking account of the apparent tendency of plasma to move about such as in cooling flows, what is the scale on which the ratio of the mean densities of seen baryons (in stars and plasma) and dark matter approaches the cosmic value? And what is this cosmic value?

The clustering phenomenon provides us with a uniquely important laboratory for assessing the roles of heredity and environment in the development of galaxies. The range of densities of galaxies is enormous, from $\sim 10^5$ times the large-scale mean $<n>$ in the cores of the richest clusters to well below $<n>$ in the neighborhoods of the great voids. We have seen an impressive variety of correlations of

environment (measured on scales from ~ 100 kpc to perhaps 3 Mpc) and
morphologies observed at X-ray, optical and radio wavelengths.
Particularly dramatic are the deformed radio lobes in clusters, though
Fanti showed that it is less clear how we are to apportion the effect
among the roles of the internal state of the galaxy, the galaxy's
motion, inhomogeneity and turbulence in the intracluster plasma,
buoyancy, magnetic fields in the ICM, and so on. We saw fascinating
evidence of HI deficiency and deformations in galaxies in dense
environments, bringing to mind van den Bergh's anemic spirals and the
original Baade-Spitzer process. Perhaps we are seeing evidence of a
similar process in Dickel's HI clouds seen around some groups of
galaxies. And it is fascinating to contrast this with the presence of
distinctly non-anemic spirals in seemingly equally destructive
environments!

There is a pronounced correlation between the ambient density of
galaxies and the frequency distribution of galaxy morphological types.
To what extent can environment induce evolution from one morphological
type to another? I did not hear much enthusiasm for the transformation
of spirals into ellipticals, but people did seem willing to suppose
that an Sa galaxy could turn into an SO (though if that is the only
source of SO galaxies the operation would have to work in low as well
as high density environments). What is the significance of the discs
so characteristic of large late-type galaxies? We heard evidence that
discs grow slowly, though that does require rather nice control on the
perpendicular component of the angular momentum of the gas fed to the
disc. In any case, discs do prefer low density environments. If, as
has been argued, such morphological features are largely hereditary it
conforms with scenarios for cosmogony in which galaxies are fragments
of pre-existing protoclusters, for with the high variability of local
conditions possible at the time protoclusters form one could easily
imagine high variability in the amount of angular momentum given to a
nascent galaxy, or in the factor by which the protogalaxy collapses
before fragmenting into stars. Of course, we must balance the
morphology correlation against the remarkably small scatter in such
global properties of galaxies as the one or two parameter sequence for
ellipticals, or the $L-v_c$ relation for spirals, which do not seem to
depend much on present ambient conditions. That might suggest that
galaxies formed under conditions a lot more uniform than what they now
find.

The cD pehnomenon is a striking case of the morphology puzzle. We
see a correlation between two distinctive features, a dense environment
and a centrally situated galaxy with an oriented amorphous optical
envelope. Does this tell us that cD galaxies are primitive objects,
nuclei around which other galaxies formed, or does the envelope represent debris that would naturally tend to arrange itself symmetrically
around the bottom of the gravitational potential well? The optical
envelopes of giant E and cD galaxies surely do grow at least in part
by cannibalism, but we heard no consensus on whether it is the dominant
process. Arguing against it in van den Bergh's opinion is the puzzle

of where the giant envelope of globular star clusters around M87 came
from. The origin of the giant envelope of dark matter around this
galaxy is another puzzle: could it have been added after the galaxy
and the Virgo cluster had formed? It seems hard to imagine if the dark
matter is some sort of exotic weakly interacting particles, much easier
if it is very low mass stars produced somehow in an earlier strong
cooling flow.

The great clusters of galaxies have been a key factor in the study
of the universe at high redshift because they serve as distinctive
markers. It is impressive to see the progress described by Oke and by
Tammann on the extension of the Hubble diagram to redshifts approaching
unity. Here relativistic effects, or whatever it is that controls the
universe, are no longer small corrections: the universe observed at a
redshift of unity surely looks a lot different from what we see at low
z. But, as Ellis and Thompson and MacGillivray emphasized, measuring
the state of evolution of the clustering pattern and the evolution of
field galaxies or even lesser cluster members at high redshifts will be
an enormous job that is only recently well begun.

The quasar phenomenon does show some correlation with redshift: we
learned from Yee that the degree of clustering of galaxy-like images
around quasars does vary with the quasar redshift about as expected if
quasar redshifts were cosmological, and from Burbidge of the difficulty
in translating that into a geometrical demonstration that the quasar
redshifts really are cosmological. The clustering of quasars has been
relatively poorly measured because of the difficulty of understanding
selection effects. Osmer's program seems particularly promising and
we will follow its progress with great interest. It does seem that
quasars become scarce at $z \gtrsim 3.5$. On the cosmological interpretation
of z something happened then, when the universe was some two billion
years old. Is this the epoch of galaxy formation, or is it the time
required to accumulate the mass to build the engine to power the quasar?
If the former, why have people not detected young galaxies not endowed
with quasars?

In the course of this meeting we have been led to consider a host
of such questions. I have mentioned those that I remember and found
memorable; another of us assigned this job surely would have come up
with a rather different list. I emphasize also that I capriciously
associated names of speakers with some questions, left others anonymous
(although easily deduced from the papers in this volume) and that that
bears absolutely no correlation to the merits of the questions. The
range of questions, so broad in the variety of disciplines from which
they arise and so deep in the fundamental problems they address, is a
wonderful testimony to the flowering of this subject.

REFERENCES

1. de Vaucouleurs, G.: 1958, A. J. 63, p. 253.

2. Abell, G. O.: 1958, Ap. J. Suppl. 3, p. 211.
3. Zwicky, F.: 1938, P.A.S.P. 50, p. 218.
4. Shane, C. D., Wirtanen, C. A. and Steinlin, U.: 1959, A. J. 64, p. 197.
5. Neyman, J.: 1962, in "Problems of Extragalactic Research," ed. G. C. McVittie (New York:Macmillan) p. 294, and earlier references therein.
6. This work was supported in part by the National Science Foundation of the U.S.A.

INDEX OF NAMES

HARRIS, W.E., 139,140,142,298,304
HARRISON, E.R., 95,97,487,496
HART, L., 35,41,75,76,229,240,273,274,385
HARTQUIST, T.W., 490,496
HAUCK, B., 603
HAUSER, M.G., 591,593
HAUSMAN, M.A., 123,130,175,176,313,317,322,338,352,427,428,531,550,551
HAWARDEN, T.G., 241
HAWLEY, D.L., 65,66,90,92
HAYES, D.J., 142,385,386
HAYLI, A., 130,428
HAYNES, M.P., 14,56,59,60,80,85,214,219,226,228,229,230,231,232,239,240,
          241,243,244,246,247,248,251,252,257,259
HAYWOOD, S. Jr., 523-524
HEATHCOTE, S.R., 600
HECKMAN, T.M., 123,130,227,240,557
HECKMANN, O., 531,550
HEESCHEN, 199,496
HEGYI, D.J., 425,428
HEILES, C., 47,298,304
HEILIGMAN, G.M., 369,371
HEISLER, J., 362,365
HELOU, G., 80,85,229,231,239,240,241,259,
HELOU, R.D., 264
HENRY, J.P., 113,116,336,339
HENRY, P.S., 40,41
HERTZ, P., 298,304
HERZOG, E., 14,172,208,291,292,379
HEUDIER, J.L., 600
HEWITT, A., 554,557
HEWITT, J.N., 223,224,226,240
HICKSON, P., 354,355,363,365,367-371,372,376,379,380,389,393,394
HILL, J.M., 93,97
HILLDROP, K.C., 296
HINTZEN, P., 58,60,97
HLIVAK, R.J., 114,116
HOAG, A.A., 581
HODGE, P.W., 473,474
HOESSEL, J., 52,54,100,101,119,120,123,124,130,131,137,193,196,199,352,
          531,539,540,541
HOFFMAN, G.L., 31,33,41
HOFFMAN, Y., 240,425,428,479,483
HOFMANN, 496
HOGAN, C.J., 493,495,496

LIST OF CONTRIBUTORS

AHMAD, F., Department of Physics, University of Kashmir, Srinagar, 190006, India.

ANDERNACH, H., Max-Planck-Institut für Radioastronomie, Auf dem Hügel 69, D-5300 Bonn 1, F.R.G.

ANDERSON, E.R., Department of Astronomy, University of Toronto, Toronto, Ontario, M5S 1A7, Canada.

APPLETON, P.N., Department of Astronomy, University of Manchester, Oxford Road, Manchester, U.K.

BAIER, F.W., Zentralinstitut für Astrophysik R.-Luxemburg-Strasse 17a, DDR-15 Potsdam-Babelsberg, G.D.R.

BAIESI-PILLASTRINI, G.C., Via Garzoni 211, 40138 Bologna, Italy.

BALKOWSKI, C., Observatoire de Paris, 92195 Meudon, France.

BARROW, J.D., Astronomy Centre, University of Sussex, Falmer, Brighton BN1 9QH, England.

BATUSKI, D.J., Department of Physics and Astronomy, University of New Mexico, Alburquerque, New Mexico 87131, U.S.A.

BEARD, S.M., Department of Astronomy, Royal Observatory, Blackford Hill, Edimburgh, U.K.

BHAVSAR, S.P., Astronomy Centre, University of Sussex, Falmer, Brighton BN1 9QH, England.

BIERMANN, P., Max-Planck-Institute für Radioastronomie, Auf dem Hügel 69, D-5300 Bonn 1, F.R.G.

BIJLEVELD, W., Sterrewacht, P.O. Box 9513, NL-2300 RA Leiden, The Netherlands.

BLUMENTAL, G.R., Board of Studies in Astronomy and Astrophysics, Lick Observatory, University of California, Santa Cruz, CA 95064, U.S.A.

BOHNENSTENGEL, H.-D., Hochschule der Bundeswehr Hamburg, FB/MB-Mathematik, D-2000 Hamburg 70, F.R.G.

BONOMETTO, S., Istituto di Fisica "G. Galilei", Via Marzolo 8, 35100 Padova, Italy.

BOTTINELLI, L., Observatoire de Meudon, Radioastronomie, 92195 Meudon Principal CEDEX, France.

BRANDUARDI-RAYMONT, G., Mullard Space Science Lab., Holmbury St. Mary, Dorking, Surrey, U.K.

BURNS, J.O., Department of Physics and Astronomy, University of
    New Mexico, Albuquerque, NM 87131, U.S.A.

BUSCH, A., Department of Astronomy, David Dunlap Observatory,
    University of Toronto, Toronto, Ontario, Canada, M5S
    1A1.

CAPACCIOLI, M., Osservatorio Astronomico, Vicolo dell'Osservatorio
    5, 35100 Padova, Italy.

CAPELATO, H., Instituto Astronomico e Geofisico, Dep.to de
    Astronomia, Caixa Postal 30627, 01000 São Paulo, Brasil.

CARTER, D., Mount Stromlo Observatory, Private Bag, Woden Post
    Office, ACT 2606, Canberra, Australia.

CAVALIERE, A., Osservatorio Astronomico, Vicolo dell'Osservatorio
    5, 35100 Padova, Italy.

CECCARELLI, C., Università di Roma, Istituto di Fisica, Piazzale
    A. Moro 2, 00185 Roma, Italy.

CIARDULLO, R. Space Telescope Science Institute, John Hopkins
    University, Department of Physics, Homewood Campus,
    Baltimore MD 21218, U.S.A.

COOKE, J.A., Department of Astronomy, Edinburgh University, Edinburgh
    EH9 3H1, Scotland, U.K.

CORBYN, P., Physics Department, Queen Mary College, Mile End Road,
    London E1, England.

CRUDDACE, R.G., Naval Research Laboratory, Washington DC 20375,
    U.S.A.

CURRIE, M.J., Royal Greenwich Observatory, Herstmonceux Castle,
    Hailsham, East Sussex BN27 1RP, England.

DALL'OGLIO, G., Istituto di Fisica della Atmosfera del C.N.R.,
    P.le Don Sturzo 31, Roma, Italy.

DAVIES, R.D., University of Manchester, Nuffield Radio Astronomy
    Laboratories, Jodrell Bank, Macclesfield, Cheshire SK11
    9DL, England.

DAVIS, R.J., Center for Astrophysics, 60 Garden Street, Cambridge,
    MA 02138 U.S.A.

DAWE, J.A., Royal Greenwich Observatory, Herstmonceux Castle,
    Hailsham, East Sussex BN27 1RP, England.

DE BERNARDIS, P., Istituto di Fisica, Università di Roma, P.le
    A. Moro 2, 00185 Roma, Italy.

DE VAUCOULEURS, G., Department of Astronomy, The University of
    Texas, Austin, Texas 78712, U.S.A.

DES FORÊTS, G., Observatoire de Meudon, 92190 Meudon, France.

DICKEL, J.R., University of Illinois at Urbana–Champaign, Room 341, Astronomy Building, 1011 West Springfield Ave., Urbana, IL 61801, U.S.A.

DICKENS, R.J., Royal Greenwich Observatory, Hailsham, East Sussex BN27 1RT, U.K.

DODD, R.J., Carter Observatory, P.O. Box 2909, Wellington, New Zealand

DRESSLER, A., Mount Wilson and Las Campanas Observatories, 813 Santa Barbara Street, Pasadena, California 91101, U.S.A.

ELLIS, R., Physics Department, Durham University, South Road, Durham, England.

EMERSON, D., Department of Astronomy, Edinburgh University, Edinburgh EH9 3HI, Scotland, U.K.

FAIRALL, A.P., Department of Astronomy, University of Cape Town, Rondebosch, 7700 South Africa.

FANG, L.Z., Astrophysics Research Division, University of Science and Technology of China, Hefei, Anhui, Peoples Republic of China.

FANTI, R., Istituto di Fisica "A. Righi", Via Irnerio 46, 40126 Bologna, Italy.

FERETTI, L., Istituto di Radioastronomia, Via Irnerio 46, 40126 Bologna, Italy.

FIALA, N., Institut für Astronomie der Universität Wien, Türkenschanz-strasse 17, A–1180 Wien, Austria.

FLICHE, H.H., Centre de Physique – C.N.R.S., Luminy Case 907, F 13288 Marseille, Cedex 9, France.

FLIN, P., Obserwatorium Astronomiczne Uniwersytetu, ul. Orla 171, 30-244 Krakow, Poland.

FOCARDI, P., Dipartimento di Astronomia, C.P. 516, 40100 Bologna, Italy.

FONG, R., Physics Department, University of Durham, South Road, Durham, OH1 3LE, England.

FORD, H., Space Telescope Science Institute, John Hopkins University, Department Of Physics, Homewood Campus, Baltimore, MD 21218, U.S.A.

FORMAN, W., Harvard–Smithsonian Center for Astrophysics, 60 Garden Street, Cambridge, MA 02138, U.S.A.

FOSTER, P.A., Department of Astronomy, The University, Manchester M13 9PL, England.

FRITZ, G., Naval Research Laboratory, Washington DC 20375, U.S.A.

FRÖLICH, H.-E., Zentralinstitut für Astrophysik, Rosa – Luxemburg Strasse 17a, DDR-15 Potsdam, G.D.R.

GAVAZZI, G., Istituto di Fisica Cosmica - C.N.R., Via Bassini 15, 20133 Milano, Italy.

GELLER, M., Harvard-Smithsonian Center for Astrophysics, 60 Garden Street, Cambridge, MA 02138, U.S.A.

GERBAL, D., LAM Observatoire de Meudon, 92195 Cedex, France.

GIOIA, I.M., Istituto di Radioastronomia, Via Irnerio 46, 40126 Bologna, Italy.

GIOVANELLI, R., National Astronomy and Ionosphere Center, Arecibo Observatory, P.O. Box 995, Arecibo PR 00613, U.S.A.

GIOVANNINI, G., Istituto di Radioastronomia, Via Irnerio 46, 40126 Bologna, Italy.

GIURICIN, G., Osservatorio Astronomico, Via Tiepolo 11, 34131 Trieste, Italy.

GODŁOWSKI, W., Jagiellonian University Observatory, ul. Orla 171, 30-224 Krakow, Poland.

GOEKE, J.F., St. Louis University H.S. St. Louis U.S.A.

GØRSKI, K., Copernicus Astronomical Centre, Bartycka 18, 00716 Warsaw, Poland.

GRAY, P.M., Anglo-Australian Observatory, P.O. Box 296, Epping NSW 2121, Australia.

GREEN, R.F., Kitt Peak National Observatory, Box 26732, Tucson Arizona 85726-6732 U.S.A.

GREGORINI, L., Istituto di Radioastronomia, Via Irnerio 46, 40126 Bologna, Italy.

HANISCH, R.J., Radiosterrenwacht Dwingeloo, Postbus 2, 7990 AA Dwingeloo, The Netherlands.

HANSEN, L., Copenhagen University Observatory, Oster Voldgade 3, DK-1350 Copenhagen K., Denmark.

HARMS, R.J., Cass C-011, UCSD, La Jolla, CA 92093, U.S.A.

HICKSON, P., Department of Geophysics and Astronomy, University of British Columbia, 2219 Main Mall, Vancouver, B.C. V6T 1W5, Canada.

HUCHRA, J.P., Center for Astrophysics, 60 Garden Street, Cambridge, MA 02138, U.S.A.

HUCHTMEIER, W.K., Max-Planck-Institut für Radioastronomie , Auf del Hügel 69, D-5300 Bonn 1, F.R.G.

ILLINGWORTH, G. , Kitt Peak National Observatory, P.O. Box 26732, Tucson, AZ 85726, U.S.A.

JAFFE, W., Space Telescope Science Institute, Homewood Campus, Baltimore, Maryland 21218, U.S.A.

JONES, C., Harvard–Smithsonian Center for Astrophysics, 60 Garden Street, Cambridge, MA 02138, U.S.A.

JØRGENSEN, H.E., Copenhagen University Observatory, Øster Voldgade 3, DK–1350 Copenhagen K, Denmark.

JUSZKIEWICZ, R., Copernicus Astronomical Centre, Bartycka 18, 00176 Warsaw, Poland.

KELLY, B.D., Royal Observatory, Blackford Hill, Edinburgh, EH9 3HJ, Scotland, U.K.

KOO, D.C., Carnegie Institution of Washington, U.S.A.

KOWALSKI, M.P., Naval Research Laboratory, Washington DC 20375, U.S.A.

KRISS, G.A., Department of Astronomy, University of Michigan, Ann Arbor, MI 48109, U.S.A.

KRON, R.G., Yerkes Observatory, The University of Chicago, William Bay, Wisconsin 53191, U.S.A.

KRONBERG, P.P., University of Toronto, Toronto Canada M5S 1A1.

KOTANYI, C., European Southern Observatory, Karl–Schwarzschild–Strasse 2, D–8046 Garching bei München, F.R.G.

LASENBY, A.N., University of Manchester, Nuffield Radio Astronomy Laboratories, Jodrell Bank, Macclesfield, Cheshire SK11 9DL, England.

LATHAM, D.W., Harvard–Smithsonian Center for Astrophysics, 60 Garden Street, Cambridge, MA 02138, U.S.A.

LUCCHIN, F., Istituto di Fisica "G. Galilei", Via Marzolo 8, 35100 Padova, Italy.

LUCEY, J., Anglo–Australian Observatory, P.O. Box 296, Epping, NSW 2121, Australia.

MACGILLIVRAY, H.T., Royal Observatory, Blackford Hill, Edinburgh, EH9 3HJ, Scotland, U.K.

MADORE, B.F., Department of Astronomy David Dunlap Observatory, University of Toronto, Toronto, Ontario, Canada, M5S 1A1.

MAMON, G.A., Physical Deparment, Princeton University, Princeton, NJ 08540, U.S.A.

MANDOLESI, N., Istituto T.E.S.R.E. C.N.R., Bologna, Italy.

MARANO, B., Dipartimento di Astronomia, Università di Bologna, Via Zamboni 33, 40126 Bologna, Italy.

MARDIROSSIAN, F., Osservatorio Astronomico, Via Tiepolo 11, 34131 Trieste, Italy.

MASI, S., Istituto di Fisica, Università di Roma, P.le A. Moro 2, 00185 Roma, Italy.

MATHEZ, G.,LAM, Observatoire de Meudon, 92195 CEDEX, France.

MAZURE, A.,LAM, Observatoire de Meudon, 92195 CEDEX, France.

MELCHIORRI , B., Istituto di Fisica dell'Atmosfera del C.N.R.,
      P.le Don Sturzo 31, Roma, Italy.

MELCHIORRI, F., Istituto di Fisica, Università di Roma, P.le A.
      Moro 2, 00185 Roma, Italy.

MEZZETTI, M., Osservatorio Astronomico, Via Tiepolo 11, 34131
      Trieste, Italy.

MONIN, J.L., LAM, Observatoire De Meudon, 92195 CEDEX, France.

MORENO, G., Istituto di Fisica "G. Marconi", P.le A. Moro 2,
      00100 Roma, Italy.

MORIGI, G., Istituto T.E.S.R.E., C.N.R., Bologna, Italy.

MOSS, C., Specola Vaticana, I-00120 Città del Vaticano, (Roma), Italy.

NANNI, D., Laboratori Nazionali di Fisica Nucleare, Casella Postale
      13, 0044 Frascati (Roma), Italy.

NINKOV, L., Department of Geophysics and Astronomy, University
      of British Columbia, 2219 Main Mall, Vancouver, B.C.
      V6T 1W5, Canada.

NØRGAARD-NIELSEN, H.U., Copenhagen University Observatory, Østervold-
      gade 3, DK-1350 Copenhagen K, Denmark.

NORMAN, C., European Southern Observatory, Karl-Schwarzschild-
      Strasse 2, D-8046 Garching bei München, F.R.G.

OCCHIONERO, F., Istituto di Astronomia, Università di Roma, P.le
      A. Moro 2, 00185 Roma, Italy.

OKE, J.B., 105-24 California Institute of Technology, Pasadena,
      CA 91125, U.S.A.

OLEAK, H., Zentralinstitut für Astrophysik,Rosa-Luxemburg-Strasse
      17a, DDR-15 Potsdam-Babelsberg, G.D.R.

OORT, J.H., Sterrewacht Leiden, Huygens Laboratorium, Wassenasarseweg
      78, Postbus 9513, 2300 RA Leiden, The Netherlands.

OSMER, P.S., Cerro Tololo Inter-American Observatory, Casilla
      603, La Serena, Chile.

OWEN, F.N., National Radio Astronomy Observatory, P.O. Box 0,
      Socorro, NM 87801, U.S.A.

PADRIELLI, L., Istituto di Radioastronomia, Via Irnerio 46, 40126
      Bologna, Italy.

PALMER, J.B., Royal Observatory, Blackford Hill, Edinburgh EH9
      3HJ, Scotland, U.K.

PALUMBO, G.G.C., Istituto T.E.S.R.E., C.N.R., Via de' Castagnoli
      1, 40126 Bologna, Italy.

PARKER, Q.A., University Observatory, Buchanan Gardens, St. Andrews,
      Fife, Scotland, U.K.

PEACH, J.V., Department of Astrophysics, South Parks Road, Oxford,
      England.

PEEBLES, P.J.E., Joseph Henry Laboratories, Physical Department, Princeton University, Princeton, NJ 08540, U.S.A.; Dominion Astrophysical Observatory, 5071 West Saanich Road, Victoria B.C. V8X 4M6, Canada.

PHILLIPPS, S., Department of Applied Mathematics and Astronomy, University College, Cardiff, Wales, U.K.

PIETRANERA, L., Istituto di Fisica, Università di Roma, P.le A. Moro 2, 00185 Roma, Italy.

PORTER, A., 105-24 California Institute of Technology, Pasadena, CA 91125, U.S.A.

PRESTAGE, R.M., University of Edimburgh, Department of Astronomy, Royal Observatory, Edinburgh EH9 3HI, Scotland, U.K.

PRIMACK, J.R., 34-63, University of California, Division of Natural Sciences, Santa Cruz, California 95064, U.S.A.

RAKOS, K.D., Institut für Astronomie, Universitäts-Sternwarte, Türkenschanzstrasse 17, A-1180 Wien, Austria.

REES, M., Institute of Astronomy, Madingley road, Cambridge CB3 OHA, England.

REPHAELI, Y., Department of Physics and Astronomy, Tel Aviv University, Tel Aviv 69978, Israel.

RICHTER, O.G., European Southern Observatory, Karl-Schwarzschild-Strasse 2, D-8046 Garching bei München, F.R.G.

RIVOLO, A.R., Space Telescope Science Institute, J.H.U./Homewood Campus, Baltimore, MD, 21218, U.S.A.

ROOD, H.J., Box 1330, Princeton, NJ 08542, U.S.A.

SADLER, E.M., European Southern Observatory, Karl-Schwarzschild-Strasse 2, D-8046 Garching bei München, F.R.G.

SANTANGELO, P., Istituto di Astronomia, Università di Roma, P.le A. Moro 2, 00187 Roma, Italy.

SCHMIDT, K.-H., Zentralinstitut für Astrophysik, Rosa-Luxemburg-Strasse 17a DDR-15 Potsdam, G.D.R.

SCHMIDT, R., Zentralinstitut für Astrophysik,Rosa-Luxemburg- Strasse 17a DDR-15 Potsdam, G.D.R.

SHANKS, T., Physics Department, University of Durham, South Road, Durham, OH1 3LE, England.

SHAPIRO, P., Department of Astronomy, University of Texas, Austin, TX 78712, U.S.A.

SHAYA, E.J., Institute for Astronomy, 2680 Woodlawn Drive, Honolulu, Hawaii 96822, U.S.A.

SMITH, H. Jr., Department of Astronomy, 211 Space Sciences Research Building, University of Florida, Gainesville, Florida 32611, U.S.A.

SOURIAU, J.M., Centre de Physique Theorique - C.N.R.S., Luminy Case 907, F 13288 Marseille, CEDEX 9, France.

STAVELEY-SMITH, L., Nuffield Radio Astronomy Laboratories, Jodrell Bank, Macclesfield, Cheshire, U.K.

STEVENSON, P., Physics Department, University of Durham, South Road, Durham, OH1 3LE, England.

TAMMANN, G.A., Astronomisches Institut der Universität Basel, Venusstrasse 7, CH-4102 Binningen, Switzerland.

TANZELLA - NITTI, G., Osservatorio Astronomico di Torino, 10025 Pino Torinese, Italy.

TARQUINI, G., Istituto di Astronomia, Università di Roma, P.le A. Moro 2, 00187 Roma, Italy.

TEAGUE, P.F., Mount Stromlo Observatory, Private Bag, Woden Post Office, ACT 2606, Canberra, Australia.

THOMPSON, L., Institute for Astronomy, 2680 Woodlawn Drive, Honolulu, Hawaii 96822, U.S.A.

TREVESE, D., Osservatorio Astronomico di Roma-Monte Mario, V.Le del Parco Mellini 84, 00136 Roma, Italy.

TUCKER, W., Harward-Smithsonian Center for Astrophysics, 60 Garden Street, Cambridge, MA 02138, U.S.A.

TULLY, B., Institute for Astronomy, 2680 Woodlawn Drive, Honolulu, Hawaii 96822, U.S.A.

TURNER, E.L., Princeton University Observatory, Peyton Hall, Princeton, NJ 08544, U.S.A.

ULMER, M.L., Department of Physics and Astronomy, Northwestern University, 2131 Sheridan Road, Evanston, Illinois 60201, U.S.A.

VALDARNINI, R., International School for Advanced Studies, Strada Costiera 11, 34100 Trieste, Italy.

VALENTIJN, E., European Southern Observatory Karl-Schwarzschild-Strasse 2, D-8046 Garching bei München, F.R.G.

VAN DEN BERGH, S., Dominion Astrophysical Observatory, 5071 West Saanich Road, Victoria B.C., V8X 4M6, Canada.

VAN GORKOM, J.H., National Radio Astronomy Observatory, P.O. Box 0, Socorro, NM 87801, U.S.A.

VAN WOERDEN, H., Kapteyn Laboratory, Pstbus 800, 9700 AV Groningen, The Netherlands.

VETTOLANI, G., Istituto di Radioastronomia, Via Irnerio 46, 40126 Bologna, Italy.

VIGNATO, A., Osservatorio Astronomico di Roma-Monte Mario, V.le del Parco Mellini 84, 00136 Roma, Italy.

VITTORIO, N., Istituto di Astronomia, Università di Roma, P.le A. Moro 2, 00185 Roma, Italy.

VOGLIS, N., Department of Astronomy, University of Athens, Panepistimiopolis, Athens (621), Greece.

WARMELS, R.H., Kapteyn Astronomical Institute, Postbox 800, 9700AV Groningen, The Netherlands.

WHITE, R.A., Goddard Space Flight Center, Greenbelt, Maryland 20771, U.S.A.

WHITTLE, M., Steward Observatory, University of Arizona, Tucson, Arizona 85721, U.S.A.

WILLIAMS, B., National Radio Astronomy Observatory, Edgemont Road, Charlottesville, Virginia 22903, U.S.A.

WINKLER, H., Department of Astronomy, University of Cape Town, Rondebosch 7700 Cape, South Africa.

WOOD, K.S., Naval Research Laboratory Washington, DC 20375, U.S.A.

YAHILA, A., State University of New York, Stony Brook, NY 11794, U.S.A.

YEE, H.K.C., Dominion Astrophysical Observatory, 5071 West Saanich Road, Victoria, BC V8X 4M6, Canada.

ASTROPHYSICS AND SPACE SCIENCE LIBRARY

Edited by

J. E. Blamont, R. L. F. Boyd, L. Goldberg, C. de Jager, Z. Kopal, G. H. Ludwig, R. Lüst,
B. M. McCormac, H. E. Newell, L. I. Sedov, Z. Švestka, and W. de Graaff

1. C. de Jager (ed.), *The Solar Spectrum, Proceedings of the Symposium held at the University of Utrecht, 26–31 August, 1963.* 1965, XIV + 417 pp.
2. J. Orthner and H. Maseland (eds.), *Introduction to Solar Terrestrial Relations, Proceedings of the Summer School in Space Physics held in Alpbach, Austria, July 15–August 10, 1963 and Organized by the European Preparatory Commission for Space Research.* 1965, IX + 506 pp.
3. C. C. Chang and S. S. Huang (eds.), *Proceedings of the Plasma Space Science Symposium, held at the Catholic University of America, Washington, D.C., June 11–14, 1963.* 1965, IX + 377 pp.
4. Zdeněk Kopal, *An Introduction to the Study of the Moon.* 1966, XII + 464 pp.
5. B. M. McCormac (ed.), *Radiation Trapped in the Earth's Magnetic Field. Proceedings of the Advanced Study Institute, held at the Chr. Michelsen Institute, Bergen, Norway, August 16– September 3, 1965.* 1966, XII + 901 pp.
6. A. B. Underhill, *The Early Type Stars.* 1966, XII + 282 pp.
7. Jean Kovalevsky, *Introduction to Celestial Mechanics.* 1967, VIII + 427 pp.
8. Zdeněk Kopal and Constantine L. Goudas (eds.), *Measure of the Moon. Proceedings of the 2nd International Conference on Selenodesy and Lunar Topography, held in the University of Manchester, England, May 30–June 4, 1966.* 1967, XVIII + 479 pp.
9. J. G. Emming (ed.), *Electromagnetic Radiation in Space. Proceedings of the 3rd ESRO Summer School in Space Physics, held in Alpbach, Austria, from 19 July to 13 August, 1965.* 1968, VIII + 307 pp.
10. R. L. Carovillano, John F. McClay, and Henry R. Radoski (eds.), *Physics of the Magnetosphere, Based upon the Proceedings of the Conference held at Boston College, June 19–28, 1967.* 1968, X + 686 pp.
11. Syun-Ichi Akasofu, *Polar and Magnetospheric Substorms.* 1968, XVIII + 280 pp.
12. Peter M. Millman (ed.), *Meteorite Research. Proceedings of a Symposium on Meteorite Research, held in Vienna, Austria, 7–13 August, 1968.* 1969, XV + 941 pp.
13. Margherita Hack (ed.), *Mass Loss from Stars. Proceedings of the 2nd Trieste Colloquium on Astrophysics, 12–17 September, 1968.* 1969, XII + 345 pp.
14. N. D'Angelo (ed.), *Low-Frequency Waves and Irregularities in the Ionosphere. Proceedings of the 2nd ESRIN-ESLAB Symposium, held in Frascati, Italy, 23–27 September, 1968.* 1969, VII + 218 pp.
15. G. A. Partel (ed.), *Space Engineering. Proceedings of the 2nd International Conference on Space Engineering, held at the Fondazione Giorgio Cini, Isola di San Giorgio, Venice, Italy, May 7–10, 1969.* 1970, XI + 728 pp.
16. S. Fred Singer (ed.), *Manned Laboratories in Space. Second International Orbital Laboratory Symposium.* 1969, XIII + 133 pp.
17. B. M. McCormac (ed.), *Particles and Fields in the Magnetosphere. Symposium Organized by the Summer Advanced Study Institute, held at the University of California, Santa Barbara, Calif., August 4–15, 1969.* 1970, XI + 450 pp.
18. Jean-Claude Pecker, *Experimental Astronomy.* 1970, X + 105 pp.
19. V. Manno and D. E. Page (eds.), *Intercorrelated Satellite Observations related to Solar Events. Proceedings of the 3rd ESLAB/ESRIN Symposium held in Noordwijk, The Netherlands, September 16–19, 1969.* 1970, XVI + 627 pp.
20. L. Mansinha, D. E. Smylie, and A. E. Beck, *Earthquake Displacement Fields and the Rotation of the Earth, A NATO Advanced Study Institute Conference Organized by the Department of Geophysics, University of Western Ontario, London, Canada, June 22–28, 1969.* 1970, XI + 308 pp.
21. Jean-Claude Pecker, *Space Observatories.* 1970, XI + 120 pp.
22. L. N. Mavridis (ed.), *Structure and Evolution of the Galaxy. Proceedings of the NATO Advanced Study Institute, held in Athens, September 8–19, 1969.* 1971, VII + 312 pp.

23. A. Muller (ed.), *The Magellanic Clouds. A European Southern Observatory Presentation: Principal Prospects, Current Observational and Theoretical Approaches, and Prospects for Future Research, Based on the Symposium on the Magellanic Clouds, held in Santiago de Chile, March 1969, on the Occasion of the Dedication of the European Southern Observatory.* 1971, XII + 189 pp.

24. B. M. McCormac (ed.), *The Radiating Atmosphere. Proceedings of a Symposium Organized by the Summer Advanced Study Institute, held at Queen's University, Kingston, Ontario, August 3 14, 1970.* 1971, XI + 455 pp.

25. G. Fiocco (ed.), *Mesospheric Models and Related Experiments. Proceedings of the 4th ESRIN-ESLAB Symposium, held at Frascati, Italy, July 6 10, 1970.* 1971, VIII + 298 pp.

26. I. Atanasijević, *Selected Exercises in Galactic Astronomy.* 1971, XII + 144 pp.

27. C. J. Macris (ed.), *Physics of the Solar Corona. Proceedings of the NATO Advanced Study Institute on Physics of the Solar Corona, held at Cavouri-Vouliagmeni, Athens, Greece, 6 17 September 1970.* 1971, XII + 345 pp.

28. F. Delobeau, *The Environment of the Earth.* 1971, IX + 113 pp.

29. E. R. Dyer (general ed.), *Solar-Terrestrial Physics/1970. Proceedings of the International Symposium on Solar-Terrestrial Physics, held in Leningrad, U.S.S.R., 12 19 May 1970.* 1972, VIII + 938 pp.

30. V. Manno and J. Ring (eds.), *Infrared Detection Techniques for Space Research. Proceedings of the 5th ESLAB-ESRIN Symposium, held in Noordwijk, The Netherlands, June 8 11, 1971.* 1972, XII + 344 pp.

31. M. Lecar (ed.), *Gravitational N-Body Problem. Proceedings of IAU Colloquium No. 10, held in Cambridge, England, August 12 15, 1970.* 1972, XI + 441 pp.

32. B. M. McCormac (ed.), *Earth's Magnetospheric Processes. Proceedings of a Symposium Organized by the Summer Advanced Study Institute and Ninth ESRO Summer School, held in Cortina, Italy, August 30 September 10, 1971.* 1972, VIII + 417 pp.

33. Antonin Rükl, *Maps of Lunar Hemispheres.* 1972, V + 24 pp.

34. V. Kourganoff, *Introduction to the Physics of Stellar Interiors.* 1973, XI + 115 pp.

35. B. M. McCormac (ed.), *Physics and Chemistry of Upper Atmospheres. Proceedings of a Symposium Organized by the Summer Advanced Study Institute, held at the University of Orléans, France, July 31 August 11, 1972.* 1973, VIII + 389 pp.

36. J. D. Fernie (ed.), *Variable Stars in Globular Clusters and in Related Systems. Proceedings of the IAU Colloquium No. 21, held at the University of Toronto, Toronto, Canada, August 29 31, 1972.* 1973, IX + 234 pp.

37. R. J. L. Grard (ed.), *Photon and Particle Interaction with Surfaces in Space. Proceedings of the 6th ESLAB Symposium, held at Noordwijk, The Netherlands, 26 29 September, 1972.* 1973, XV + 577 pp.

38. Werner Israel (ed.), *Relativity, Astrophysics and Cosmology. Proceedings of the Summer School, held 14 26 August, 1972, at the BANFF Centre, BANFF, Alberta, Canada.* 1973, IX + 323 pp.

39. B. D. Tapley and V. Szebehely (eds.), *Recent Advances in Dynamical Astronomy. Proceedings of the NATO Advanced Study Institute in Dynamical Astronomy, held in Cortina d'Ampezzo, Italy, August 9 12, 1972.* 1973, XIII + 468 pp.

40. A. G. W. Cameron (ed.), *Cosmochemistry. Proceedings of the Symposium on Cosmochemistry, held at the Smithsonian Astrophysical Observatory, Cambridge, Mass., August 14 16, 1972.* 1973, X + 173 pp.

41. M. Golay, *Introduction to Astronomical Photometry.* 1974, IX + 364 pp.

42. D. E. Page (ed.), *Correlated Interplanetary and Magnetospheric Observations. Proceedings of the 7th ESLAB Symposium, held at Saulgau, W. Germany, 22 25 May, 1973.* 1974, XIV + 662 pp.

43. Riccardo Giacconi and Herbert Gursky (eds.), *X-Ray Astronomy.* 1974, X + 450 pp.

44. B. M. McCormac (ed.), *Magnetospheric Physics. Proceedings of the Advanced Summer Institute, held in Sheffield, U.K., August 1973.* 1974, VII + 399 pp.

45. C. B. Cosmovici (ed.), *Supernovae and Supernova Remnants. Proceedings of the International Conference on Supernovae, held in Lecce, Italy, May 7 11, 1973.* 1974, XVII + 387 pp.

46. A. P. Mitra, *Ionospheric Effects of Solar Flares.* 1974, XI + 294 pp.

47. S.-I. Akasofu, *Physics of Magnetospheric Substorms.* 1977, XVIII + 599 pp.

48. H. Gursky and R. Ruffini (eds.), *Neutron Stars, Black Holes and Binary X-Ray Sources.* 1975, XII + 441 pp.
49. Z. Švestka and P. Simon (eds.), *Catalog of Solar Particle Events 1955–1969. Prepared under the Auspices of Working Group 2 of the Inter-Union Commission on Solar-Terrestrial Physics.* 1975, IX + 428 pp.
50. Zdeněk Kopal and Robert W. Carder, *Mapping of the Moon.* 1974, VIII + 237 pp.
51. B. M. McCormac (ed.), *Atmospheres of Earth and the Planets. Proceedings of the Summer Advanced Study Institute, held at the University of Liège, Belgium, July 29–August 8, 1974.* 1975, VII + 454 pp.
52. V. Formisano (ed.), *The Magnetospheres of the Earth and Jupiter. Proceedings of the Neil Brice Memorial Symposium, held in Frascati, May 28–June 1, 1974.* 1975, XI + 485 pp.
53. R. Grant Athay, *The Solar Chromosphere and Corona: Quiet Sun.* 1976, XI + 504 pp.
54. C. de Jager and H. Nieuwenhuijzen (eds.), *Image Processing Techniques in Astronomy. Proceedings of a Conference, held in Utrecht on March 25–27, 1975.* XI + 418 pp.
55. N. C. Wickramasinghe and D. J. Morgan (eds.), *Solid State Astrophysics. Proceedings of a Symposium, held at the University College, Cardiff, Wales, 9–12 July 1974.* 1976, XII + 314 pp.
56. John Meaburn, *Detection and Spectrometry of Faint Light.* 1976, IX + 270 pp.
57. K. Knott and B. Battrick (eds.), *The Scientific Satellite Programme during the International Magnetospheric Study. Proceedings of the 10th ESLAB Symposium, held at Vienna, Austria, 10–13 June 1975.* 1976, XV + 464 pp.
58. B. M. McCormac (ed.), *Magnetospheric Particles and Fields. Proceedings of the Summer Advanced Study School, held in Graz, Austria, August 4–15, 1975.* 1976, VII + 331 pp.
59. B. S. P. Shen and M. Merker (eds.), *Spallation Nuclear Reactions and Their Applications.* 1976, VIII + 235 pp.
60. Walter S. Fitch (ed.), *Multiple Periodic Variable Stars. Proceedings of the International Astronomical Union Colloquium No. 29, held at Budapest, Hungary, 1–5 September 1976.* 1976, XIV + 348 pp.
61. J. J. Burger, A. Pedersen, and B. Battrick (eds.), *Atmospheric Physics from Spacelab. Proceedings of the 11th ESLAB Symposium, Organized by the Space Science Department of the European Space Agency, held at Frascati, Italy, 11–14 May 1976.* 1976, XX + 409 pp.
62. J. Derral Mulholland (ed.), *Scientific Applications of Lunar Laser Ranging. Proceedings of a Symposium held in Austin, Tex., U.S.A., 8–10 June, 1976.* 1977, XVII + 302 pp.
63. Giovanni G. Fazio (ed.), *Infrared and Submillimeter Astronomy. Proceedings of a Symposium held in Philadelphia, Penn., U.S.A., 8–10 June, 1976.* 1977, X + 226 pp.
64. C. Jaschek and G. A. Wilkins (eds.), *Compilation, Critical Evaluation and Distribution of Stellar Data. Proceedings of the International Astronomical Union Colloquium No. 35, held at Strasbourg, France, 19–21 August, 1976.* 1977, XIV + 316 pp.
65. M. Friedjung (ed.), *Novae and Related Stars. Proceedings of an International Conference held by the Institut d'Astrophysique, Paris, France, 7–9 September, 1976.* 1977, XIV + 228 pp.
66. David N. Schramm (ed.), *Supernovae. Proceedings of a Special IAU-Session on Supernovae held in Grenoble, France, 1 September, 1976.* 1977, X + 192 pp.
67. Jean Audouze (ed.), *CNO Isotopes in Astrophysics. Proceedings of a Special IAU Session held in Grenoble, France, 30 August, 1976.* 1977, XIII + 195 pp.
68. Z. Kopal, *Dynamics of Close Binary Systems,* XIII + 510 pp.
69. A. Bruzek and C. J. Durrant (eds.), *Illustrated Glossary for Solar and Solar-Terrestrial Physics.* 1977, XVIII + 204 pp.
70. H. van Woerden (ed.), *Topics in Interstellar Matter.* 1977, VIII + 295 pp.
71. M. A. Shea, D. F. Smart, and T. S. Wu (eds.), *Study of Travelling Interplanetary Phenomena.* 1977, XII + 439 pp.
72. V. Szebehely (ed.), *Dynamics of Planets and Satellites and Theories of Their Motion. Proceedings of IAU Colloquium No. 41, held in Cambridge, England, 17–19 August 1976.* 1978, XII + 375 pp.
73. James R. Wertz (ed.), *Spacecraft Attitude Determination and Control.* 1978, XVI + 858 pp.

74. Peter J. Palmadesso and K. Papadopoulos (eds.), *Wave Instabilities in Space Plasmas. Proceedings of a Symposium Organized Within the XIX URSI General Assembly held in Helsinki, Finland, July 31–August 8, 1978.* 1979, VII + 309 pp.

75. Bengt E. Westerlund (ed.), *Stars and Star Systems. Proceedings of the Fourth European Regional Meeting in Astronomy held in Uppsala, Sweden, 7–12 August, 1978.* 1979, XVIII + 264 pp.

76. Cornelis van Schooneveld (ed.), *Image Formation from Coherence Functions in Astronomy. Proceedings of IAU Colloquium No. 49 on the Formation of Images from Spatial Coherence Functions in Astronomy, held at Groningen, The Netherlands, 10–12 August 1978.* 1979, XII + 338 pp.

77. Zdeněk Kopal, *Language of the Stars. A Discourse on the Theory of the Light Changes of Eclipsing Variables.* 1979, VIII + 280 pp.

78. S.-I. Akasofu (ed.), *Dynamics of the Magnetosphere. Proceedings of the A.G.U. Chapman Conference 'Magnetospheric Substorms and Related Plasma Processes' held at Los Alamos Scientific Laboratory, N.M., U.S.A., October 9–13, 1978.* 1980, XII + 658 pp.

79. Paul S. Wesson, *Gravity, Particles, and Astrophysics. A Review of Modern Theories of Gravity and G-variability, and their Relation to Elementary Particle Physics and Astrophysics.* 1980, VIII + 188 pp.

80. Peter A. Shaver (ed.), *Radio Recombination Lines. Proceedings of a Workshop held in Ottawa, Ontario, Canada, August 24–25, 1979.* 1980, X + 284 pp.

81. Pier Luigi Bernacca and Remo Ruffini (eds.), *Astrophysics from Spacelab*, 1980, XI + 664 pp.

82. Hannes Alfvén, *Cosmic Plasma*, 1981, X + 160 pp.

83. Michael D. Papagiannis (ed.), *Strategies for the Search for Life in the Universe*, 1980, XVI + 254 pp.

84. H. Kikuchi (ed.), *Relation between Laboratory and Space Plasmas*, 1981, XII + 386 pp.

85. Peter van der Kamp, *Stellar Paths*, 1981, xxii + 155 pp.

86. E. M. Gaposchkin and B. Kołaczek (eds.), *Reference Coordinate Systems for Earth Dynamics*, 1981, XIV + 396 pp.

87. R. Giacconi (ed.), *X-Ray Astronomy with the Einstein Satellite. Proceedings of the High Energy Astrophysics Division of the American Astronomical Society Meeting on X-Ray Astronomy held at the Harvard-Smithsonian Center for Astrophysics, Cambridge, Mass., U.S.A., January 28–30, 1980.* 1981, VII + 330 pp.

88. Icko Iben Jr. and Alvio Renzini (eds.), *Physical Processes in Red Giants. Proceedings of the Second Workshop, held at the Ettore Majorana Centre for Scientific Culture, Advanced School of Agronomy, in Erice, Sicily, Italy, September 3–13, 1980.* 1981, XV + 488 pp.

89. C. Chiosi and R. Stalio (eds.), *Effect of Mass Loss on Stellar Evolution. IAU Colloquium No. 59 held in Miramare, Trieste, Italy, September 15–19, 1980.* 1981, XXII + 532 pp.

90. C. Goudis, *The Orion Complex: A Case Study of Interstellar Matter*, 1982 (forthcoming).

91. F. D. Kahn (ed.), *Investigating the Universe. Papers Presented to Zdeněk Kopal on the Occasion of his retirement, September 1981.* 1981, X + 458 pp.

92. C. M. Humphries (ed.), *Instrumentation for Astronomy with Large Optical Telescopes, Proceedings of IAU Colloquium No. 67.* 1982 (forthcoming).

93. R. S. Roger and P. E. Dewdney (eds.), *Regions of Recent Star Formation, Proceedings of the Symposium on "Neutral Clouds Near HII Regions - Dynamics and Photochemistry", held in Penticton, B.C., June 24–26, 1981.* 1982, XVI + 496 pp.

94. O. Calame (ed.), *High-Precision Earth Rotation and Earth-Moon Dynamics. Lunar Distances and Related Observations*, 1982, xx + 354 pp.

95. M. Friedjung and R. Viotti (eds.), *The Nature of Symbiotic Stars*, xx + 310 pp.

96. W. Fricke and G. Teleki (eds.), *Sun and Planetary System*, xiv + 538 pp.

97. C. Jaschek and W. Heintz (eds.), *Automated Data Retrieval in Astronomy*, xx + 324 pp.

98. Z. Kopal and J. Rahe (eds.), *Binary and Multiple Stars as Tracers of Stellar Evolution*, 1982, XXX + 504 pp.

99. A. W. Wolfendale (ed.), *Progress in Cosmology*, 1982, VI + 360 pp.

100. W. L. H. Shuter (ed.), *Kinematics, Dynamics and Structure of the Milky Way*, 1983, XII + 392 pp.

101. M. Livio and G. Shaviv (eds.), *Cataclysmic Variables and Related Objects*, 1983, XII + 352 pp.

102. P. B. Byrne and M. Rodonò (eds.), *Activity in Red-Dwarf Stars*, 1983, XXVI + 670 pp.
103. A. Ferrari and A. G. Pacholczyk (eds.), *Astrophysical Jets*, 1983, XVI + 328 pp.
104. R. L. Carovillano and J. M. Forbes (eds.), *Solar-Terrestrial Physics*, 1983, XVIII + 860 pp.
105. W. B. Burton and F. P. Israel (eds.), *Surveys of the Southern Galaxy*, 1983, XIV + 310 pp.
106. V. V. Markellos and Y. Kozai (eds.), *Dynamical Trapping and Evolution on the Solar System*, 1983, XVI + 424 pp.
107. S. R. Pottasch, *Planetary Nebulae*, 1984, X + 322 pp.
108. M. F. Kessler and J. P. Phillips (eds.), *Galactic and Extragalactic Infrared Spectroscopy*, 1984, XII + 472 pp.
109. C. Chiosi and A. Renzini (eds.), *Stellar Nucleosynthesis*, 1984, XIV + 398 pp.
110. M. Capaccioli (ed.), *Astronomy with Schmidt-type Telescopes*, 1984, XXII + 620 pp.